A Foundation for PROPs, Algebras, and Modules

Mathematical
Surveys
and
Monographs

Volume 203

A Foundation for PROPs, Algebras, and Modules

Donald Yau
Mark W. Johnson

American Mathematical Society
Providence, Rhode Island

2010 *Mathematics Subject Classification.* Primary 18D99;
Secondary 55U40, 13D10, 81T30.

For additional information and updates on this book, visit
www.ams.org/bookpages/surv-203

Library of Congress Cataloging-in-Publication Data

Yau, Donald Y. (Donald Ying), 1977–
A foundation for PROPs, algebras, and modules / Donald Yau, Mark W. Johnson.
pages cm. — (Mathematical surveys and monographs ; volume 203)
Includes bibliographical references and index.
ISBN 978-1-4704-2197-7 (alk. paper)
1. Permutation groups. 2. Categories (Mathematics) 3. Modules (Algebra) I. Johnson, Mark W., 1973– II. Title

QA175.Y38 2015
512′.21—dc23

2014049153

♾ The paper used in this book is acid-free and falls within the guidelines
established to ensure permanence and durability.
Visit the AMS home page at `http://www.ams.org/`

10 9 8 7 6 5 4 3 2 1 20 19 18 17 16 15

The first author dedicates this book to Eun Soo and Jacqueline.

The second author would like to dedicate this book to his lovely wife Terri and to his nearly perfect children Lizzie and Ben.

Contents

Preface

The purpose of this monograph is to introduce and study a unifying object we call a generalized PROP, which includes the colored version of an operad, a PROP, a wheeled PROP, or any variant as a special case. Before we describe the topics discussed in this monograph, let us briefly review operads and PROPs.

Operads are an efficient machinery for organizing operations and the relations between them. Operads were introduced in homotopy theory by May [**May72**] to describe spaces with the weak homotopy types of iterated loop spaces. An earlier motivating example for the concept of an operad was Stasheff's A_∞-spaces [**Sta63**].

Briefly, an operad O has objects $\mathsf{O}(n)$ for $n \geq 0$ and structure maps

$$\gamma: \mathsf{O}(n) \otimes \mathsf{O}(k_1) \otimes \cdots \otimes \mathsf{O}(k_n) \longrightarrow \mathsf{O}(k_1 + \cdots + k_n)$$

that satisfy some associativity, equivariance, and unity conditions. The prototypical example of an operad is called the endomorphism operad of an object A with

$$\mathsf{E}_A(n) = \mathrm{Hom}(A^{\otimes n}, A).$$

Here the elements are normally called n-ary operations, and the structure map γ comes from using the outputs of choices of k_j-ary operations as the inputs of an n-ary operation, thereby producing a single operation with $\Sigma_j k_j$ inputs. As expected, an operad map $\mathsf{O} \longrightarrow \mathsf{E}_A$ then has entries $\mathsf{O}(n) \longrightarrow \mathsf{E}_A(n)$ compatible with the structure maps, and in this way $\mathsf{O}(n)$ can be used to parametrize n-ary operations on A. As a consequence, this special case of a map into the endomorphism operad of A earns the name of an O-algebra structure on A. In addition, relations described in terms of the structure map of O must also remain present among the families of operations in the image of such a map, due to the compatibility of a morphism $\mathsf{O} \longrightarrow \mathsf{E}_A$ with structure maps γ.

There are many important uses of operads in homotopy theory and algebra, so we will mention a few of them. Besides the study of iterated loop spaces, operads are used in the algebraic classification of homotopy types [**Man06, Smi82, Smi01**]. The singular cochain complex of a space is an E_∞-algebra, which is a homotopy version of a commutative algebra. For certain nice spaces, this E_∞-algebra determines the weak homotopy type. Another operadic link between topology and algebra is the solution of Deligne's Conjecture [**Kau07, MS02**]. It says that the Hochschild cochain complex of an associative algebra is an algebra over a suitable chain version of May's little 2-cubes operad. Furthermore, Stasheff's work on homotopy associative H-spaces can be generalized to other homotopy invariant structures. Given any reasonably nice operad O, Boardman and Vogt [**BV73**] constructed an operad $W\mathsf{O}$ that is weakly equivalent to O such that $W\mathsf{O}$-algebras are homotopy invariant. Other applications of operads are discussed in [**KM95, Mar08, MSS02, Smi01**].

In many algebraic situations, one encounters not only n-ary operations but also operations with multiple inputs and multiple outputs. The simplest example is a bialgebra, which has a multiplication and a comultiplication. PROPs are a machinery that can be used similarly to organize operations with multiple inputs and multiple outputs. PROPs were introduced by Mac Lane [**Mac63, Mac65**] to describe the structure on the iterated bar constructions on a commutative differential graded Hopf algebra. Briefly, a PROP P has objects

$$\mathsf{P}\binom{n}{m}$$

for $n, m \geq 0$ and structure maps

$$\mathsf{P}\binom{n}{m} \otimes \mathsf{P}\binom{q}{p} \longrightarrow \mathsf{P}\binom{n+q}{m+p} \quad \text{(horizontal composition)}$$

and

$$\mathsf{P}\binom{n}{m} \otimes \mathsf{P}\binom{m}{l} \longrightarrow \mathsf{P}\binom{n}{l} \quad \text{(vertical composition)}$$

that satisfy some associativity, bi-equivariance, unity, and compatibility conditions. The prototypical example here is the endomorphism PROP of an object A with

$$\mathsf{E}_A\binom{n}{m} = \mathrm{Hom}(A^{\otimes m}, A^{\otimes n}).$$

As above, the object $\mathsf{P}\binom{n}{m}$ parametrizes operations with m inputs and n outputs via an entry $\mathsf{P}\binom{n}{m} \longrightarrow \mathsf{E}_A\binom{n}{m}$ of a map $\mathsf{P} \longrightarrow \mathsf{E}_A$, so such a map is again called a P-algebra structure on A.

There are numerous applications in mathematics and physics of PROPs and variants, such as the smaller half-PROPs and properads and the bigger wheeled PROPs. For example, these objects are used prominently in deformation theory [**FMY09, MV09**], graph cohomology [**MV09, Mer09**], homotopy invariant structures [**JY09**], Batalin-Vilkovisky structures [**Mer10a**], the Master Equation [**MMS09**], deformation quantization [**Mer08, Mer10b**], Poisson structures [**Str10**], string topology [**Cha05, CG04, CV06**], and field theories [**JY09, Ion07, Seg01, Seg04**].

There are close relationships between operads and PROPs. Their definitions are formally similar to each other, and their algebras are both given by morphisms into the endomorphism objects. In fact, every PROP P has an underlying operad with

$$\mathsf{P}(m) = \mathsf{P}\binom{1}{m}.$$

Conversely, every operad O generates a PROP O' such that O-algebras are exactly O'-algebras [**BV73**]. There is a conceptual description of an operad as a monoid in the monoidal category of Σ-modules [**May97**]. There is a conceptually similar, but more complicated, description of a PROP as a 2-monoid in the category of Σ-bimodules [**JY09**]. Furthermore, the homotopy theory of PROPs is, in a precise sense, a homotopy refinement of the homotopy theory of operads [**JY09**].

In general, PROPs and wheeled PROPs are harder to deal with than operads, because they are much bigger. The operations in an operad are parametrized by level trees, which have nice combinatorial properties that allow one to do induction on the internal edges. For example, the Boardman-Vogt W-construction [**BV73, BM06, BM07, Vog03**] for an operad uses level trees in an essential way. On the other hand, the operations in a PROP are parametrized by directed cycle-free graphs, which may have multiple connected components. There are many more such graphs than level trees. Moreover, induction on the internal edges in directed cycle-free graphs is usually not possible because there can be many edges between two vertices, and one cannot generally shrink away an internal edge without the chance of producing a cycle in the resulting graph. Going even further, the operations in a wheeled PROP are parametrized by directed graphs, which may have multiple connected components along with directed cycles and loops.

As discussed by Markl in [**Mar08**], operads, PROPs, and wheeled PROPs can all be described using collections of graphs he called pasting schemes, in these cases consisting of the level trees, the directed cycle-free graphs, and the directed graphs, respectively. Since the definition of a pasting scheme was left ambiguous there, part of our aim in the first half of this monograph is to make precise this notion of pasting scheme. The main motivation is that there should be a variant of PROPs associated to any reasonable pasting scheme. By choosing the right pasting schemes, one can obtain colored versions of (wheeled) operads, (wheeled) properads, (wheeled) PROPs, dioperads, and half-PROPs, among others. Unfortunately, the related cyclic and modular operads would require more cumbersome versions of the underlying graph theory, so we have elected not to complicate our approach throughout in order to include those structures, which are mentioned only at the very end of the first chapter.

In this monograph, we introduce and study this unifying object, called a generalized PROP. This monograph is divided into two parts. The first part describes the theory of pasting schemes in careful detail, which requires a new definition of graph, a new description of graph substitution, a careful description of graph operations, a theory of generating sets for graph groupoids, and notions of intersections and free products of graph groupoids. This part is somewhat technical, but the point is to reduce the technical issues in the subsequent theory to the underlying questions about graphs by taking all aspects of the theory of pasting schemes seriously. The second part of this monograph contains categorical properties of generalized PROPs along with their algebras and modules. In this second part, we work over an arbitrary symmetric monoidal (closed) category with enough limits and colimits. In future work, we plan to investigate questions related to the homotopy theory of all of these objects, including constructive approaches to cofibrant replacements, where possible. The graph theory built up in the first part of this monograph and the equivalence established here between the biased and unbiased versions of (wheeled) properads are also used in [**HRY**] to develop a theory of higher (wheeled) properads.

Several other projects have worked to provide a unifying view of a variety of operational structures, (e.g., [**Get09**], [**KW**], or [**BM14**]), while [**BB**, Subsec. 15.4] even does so using a version of graphs which they show to be equivalent to our presentation here. We hope the present monograph will serve as a fully detailed reference for at least one such unifying approach. A brief description of each chapter follows.

The first chapter introduces the new definition of a wheeled graph. First, a basic graph consists of a partitioned finite set of flags equipped with an involution, so the partitions correspond to vertices, the flags correspond to half-edges, and the involution pairs two half-edges together to form edges. Flags fixed by the involution are called legs and represent either inputs or outputs of the whole graph. Unfortunately, some additional structure must also be included to deal properly with exceptional graphs, which contain no vertices, and so any graph can have an exceptional part. A wheeled graph is then a basic graph together with three extra pieces of structure, called a direction, a coloring, and a listing. The listing is a new feature, introduced so the inputs or outputs of any vertex, or of the full graph, may be expressed as a (finite) ordered sequence of colors, which is vital to defining graph substitution in full detail later. A series of small examples is included to clarify the many definitions.

The second chapter is devoted to understanding the various technical properties which form the distinctions between the pasting schemes of interest. Connected and simply-connected graphs are defined without recourse to any geometric realization, exploiting a careful presentation of the notion of paths in a directed graph. Wheel-free graphs, half-graphs, dioperadic graphs, and several variants of trees, including level trees, simple trees, special trees, and wheeled trees are also discussed, including some pictures.

Some basic graph operations are the topic of the third chapter. In Chapter 6, a more general viewpoint is taken to characterize all graph operations compatible with the fundamental operation of graph substitution, but a few key operations must be introduced much earlier. These include relabeling operations, which shuffle the ordered sequences of inputs and outputs, and a disjoint union operation. There is also a grafting operation where one matches the inputs of one graph with the outputs of another, creating a series of new internal edges as a result. Also included is a partial grafting variant, that can be used to describe the comp-i operations of Gerstenhaber and the related j-comp-i operations. Finally, a contraction operation, which connects two former legs to construct a new internal edge is described, followed by a discussion of invertible graph operations. Along with each of these operations, an example involving a graph with one or two vertices is included, and these graphs will be shown in Chapter 6 to generate the associated operations via the fundamental operation of graph substitution.

In Chapter 4, we present two different notions of isomorphism and describe the main examples of graph groupoids. Since our notion of listing is new, in some instances we want to insist it is preserved by isomorphism, so we define what we call strict isomorphisms, and in other instances we want to relax this constraint, so we define weak isomorphisms. A variety of results concerning both strict automorphism groups and weak automorphism groups of graphs is included, and the strict automorphisms are shown to be quite rigid. For example, the strict automorphism group of any simply-connected graph is trivial.

Chapter 5 is devoted to a careful construction of the fundamental operation of graph substitution, as well as verifying that it is unital, associative, and natural with respect to both types of isomorphisms. The basic idea of substitution is to cut a small hole around any vertex and to insert a shrunken copy of another graph with the same ordered sequence of inputs and outputs as the vertex removed. Unfortunately, the exceptional parts of the graphs inserted cause a variety of technical problems. Thus, we introduce a new object called a pre-graph which is discussed only in this chapter, and building the associated graph of a pre-graph becomes one key technical complication. Forming the substitution pre-graph is relatively straightforward, and we show the process is associative and nearly unital. However, there are several choices for where to apply the associated graph construction, and verifying they all produce the same eventual substitution graph is also a key technical issue, which is necessary in order to verify the associativity of the full graph substitution operation.

With the formal properties of graph substitution now established, the sixth chapter starts by verifying the consequence that any operation compatible with graph substitution can be described as graph substitution into a fixed graph. This includes all of the major operations of Chapter 3, which establishes graph substitution as a strong unifying principle. For example, grafting can be viewed as graph substitution into a graph with two vertices, where the inputs and outputs of these two vertices match those of the two graphs to be grafted. Contraction becomes graph substitution into a one vertex graph with a directed loop, and so on. Also included is verification that graph substitution preserves the key technical properties of being either (simply-)connected or wheel-free. Since these properties are easily established for many of our representing graphs, it follows immediately that many operations preserve these properties as well. The chapter ends with an extensive series of technical lemmas, which we refer to as the calculus of graph substitution. In all cases, there is a result involving graph substitutions with relatively few vertices, and in most cases this is paired with a corresponding statement about the interaction of two graph operations. It is convenient for various purposes later to have these results collected somewhere, and at this point in the presentation, they also serve as a list of examples of graph substitutions computed explicitly. Finally, they set the stage for the Reidemeister theory of graphs and strong generating sets introduced in Chapter 7, as well as the examples of compatible pairs of strong generating sets in Chapter 8.

Chapter 7 is dedicated to determining strong generating sets for all of the major examples of graph groupoids. The consequence later will be that an abstract definition in terms of an algebra over a certain monad associated to a pasting scheme will be reduced to instead requiring a series of structure maps and prescribed relations among them. This material, while also somewhat technical, will apply not only to generalized PROPs but also to the algebras and modules over them. As a consequence, each strong generating set will be used three times in Part 2, and the variety inherent in these methods of decomposing graphs while performing only operations whose generating graphs are included within the pasting scheme itself is quite attractive. The main idea is to introduce an analog of Reidemeister moves from knot theory, so a strong generating set is required to satisfy an analog of Reidemeister's Theorem, connecting any two finite strings of possible graph substitutions with the same composite by a finite string of relaxed moves. We establish strong generating sets for wheeled graphs, wheel-free graphs, level trees, unital trees, wheeled trees, simply-connected graphs, and half-graphs, in addition to the connected graphs and connected wheel-free graphs which appear to be more of a surprise.

Chapter 8 presents the definition of a pasting scheme, implied but never stated by Markl, as a graph groupoid closed under graph substitution and containing the units thereof. A large number of examples is now available, with strong generating sets established earlier in most cases. Next is a discussion of free products of pasting schemes, and the technical conditions necessary to say the union of strong generating sets remains a strong generating set for the free product of pasting schemes. The chapter also includes a discussion of the pasting schemes where each graph contains a single vertex, which implies the strict isomorphism classes of graphs can be used to define the morphisms in a category where graph substitution defines the composition law.

The ninth chapter introduces the notion of orthogonal pasting schemes, the Kontsevich groupoid associated to a pair of pasting schemes, and the notion of well-matched pasting schemes. The point here is to understand the free construction, or induction functor, for moving from a small pasting scheme up to a larger pasting scheme. Once again, this technical theory will be applied in the context of generalized PROPs as well as that of algebras and modules. The Kontsevich groupoid generalizes an idea underlying Kontsevich's suggestion of using half-PROPs to establish results about dioperads and PROPs, and over a dozen examples are computed explicitly. The idea behind well-matched pasting schemes, studied in some form in [**MV09**], is that the free functor has a particularly convenient presentation. In fact, one of our important non-examples is established in [**MV09**], but we also provide two more non-examples, to contrast with a variety of examples. This discussion ends the foundational work of Part 1.

Part 2 begins with a review of relevant categorical background, including symmetric monoidal categories, unordered tensor products, as well as monads and their algebras. Then a notion of a pointed extension of a monad is presented, which can be used to describe analogs of classical modules over a ring, where algebras over a monad play the role of the ring. Here the technical issue is that we are able to produce a new monad whose algebras will be the analog of modules in question, which will allow us later to produce free module constructions as well as (co)limits of modules over a generalized PROP. The next step is to proceed with the definition of generalized PROPs associated to a pasting scheme. Given a pasting scheme $\mathcal{G}$, there is a monad $F_{\mathcal{G}}$ on the category of appropriately colored objects in a symmetric monoidal $\mathcal{E}$ whose monadic multiplication is induced by graph substitution. A $\mathcal{G}$-PROP is then defined simply as an $F_{\mathcal{G}}$-algebra, and a few simple examples are included to illustrate the basic theory.

Chapter 11 begins with the Biased Definition Theorem and Biased Morphism Theorem, which are the first main results exploiting our strong generating sets from Chapter 7. Essentially, this is a blueprint for how to turn a strong generating set into a characterization of the $\mathcal{G}$-PROPs formally defined as algebras over a monad instead using a collection of structure maps and relations among them. The remainder of the chapter consists of an extensive list of major examples, including Markl Non-Unital Operads, May Operads, Dioperads, Half-PROPs, Properads, PROPs, Wheeled PROPs, Wheeled Properads, and Wheeled Operads. In each case, a complete definition of the relevant structure is carefully stated, which in several cases seems to be new in the literature, followed by a description of how the strong generating set translates into the indicated structure maps and required commutative diagrams, before stating the characterizations formally.

The discussion in Chapter 12 centers around the relationships between the generalized PROPs associated to different pasting schemes and base categories. Given two pasting schemes with one contained in the other, we give a detailed construction of the free-forgetful adjunction between their categories of generalized PROPs and study the left adjoint. This left adjoint contains all such free functors in the operad/PROP literature as special cases. For example, the free wheeled PROP generated by a wheeled operad and the free PROP generated by a Σ-bimodule are both special cases of this construction. In the situation that the pasting schemes are well-matched, we have a simpler construction reminiscent of the monad associated to a pasting scheme, but in this case depending heavily upon the Kontsevich groupoid. We also discuss the change-of-base functors on the category of generalized PROPs for a fixed pasting scheme, for example, to understand when functors or adjoint pairs extend to $\mathcal{G}$-PROP categories.

In Chapter 13 we introduce and discuss algebras over a generalized PROP. This begins by defining endomorphism objects E_A associated to a colored object A for a given pasting scheme, which is subtle whenever contractions might be involved. As usual, we define an algebra over a $\mathcal{G}$-PROP P as a morphism $\mathsf{P} \longrightarrow \mathsf{E}_A$. Then we use the relative endomorphism object associated to a morphism of colored objects, which involves a pullback construction, to characterize a morphism of algebras. There are also various results about the category of algebras under a change of pasting scheme, $\mathcal{G}$-PROP P or base category. Finally, the chapter closes with presentations of P-algebras in terms of each of our strong generating sets, for the same list of examples considered in detail for generalized PROPs.

The focus in Chapter 14 is on two conceptual characterizations of generalized PROPs in terms of more familiar objects. First, we associate to any pasting scheme $\mathcal{G}$ a colored operad whose algebras are exactly the $\mathcal{G}$-PROPs. The essence of this construction is that graph substitution can be repackaged as a colored operad composition, or equivalently as the composition in a multicategory. Generalized PROPs associated to a pasting scheme $\mathcal{G}$ are also characterized as enriched multicategorical functors from a fixed small enriched multicategory into the base category.

Chapter 15 is devoted to the study of modules over a generalized PROP, which is a new definition in this generality. This builds upon the theory of pointed extensions of a monad developed early in Chapter 10 for precisely this purpose. The fundamental viewpoint is that a classical bimodule over a ring can be viewed as involving multiplications of a number of factors where a single factor is the module and all others are the ring. The key technical mechanism here involves pointed graphs, which is why we worked out the implications of strong generating sets for pointed graphs and their graph substitutions. In essence, we take the construction of the monad $F_{\mathcal{G}}$ associated to a pasting scheme and insert pointed graphs in appropriate places to construct a pointed extension, which leads to the definition of modules. The theory from Chapter 10 still provides a new monad with these as its algebras, so a free object construction and the existence of (co)limits. We then provide a series of results about changes of base category, pasting scheme, or $\mathcal{G}$-PROP P. Finally, the chapter ends with characterizations of modules in concrete terms for each of our usual examples, once again exploiting our strong generating sets, and verifying that our definition coincides with others in the literature even if it is presented differently.

Finally, Chapter 16 is devoted to the study of May modules over an algebra over an operad. In this case, the initial definition is given in concrete terms, but then shown to agree with a notion of module associated to a pointed extension of a monad once again. When using the material of Chapter 14 to view a $\mathcal{G}$-PROP P as the algebras over the specific colored operad indicated there, the result is to recover the modules over P in the sense of Chapter 15 as May modules. This point of view fits in well with deformation theory (e.g., [**FMY09**]) and provides closer connections to a variety of previous work.

Organizational Graphics

Diagram of the inclusions of pasting schemes:

(0.1)

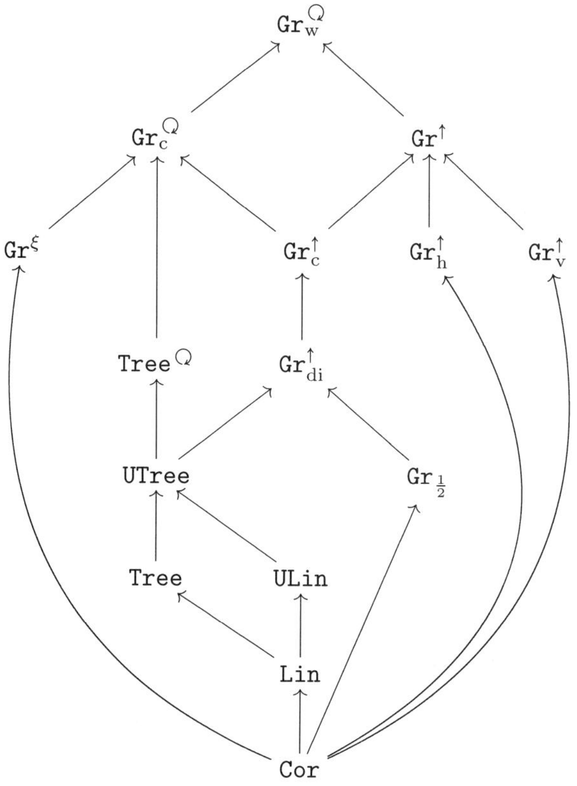

Structure	**Pasting Scheme**	**Gen. Set**	**Key Refs.**
Markl non-unital operads	Level trees $\mathtt{Tree}$	$\mathcal{T}^{Tree}$	Thm 7.35 Nota 8.10 Sec 11.3
May operads	Unital trees $\mathtt{UTree}$	$\mathcal{T}^{UTree}$	Thm 7.41 Nota 8.10 Sec 11.4
Dioperads	Simply connected graphs $\mathtt{Gr}^{\uparrow}_{\mathrm{di}}$	$\mathcal{T}^{di}$	Thm 7.57 Nota 8.9 Sec 11.5
Half-PROPs	Half-graphs $\mathtt{Gr}_{\frac{1}{2}}$	$\mathcal{T}^{1/2}$	Thm 7.64 Nota 8.9 Sec 11.6
Properads	Connected wheel-free graphs $\mathtt{Gr}^{\uparrow}_{\mathrm{c}}$	$\mathcal{T}^{\uparrow}_{\mathrm{c}}$, $\mathcal{T}^{\uparrow}_{\mathrm{c},2}$	Thms 7.67, 7.81 Nota 8.9 Sec 11.7
PROPs	Wheel-free graphs $\mathtt{Gr}^{\uparrow}$	$\mathcal{T}^{\uparrow}$	Thm 7.27 Nota 8.9 Sec 11.8
Wheeled PROPs	Wheeled graphs $\mathtt{Gr}^{\circlearrowleft}_{\mathrm{w}}$	$\mathcal{T}^{\circlearrowleft}$	Prop 7.22 Nota 8.4 Sec 11.9
Wheeled Properads	Connected wheeled graphs $\mathtt{Gr}^{\circlearrowleft}_{\mathrm{c}}$	$\mathcal{T}^{\circlearrowleft}_{\mathrm{c}}$	Thm 7.88 Nota 8.9 Sec 11.10
Wheeled Operads	Wheeled trees $\mathtt{Tree}^{\circlearrowleft}$	$\mathcal{T}^{\circlearrowleft Tree}$	Thm 7.53 Nota 8.10 Sec 11.11
Enriched Categories	Linear (incl. excep.) graphs $\mathtt{ULin}$		Ex 8.7 Subsec 10.5.1
Contraction PROPs	Contracted Corollas $\mathtt{Gr}^{\xi}$		Nota 8.4 Subsec 10.5.2
hPROPs	Unions of Corollas $\mathtt{Gr}^{\uparrow}_{\mathrm{h}}$		Nota 8.4 Subsec 10.5.3
vPROPs	Grafted Corollas $\mathtt{Gr}^{\uparrow}_{\mathrm{v}}$		Nota 8.4 Subsec 10.5.3
Bimodules	Permuted Corollas $\mathtt{Cor}$		Nota 8.4 Ex 10.44

Acknowledgments. The authors would like to express their thanks for helpful discussions during this project to J. Peter May, Sinan Yalin, Martin Frankland, Matt Ando, Michael Batanin, Clemens Berger, and the anonymous referees.

DONALD YAU
MARK W. JOHNSON

List of Notations

Notation	Page	Description
$\mathfrak{C}$	3	set of colors
$\lvert S\rvert$	3	cardinality of a set
$\underline{c}$, $\underline{d}$	3	profiles
$(\underline{c}, \underline{d})$	4	concatenation of profiles
Σ_n	4	symmetric group on n letters
$\sigma\underline{d}$	4	permutation of a profile
$\mathcal{P}(\mathfrak{C})$	4	groupoid of profiles
$[\underline{c}]$	4	orbit type of a profile
$\Sigma_{[\underline{c}]}$, $\Sigma_{\underline{c}}$	4	orbit subgroupoid of a profile
$\sigma\langle i_1, \ldots, i_n\rangle$	4	associated block permutation
$\underline{c}'$	4	k-segment within $\underline{c}$
$\underline{c} \circ_{\underline{c}'} \underline{d}$ or $\underline{c} \circ_l \underline{d}$	4	profile with segment replaced
$\mathfrak{F}$	6	set of flags
G, H, K	6	graphs
$\mathrm{Flag}(G)$	6	the (finite) set of flags of a graph
ι_G	6	the involution of a graph
π_G	6	the free involution of the exceptional legs of a graph
$\mathrm{Vt}(G)$	8	the set of vertices of a graph
$\mathrm{Leg}(G)$	8	the set of legs of a graph
$\mathrm{Flag}_{\mathrm{e}}(G)$	8	the exceptional flags of a graph
$\mathrm{Flag}_{\mathrm{o}}(G)$	8	the ordinary flags of a graph
$\mathrm{Flag}_{\mathrm{i}}(G)$	8	the internal flags of a graph
$\mathrm{Flag}_{\mathrm{ie}}(G)$	8	the internal exceptional flags (in exceptional loops)
$\mathrm{Leg}_{\mathrm{e}}(G)$	8	the exceptional legs of a graph
$\mathrm{Leg}_{\mathrm{o}}(G)$	8	the ordinary legs of a graph
$\mathrm{Edge}_{\mathrm{i}}(G)$	8	the internal edges of a graph
$\mathrm{Edge}_{\mathrm{e}}(G)$	8	the exceptional edges of a graph
κ, κ_G	9	a coloring for a graph
δ, δ_G	10	a direction for a graph
$\mathrm{in}(v)$, $\mathrm{in}(G)$	10	the inputs of a vertex or a full graph
$\mathrm{out}(v)$, $\mathrm{out}(G)$	10	the outputs of a vertex or a full graph
ℓ_v, ℓ_G	11	a listing for a vertex or a full graph
$C_{(\underline{c};\underline{d})}$	13	the $(\underline{c};\underline{d})$-corolla
C_v	13	the corolla associated to a vertex
$\uparrow_{\underline{c}}$	14	the $\underline{c}$-exceptional edge
$\circlearrowleft_{\underline{c}}$	14	the $\underline{c}$-exceptional loop

$C^{j,i}_{(\underline{a};\underline{b};\underline{c};\underline{d})}$	29	a basic dioperadic graph
$C_{(\underline{c};d)} \circ_i C_{(\underline{b};c_i)}$	30	a simple tree
$T\big(\{\underline{b}^i\};\underline{c};d\big)$	32	a special tree
$\ell_{\sigma G \tau}$	36	the permuted listing for an input and output relabeling
$\sigma G \tau$	36	the input and output relabeling of a graph
$\sigma C_{(\underline{c};\underline{d})}\tau$	37	a permuted corolla
$\coprod_{j=1}^{r} G_j$, $G_1 \sqcup G_2$	37	a disjoint union of graphs
G_{ord}	39	the ordinary part of a graph
$G_1 \boxtimes G_2$	39	the grafting of two graphs with matching profiles
$C_{(\underline{b};\underline{c};\underline{d})}$	40	a grafted corollas
$G_1 \boxtimes^{\underline{c}'}_{\underline{b}'} G_2$	41	the partial grafting of two graphs with matching segments
$C^{l_c,l_b,k}_{\underline{a},\underline{b},\underline{c},\underline{d}}$	42	a partially grafted corollas
$G_1 \circ_i G_2$	43	a partial grafting with the single output matching the ith input
$(G_1)_j{\circ_i}(G_2)$	44	a partial grafting with the jth output matching the ith input
$\xi^i_j G$	44	a contraction with the jth input connected to the ith output
$\xi^i_j C_{(\underline{c};\underline{d})}$	45	a contracted corolla
$\mathrm{Aut}_{\mathrm{str.}}(G)$	48	the strict automorphism group of a graph
$\mathrm{Aut}_{\mathrm{w.}}(G)$	53	the weak automorphism group of a graph
$\mathtt{Gr}^{\circlearrowright}_{\mathrm{str}}\binom{\underline{d}}{\underline{c}}$	55	the groupoid of $(\underline{c};\underline{d})$-wheeled graphs and strict isomorphisms
$\mathtt{Gr}^{\circlearrowright}_{\mathrm{w}}$	55	the groupoid of wheeled graphs and weak isomorphisms
$\mathtt{Cor}$	55	the groupoid of permuted corollas and weak isomorphisms
$\mathtt{Gr}^{\circlearrowright}_{\mathrm{c}}$	55	the groupoid of connected wheeled graphs and weak isomorphisms
$\mathtt{Gr}^{\xi}$	55	the groupoid of possibly repeatedly contract -ed corollas and weak isomorphisms
$\mathtt{Tree}^{\circlearrowright}$	55	the groupoid of wheeled trees and weak isomorphisms
$\mathtt{Gr}^{\uparrow}$	56	the groupoid of wheel-free graphs and weak isomorphisms
$\mathtt{Gr}^{\uparrow}_{\mathrm{c}}$	56	the groupoid of connected wheel-free graphs and weak isomorphisms
$\mathtt{Gr}^{\uparrow}_{\mathrm{di}}$	56	the groupoid of simply-connected graphs and weak isomorphisms
$\mathtt{Gr}_{\frac{1}{2}}$	56	the groupoid of half-graphs and weak isomorphisms
$\mathtt{Tree}$	56	the groupoid of level trees and weak isomorphisms

$\mathtt{UTree}$	56	the groupoid of level trees and exceptional edges with weak isomorphisms
$\mathtt{Gr}^{\uparrow}_{\mathrm{h}}$	56	the groupoid of finite unions of corollas and their input and output relabelings with weak isomorphisms
$\mathtt{Gr}^{\uparrow}_{\mathrm{v}}$	56	the groupoid of wheeled graphs weakly isomorphic to iterated graftings of corollas with weak isomorphisms
$\mathtt{Lin}$	56	the groupoid of iterated graftings of corollas with unique inputs and outputs with weak isomorphisms
$G(H_v)$	58	the (pre-)graph substitution where each v is replaced by H_v
$\widehat{G}$, $\widehat{H}$, $\widehat{K}$	61	pre-graphs
$\mathrm{Ambig}(\widehat{K})$	61	the set of ambiguous flags in a pre-graph
$\mathrm{Arm}(\widehat{K})$	61	the set of arms, or ambiguous flags paired with ordinary flags, in a pre-graph
$\mathrm{Ambig}^{\uparrow}(\widehat{K})$	65	the set of ambiguous components in a pre-graph
$\mathrm{Ambig}^{\circlearrowright}(\widehat{K})$	65	the set of ambiguous loops in a pre-graph
$\mathrm{Leg}_{\mathrm{q}}(\widehat{K})$	65	the set of quasi-legs in a pre-graph
$\mathrm{Edge}_{\mathrm{q}}(\widehat{K})$	65	the set of quasi-edges in a pre-graph
G_{in}	79	the input extension of a graph
G_{out}	80	the output extension of a graph
$\sigma^{(i)}$	84	a permutation of a profile with an entry removed
$\lambda \circ_{\underline{b}'} \sigma$	90	outer permutation for a partial grafting
$\mathcal{H}$	109	a graph simplex
$\mathrm{sub}(\mathcal{H})$	109	the substitution of a graph simplex
$\mathcal{T}$	110	a set of graphs
$\mathcal{T}_*$	111	the set of all pointed versions of graphs in $\mathcal{T}$
$\mathcal{T}^{\circlearrowright}$	113	the generating graphs for wheeled graphs
$\mathcal{W}^{\circlearrowright}$	113	the strong set of moves for wheeled graphs
$\mathcal{T}^{\uparrow}$	115	the generating graphs for wheel-free graphs
$\mathcal{W}^{\uparrow}$	115	the strong set of moves for wheel-free graphs
$\mathcal{T}^{Tree}$	118	the generating graphs for level trees
$\mathcal{W}^{Tree}$	118	the strong set of moves for level trees
$\mathcal{T}^{UTree}$	119	the generating graphs for unital trees
$\mathcal{W}^{UTree}$	119	the strong set of moves for unital trees
$\mathcal{T}^{\circlearrowright Tree}$	122	the generating graphs for wheeled trees
$\mathcal{W}^{\circlearrowright Tree}$	122	the strong set of moves for wheeled trees
$\mathcal{T}^{di}$	123	the generating graphs for simply-connected graphs
$\mathcal{W}^{di}$	123	the strong set of moves for simply-connected graphs
$\mathcal{T}^{1/2}$	124	the generating graphs for half-graphs
$\mathcal{W}^{1/2}$	124	the strong set of moves for half-graphs

$\mathcal{T}_c^{\uparrow}$	125	the generating graphs for connected wheel-free graphs
$\mathcal{W}_c^{\uparrow}$	125	the strong set of moves for connected wheel-free graphs
$\mathcal{T}_{c,2}^{\uparrow}$	128	the alternative generating graphs for connected wheel-free graphs
$\mathcal{W}_{c,2}^{\uparrow}$	128	the alternative strong set of moves for connected wheel-free graphs
$\mathcal{T}_c^{\circlearrowleft}$	129	the generating graphs for connected wheeled graphs
$\mathcal{W}_c^{\circlearrowleft}$	129	the strong set of moves for connected wheeled graphs
$\mathtt{G}_S$	134	graphs in $\mathtt{G}$ whose vertices all have profiles in S
$\mathcal{G} = (S, \mathtt{G})$ or $\mathcal{G}'$	134	pasting schemes
$\mathcal{G} \leq \mathcal{G}'$	134	partial order on pasting schemes
$Min(S)$	135	minimal pasting scheme, solely of permuted corollas
$\mathtt{Lin}$	136	linear pasting scheme
$c \in S$	136	color occurring in a pasting scheme
$\mathtt{ULin}$	136	unital linear pasting scheme
$\mathfrak{M}$	138	a multicategory
$\mathcal{G}_1 * \mathcal{G}_2$	139	free product of pasting schemes
$\overline{\mathtt{Gr}_h^{\uparrow}}$	139	pasting scheme of relabeled unions of corollas and $\underline{c}$-exceptional edges
$\mathtt{Gr}_{\mathrm{ord}}^{\circlearrowleft}$	139	pasting scheme of ordinary wheeled graphs
$\mathtt{Cor}_S^{\mathrm{in/out}}$	139	the minimal unital pasting scheme
$D(\mathcal{H})$	141	the deviation of a weakly separating graph simplex
$\mathcal{C}(\mathcal{G})$	145	the associated category of a monogenic pasting scheme
$\mathtt{G}_{cor}$	148	the subgroupoid of permuted corollas in $\mathtt{G}$
$\mathtt{G}_{ecor}$	148	the extended corollas in $\mathtt{G}$
$\mathtt{G}_{ntriv}$	148	subgroupoid of wheeled graphs without exceptional legs
$\mathtt{Kont}(\mathcal{G}, \mathcal{G}')$	149	the Kontsevich groupoid of $\mathcal{G} \leq \mathcal{G}'$
I	160	the unit in a (symmetric) monoidal category
$\mathcal{E}$ or $\mathcal{D}$	161	(symmetric) monoidal categories
$\bigotimes_\sigma A_x$	164	ordered tensor product
$\bigotimes_{x \in X} A_x$, $\bigodot_{i=1}^n A_i$	165	unordered tensor product
T or (T, μ, ν)	165	a monad
(X, γ)	166	an algebra over a monad
$\mathbf{Alg}(T)$	167	the category of algebras over a monad
$T(?,?)$	169	a pointed extension of a monad
$\overline{T}_X$	169	source of multiplication in a pointed extension of a monad
$\mathrm{Module}_T(X)$	170	the category of modules over X with

		respect to T
$\mathcal{F}_X$	172	the monad for X-modules
$\mathcal{E}^{\mathcal{D}}$	176	a category of diagrams in $\mathcal{E}$
$dis(S)$	177	a discrete subcategory
$\mathcal{E}^{dis(S)}$	177	the category of S-colored objects
$\mathcal{E}^S$	177	the category of S factor bimodules
$\mathcal{E}^{\mathcal{P}(\mathfrak{C})^{op}\times\mathcal{P}(\mathfrak{C})}$	177	the full category of bimodules
P, Q	177	bimodules
$\mathsf{P}[G]$	179	P-decorated graph
$F_{\mathcal{G}}$	181	the monad associated to a pasting scheme
$\eta_{[G]}$	181	the inclusion of a summand in $F_{\mathcal{G}}$
μ_{P}	182	the multiplication of the monad $F_{\mathcal{G}}$
ν_{P}	183	the unit of the monad $F_{\mathcal{G}}$
(P,γ)	184	a $\mathcal{G}$-PROP
$\mathtt{PROP}^{\mathcal{G}}$	184	the category of $\mathcal{G}$-PROPs in $\mathcal{E}$
$\mathbf{1}_c$	186	a vertical unit of a $\mathcal{G}$-PROP
$\gamma_{\mathcal{H}}$	192	a structure map of a $\mathcal{T}$-algebra
$\circ_i$	195	a comp-i operation in a Markl non-unital operad
${}_j\circ_i$	199	a j-comp-i operation in a dioperad
${}_j\circ$	202	a j-comp operation in a half-PROP
$\boxtimes^{\underline{c}'}_{\underline{b}'}$	204	a properadic composition
$\boxtimes^{\underline{c}'}_{\underline{b}'}(\tau;\sigma)$	207	an extended properadic composition
$\otimes_h$	209	the horizontal composition in a PROP
$\otimes_v$	209	the vertical composition in a PROP
$\mathbf{1}_{\varnothing}$	209	the empty (or horizontal) unit in a PROP
ξ^i_j	213	a contraction in a wheeled PROP
P_w, P_o	218	the wheeled and operadic parts of a wheeled operad
ρ	218	the right P_o-action on P_w in a wheeled operad
U	223	a forgetful functor from $\mathcal{G}'$-PROP to $\mathcal{G}$-PROP for $\mathcal{G} \leq \mathcal{G}'$
L	223	the left adjoint of some U as above
$\mathcal{D}\binom{\underline{d}}{\underline{c}}$	228	an extension category for $\mathcal{G} \leq \mathcal{G}'$
$\mathcal{M}$	240	a symmetric monoidal $\mathcal{E}$-category
$X_{\underline{c}}$	241	tensor extension of a colored object
$f_{\underline{c}}$	241	tensor extension of a morphism of colored objects
$\mathsf{E}_{X,Y}$	241	mixed endomorphism object of two colored objects
E_X	241	endomorphism object of a colored object
$\mathsf{E}_f \in \mathbf{\Sigma}_S$	241	relative endomorphism object of a morphism of colored objects
f_*	241	postcomposition map in an endomorphism object
f^*	241	precomposition map in an endomorphism

		object
E^c_X	245	a coendomorphism operad
$\mathbf{Alg}_{\mathcal{M}}(\mathsf{P})$, $\mathbf{Alg}(\mathsf{P})$	251	the category of algebras over a generalized PROP
$\overline{\mathsf{U}}_{\mathcal{G}}$	263	the S-colored operad associated to the pasting scheme $\mathcal{G}$
$\mathsf{U}_{\mathcal{G}}$	264	the **Set**-valued operad associated to the pasting scheme $\mathcal{G}$
s, t	264	pairs of profiles in S
$x_{[i,j]}$, $f(x_{[i,j]})$	266	strings of objects in a multicategory
$\mathcal{C}(x_{[1,m]}, y)$	266	object of multi-morphisms in an $\mathcal{E}$-multicategory
$[X, Y]$	267	an internal hom object
$\mathtt{Fun}_{\mathcal{E}}(\mathcal{C}, \mathcal{E})$	269	the category of $\mathcal{E}$-multicategorical functors
$(\mathsf{P}, M)[G, v]$	274	a pointed decorated graph
$F_{\mathcal{G}}(-,-)$	274	the pointed extension of the monad associated to a pasting scheme
$\nu_{\mathsf{P},M}$	274	the unit of $F_{\mathcal{G}}(-,-)$
$\mu_{\mathsf{P},M}$	274	the multiplication of $F_{\mathcal{G}}(-,-)$
(M, λ)	276	a module over a generalized PROP
$\mathbf{Mod}(\mathsf{P})$	277	the category of modules over the $\mathcal{G}$-PROP P
Module(P)	284	the category of biased modules over the biased generalized PROP P
$\circ^l_i$	284	the left comp-i action of a biased module over a Markl non-unital operad
$\circ^r_i$	284	the right comp-i action of a biased module over a Markl non-unital operad
γ^l	285	the left action map of a biased module over a May operad
γ^r	285	the right action map of a biased module over a May operad
${}_j\circ_i{}^l$	285	the left j-comp-i action of a biased module over a dioperad
${}_j\circ_i{}^r$	285	the right j-comp-i action of a biased module over a dioperad
${}_j\circ^l$	286	the left j-comp action of a biased module over a half-PROP
${}_j\circ^r$	286	the right j-comp action of a biased module over a half-PROP
${}^l\boxtimes^{\underline{c}'}_{\underline{b}'}$	287	the left properadic action on a biased module
${}^r\boxtimes^{\underline{c}'}_{\underline{b}'}$	287	the right properadic action on a biased module
$\otimes^l_h$, $\otimes^r_h$	287	the left and right horizontal action map of a biased module over a PROP
$\otimes^l_v$, $\otimes^r_v$	287	the left and right vertical action map of a biased module over a PROP

ρ^l	290	the left action map of a biased module over a wheeled operad
ρ^r	290	the right action map of a biased module over a wheeled operad
$X \otimes Y$	292	tensoring groupoid-indexed functors of opposite variance
$X \otimes_{\mathtt{G}} Y$	292	tensoring over a groupoid
$F_{\mathsf{O}}(A)$	293	the monad for algebras A over an operad O
$F_{\mathsf{O}}(A, M)$	293	the pointed extension of a monad for defining May modules
$\mathcal{F}(A, M)$	296	the associated monad for May modules
$\eta_{A,M}$	296	the (quotient) map $F_{\mathsf{O}}(A, M) \longrightarrow \mathcal{F}(A, M)$
$\mathrm{Mod}_{\mathsf{O}}(A)$	298	the category of May modules

Part 1

Wheeled Graphs and Pasting Schemes

CHAPTER 1

Wheeled Graphs

The goal in Part I is to understand the fundamental operation of graph substitution and the groupoids of graphs that are closed under this operation, called the pasting schemes. Unfortunately, this construction is quite technical once exceptional graphs, with no vertices, are taken seriously. As a consequence, Part I is an extensive technical study of the appropriate notion of wheeled graph.

The most basic method of classifying graphs will be to look at their inputs and outputs, which will be viewed as ordered sequences of colors. As a consequence, we begin by establishing notation for such ordered finite sequences, here called profiles, together with the permutations that would naturally act upon them.

Section 1.2 provides the detailed definition of wheeled graphs, which are different from the traditional objects of study in a graph theory course. In particular, directed cycles are allowed, as are edges incident to no vertex. Then section 1.3 provides a collection of examples of graphs to be considered throughout this monograph.

1.1. Profiles

In this section, we first recall the basic definitions regarding colors and profiles as presented in [**JY09**].

1.1.1. Colors. Fix a non-empty set $\mathfrak{C}$ once and for all, whose elements will be called **colors** These colors will be used to parametrize the inputs and outputs of the operations making up our generalized PROPs, as well as the entries of related constructions like algebras and modules. The primary reason is to provide a mechanism for disallowing certain combinations of operations, by only allowing combinations where the colors match. Throughout, in order to reduce to the uncolored case, simply choose $\mathfrak{C}$ to consist of a single element.

The cardinality of a set S will be denoted by $|S|$.

1.1.2. Profiles and Shuffles. Given an operation, to describe the numbers of inputs and outputs as well as their associated colors, we use the following concept.

DEFINITION 1.1. **$\mathfrak{C}$-profiles** are finite, possibly empty, sequences of colors. If $\mathfrak{C}$ is clear from the context, we will often just say **profiles**.

We use a normal letter, possibly with a subscript (e.g., d_i), to denote a color and an underlined letter (e.g., $\underline{d}$) to denote a $\mathfrak{C}$-profile. If

$$\underline{d} = (d_1, \ldots, d_n),$$

then we write $|\underline{d}| = n$ and sometimes $\underline{d} = d_{[1,n]}$. Given $\underline{d}$ as above with $n \geq 1$, we write

$$\underline{d} \smallsetminus d_i = (d_1, \ldots, d_{i-1}, d_{i+1}, \ldots, d_n),$$

if $1 \le i \le n$, and likewise for $\underline{d} \setminus \{d_i, d_k\}$ if $1 \le i < k \le n$. The **empty profile** is denoted by $\varnothing$.

Suppose $\underline{c} = c_{[1,m]}$ and $\underline{d} = d_{[1,n]}$ are both profiles. Then their **concatenation** is the profile

$$(\underline{c}, \underline{d}) = (c_1, \ldots, c_m, d_1, \ldots, d_n),$$

with the indicated notation, so ordered pairs of profiles will instead be written $(\underline{c}; \underline{d})$. Notice the empty profile clearly serves as a two-sided unit for concatenation.

Permutations $\sigma \in \Sigma_{|\underline{d}|}$ act on a profile $\underline{d}$ from the left by permuting the $|\underline{d}|$ colors, so

$$\sigma \underline{d} = (d_{\sigma(1)}, d_{\sigma(2)}, \ldots d_{\sigma(n)}).$$

Define a category $\mathcal{P}(\mathfrak{C})$ with objects the profiles and where a **morphism**

$$\underline{c} \longrightarrow \underline{d} \in \mathcal{P}(\mathfrak{C})$$

is a permutation σ such that $\sigma(\underline{c}) = \underline{d}$. Such a morphism exists if and only if $\underline{d}$ is in the orbit of $\underline{c}$, which defines a groupoid structure on $\mathcal{P}(\mathfrak{C})$. The **orbit type** of a profile $\underline{c}$, or equivalently the set of objects of the same connected component in the groupoid, will be denoted by $[\underline{c}]$ when necessary. The full (orbit) subcategory of $\mathcal{P}(\mathfrak{C})$ with objects $[\underline{c}]$ is equivalent to $\Sigma_{|\underline{c}|}$ (the difference being that when viewed as a category, the symmetric group $\Sigma_{|\underline{c}|}$ has a single object) and will be denoted $\Sigma_{[\underline{c}]}$ (or even $\Sigma_{\underline{c}}$). As with any groupoid, notice $\mathcal{P}(\mathfrak{C})$ splits as a disjoint union of its orbit subcategories.

LEMMA 1.2. *There is an isomorphism of categories* $\mathcal{P}(\mathfrak{C}) \cong \coprod_{[\underline{c}]} \Sigma_{[\underline{c}]}$.

EXAMPLE 1.3. In the simplest case $\mathfrak{C} = \{*\}$, a profile is really just a non-negative integer, and $\mathcal{P}(\mathfrak{C})$ is the groupoid consisting of the non-negative integers with the permutation groups as endomorphisms, while the concatenation of two profiles represents the sum of these numbers.

NOTATION 1.4. At various points we will want to talk about block permutations, where a chosen permutation tells us how to pass strings past each other, but we do not change the order within any string. In such cases, if σ is a permutation of n letters, we will write $\sigma\langle i_1, \ldots, i_n\rangle$ for the associated block permutation of $i_1 + \cdots + i_n$ letters, where the i_j indicate the lengths of the individual strings within which the operation preserves order.

The most important instance of this concept will come from the following situation, relevant to indexing partial graftings.

DEFINITION 1.5. For a non-empty profile $\underline{c}$ and $k > 0$, a k-**segment** of $\underline{c}$ is a profile $\underline{c}' = (c_l, c_{l+1}, \ldots, c_{l+k-1})$ for some $1 \le l \le |\underline{c}| - k + 1$.

DEFINITION 1.6. Suppose $\underline{e}' = e_{[l,l+k-1]} \subset \underline{e}$ is a k-segment, and $\underline{c}$ is another profile. Define the profile

$$\underline{e} \circ_{\underline{e}'} \underline{c} = \left(e_{[1,l-1]}, \underline{c}, e_{[l+k,|\underline{e}|]}\right).$$

If $\underline{e}' = (e_l)$ is a 1-segment, we will also write $\underline{e} \circ_l \underline{c}$ for $\underline{e} \circ_{\underline{e}'} \underline{c}$.

Later, we will require the following generalized commutativity statements for this notation.

LEMMA 1.7. *There is a unique block permutation σ associated to choices of l_b, l_d, $|\underline{b}| - (l_b + |\underline{b}'|)$, and $|\underline{d}| - (l_d + |\underline{d}'|)$ satisfying*

$$\underline{b} \circ_{\underline{b}'} (\underline{d} \circ_{\underline{d}'} \underline{f}) = \sigma[\underline{d} \circ_{\underline{d}'} (\underline{b} \circ_{\underline{b}'} \underline{f})]$$

for all such choices of $\underline{b}$, $\underline{d}$, and $\underline{f}$. On the other hand, if $\underline{e}'$ is a segment which occurs before the segment $\underline{e}''$ in $\underline{e}$, then we have the equation

$$(\underline{e} \circ_{\underline{e}'} \underline{c}) \circ_{\underline{e}''} \underline{a} = (\underline{e} \circ_{\underline{e}''} \underline{a}) \circ_{\underline{e}'} \underline{c}$$

for any $\underline{a}$ and $\underline{c}$.

PROOF. The left side generically looks like

$$(b_1, \ldots, b_{l_b - 1}, d_1, \ldots, d_{l_d - 1}, \underline{f}, d_{l_d + |\underline{d}'|}, \ldots, d_{|\underline{d}|}, b_{l_b + |\underline{b}'|}, \ldots b_{|\underline{b}|})$$

and before applying σ the right side generically looks like

$$(d_1, \ldots, d_{l_d - 1}, b_1, \ldots, b_{l_b - 1}, \underline{f}, b_{l_b + |\underline{b}'|}, \ldots, b_{|\underline{b}|}, d_{l_d + |\underline{d}'|}, \ldots d_{|\underline{d}|}).$$

Thus, σ is a block permutation that (if both are non-empty) moves the first $l_d - 1$ entries to come after the next $l_b - 1$ entries, as well as (if both are non-empty) moving the last $|\underline{d}| - (l_d + |\underline{d}'|) + 1$ entries to come before the previous block of $|\underline{b}| - (l_b + |\underline{b}'|) + 1$ entries. This depends solely on the indicated lengths, since we are treating $\underline{f}$ as a single invariant block.

For the second claim, in either case the result looks generically like

$$(e_1, \ldots, e_{l_e - 1}, \underline{c}, e_?, \ldots, e_?, \underline{a}, e_?, \ldots, e_{|\underline{e}|})$$

where each of the three $\underline{e}$ segments may be empty. □

1.1.3. Pairs of Profiles. Our fundamental context throughout will involve indexing on pairs of profiles, which we introduce now. The primary motivation is to consider an $X \in \mathcal{E}^{\mathfrak{C}}$ where $\mathcal{E}$ has a symmetric monoidal structure, and consider the collections of morphisms $\mathcal{E}(\overline{X}_{\underline{c}}, \overline{X}_{\underline{d}})$, with

$$\overline{X}_{\underline{c}} = X_{c_1} \otimes X_{c_2} \otimes \cdots \otimes X_{c_m}.$$

This collection of morphisms is viewed as the natural home of any algebraic structure maps which one might impose upon the entries of X, and it becomes convenient to keep all of these possible structure maps in sight via this presentation.

Given $\mathcal{P}(\mathfrak{C})$ above, the opposite category $\mathcal{P}(\mathfrak{C})^{op}$ is regarded as the category of profiles in which permutations act on the profiles from the right. An element of the product category $\mathcal{P}(\mathfrak{C})^{op} \times \mathcal{P}(\mathfrak{C})$ is written as either $\binom{\underline{d}}{\underline{c}}$ or $(\underline{c}; \underline{d})$, so for appropriate choices $(\tau; \sigma)$ induces a morphism $(\underline{c}; \underline{d}) \longrightarrow (\underline{c}\tau; \sigma\underline{d})$. The notation $\Sigma_{\underline{c};\underline{d}}$ will be used to indicate the orbit subgroupoid containing the object $(\underline{c}; \underline{d})$, which is isomorphic to $\Sigma^{op}_{[\underline{c}]} \times \Sigma_{[\underline{d}]}$ and equivalent to $\Sigma^{op}_{|\underline{c}|} \times \Sigma_{|\underline{d}|}$.

LEMMA 1.8. *There is an isomorphism of categories*

$$\mathcal{P}(\mathfrak{C})^{op} \times \mathcal{P}(\mathfrak{C}) \cong \coprod_{[\underline{c}],[\underline{d}]} \Sigma_{\underline{c};\underline{d}}.$$

1.2. Defining Wheeled Graphs

To discuss our graph related constructions later, especially graph substitutions (section 5.4), we need a precise definition of a generalized graph. The focus in this presentation is on the collection of 'half-edges', or 'flags', so a vertex corresponds to a (possibly empty) set of flags, consisting of those flags which are incident to that vertex. In order to avoid pathological half-edges, any two vertices must then be disjoint when considered as sets of flags, and flags are usually treated in pairs that define edges. Inputs and outputs for the graph will come from flags where one end touches a vertex but the other end remains unpaired, called legs. There is also the potentially surprising possibility of an edge not attached to any vertex, which may be either a pair of legs or simply a closed loop on its own. To some extent, dealing properly with these exceptional edges and exceptional loops is the primary reason for our technical difficulties throughout Part I. In particular, disconnected graphs as well as graphs containing cycles, directed or undirected, are included here.

1.2.1. Basic Graphs. Here we discuss the precise definition of a generalized graph, with additional structures we will often impose introduced a bit later.

DEFINITION 1.9. A **partition** of a finite set X is given by an equality of sets $X = \coprod_{\alpha \in A} Y_\alpha$ with A finite. Each Y_α will be called a cell of the partition, so if $|A| = n$, this may be called an n-celled partition.

Notice finiteness of A is not a consequence of the finiteness of X since any individual Y_α may be empty.

DEFINITION 1.10. An **involution** on a set X is simply a function $\iota : X \longrightarrow X$ such that $\iota \circ \iota$ is the identity map. A **fixed point** of the involution will refer to an $x \in X$ with $\iota x = x$. An **isolated cell** of a partitioned set with involution will refer to a cell Y_α with $\iota Y_\alpha \subset Y_\alpha$.

Notice the containment in the definition of an isolated cell becomes an equality by the nature of an involution, and also implies $\iota(X \smallsetminus Y_\alpha) = X \smallsetminus Y_\alpha$. Also notice a choice of involution is equivalent to the choice of an action by Σ_2, the cyclic group with two elements, which is particularly evident when interpreting the terminology.

We will detail the geometric intuition after presenting the formal definition of a graph. Once again, the key feature is to look at 'half-edges' or 'flags' and allow pairing of flags to build edges. To avoid set-theoretic issues, as with colors, we will choose once and for all an infinite set $\mathfrak{F}$ from which our flags will be chosen.

DEFINITION 1.11. A **generalized graph**, or simply a **graph**, G will refer to a finite set $\mathrm{Flag}(G) \subset \mathfrak{F}$ with involution ι_G, together with a partition and the choice of an isolated cell G_0, such that there is a free involution π_G on the set of fixed points of ι_G within G_0.

If $G_0 = \varnothing$, the graph G is called an **ordinary graph** and if $\mathrm{Flag}(G)$ consists solely of the non-empty (isolated) cell G_0, the graph G is called **exceptional**. As such, G_0 will be referred to as the **exceptional cell** of G and all other cells as **ordinary cells**, or **vertices**, of the graph G. The fixed points of the involution ι will be called the **legs** of the graph, and the orbits of ι together with those of π will be called the **edges** of the graph. Non-trivial orbits of ι will be called the **internal edges** of the graph, so the flags which make up a non-trivial orbit of ι

will be called **internal flags**. The orbits of π will be called **exceptional edges**, while the non-trivial orbits of ι within G_0 will be called **exceptional loops**. Notice exceptional loops remain internal edges, while exceptional edges do not.

The geometric intuition is suggested by the terminology. Each ordinary cell Y_α denotes a vertex in the traditional sense, depicted as a dot, with the flags which are elements of Y_α corresponding to half-edges incident to that vertex. **Isolated vertices**, or dots with no incident half-edges, correspond to empty ordinary cells Y_α in the partition, so ordinary graphs can contain isolated vertices. Internal edges of the graph are formed by connecting pairs of flags which make up a non-trivial orbit of the involution ι. For example, a pair $f \neq \iota(f)$ in the same ordinary cell corresponds to a small loop at that vertex. Similarly, if e and f lie in different ordinary cells of the partition with $\iota(e) = f$, then the pair creates an internal edge connecting these two vertices. Legs have at most one end incident to a vertex (none for exceptional legs), so should be thought of as possible inputs or outputs of graphs as detailed later.

The difficult part to understand is the role of flags in the exceptional cell G_0, where there are no vertices. Once again, non-trivial orbits of the involution ι within G_0 correspond to a pairing of half-edges, but in this case they are glued at both ends to produce exceptional loops incident to no vertex, so visually floating freely. On the other hand, fixed points of ι within the exceptional cell correspond to free-standing legs. The role of the free involution π is to pair up these free-standing half-edges, which then form exceptional edges. The primary role of exceptional edges later will be as placeholders for identity maps, bypassing operations by doing nothing, in some sense.

EXAMPLE 1.12. Consider the following indicative example of a graph, which will be expanded upon below.

(1.1)

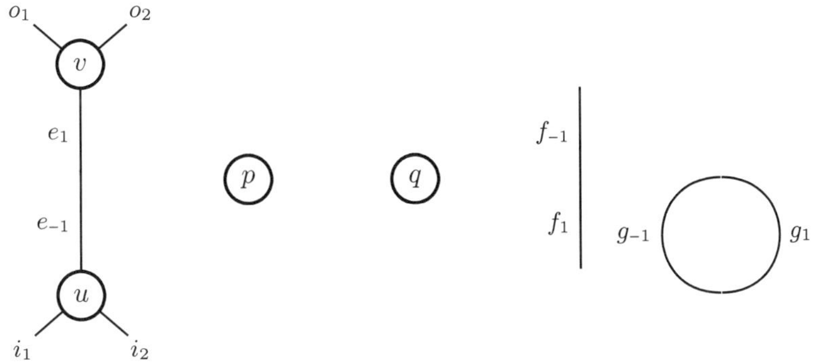

Our convention is to read (directed) graphs from bottom to top. Inputs, outputs, and (directed) internal edges will be represented by line segments (arrows), and vertices will be represented by circled letters such as p or q above.

As a partitioned set, we have $\{\{i_1, i_2, e_{-1}\}, \{e_1, o_1, o_2\}, \varnothing, \varnothing, \{f_{\pm 1}, g_{\pm 1}\}\}$ with the last representing the exceptional cell. In addition, $\iota(e_j) = e_{-j}$, $\iota(g_j) = g_{-j}$ and ι fixes all other elements. This leaves $\pi(f_j) = f_{-j}$ as the only remaining bit of structure.

Notice the two flags forming a 2-cycle of the involution ι are connected in the non-vertex end, or in both ends if the flags are exceptional. The two exceptional legs f_1 and f_{-1} are also connected to form an edge, since they are paired by the free involution π on the exceptional legs. In the circle, the left (resp., right) half is the flag g_{-1} (resp., g_1).

Our convention will be to list the exceptional cell last when presenting examples of graphs.

The definition of a graph may be presented alternatively as follows:

LEMMA 1.13. *A* ***graph*** *G is equivalent to a quadruple*

$$(\mathrm{Flag}(G), \iota_G, \mathrm{Vt}(G), \pi_G)$$

consisting of:

- $\mathrm{Flag}(G) \subset \mathfrak{F}$*, a finite, possibly empty, set whose elements are called* ***flags,***
- *an involution*
$$\iota_G\colon \mathrm{Flag}(G) \longrightarrow \mathrm{Flag}(G),$$
- $\mathrm{Vt}(G)$*, a finite, possibly empty, set of disjoint, possibly empty, subsets of* $\mathrm{Flag}(G)$ *whose elements are called* ***vertices****, and*
- *a free involution π_G on the set of fixed points of ι_G which do not appear in any vertex*

such that defining the subset

$$\mathrm{Flag}_{\mathrm{e}}(G) = \mathrm{Flag}(G) \smallsetminus \coprod_{v \in \mathrm{Vt}(G)} v \tag{1.2}$$

of $\mathrm{Flag}(G)$*, whose elements are called* ***exceptional flags****, we require*

$$\iota_G\left(\coprod_{v \in \mathrm{Vt}(G)} v\right) \subseteq \coprod_{v \in \mathrm{Vt}(G)} v \quad \textit{and} \quad \iota_G\left(\mathrm{Flag}_{\mathrm{e}}(G)\right) \subseteq \mathrm{Flag}_{\mathrm{e}}(G).$$

Notice the containments above both become identities by the involution property.

REMARK 1.14. The description of a graph in Lemma 1.13 is more in keeping with other sources, such as [**Mar08**] and [**MMS09**]. However, the description in terms of partitioned sets is conceptually simpler as well as having the advantage of including the exceptional edges and exceptional loops within the definition of a graph. This improvement is important for our effort to clarify technical issues by reducing wherever possible to arguments at the level of graphs.

NOTATION 1.15. $\mathrm{Flag}(G)$ will denote the set of flags in G. $\mathrm{Vt}(G)$ will denote the set of vertices, or ordinary cells, of G. $\mathrm{Leg}(G)$ will denote the set of fixed points of ι_G. As a first variation on these notations, $\mathrm{Flag}_{\mathrm{e}}(G)$ will denote the elements of G_0, called the **exceptional flags**. On the other hand, $\mathrm{Flag}_{\mathrm{o}}(G)$ will denote the **ordinary flags**, or those flags which are incident to some vertex, or even those flags which appear in some ordinary cell of the partition. Then $\mathrm{Leg}_{\mathrm{e}}(G)$ will denote the **exceptional legs**, or $\mathrm{Leg}(G) \cap \mathrm{Flag}_{\mathrm{e}}(G)$, which forms the domain of π by definition. The other legs will be called **ordinary legs** and denoted by $\mathrm{Leg}_{\mathrm{o}}(G)$. Recall flags which are not part of any leg are called internal flags and denoted $\mathrm{Flag}_{\mathrm{i}}(G)$, with the subset consisting of those in exceptional loops denoted $\mathrm{Flag}_{\mathrm{ie}}(G)$. Finally, let $\mathrm{Edge}_{\mathrm{i}}(G)$ denote the set of 2-element orbits of ι, which are called internal edges and let $\mathrm{Edge}_{\mathrm{e}}(G)$ denote the set of orbits of π, which are called **exceptional edges**.

EXAMPLE 1.16. The **empty graph** $\varnothing$ is the partition of the empty set with a single empty cell chosen as the exceptional cell. Notice this is an ordinary graph according to our conventions. Alternatively, it is the unique graph with both $\mathrm{Flag}(\varnothing)$ and the indexing set $\mathrm{Vt}(\varnothing)$ empty.

EXAMPLE 1.17. The graph of example 1.12 may also be described explicitly as follows.

- $\mathrm{Flag}(G) = \{i_1, i_2, o_1, o_2, e_{-1}, e_1, f_{-1}, f_1, g_{-1}, g_1\}$.
- $\iota(i_j) = i_j$, $\iota(o_j) = o_j$, $\iota(f_{-1}) = f_{-1}$, $\iota(f_1) = f_1$, $\iota(e_j) = e_{-j}$, and $\iota(g_j) = g_{-j}$.
- $\mathrm{Vt}(G) = \{\{i_1, i_2, e_{-1}\}, \{o_1, o_2, e_1\}, \varnothing, \varnothing\}$.
- $\pi(f_j) = f_{-j}$

In particular, the graph G has four vertices, two of which are isolated. Moreover, we have:

- $\mathrm{Flag}_{\mathrm{o}}(G) = \{i_1, i_2, o_1, o_2, e_{\pm 1}\}$.
- $\mathrm{Flag}_{\mathrm{i}}(G) = \{e_{\pm 1}, g_{\pm 1}\}$.
- $\mathrm{Edge}_{\mathrm{i}}(G) = \{\{e_{\pm 1}\}, \{g_{\pm 1}\}\}$.
- $\mathrm{Flag}_{\mathrm{e}}(G) = \{f_{\pm 1}, g_{\pm 1}\}$.
- $\mathrm{Flag}_{\mathrm{ie}}(G) = \{g_{\pm 1}\}$.
- $\mathrm{Leg}(G) = \{i_1, i_2, o_1, o_2, f_{\pm 1}\}$.
- $\mathrm{Leg}_{\mathrm{o}}(G) = \{i_1, i_2, o_1, o_2\}$.
- $\mathrm{Leg}_{\mathrm{e}}(G) = \{f_{\pm 1}\}$.

In what follows, we will describe various graphs concretely using our definition, but the reader who is interested will notice that translating these into the language of Lemma 1.13 is straightforward, as in the example above.

We will be quite interested in the groupoid of graphs later, so we establish the following definition now.

DEFINITION 1.18. An **isomorphism** of graphs is a bijection of partitioned sets preserving the choice of exceptional cell and compatible with both involutions.

In particular, an isomorphism of graphs induces a bijection of vertices, edges, legs, and preserves the notions of internal, ordinary and exceptional.

REMARK 1.19. The collection of isomorphism classes of graphs forms a set, but unless we restrict the choices of flags by introducing $\mathfrak{F}$, the full collection of all graphs is too large to be a set. This restriction does not alter the set of isomorphism classes of graphs, as given any graph one could rename all of the flags by elements of $\mathfrak{F}$ and thereby produce an isomorphic graph. As such, this restriction that flags come from $\mathfrak{F}$ is fairly innocuous and allows us to avoid tripping over set-theoretic issues throughout the presentation.

At various points we will want to treat the sets of flags coming from a finite set of graphs as if they were disjoint. Rather than clutter the presentation with such details, we will assume the interested reader has replaced $\mathfrak{F}$ with $\mathfrak{F}' = \mathfrak{F} \times \mathbb{N}$ and manipulated the index in $\mathbb{N}$ in order to produce disjoint sets with 'the same flag names'.

1.2.2. Structures on Graphs. Now some additional layers of structure one might naturally impose on a graph.

DEFINITION 1.20. A **coloring** for a graph G is a function

$$\kappa : \mathrm{Flag}(G) \longrightarrow \mathfrak{C}$$

which is constant on the orbits of both involutions.

The point here is that colors are really defined for edges, hence the compatibility with involutions that is required. However, we will later want the flexibility of saying the colors are defined for individual flags, particularly when involved in the details of defining graph substitution.

DEFINITION 1.21. A **direction** for a graph G is a function

$$\delta : \mathrm{Flag}(G) \longrightarrow \{-1, 1\}$$

such that

- if $\iota x \neq x$, then $\delta(\iota x) = -\delta(x)$ and similarly,
- if πx is defined then $\delta(\pi x) = -\delta(x)$.

The point here is that we think of an ordinary edge or an exceptional loop as directed from the negative flag toward the positive flag. Unfortunately, it will be important later to consider exceptional edges as directed from positive flag to negative flag, which will also be true of full graphs. A directed internal edge will in general be denoted by a pair such as (f_{-i}, f_i). One can represent such a directed internal edge as follows:

(1.3)

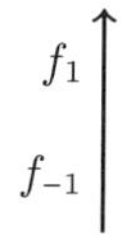

DEFINITION 1.22. Given a vertex v of a graph G with a direction, the **inputs of the vertex**, or $\mathrm{in}(v)$, will refer to those $x \in v$ with $\delta(x) = 1$ and the **outputs of the vertex**, or $\mathrm{out}(v)$, will refer to those $x \in v$ with $\delta(x) = -1$. Similarly, the legs of G with $\delta(x) = 1$ will be called **inputs of the graph**, denoted $\mathrm{in}(G)$, while the legs with $\delta(x) = -1$ will be called **outputs of the graph**, denoted $\mathrm{out}(G)$.

Clearly, when $\mathfrak{C}$ has a unique element (the one-color case) there is a unique coloring of each graph. In this case, the term **$(m;n)$-graph** will be used to indicate a graph G with $|\mathrm{in}(G)| = m$ and $|\mathrm{out}(G)| = n$. A directed $(m;n)$-graph can be represented as

(1.4)

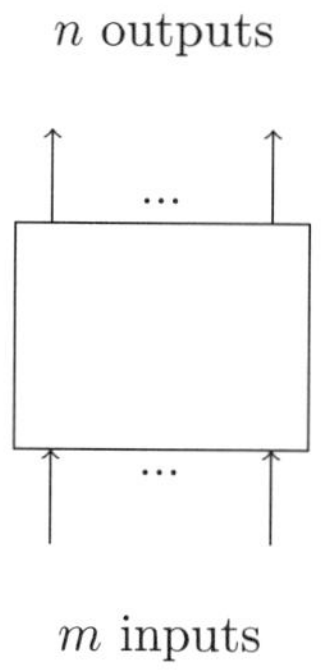

NOTATION 1.23. Let $\mathrm{in}_\mathrm{e}(G)$ denote the set of **exceptional inputs** of the graph, or $\mathrm{Leg}_\mathrm{e}(G) \cap \mathrm{in}(G)$, and $\mathrm{in}_\mathrm{o}(G)$ will denote the set of **ordinary inputs** of the graph, or $\mathrm{Leg}_\mathrm{o}(G) \cap \mathrm{in}(G)$. The obvious variants define the **exceptional outputs** $\mathrm{out}_\mathrm{e}(G)$ and **ordinary outputs** $\mathrm{out}_\mathrm{o}(G)$.

EXAMPLE 1.24. One directed version of Example 1.12 is the following directed (3; 3)-graph

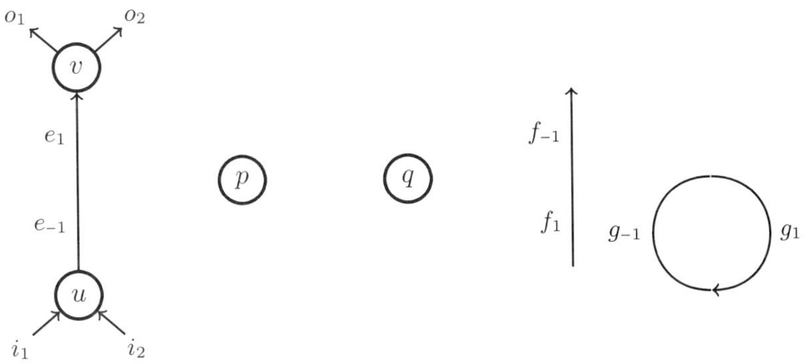

with

$$\delta(o_j) = -1, \quad \delta(i_j) = 1, \quad \delta(e_j) = j, \quad \delta(f_j) = j \quad \text{and} \quad \delta(g_j) = j.$$

In particular, we have:

$$\text{in}_{\text{o}}(G) = \{i_1, i_2\}, \quad \text{out}_{\text{o}}(G) = \{o_1, o_2\}, \quad \text{in}_{\text{e}}(G) = \{f_1\} \quad \text{and} \quad \text{out}_{\text{e}}(G) = \{f_{-1}\}.$$

For ease of reference later, we compile the following observations which are immediate from the definitions.

LEMMA 1.25. *Let G be a directed graph. Then we have the equalities*

$$\begin{aligned} |\text{in}_{\text{e}}(G)| &= |\text{out}_{\text{e}}(G)|, \\ \text{Flag}(G) &= \text{Flag}_{\text{o}}(G) \sqcup \text{Flag}_{\text{e}}(G), \\ \text{Leg}(G) &= \text{Leg}_{\text{o}}(G) \sqcup \text{Leg}_{\text{e}}(G), \\ \text{Leg}_{\text{e}}(G) &= \text{in}_{\text{e}}(G) \sqcup \text{out}_{\text{e}}(G), \\ \text{Leg}_{\text{o}}(G) &= \text{Leg}(G) \cap \text{Flag}_{\text{o}}(G) = \text{in}_{\text{o}}(G) \sqcup \text{out}_{\text{o}}(G), \quad \textit{and} \\ \text{Flag}(G) \smallsetminus \text{Leg}_{\text{e}}(G) &= \text{Flag}_{\text{o}}(G) \sqcup \text{Flag}_{\text{ie}}(G) = \text{Flag}_{\text{i}}(G) \sqcup \text{Leg}_{\text{o}}(G). \end{aligned}$$

The last bit of structure we impose is a bit abstract for now, but will be used together with a coloring and a direction to provide input and output profiles for each vertex and for the full graph.

DEFINITION 1.26. A **listing** for a graph G with direction is a choice for each vertex and for the full graph $u \in \{G\} \bigcup \text{Vt}(G)$ of a bijection of pairs of sets

$$\ell_u : (\text{in}(u), \text{out}(u)) \longrightarrow (\{1, \dots, |\text{in}(u)|\}, \{1, \dots, |\text{out}(u)|\}).$$

If $f \in \text{in}(u)$ and $\ell_u(f) = k$, we will write $f = i_k^u$ or simply i_k. Likewise, if $f \in \text{out}(u)$ and $\ell_u(f) = k$, we will write $f = o_k^u$ or simply o_k.

Notice a listing requires first choosing a direction, but does not require a choice of coloring. Essentially, each ℓ_u picks a linear ordering on the inputs and outputs of either a vertex or of the full graph. Rather than leaving this ordering abstract, it will facilitate clarity later that we are specifying indices precisely.

REMARK 1.27. Henceforth, we will generally intend the ordered set, or profile, rather than the underlying set when we write in(v), out(v), etc.

REMARK 1.28. Be aware this decision to include a listing is not made consistently throughout the literature, but imposing the structure now rather than leaving it ambiguous until later is a deliberate choice for us. For example, in [**Val07**, 2nd Definition on page 4868] certain graphs which are essentially the connected and wheel-free directed graphs in our terminology, are equipped with labelings for the input and output flags at each vertex.

Many authors define the key operation of graph substitution without forcing this listing information at the vertices to decide if the substitution is defined. That is not wholly unreasonable, as they are really asking if some choice of listing exists that would allow them to define the graph substitution. However, there may actually be a choice between several possibilities, particularly when a color is repeated several times as either an input or output. Most authors deal with this question later by using more equivariant structures on their bimodules, essentially making all possible consistent choices at once and then coequalizing them. Since part of our goal is to treat as much of the structure as possible as a consequence of the nature of graphs, with graph substitution the only real source of operations, we prefer to work with slightly more structured graphs in order to eliminate confusion later.

DEFINITION 1.29. A **wheeled graph** G is the combination of a graph together with a choice of a coloring κ, a direction δ, and a listing ℓ. If $\underline{c} = \text{in}(G)$ and $\underline{d} = \text{out}(G)$, then G will be called a **$(\underline{c};\underline{d})$-wheeled graph.**

The term **$(m;n)$-wheeled graph** will be used when $\mathfrak{C}$ has a unique element with $|\text{in}(G)| = m$ and $|\text{out}(G)| = n$.

EXAMPLE 1.30. One wheeled version of the directed $(3;3)$-graph in Example 1.24 is the following $(b, c_1, c_2; d_2, b, d_1)$-wheeled graph

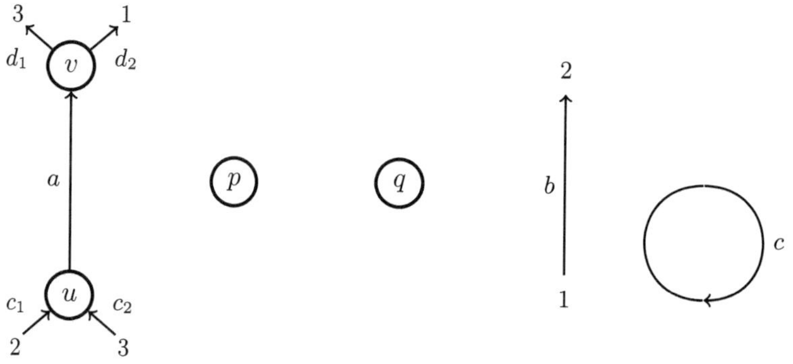

In this case the coloring is defined as

$$\kappa(i_j) = c_j, \quad \kappa(o_j) = d_j, \quad \kappa(e_j) = a, \quad \kappa(f_j) = b \quad \text{and} \quad \kappa(g_j) = c$$

where a, b, c, c_1, c_2, d_1 and d_2 are arbitrary colors. The listing is defined by

$$\ell_G(i_1) = 2, \quad \ell_G(i_2) = 3, \quad \ell_G(f_1) = 1, \quad \ell_G(o_1) = 3, \quad \ell_G(o_2) = 1$$
$$\text{and} \quad \ell_G(f_{-1}) = 2,$$

as well as

$$\ell_{i_1,i_2,e_{-1}}(i_1) = 2, \quad \ell_{i_1,i_2,e_{-1}}(i_2) = 1, \quad \ell_{o_1,o_2,e_1}(o_1) = 2, \quad \text{and} \quad \ell_{o_1,o_2,e_1}(o_2) = 1.$$

The numbers 1, 2 and 3 refer to the inputs and the outputs of the wheeled graph. In order to avoid overloading the above picture with symbols, we left out the flag symbols and the labels of the incoming and the outgoing flags at the two vertices. We will often choose more convenient names for the vertices as well, possibly by choosing an ordering of the vertices.

1.3. Basic Examples of Wheeled Graphs

Here we discuss in detail the crucial wheeled graphs called corollas, the two types of exceptional wheeled graphs, and a collection of other small examples of wheeled graphs.

1.3.1. Corollas.

We start by discussing a simple class of wheeled graphs called corollas, which later serve as the units for our key operation of graph substitution. Mild variations of corollas will be used together with graph substitution to yield certain graph operations, including input/output relabeling, union, (partial) grafting, and contraction.

DEFINITION 1.31. Let $\underline{c} = c_{[1,m]}$ and $\underline{d} = d_{[1,n]}$ be $\mathfrak{C}$-profiles. The **$(\underline{c};\underline{d})$-corolla** $C_{(\underline{c};\underline{d})}$ begins with the partitioned set $\{\{i_1,\ldots,i_m,o_1,\ldots,o_n\},\varnothing\}$ with the empty cell chosen as the exceptional cell and ι the identity map (so no need to define π since there are no exceptional legs). The direction is defined by $\delta(i_j) = 1$ and $\delta(o_j) = -1$, so these represent the inputs and outputs, respectively, of both the unique vertex and the entire graph. The coloring and listing are as indicated by the subscripts and the pair of profiles $(\underline{c};\underline{d})$, so

$$\kappa(i_k) = c_k, \quad \kappa(o_k) = d_k, \quad \ell_w(i_k) = k = \ell_G(i_k) \quad \text{and} \quad \ell_w(o_k) = k = \ell_G(o_k)$$

where w is the unique vertex. A **corolla** is a $(\underline{c};\underline{d})$-corolla for some pair $(\underline{c};\underline{d})$ of profiles.

In other words, the $(\underline{c};\underline{d})$-corolla $C_{(\underline{c};\underline{d})}$ is an ordinary wheeled graph that has exactly one vertex w and no internal edges. It has $m = |\underline{c}|$ inputs that are adjacent to w, are labeled $1,\ldots,m$ from left to right, and have colors $c_1,\ldots,c_m$. Moreover, it has $n = |\underline{d}|$ outputs that are adjacent to w, are labeled $1,\ldots,n$ from left to right, and have colors $d_1,\ldots,d_n$. The incoming and outgoing flags at the vertex w are also labeled from left to right. Graphically, the corolla $C_{(\underline{c};\underline{d})}$ is represented as:

(1.5)
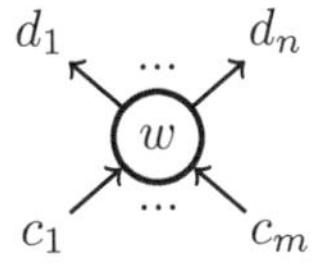

Note that if $\underline{c} = \underline{d} = \varnothing$, then the $(\varnothing;\varnothing)$-corolla, or $C_{(\varnothing;\varnothing)}$,

(1.6)

has an isolated vertex and no flags.

Suppose G is a wheeled graph and $v \in \mathrm{Vt}(G)$ has $\binom{\mathrm{out}(v)}{\mathrm{in}(v)} = \binom{\underline{d}}{\underline{c}}$. Then we define

$$C_v = C_{(\underline{c};\underline{d})}, \tag{1.7}$$

and call C_v the **corolla associated to** v.

1.3.2. Exceptional Edges. We now discuss one of the two basic types of exceptional wheeled graphs, which will serve as units for the operation of grafting, among other things.

DEFINITION 1.32. Let $\underline{c} = c_{[1,m]}$ be a non-empty profile. Define the **$\underline{c}$-exceptional edge**, an exceptional $(\underline{c};\underline{c})$-wheeled graph, denoted

$$\uparrow_{\underline{c}}, \tag{1.8}$$

as the partitioned set $\{\{i_1,\ldots,i_m,o_1,\ldots,o_m\}\}$ with the unique cell chosen as the exceptional cell, ι the identity, $\pi(i_j) = o_j$, and vice-versa. The direction is then chosen as $\delta(i_j) = 1$ and $\delta(o_j) = -1$, with the coloring and listing as suggested by the subscripts and the profile $\underline{c}$, so

$$\kappa(i_j) = c_j = \kappa(o_j) \quad \text{and} \quad \ell_G(i_j) = j = \ell_G(o_j).$$

Note that

$$\mathrm{Flag}(\uparrow_{\underline{c}}) = \mathrm{Leg}_{\mathrm{e}}(\uparrow_{\underline{c}}),$$

i.e., $\uparrow_{\underline{c}}$ consists of only exceptional legs. Graphically, the $\underline{c}$-exceptional edge $\uparrow_{\underline{c}}$ can be represented as

$$\uparrow_{\underline{c}} = \uparrow_{c_1} \cdots \uparrow_{c_m}$$

in which $\uparrow_{c_j}$ consists of the two exceptional legs i_j and o_j. For a single color c, we also say the **c-colored exceptional edge** or the c-exceptional edge

$$\uparrow_c. \tag{1.9}$$

The exceptional wheeled graphs $\uparrow_{\underline{c}}$ will be used to index vertical composition units in $\mathfrak{C}$-colored generalized PROPs (see subsection 8.1.5).

1.3.3. Exceptional Loops. In wheeled PROPs, wheeled properads, and wheeled operads, there is an important operation called contraction. Appropriate contractions, as discussed in section 6.1.5, when applied to the $\underline{c}$-exceptional edge $\uparrow_{\underline{c}}$ give rise to exceptional loops.

DEFINITION 1.33. For a non-empty profile $\underline{c}$, define the **$\underline{c}$-exceptional loop**, an exceptional $(\varnothing;\varnothing)$-wheeled graph, denoted

$$\circlearrowright_{\underline{c}}, \tag{1.10}$$

as the variant of the $\underline{c}$-exceptional edge (Definition 1.32) in which we instead have $\iota(i_j) = o_j$ and vice-versa, so there is no longer a need to define π or a listing.

Note that

$$\mathrm{Flag}\left(\circlearrowright_{\underline{c}}\right) = \mathrm{Flag}_{\mathrm{ie}}\left(\circlearrowright_{\underline{c}}\right),$$

i.e., every flag in $\circlearrowright_{\underline{c}}$ is both internal and exceptional. Graphically, the $\underline{c}$-exceptional loop can be represented as

$$\circlearrowright_{\underline{c}} = \circlearrowright_{c_1} \cdots \circlearrowright_{c_m}$$

in which $\circlearrowright_{c_j}$ consists of the internal exceptional flags i_j and o_j. For a single color c, we also say the **c-colored exceptional loop** or the c-exceptional loop

$$\circlearrowright_c. \tag{1.11}$$

REMARK 1.34. Even though we will occasionally refer to the $\underline{c}$-exceptional loop, it is the case that a listing fails to define an ordering on the collection of exceptional loops in a graph (with more than one). One could impose more structure, namely a choice of ordering on exceptional loops as part of a listing, but such an ordering is irrelevant for the algebraic structures to be indexed by these graphs, in part due to their nature as disconnected pieces with no inputs or outputs.

1.3.4. Other Basic Graphs. Here is a collection of small graphs which will be used in later examples, which are included here to further illustrate the basic definitions.

EXAMPLE 1.35. The **k-vertex cycle** will refer to the following graph with direction

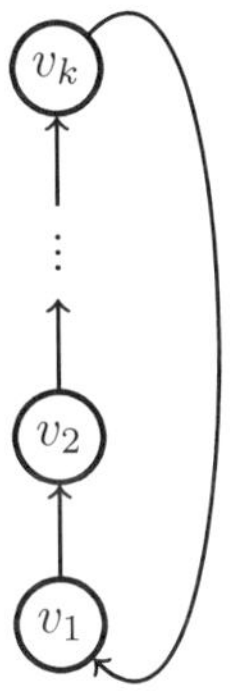

defined as follows. Take the partitioned set

$$\{\{e_1^k, e_{-1}^1\}, \dots \{e_1^{k-2}, e_{-1}^{k-1}\}, \{e_1^{k-1}, e_{-1}^k\}, \varnothing\}$$

with the empty exceptional cell, $\iota(e_j^i) = e_{-j}^i$, $\delta(e_j^i) = j$, and notice there is no need to define π or ℓ, while one could choose $\kappa(e_j^i)$ in various ways.

EXAMPLE 1.36. Let a, b, c and c_1 be arbitrary colors. The **two vertex linear graph with exceptional edge and loop** G is the $(c, c_1; c, c_1)$-wheeled graph

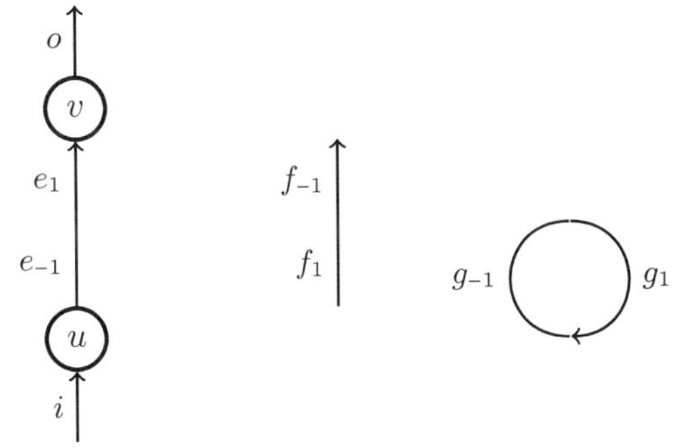

defined as follows. Take the partitioned set

$$\{\{i, e_{-1}\}, \{e_1, o\}, \{f_1, f_{-1}, g_1, g_{-1}\}\}$$

with the last chosen as exceptional cell, with involution fixing all but e_j and g_j, while

$$\iota(e_j) = e_{-j}, \quad \iota(g_j) = g_{-j} \quad \text{and} \quad \pi(f_j) = f_{-j}.$$

For the direction,

$$\delta(i) = 1 = \delta(e_1) = \delta(f_1) = \delta(g_1) \quad \text{and} \quad \delta(o) = -1 = \delta(e_{-1}) = \delta(f_{-1}) = \delta(g_{-1}).$$

For the coloring,

$$\kappa(i) = c = \kappa(o), \quad \kappa(f_j) = c_1, \quad \kappa(e_j) = a \quad \text{and} \quad \kappa(g_j) = b.$$

Finally, for the listing,

$$\ell_G(i) = 1 = \ell_G(o) \quad \text{and} \quad \ell_G(f_1) = 2 = \ell_G(f_{-1}).$$

EXAMPLE 1.37. The **two-vertex cycle with exceptional edge and loop** G' is the $(c_1; c_1)$-wheeled graph, which contains a 2-vertex cycle

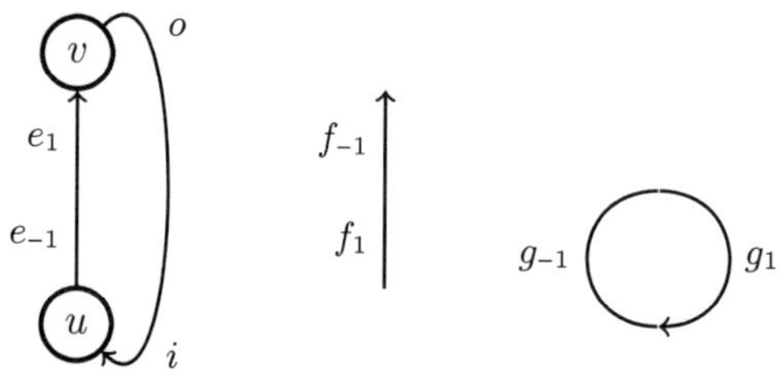

and G' is obtained from G in Example 1.36 by instead declaring $\iota(i) = o$ and $\iota(o) = i$ (so $\ell_G(f_1) = 1$ and $\ell_G(f_{-1}) = 1$ is now forced).

EXAMPLE 1.38. The **butterfly net** graph G is the $(\varnothing; \varnothing)$-wheeled graph

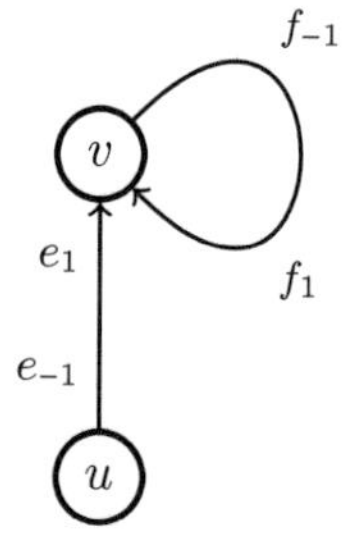

defined as follows. Take the partitioned set

$$\{\{e_{-1}\}, \{e_1, f_1, f_{-1}\}, \varnothing\}$$

with empty exceptional cell, with involution $\iota(e_j) = e_{-j}$ and $\iota(f_j) = f_{-j}$ and no need to define π. For the direction,

$$\delta(e_j) = j = \delta(f_j),$$

while

$$\ell_v(e_1) = 1 \quad \text{and} \quad \ell_v(f_1) = 2,$$

and the coloring arbitrary provided

$$\kappa(e_1) = \kappa(e_{-1}) \quad \text{and} \quad \kappa(f_1) = \kappa(f_{-1}).$$

EXAMPLE 1.39. The **upward ray** graph G is the $(\varnothing; c)$-wheeled graph

defined as follows. Take the partitioned set

$$\{\{o\}, \varnothing\}$$

with empty exceptional cell, ι the identity and no need to define π. Then

$$\delta(o) = -1 \quad \text{and} \quad \kappa(o) = c$$

with ℓ forced.

EXAMPLE 1.40. The **bat and ball** graph G is the $(\varnothing; \varnothing)$-wheeled graph

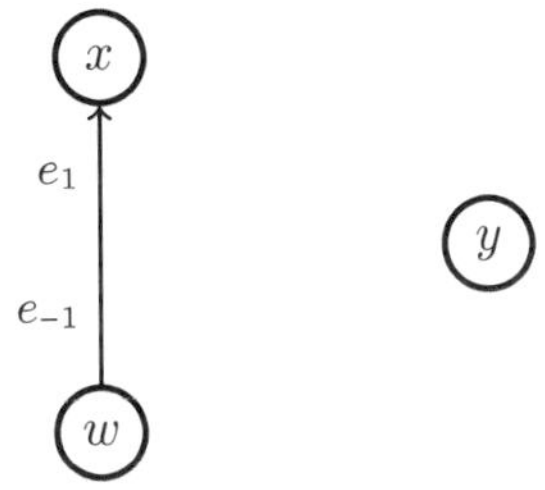

given by the partitioned set

$$\{\{e_{-1}\}, \{e_1\}, \varnothing, \varnothing\}$$

with empty exceptional cell,

$$\iota(e_j) = e_{-j}, \quad \delta(e_j) = j \quad \text{and} \quad \kappa(e_1) = \kappa(e_{-1}).$$

EXAMPLE 1.41. The **walnut** graph G is the $(\varnothing; \varnothing)$-wheeled graph

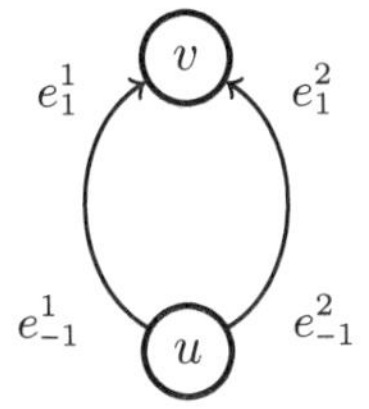

given by the partitioned set

$$\{\{e^1_{-1}, e^2_{-1}\}, \{e^1_1, e^2_1\}, \varnothing\}$$

with empty exceptional cell,

$$\iota(e^k_j) = e^k_{-j}, \quad \delta(e^k_j) = j, \quad \kappa(e^k_1) = \kappa(e^k_{-1}) \quad \text{and} \quad \ell_u(e^k_j) = k.$$

EXAMPLE 1.42. The **growth chart graph** G will denote the $(c,c;c)$-wheeled graph

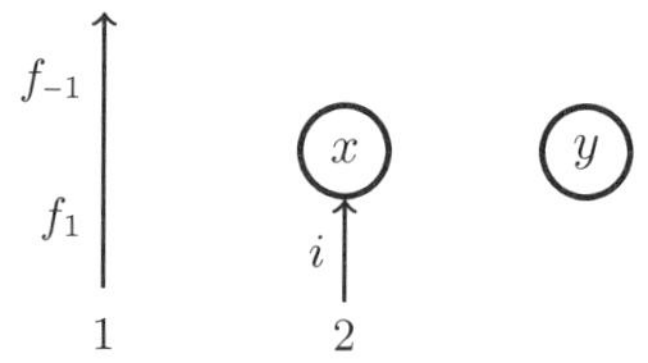

defined by the partitioned set

$$\{\{i\}, \varnothing, \{f_1, f_{-1}\}\}$$

with last the exceptional cell, ι the identity,

$$\pi(f_j) = f_{-j}, \quad \delta(i) = 1, \quad \delta(f_j) = j, \quad \kappa(f_1) = c = \kappa(f_{-1}) = \kappa(i),$$
$$\ell_G(f_1) = 1 \quad \text{and} \quad \ell_G(i) = 2.$$

1.4. Technical Variants of Wheeled Graphs

There are a few variants of the definition of graph that occur for technical reasons in very specific contexts. We separate those from the general definition by including them here, along with some indication of where they become important.

1.4.1. Pointed Graphs. As one might expect, a pointed graph simply involves the choice of a distinguished vertex within the graph, and only isomorphisms preserving distinguished vertices should be considered.

DEFINITION 1.43. A **pointed graph** will denote an ordered pair $(G, \overline{v})$ consisting of a graph and a distinguished vertex. An **isomorphism of pointed graphs** will then denote an isomorphism of graphs $\psi : G \longrightarrow G'$ with $\psi(\overline{v}) = \overline{v}'$.

Of course, this excludes the exceptional graphs without isolated vertices, whose sets of vertices are empty. Notice the graphs with a single vertex may be viewed canonically as pointed graphs. This language will be used heavily when discussing modules over a generalized PROP in Chapter 15.

1.4.2. Ordered Graphs. The natural description of an ordered graph is as a wheeled graph together with a choice of ordering of the vertices.

DEFINITION 1.44. An **ordered graph** will consist of an ordered pair (G, φ) with

$$\varphi : \mathrm{Vt}(G) \longrightarrow \{1, 2, \ldots, |\mathrm{Vt}(G)|\}$$

a bijection. An **isomorphism of ordered graphs** will denote an isomorphism of graphs $\psi : G \longrightarrow G'$ such that $\varphi'\psi = \varphi$.

As above, there is a unique choice of ordering for any graph with a single vertex, but in this case exceptional graphs without isolated vertices are included with an empty set of vertices trivially ordered.

One could work extensively with ordered graphs as an alternative to using the unordered tensor product when defining decorated graphs in Chapter 10 or algebras over a generalized PROP in Chapter 13. They will also be used in Chapter 14 when building a multi-category of wheeled graphs with graph substitution used to define the composition operation.

1.4.3. Multi-stage Graphs. These are generalizations of wheeled graphs where the profiles of the whole wheeled graph are partitioned further into finite ordered sets of $\mathfrak{C}$-profiles and the vertices are ordered. Since we will not use such a construct anywhere, we avoid a formal definition in favor of an illustrative example.

EXAMPLE 1.45. Consider H the wheeled graph of Ex. 1.30 with the vertices ordered by height.

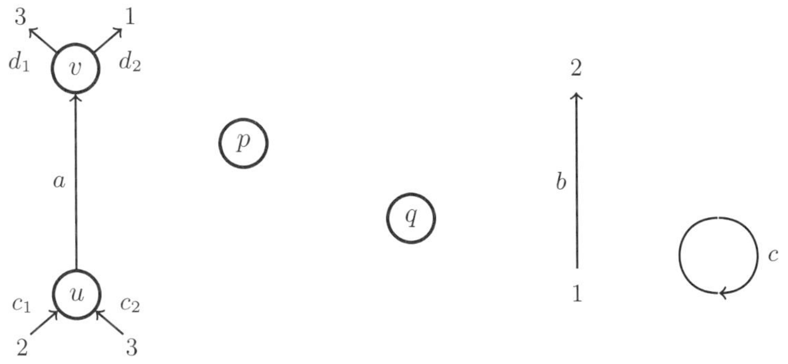

Rather than considering this as a $(b, c_1, c_2; d_2, b, d_1)$-wheeled graph, we could instead think of it as a $((\{c_1, c_2\}; \{d_2, d_1\}), (\{b\}; \{b\}))$-multi-stage graph. Now given G a two vertex multi-stage graph whose first vertex has profiles $(\{c_1, c_2\}; \{d_2, d_1\})$ and whose second vertex has profiles $(\{b\}; \{b\})$, we can form a graph substitution $G(H)$.

This is somewhat different than a related idea of forming $\mathcal{P}(\mathfrak{C})$-colored wheeled graphs, where both the vertices and the full graphs would have inputs/outputs consisting of profiles of $\mathfrak{C}$-profiles. The primary difference is that for multi-stage graphs, the input/output information for a full graph is inherently different from that of an individual vertex. This reflects the fact that there exists a universal operad for describing $\mathcal{P}(\mathfrak{C})$-colored wheeled graphs (see Chapter 14), while the collection of multi-stage graphs would instead require a universal PROP to describe.

One possible use for multi-stage graphs would be tracking processes where all input or output does not occur at the same time, but instead occurs in discrete stages.

1.4.4. Cyclic or Modular Graphs. As discussed in detail in [**MSS02**, Sec. II.5.2], one should essentially ignore the distinction between inputs and outputs, or even the direction, if working with cyclic operads. One natural extension of this idea would come from working with graphs without listings or directions at all, but then there are problems with defining carefully when graph substitutions make sense. This issue is largely irrelevant for planar graphs, such as trees, where orderings can be imposed more naturally. However, for general graphs this added flexibility can cause serious confusion at the level of graphs, an issue which many authors instead address at the level of operational structures. Although this variant could lead to other cyclic structures, this change would lead us away from the core examples of interest to us, as well as our fundamental principle of working only with the indexing graphs wherever possible. As a consequence, allowing such additional flexibility throughout this document was not appealing.

Similarly, when considering modular operads following [**MSS02**, Sec. II.5.3], one should work with graphs equipped with a genus map from the vertices to the non-negative integers satisfying a stability condition. In addition, the underlying equivariant objects of a modular operad differ from those we consider later. Once again, imposing generalizations of stable genus maps in more complex graphs than trees would require a substantial digression from our main line of reasoning, so was avoided throughout.

CHAPTER 2

Special Sets of Graphs

We will need to discuss a variety of technical conditions in order to talk about the different pasting schemes, or types of generalized PROPs, of strongest interest. This requires us to make careful choices when introducing these conditions and to study them in some detail. This chapter begins with a careful presentation of the idea of a path in a wheeled graph. As a consequence, we are able to introduce (simply-)connected and wheel-free graphs, as well as level trees, economically. Other important sets of graphs studied repeatedly later are also introduced, including half-graphs, dioperadic graphs, simple trees, wheeled trees, and special trees. There are also several technical results used to break the set of wheeled trees into three disjoint parts, setting the stage for bootstrapping up to understanding wheeled trees later, especially in Chapter 7.

2.1. Paths and Directed Paths

We would like to introduce notions of (simply-)connected graphs and wheel-free graphs without reference to any topological realization operation. This depends upon first providing a careful description of paths in a graph, in particular the special case of internal paths in a graph. This is unfortunately quite technical, but the same technical background is used again in later chapters, especially when describing the key operation of graph substitution.

2.1.1. Paths. We would like to define a path as a sequence of edges that meet at vertices, but the statements are complicated by the technical differences among the sets of internal edges, exceptional edges or loops, and ordinary legs.

DEFINITION 2.1. A **mono-edge** in a graph G will refer to one of the following:

- a single exceptional edge,
- a single exceptional loop,
- an ordinary internal edge, or
- an ordinary leg.

The following definition is complicated by the desire to allow paths to possibly include legs at the ends, as well as exceptional edges or loops.

DEFINITION 2.2. Given a natural number r together with a pair of binary variables s, $t \in \{0,1\}$, a **path** in a graph G of **length** r

$$P = \left((e^j)_{j=1}^r, (v_j)_{j=s}^{r-t}\right)$$

consists of an r-tuple of mono-edges of G, together with an $(r-s-t+1)$-tuple of vertices, which is empty if $r+1 = s+t$, satisfying:

- $v_i \neq v_j$ for $i < j$ except possibly when $j - i = r$, and
- each e^j is adjacent to both v_{j-1} and v_j whenever these were specified.

A **sub-path** of P is defined by the choice of some $1 \leq i < k \leq r$, so only the mono-edges e^j with $j \in [i,k]$ and incident vertices are considered.

If chosen, the vertices v_0 and v_r are called the **initial vertex** and the **terminal vertex**, respectively, of the path and either will be called an **end vertex**.

Notice an ordinary leg can only be adjacent to a single vertex, so ordinary legs can only occur at either end of a path, if at all. A path starting with an ordinary leg has $s = 1$ (so no v_0), and a path ending with an ordinary leg has $t = 1$ (so no v_r). Only those paths without ordinary legs ($s = 0 = t$) can satisfy $j - i = r$, hence be allowed to satisfy $v_0 = v_r$. Also notice $r = 1$ with e^1 an exceptional edge or exceptional loop and $s = 1 = t$ is one possibility.

Keep in mind that a path is defined in a graph, with no reference to a coloring, direction or listing. Also notice that choosing a pair of sub-paths of P, so choosing $i < k$ and $i' < k'$, it might happen that $k = i'$, in which case their **concatenation** refers to the sub-path associated to $i < k'$.

DEFINITION 2.3. An **internal path** in a graph G is a path in G where each mono-edge e^j is an internal edge whose adjacent vertices are also included in the path.

A **cycle** in G is an internal path in G whose initial vertex is not different from its terminal vertex, while a **trail** is any path in G which is not a cycle in G.

Notice a single exceptional loop is included as a cycle with this definition, since there are no initial or terminal vertices to differ. Also, a trail may be an internal path or not, but cannot be an internal path whose ends coincide or a single exceptional loop. Finally, an internal path is either a single exceptional loop (with $r = s = t = 1$) or every e^j is an ordinary internal edge and P has both end vertices (with $s = t = 0$).

REMARK 2.4. Given an ordinary cycle, a cyclic permutation of the e^j and the v_j yields another cycle. The resulting cycle consists of the same sets of internal edges and vertices. Therefore, it is reasonable to identify a cycle with any of its cyclic permutations.

2.1.2. Directed Paths. Now we would like to define directed paths. Our mechanism is to say that an ordinary internal edge has initial and terminal vertices, while an ordinary input leg has a terminal vertex and an ordinary output leg has an initial vertex. Exceptional edges and exceptional loops are directed by default, so no additional condition is necessary.

DEFINITION 2.5. Let G be a graph with a direction. The **initial vertex** of a mono-edge will refer to the vertex adjacent to the flag with $\delta = -1$ (if these both exist) and the **terminal vertex** of a mono-edge will refer to the vertex adjacent to the flag with $\delta = 1$ (if these both exist).

A **directed path** in G is a path P as in Definition 2.2 such that each ordinary mono-edge e^j has initial vertex v_{j-1} and terminal vertex v_j whenever either has been chosen.

As expected, a **directed internal path** in G is an internal path P as in Definition 2.3 which is also a directed path. A **directed cycle**, also called a **wheel**, is a directed internal path whose underlying path is a cycle. An **undirected cycle** is a cycle that is not the underlying path of any directed path.

Thus, a cycle in a graph with a direction is either a wheel or an undirected cycle, but not both. The simplest example of an undirected cycle is a cycle containing both vertices of the walnut graph 1.41, while any cycle in the k-vertex cycle graph 1.35 is a wheel. As above, a wheel (or an undirected cycle) and its cyclic permutations are identified. Intuitively, a wheel is a loop which one could travel around indefinitely, even though there are only a finite number of internal edges in the wheel. Since a single exceptional loop is always a directed cycle, it constitutes a wheel.

In the following definition and lemma, keep in mind that the larger path P is not directed, but will essentially be decomposed as a concatenation of maximal directed pieces.

DEFINITION 2.6. Suppose P is an internal path in a graph G with a direction.

A **positively directed sub-path** in P is an internal sub-path which is also a directed path

$$v_i \xrightarrow{e^{i+1}} v_{i+1} \xrightarrow{e^{i+2}} \cdots \xrightarrow{e^{i+k}} v_{i+k},$$

in which the indices are taken modulo r if P is a cycle. Such a directed sub-path will be called **positive**, with initial vertex v_i and terminal vertex v_{i+k}.

A **negatively directed sub-path** in P is an internal sub-path where each internal edge e^j has initial vertex v_j and terminal vertex v_{j-1}

$$v_j \xrightarrow{e^{j}} v_{j-1} \xrightarrow{e^{j-1}} \cdots \xrightarrow{e^{j-l+1}} v_{j-l},$$

in which the indices are taken modulo r if P is a cycle. Such a directed sub-path will be called **negative**, with initial vertex v_j and terminal vertex v_{j-l}.

A **directed sub-path** in P means either a positively directed sub-path or a negatively directed sub-path, and will be called **maximal** if there is no *directed* sub-path in P of strictly greater length that contains it.

LEMMA 2.7. *Suppose P is an internal path in a graph G with a direction.*

(1) *Any two distinct maximal directed sub-paths in P are disjoint, except possibly at end vertices.*
(2) *Every internal edge e^j is contained in a unique maximal directed sub-path.*
(3) *Every directed sub-path in P is contained in a unique maximal directed sub-path.*
(4) *P can be written as a concatenation of maximal directed sub-paths of alternating directions.*

PROOF. The first assertion is immediate from maximality. For the next two assertions, simply take the directed sub-path with the largest length containing the internal edge or the directed sub-path in question.

For the final claim, begin with the maximal directed sub-path containing e^1. If this is not all of P, then let e^{k_1} denote the first entry of P it does not contain, whose maximal directed sub-path must then be oppositely directed. Proceeding in this manner produces a sequence of maximal directed sub-paths of P with alternating directions whose concatenation is all of P by finiteness. □

The following technical observation will be important later, when we discuss wheeled trees.

LEMMA 2.8. *An undirected cycle in a graph with a direction must contain a vertex with at least two outgoing flags.*

PROOF. Suppose P is an undirected cycle and $P^1, \ldots, P^k$ are the maximal directed sub-paths in the decomposition of Lemma 2.7(4). By the undirected cycle assumption, we have $k \geq 2$. Identifying P with a cyclic permutation if necessary, we can assume without loss of generality that P^1 is positively graded and P^k is negatively graded, while the initial vertex of P^1 and the terminal vertex of P^k must coincide in order to produce the cycle P. This initial vertex of P^1 must then have at least two outgoing flags, because these distinct maximal directed sub-paths do not have any common internal edges. □

2.2. Connected Graphs

The next two sections are devoted to studying the existence of wheels, undirected cycles, or sufficient paths to connect different flags within a graph. These technical properties will be used to sort graphs into various different groupoids exploited to model various structures as generalized PROPs. First we discuss connectedness for graphs.

2.2.1. Flag-Connected Graphs. To define connected graphs, we have to eliminate isolated vertices and otherwise exploit the following concept.

DEFINITION 2.9. Let G be a graph, and let f and g be two distinct flags in G. We say that f and g are **flag-connected** if there exists a path P in G containing both flags.

A graph G is **flag-connected** if any two distinct flags in it are flag-connected.

Intuitively, in a flag-connected graph, one can travel from any flag to any other flag by moving along paths.

EXAMPLE 2.10. An exceptional edge or exceptional loop constitutes a path (with $r = s = t = 1$), so is flag-connected. On the other hand, this is the only path which can contain a given exceptional flag. In addition, every graph with no flags is trivially flag-connected, so a finite number of isolated vertices is flag-connected.

2.2.2. Connected Graphs.

DEFINITION 2.11. A non-empty graph G is **connected** if either G is an isolated vertex (below Definition 1.11), or G is flag-connected and has no isolated vertices.

In particular, if G is connected it is either ordinary or exceptional, but cannot contain both a non-trivial ordinary part and a non-trivial exceptional part. In fact, G connected can only contain a single exceptional loop or exceptional edge, or a single isolated vertex, and otherwise G must be ordinary, flag-connected, and contain no isolated vertices.

Note that connectedness is defined for a graph with no reference to a direction, a coloring, or a listing.

EXAMPLE 2.12. Examples of connected wheeled graphs include every corolla (Definition 1.31), the c-colored exceptional edge $\uparrow_c$ (1.9), the c-colored exceptional loop $\circlearrowleft_c$ (1.11), the k-vertex cycle (Example 1.35), the butterfly net graph (Example 1.38), the upward ray graph (Example 1.39), and the walnut graph (Example 1.41).

EXAMPLE 2.13. Examples of wheeled graphs that are not connected include the $\underline{c}$-exceptional edge and the $\underline{c}$-exceptional loop with $|\underline{c}| \geq 2$, our generic wheeled graph (Example 1.30), the two vertex linear graph with exceptional edge and loop (Example 1.36), the two vertex cycle with exceptional edge and loop (Example 1.37), the growth chart graph (Example 1.42), and the bat and ball graph (Example 1.40).

We now define a **connected component** of a graph G associated to a given flag e to consist of all flags that are flag-connected to e with the partitioning and involutions inherited from G itself. If G has a direction or coloring, they immediately descend to the connected component. If there is a listing for G, it will remain the same for the connected component at each vertex, since all flags adjacent to the vertex in G will be flag-connected to one another, and so to e at the same time. For the listing of the full graph, simply maintain the ordering of both inputs and outputs imposed by restricting ℓ_G.

Once we discuss disjoint union of graphs, we will be able to describe any graph as a disjoint union of connected components, which can instead consist of isolated vertices as well. If one includes listings, then one may need to alter the listing on the full graph from that given by the disjoint union to recover the full graph, which will also be discussed later.

Next we have two technical observations that will be exploited later.

LEMMA 2.14. *Suppose G is a connected ordinary graph with f an ordinary leg. Then we can define a connected graph $G' = G \setminus \{f\}$ by simply removing this single flag.*

PROOF. To verify we have a wheeled graph, it suffices to describe the listings $\ell_{G'}$ and ℓ_v, for v the unique vertex containing f, by decreasing by one the listing for each leg of the same type listed after f in ℓ_G or ℓ_v. To see that G' is also connected, notice that any path containing f is unnecessary to connect flags other than f, while f itself has become irrelevant. $\square$

The following extension of Lemma 2.8 provides several situations where there must be a vertex with at least two outgoing flags, which will be important in our discussion of wheeled trees.

LEMMA 2.15. *Suppose G is a connected graph with a direction and that one of the following statements hold.*

- *G has at least two vertices without outgoing flags.*
- *G has a wheel and a vertex without outgoing flags.*
- *G has at least two wheels.*

Then there exists a vertex in G with at least two outgoing flags.

PROOF. First note that all three cases imply that G is ordinary, flag-connected, and has no isolated vertices. In each case, the approach is to decompose an internal trail from a vertex v_0 to a vertex v_r, which must exist by connectivity of G, as in Lemma 2.7(4). In this decomposition, it suffices to verify that there is a negatively directed sub-path followed by a positively directed sub-path. As a consequence of maximal sub-paths sharing no flags, the shared vertex then has at least two outgoing flags.

For the case where v_0 and v_r are distinct vertices without outgoing flags, we must have P^1 negatively directed and P^k positively directed, with $k \geq 2$, which suffices.

For the case where v_0 is on a wheel and v_r (not on the wheel) has no outgoing flags, we can choose a new trail if necessary, by moving around the wheel in the opposite direction at first, in order to assume that P^1 is negatively directed. The fact that v_r has no outgoing flags then again suffices to imply P^k is positively directed with $k \geq 2$.

If there are two different wheels, it remains possible that the vertices of the first wheel are all among the vertices of the second wheel. However, this implies at least one vertex in the first has at least two outgoing flags, one from the first wheel and one from the second. Otherwise, we can assume v_0 and v_r are each on one wheel and not on the other. In this case, we can again choose a different trail if necessary, by reversing direction around the wheel(s), in order to assume P^1 is negatively directed and P^k is positively graded, so $k \geq 2$. □

2.3. Simply-Connected and Wheel-Free Graphs

These two related concepts both depend upon the non-existence of cycles, or of wheels (also called directed cycles), within certain graphs.

2.3.1. Simply-Connected Graphs.

Keep in mind in the next definition that a single exceptional loop is a cycle.

DEFINITION 2.16. A connected graph is called **simply-connected** if it contains no cycles.

Like connectedness, being simply-connected is defined for a graph without reference to a direction, a coloring, or a listing. Intuitively, by contracting edges, a simply-connected graph can be shrunk down to a single vertex without breaking any internal edges, or it has trivial fundamental group π_1. Again, our definition based on the technical work with paths requires no reference to any form of topological realization.

EXAMPLE 2.17. Corollas, the c-colored exceptional edge $\uparrow_c$, and the upward ray graph (Example 1.39) are simply-connected. None of the other examples in section 1.3 are simply-connected.

2.3.2. Wheel-Free Graphs.

Next we have the broader class of graphs containing no wheels, although they may contain undirected cycles or fail to be connected, and so may fail to be simply-connected in two different ways.

DEFINITION 2.18. A graph with direction G is said to be **wheel-free** if it contains no wheels.

EXAMPLE 2.19. Here are some examples of wheel-free graphs.

(1) Every simply-connected graph with a direction is wheel-free.
(2) On the other hand, the walnut graph (Example 1.41) is a connected wheel-free graph that is not simply-connected.
(3) The $\underline{c}$-exceptional edge $\uparrow_{\underline{c}}$ with $|\underline{c}| \geq 2$, the growth chart graph (Example 1.42), and the bat and ball graph (Example 1.40) are wheel-free but not connected.

EXAMPLE 2.20. Here are some examples of wheeled graphs that are not wheel-free.

(1) The c-colored exceptional loop $\circlearrowleft_c$, the k-vertex cycle (Example 1.35), and the butterfly net graph (Example 1.38) are connected but not wheel-free.
(2) The $\underline{c}$-exceptional loop $\circlearrowleft_{\underline{c}}$ with $|\underline{c}| \geq 2$, our generic wheeled graph (Example 1.30), the two vertex linear graph with exceptional edge and loop (Example 1.36), and the two vertex cycle with exceptional edge and loop (Example 1.37) are neither connected nor wheel-free.

2.4. Half-Graphs and Dioperadic Graphs

In this section we discuss two types of simply-connected graphs that will be important later, in the discussions of half-PROPs and dioperads.

2.4.1. Half-Graphs. The following set of graphs will serve as the pasting scheme in the discussion of half-PROPs.

DEFINITION 2.21. A **half-graph** is a simply-connected graph with a direction G that satisfies the following conditions:

(1) G is ordinary.
(2) Every vertex in G contains at least three flags with both $\text{in}(v)$ and $\text{out}(v)$ non-empty.
(3) For every internal edge the initial vertex has a single outgoing flag, or the terminal vertex has a single incoming flag, or both.

Graphically, the last condition says that every internal edge in a half-graph must be of one of the following two forms:

Note that the definition of a half-graph requires a direction but does not refer to a coloring or a listing.

EXAMPLE 2.22. There are simply-connected graphs with direction that satisfy each combination of exactly two of the three assumptions in the definition of a half-graph.

(1) For example, consider G the exceptional edge

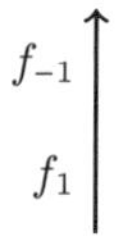

given by $\{\{f_1, f_{-1}\}\}$. It has no ordinary cell, ι is the identity and $\pi(f_i) = f_{-i}$ while $\delta(f_i) = i$. Then G is simply-connected and satisfies only the last two of the three conditions.

(2) On the other hand, modify the previous case to obtain $G' = \{\{f_{-1}, f_1\}, \varnothing\}$

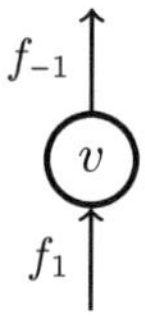

with $v = \{f_{-1}, f_1\}$ the only ordinary cell, empty exceptional cell (hence no need for π), and the same ι and δ as above. Then G' is simply-connected and satisfies only the first and the third conditions above. Note that G' becomes a corolla with one input and one output if it is given the unique listing and if its flags are assigned any colors.

(3) Finally, the graph with direction G depicted as

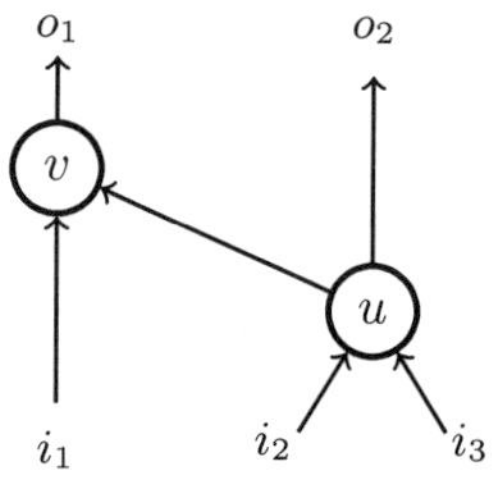

is simply-connected and satisfies only the first two of the above three conditions. More precisely, we define G as the partitioned set

$$\{\{i_2, i_3, e_{-1}, o_2\}, \{i_1, e_1, o_1\}, \varnothing\}$$

with empty exceptional cell and ι fixing all but $\iota(e_i) = e_{-i}$, in addition to

$$\delta(i_j) = 1 = \delta(e_1) \quad \text{and} \quad \delta(o_j) = -1 = \delta(e_{-1}).$$

2.4.2. Basic Dioperadic Graphs. A basic dioperadic graph is a simply-connected (wheel-free) graph with one internal edge, which can be used as the standard building block for simply-connected wheeled graphs. In a real sense, it represents the ${}_j\circ_i$ construction commonly used to construct dioperads (see subsection 3.3.5) and it will also be considered an example of a partial grafting of two corollas.

A basic dioperadic graph can be depicted as follows.

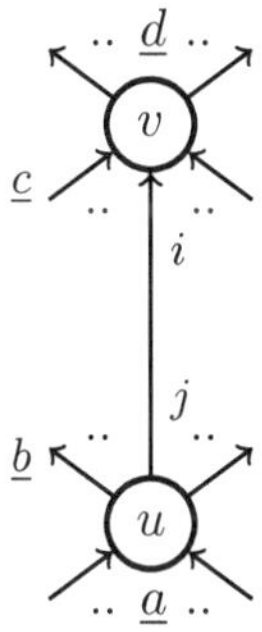

DEFINITION 2.23. A **basic dioperadic graph** is a wheeled graph

$$G = C^{j,i}_{(\underline{a};\underline{b};\underline{c};\underline{d})}$$

with $|\underline{a}| = k$, $|\underline{b}| = l$, $|\underline{c}| = m$, $|\underline{d}| = n$, and $b_j = c_i$ for some i and j, defined as the partitioned set

$$\left\{\left\{f_a^1, \ldots, f_a^k, f_b^1, \ldots, f_b^l\right\}, \left\{f_c^1, \ldots f_c^m, f_d^1, \ldots, f_d^n\right\}, \varnothing\right\}$$

with empty exceptional cell and ι fixing all but

$$\iota(f_b^j) = f_c^i \quad \text{and} \quad \iota(f_c^i) = f_b^j,$$

with direction

$$\delta(f_a^p) = 1 = \delta(f_c^r) \quad \text{and} \quad \delta(f_b^q) = -1 = \delta(f_d^s),$$

coloring $\kappa(f_t^o) = t_o$, and listing

$$\ell_u\left(f_a^o\right) = o = \ell_u\left(f_b^o\right), \quad \ell_v\left(f_c^o\right) = o = \ell_v\left(f_d^o\right),$$
$$\ell_G\left(f_a^p\right) = p + i - 1, \quad \ell_G\left(f_d^s\right) = s + j - 1,$$

and

$$\ell_G\left(f_b^q\right) = \begin{cases} q & \text{if } q < j, \\ q - 1 + n & \text{if } q > j. \end{cases}$$

$$\ell_G\left(f_c^r\right) = \begin{cases} r & \text{if } r < i, \\ r - 1 + k & \text{if } r > i. \end{cases}$$

It is immediate to check that the basic dioperadic graph is simply-connected. Using the notation of Definition 1.6, the profiles of the basic dioperadic graph $C^{j,i}_{(\underline{a};\underline{b};\underline{c};\underline{d})}$ will be

$$\binom{\underline{b} \circ_j \underline{d}}{\underline{c} \circ_i \underline{a}}.$$

In particular, the inputs $c_{[1,i-1]}$ come first, followed by the bottom inputs $\underline{a}$, and then by $c_{[i+1,m]}$. Likewise, the outputs $b_{[1,j-1]}$ come first, followed by the top outputs $\underline{d}$, and then by $b_{[j+1,l]}$. In Example 6.5, we will observe that the basic dioperadic graph can be constructed by grafting two corollas in one place, or in Example 6.7, by applying a contraction to a union of two corollas and then adjusting the listing.

Up to a change of listing, basic dioperadic graphs are the only simply-connected wheeled graphs with two vertices. They will allow us to build ordinary simply-connected wheeled graphs one internal edge at a time via graph substitution.

EXAMPLE 2.24. Some basic dioperadic graphs are half-graphs. Indeed, for $C^{j,i}_{(\underline{a};\underline{b};\underline{c};\underline{d})}$ to be a half-graph, it is necessary and sufficient that

(1) $|\underline{b}| = 1$ (which implies $j = 1$), $|\underline{a}| \geq 2$, and $|\underline{c}||\underline{d}| \geq 2$, or
(2) $|\underline{c}| = 1$ (which implies $i = 1$), $|\underline{d}| \geq 2$, and $|\underline{a}||\underline{b}| \geq 2$.

2.5. Trees

Here we introduce several classes of connected graphs that will be important in the discussion of Markl's non-unital operads, May's operads, and wheeled operads.

2.5.1. Simple Trees. Simple trees will be used to gradually build all trees used in the discussion of Markl's non-unital operads.

DEFINITION 2.25. A **simple tree** is a basic dioperadic graph of the form $C^{1,i}_{(\underline{b};c_i;\underline{c};d)}$. It is also denoted by $C_{(\underline{c};d)} \circ_i C_{(\underline{b};c_i)}$.

In the notation of Definition 1.6,the profiles of the simple tree $C_{(\underline{c};d)} \circ_i C_{(\underline{b};c_i)}$ are

$$\binom{d}{\underline{c} \circ_i \underline{b}}.$$

It can be pictorially represented as:

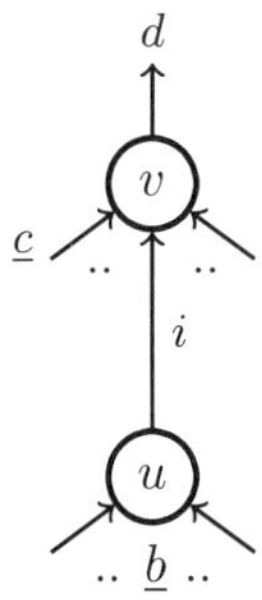

The notation $C_{(\underline{c};d)} \circ_i C_{(\underline{b};c_i)}$ for the simple tree is intended to suggest that the corolla $C_{(\underline{b};c_i)}$ is composed into the ith input of the corolla $C_{(\underline{c};d)}$. In fact, as we will see in section 3.3.4, this simple tree generates the comp-i operation in a Markl non-unital operad. In Example 6.6, we will observe that the simple tree can be constructed from two corollas by partial grafting.

EXAMPLE 2.26. The simple tree $C_{(\underline{c};d)} \circ_i C_{(\underline{b};c_i)}$ is a half-graph if and only if $|\underline{b}| \geq 2$ and $|\underline{c}| \geq 2$.

2.5.2. Level Trees. Here we discuss level trees, which will play a key role in the discussions of May's operads.

DEFINITION 2.27. A **level tree** consists of a simply-connected graph T with a direction together with a function

$$L: \mathrm{Vt}(T) \longrightarrow \{1, 2, \ldots\}$$

such that the following conditions are satisfied:

(1) Every vertex has exactly one outgoing flag.
(2) There exists a unique vertex w such that $L(w) = 1$.
(3) Given any internal edge e with initial vertex u and terminal vertex v,

$$L(u) = L(v) + 1.$$

If $L(v) = k$, then we call v a **level k vertex**. If n is the largest integer in the image of L, then we call T an **n-level tree**. A **fully n-level tree** is an n-level tree in which every input leg is adjacent to a level n vertex.

Note that to define a level tree, a direction is necessary, but a coloring or a listing are not necessary.

EXAMPLE 2.28. Here is a fully 2-level tree:

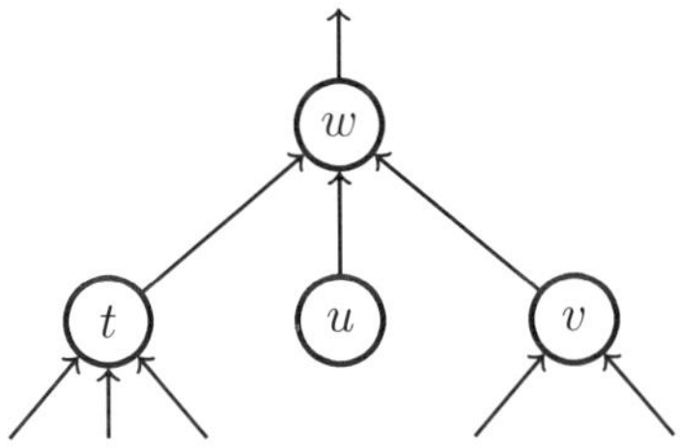

This level-tree has twelve flags, four vertices, one output, five inputs, and three internal edges. The unique level 1 vertex is w, and the three vertices under w are all level 2 vertices. As always, the five inputs have $\delta = 1$, and the unique output has $\delta = -1$. The three internal flags adjacent to the level 2 vertices have $\delta = -1$. The other three internal flags, all adjacent to w, have $\delta = 1$.

The next observation makes clear that assuming the existence of L is simply a matter of convenience.

LEMMA 2.29. *A simply-connected graph with a direction is a level tree precisely when it is ordinary and each vertex has exactly one outgoing flag.*

PROOF. Both of these properties are immediate from the definition of a level tree. On the other hand, it follows by induction on the length of the longest directed internal path through the graph that the function L is uniquely determined by the underlying simply-connected ordinary graph. □

EXAMPLE 2.30. The simple tree $C_{(\underline{c};d)} \circ_i C_{(\underline{b};c_i)}$ is a 2-level tree, but it is not a fully 2-level tree if $|\underline{c}| \geq 2$.

2.5.3. Special Trees. Here we introduce a type of fully 2-level tree that will play a crucial role in the discussion of May's operads. Suppose d is a color and $\underline{c} = c_{[1,m]}$ and $\underline{b}^i$ are profiles with $1 \leq i \leq m$, $m \geq 1$, and $|\underline{b}^i| = p_i$. The special tree

$T\left(\{\underline{b}^i\};\underline{c};d\right)$ can be pictorially represented as:

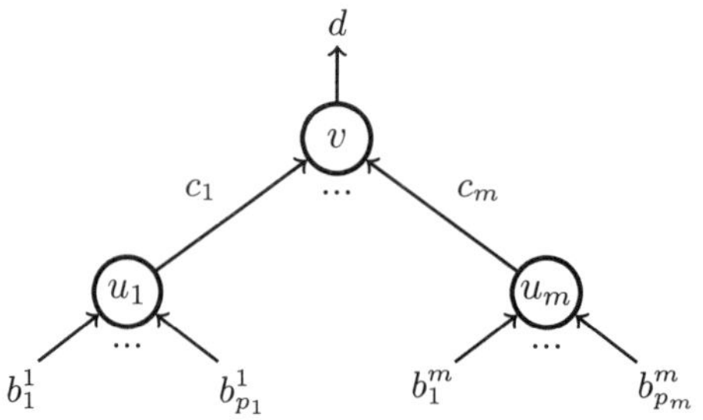

DEFINITION 2.31. Suppose d is a color while $\underline{c} = c_{[1,m]}$ and $\underline{b}^i$ are profiles with $1 \le i \le m$, $m \ge 1$, and $|\underline{b}^i| = p_i$. Define the **special tree**

$$T = T\left(\{\underline{b}^i\};\underline{c};d\right) \tag{2.1}$$

as the partitioned set

$$\{\{e, f_1^1, \ldots, f_1^m\}, \{f_{-1}^1, g_1^1, \ldots, g_{p_1}^1\}, \ldots, \{f_{-1}^m, g_1^m, \ldots, g_{p_m}^m\}, \varnothing\}$$

with empty exceptional cell, where ι fixes all except

$$\iota(f_k^j) = f_{-k}^j, \quad \text{with direction} \quad \delta(e) = -1 = \delta\left(f_{-1}^i\right) \quad \text{and} \quad \delta\left(g_j^i\right) = 1 = \delta\left(f_1^i\right),$$

coloring

$$\kappa(e) = d, \quad \kappa\left(f_{\pm 1}^i\right) = c_i \quad \text{and} \quad \kappa\left(g_j^i\right) = b_j^i$$

and listing

$$\ell_v\left(f_1^i\right) = i, \quad \ell_{u_i}\left(g_j^i\right) = j \quad \text{and} \quad \ell_G\left(g_j^i\right) = j + p_1 + \cdots + p_{i-1}.$$

It is immediate to check that the special tree is a simply-connected, fully 2-level tree, with the u_i as level 2 vertices. There is one internal edge for each $1 \le i \le m$ consisting of the internal flags $f_{\pm 1}^i$. The inputs and the incoming flags are all labeled, according to the picture, from left to right. The unique output is the flag e, and the inputs are the flags g_j^i. As we will discuss in Example 3.11, the special tree can be constructed from corollas using the operations of union and grafting.

2.5.4. Wheeled and Truncated Trees.

The graphs discussed here will be important for our presentation of wheeled operads.

DEFINITION 2.32. A **wheeled tree** is a connected graph with a direction in which each vertex has at most one outgoing flag.

Note that a wheeled tree may not have any vertices at all.

EXAMPLE 2.33. Both the exceptional edge and the exceptional loop are wheeled trees. Also, it follows from Lemma 2.29 that every level tree is a wheeled tree.

The following will be used below to divide wheeled trees into three types, two of which we can characterize here.

LEMMA 2.34. *Suppose G is a wheeled tree. Then:*

1. *G cannot contain an undirected cycle.*
2. *G can contain at most one wheel.*
3. *G can contain at most one vertex with no outgoing flags.*

(4) *G cannot simultaneously contain both a wheel and a vertex with no outgoing flags.*

PROOF. Combine Lemmas 2.8 and 2.15. □

One way to obtain a wheeled tree is to cut off the output leg of a level tree.

DEFINITION 2.35. Given a level tree G with level 1 vertex v, define a new graph G' as in Lemma 2.14, called the **truncation** of G, by removing the single flag in G that corresponds to the sole output of v.

Graphs constructed as G' above will be called **truncated trees**. It follows from Lemma 2.14 that a truncated tree is a wheeled tree but it is clearly not a level tree.

Before characterizing truncated trees, we need a technical result.

LEMMA 2.36. *If G is a wheeled tree containing a vertex w with no outgoing flags, then* $\mathrm{out}(G)$ *is empty.*

PROOF. Since G is connected and contains a vertex, we can assume there is at least one other vertex v in G that has an output leg f, as otherwise $\varnothing = \mathrm{out}(w) = \mathrm{out}(G)$. By Lemma 2.34, v must have exactly one outgoing flag. Forming $G' = G \smallsetminus \{f\}$ as in Lemma 2.14 then produces a connected graph with two vertices, the original w and now v, with no outgoing flags. As such, Lemma 2.15 implies G', and so G, contains another vertex with at least two outgoing flags, contradicting G a wheeled tree. □

The following result characterizes truncated trees using a 'glue and truncate' argument.

LEMMA 2.37. *Suppose G is a wheeled tree. Then G is a truncated tree if and only if it contains a vertex with no outgoing flags.*

PROOF. First, the unique level 1 vertex of a truncated tree has no outgoing flags. Now suppose G is a wheeled tree and w is a vertex of G with no outgoing flags. Then by Lemma 2.34, G contains neither wheels nor undirected cycles, hence G is simply-connected. Moreover, every other vertex in G has exactly one outgoing flag. We can then form a simply-connected G'' from G by adding one additional flag chosen to be the sole outgoing flag of w (with any choice of color) and forced to be the sole output leg by applying Lemma 2.36 to G. By Lemma 2.29 G'' is then a level tree, whose truncation is G, so G is a truncated tree. □

REMARK 2.38. Wheeled trees naturally break into three distinct subsets:

- the simply-connected graphs with direction where each vertex has exactly one outgoing flag,
- the simply-connected graphs with direction where there is one vertex with no outgoing flags and all other vertices have exactly one outgoing flag, and
- the wheeled trees containing exactly one wheel.

Notice the ordinary graphs of the first type are precisely the level trees and single exceptional edges by Lemma 2.29, while the ordinary graphs of the second type are precisely the truncated trees by Lemma 2.37. The remaining class of wheeled trees will be called contracted trees. We will discuss them further later, after introducing the operation of contraction.

CHAPTER 3

Basic Operations on Wheeled Graphs

In this short chapter we discuss certain basic operations on wheeled graphs. Some of them are only defined for ordinary wheeled graphs here, leaving the general case to a later chapter when we have graph substitution at our disposal to give a comprehensive formulation. For now, graph operations will simply be defined as mappings from sets of graphs to other sets of graphs, although we will consider them as functors of graph groupoids later.

It will be shown later that every operation which is compatible with graph substitution in a reasonable way is defined by graph substitution into a fixed graph. This fixed graph is produced simply by applying the operation to an appropriate collection of corollas. Thus, for each operation discussed in this chapter, we single out one 'generating graph' related to a corolla. This representability result is somewhat analogous to the Riesz Representation Theorem in Analysis. A positive linear functional and the corresponding Borel measure are analogous to a graph operation and the corresponding corolla-related graph, respectively. Integrating a function against a measure is analogous to substituting wheeled graphs into the representing corolla-related graph.

Our first operation is relabeling, which simply alters the listing of the graph, and the representing graph will be called a permuted corolla. The next is a disjoint union of a finite set of graphs, with the representing graph a disjoint union of corollas. In particular, this yields a decomposition of each graph as a disjoint union of its connected components.

The third operation considered is a grafting operation for ordinary graphs. If the input profile of G_1 coincides with the output profile of G_2, we can think of building a new graph from the two by altering ι to pair the corresponding former legs to create new internal edges. The representing graph in this case is a two-vertex graph where the input and output profiles of the lower vertex match those of G_2 and the input and output profiles of the upper vertex match those of G_1.

In fact, we can consider this same idea of gluing legs together even if not all of the profiles match but just some segment within the input profile of G_1 matches a segment within the output profile of G_2, which leads to a partial grafting operation. Again, the representing graph here has two vertices whose profiles match those of the graphs involved, while the representing graphs of both the comp-i operation $\circ_i$ and the related j-comp-i operation ${}_j\circ_i$ are easily constructed this way. One technical question is how to deal with listing the inputs and outputs of partial graftings, which was the point of introducing segment substitutions of profiles (Definition 1.6).

Our final major operation to consider will be graph contraction for ordinary graphs. This operation takes an output of G and an input of the same color, and alters ι_G to produce an internal edge by connecting these two legs. The representing graph here is like a corolla, although with one directed loop at the single vertex. The vertex profile is that of G, so the profile of the full graph will be that of G with a single input and a single output removed. Later, contracting a disjoint union will be shown to provide a constructive method that can be used inductively to build any wheeled graph.

From this point on, we almost always work with wheeled graphs. Therefore, (simply-)connected graphs, wheel-free graphs, etc., unless otherwise specified, will refer to *wheeled* graphs that are (simply-)connected, wheel-free, and so forth.

3.1. Relabeling

One piece of the structure of a wheeled graph is the listing, and our first operation simply alters the listing and nothing else.

3.1.1. Input and Output Relabeling.

DEFINITION 3.1. Suppose G is a graph with a listing, σ is a permutation on $|\mathrm{out}(G)|$ letters, and τ is a permutation on $|\mathrm{in}(G)|$ letters. Define a **permuted listing** $\ell_{\sigma G\tau}$ by $\ell_{\sigma G\tau}(e) = \tau(\ell_G(e))$ for an input leg e and $\ell_{\sigma G\tau}(f) = \sigma^{-1}(\ell_G(f))$ for an output leg f.

While our goal is to have τ acting on $\mathrm{in}(G)$ from the right and σ acting on $\mathrm{out}(G)$ from the left, when working with the listing we must reverse the actions. In order to produce the output $\sigma\underline{d} = (d_{\sigma(1)}, d_{\sigma(2)}, \cdots, d_{\sigma(n)})$, the new listing needs to list $d_{\sigma(1)}$ first, $d_{\sigma(2)}$ second, and so on.

DEFINITION 3.2. Given a $(\underline{c};\underline{d})$-wheeled graph G and permutations σ on $|\mathrm{out}(G)|$ letters and τ on $|\mathrm{in}(G)|$ letters, the **input and output relabeling of** G by σ and τ will denote the $(\underline{c}\tau;\sigma\underline{d})$-wheeled graph $\sigma G\tau$ described by the same underlying graph, coloring, and direction equipped with the permuted listing of Definition 3.1.

EXAMPLE 3.3. Consider the case where G has profiles $(\underline{c};\underline{d})$ with $|\underline{d}| = 3$ while $\sigma(1) = 3$, $\sigma(2) = 1$, and $\sigma(3) = 2$. Then $\sigma^{-1}(3) = 1$ so o_3 with color d_3 is listed first, $\sigma^{-1}(1) = 2$ so o_1 with color d_1 is listed second, and $\sigma^{-1}(2) = 3$ so o_2 with color d_2 is listed third. As a consequence, $\mathrm{out}(\sigma G\tau)$ is

$$(d_3, d_1, d_2) = (d_{\sigma(1)}, d_{\sigma(2)}, d_{\sigma(3)}) = \sigma\underline{d}$$

as expected.

This input and output relabeling will be an important part of our equivariance properties later. We would like to think of this as our first example of a **graph operation**, which will simply refer to a set-valued mapping from one set of graphs to another. Once we have discussed our two variants of isomorphisms of graphs, we will show that all operations discussed will be compatible with at least one type of isomorphism. In the present case, we will be saying any isomorphism of colored, directed graphs remains so even when the listings on the inputs and outputs are altered.

A key result later will be that given a graph operation which is compatible with graph substitution in an appropriate sense, applying the operation to a corolla, or an appropriately chosen collection of corollas in some cases, will produce a 'representing graph'. In other words, graph substitution into the representing graph defines an isomorphic graph operation. Our first example of this phenomenon follows.

3.1.2. Permuted Corollas.

DEFINITION 3.4. A **permuted corolla** is an input and output relabeling of a corolla.

In other words, a permuted corolla is a wheeled graph of the form $\sigma C_{(\underline{c};\underline{d})}\tau$ for some permutations σ and τ on $|\underline{d}|$ letters and $|\underline{c}|$ letters, respectively. Every corolla is a permuted corolla if one chooses the identity permutations. However, the permuted corolla $\sigma C_{(\underline{c};\underline{d})}\tau$ is *not* a corolla, unless both σ and τ are equal to the identity. This is evident by noticing $\ell_G \neq \ell_w$ (with w the unique vertex) if either permutation is non-trivial.

More directly, the permuted corolla $\sigma C_{(\underline{c};\underline{d})}\tau$ can be described as the partitioned set

$$\{\{i_1,\ldots,i_m,o_1,\ldots,o_n\},\varnothing\}$$

with empty exceptional cell, ι the identity,

$$\delta(i_k)=1,\quad \delta(o_k)=-1,\quad \kappa(i_k)=c_k \quad\text{and}\quad \kappa(o_j)=d_j,$$

and equipped with the listing

$$\ell_w(i_k)=k,\quad \ell_w(o_j)=j,\quad \ell_G(i_k)=\tau(k) \quad\text{and}\quad \ell_G(o_j)=\sigma^{-1}(j).$$

The permuted corolla $\sigma C_{(\underline{c};\underline{d})}\tau$ is pictorially depicted as:

(3.1)

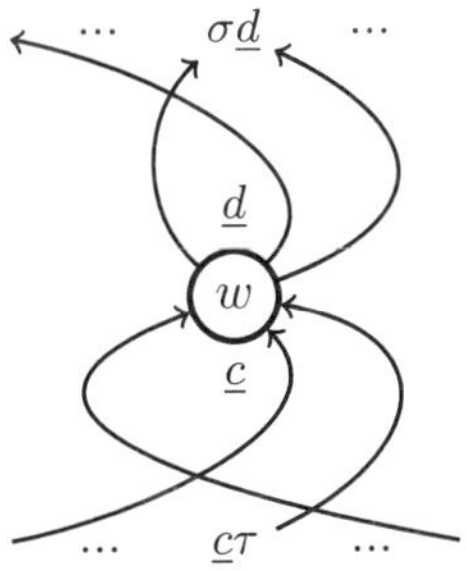

As we will discuss in section 6.1.2, the input and output relabeling $\sigma G\tau$ will agree with the graph substitution of G into the permuted corolla $\sigma C_{(\underline{c};\underline{d})}\tau$.

3.2. Disjoint Union

In some cases, the domain of our graph operation will be a product of sets of graphs, in order to allow several inputs at once. That is the case with the disjoint union operation we discuss next.

One can intuitively think of the disjoint union of graphs as arranging a finite number of graphs in sequence and then using the ordering of the original graphs together with their listings in order to define listings on the full collections of inputs and outputs.

3.2.1. Disjoint Union of Wheeled Graphs. The reader may want to revisit Remark 1.19 concerning disjoint sets of flags at this point.

DEFINITION 3.5. Let $G_1,\ldots,G_r$ be wheeled graphs. Define the **disjoint union**

$$G=\coprod_{j=1}^{r} G_j \tag{3.2}$$

as the disjoint union of partitioned sets for the ordinary part, but with the new exceptional cell chosen to be the union of all exceptional cells. This new exceptional cell remains an isolated cell when defining ι also as the union of the involutions. Similarly, π is then defined as the union of the various free involutions, while the direction, coloring, and listing remain unchanged except for the listing of the entire graph. When defining ℓ_G one uses the lexicographical ordering coming from first looking at the index j and then the order of the inputs or outputs of the individual graph G_j.

Note that if the profiles of G_j are $(\underline{c}^j;\underline{d}^j)$, then the disjoint union $\coprod G_j$ has profiles $(\underline{c};\underline{d})$, where $\underline{c} = (\underline{c}^1,\ldots,\underline{c}^r)$ and $\underline{d} = (\underline{d}^1,\ldots,\underline{d}^r)$. For example, if each G_j is the c_j-colored exceptional edge $\uparrow_{c_j}$, then their disjoint union is the $\underline{c}$-exceptional edge $\uparrow_{\underline{c}}$ with $\underline{c} = (c_1,\ldots,c_r)$. When there are just a few graphs involved, we will often write $G_1 \sqcup G_2$, etc.

The following alternative description of the disjoint union is immediate from the definition.

LEMMA 3.6. *Suppose G_j has profiles $(\underline{c}^j;\underline{d}^j)$. Then the disjoint union $G = \coprod_j G_j$ can be equivalently described as follows:*

(1) $\mathrm{Flag}(G) = \coprod_{j=1}^r \mathrm{Flag}(G_j)$.
(2) $\mathrm{Vt}(G) = \coprod_{j=1}^r \mathrm{Vt}(G_j)$.
(3) $\iota = \coprod_{j=1}^r \iota_j$, *which implies internal edges and legs are also defined by restriction to the various G_j, hence the same may be said for the remaining structure aside from ℓ_G.*
(4) ℓ_G *is defined as*

$$\ell_G(f) = \begin{cases} k + \sum_{p=1}^{j-1} |\underline{c}^p| & \text{if } f = i_k^{G_j}, \\ k + \sum_{q=1}^{j-1} |\underline{d}^q| & \text{if } f = o_k^{G_j}. \end{cases}$$

The disjoint union operation is associative, although for those who consider the disjoint union of sets to be associative only up to canonical isomorphism the more precise statement would be that disjoint union is associative up to a canonical isomorphism in the strictest sense we will consider. It is unfortunate that the disjoint union operation is not quite commutative, although changing the order of the arguments is simply an input and output relabeling of the original disjoint union. There is a special case with two graphs under which disjoint union becomes strictly commutative, namely when one graph has no inputs and the other has no outputs, since the listings for the full graph do not interact in this case.

3.2.2. Disjoint Union of Corollas. Let C_j be the corolla $C_{(\underline{c}^j;\underline{d}^j)}$ for each $j \in \{1,\ldots,r\}$. The disjoint union of corollas $G = \coprod_j C_j$ is pictorially depicted as follows:

As we will discuss in section 6.1.3, the disjoint union $\coprod_j G_j$ agrees with the graph substitution of the G_j into the disjoint union $\coprod_j C_j$ of corollas.

3.2.3. Disjoint Union Decompositions of Wheeled Graphs. We have two different ways of rewriting graphs using disjoint unions, one in terms of connected components and another separates the ordinary from the two types of exceptional pieces.

Since the collection of connected components collectively know everything about the full graph except for the listing of the inputs and outputs of the full graph, ℓ_G, the following says each graph can be written as an input and output relabeling of the disjoint union of its connected components.

LEMMA 3.7. *Every non-empty wheeled graph G can be written as*

$$G = \sigma(G_1 \sqcup \cdots \sqcup G_r)\tau$$

for some permutations σ and τ, with each G_j connected.

For our other decomposition, notice it is possible to alter ℓ_G so that all ordinary legs are listed first without altering their relative ordering, then followed by the domain of π. Therefore, the following decomposition result is also immediate.

LEMMA 3.8. *For each wheeled graph G, there exist (not necessarily unique) permutations $(\sigma;\tau) \in \Sigma_{|\underline{d}|} \times \Sigma_{|\underline{c}|}$ such that*

$$\sigma G \tau = G_{ord} \sqcup \uparrow_{\underline{a}} \sqcup \circlearrowleft_{\underline{b}},$$

where G_{ord} is an ordinary wheeled graph, $\uparrow_{\underline{a}}$ is the $\underline{a}$-exceptional edge, and $\circlearrowleft_{\underline{b}}$ is the $\underline{b}$-exceptional loop.

As discussed in Remark 1.34, permutations of $\underline{b}$ will not be noticed in our discussion.

3.3. Grafting

Recall that the composition operation in a category has domain those ordered pairs of morphisms where the source of the first morphism matches the target of the second. Our grafting operation can be thought of as requiring a similar condition, where we check that the input profile of the first graph matches the output profile of the second graph.

3.3.1. Grafting of Ordinary Wheeled Graphs. Suppose G_1 and G_2 are ordinary wheeled graphs with profiles $(\underline{c};\underline{d})$ and $(\underline{b};\underline{c})$ respectively. Since $\mathrm{in}(G_1) = \underline{c} = \mathrm{out}(G_2)$, one can imagine connecting the outputs of G_2 to the inputs of G_1 in order. The restriction here to ordinary wheeled graphs keeps the following definition relatively simple, although the general case discussed in section 6.1.4 is more complex. For example, the grafting of two exceptional edges of the same color simply produces a single exceptional edge of that color.

The reader may want to revisit Remark 1.19 concerning disjoint sets of flags at this point.

DEFINITION 3.9. Given two ordinary wheeled graphs with $\mathrm{in}(G_1) = \underline{c} = \mathrm{out}(G_2)$, define their **grafting**

$$G = G_1 \boxtimes G_2$$

as follows. The underlying partitioned set is the disjoint union of partitioned sets with an empty exceptional cell. The involution ι_G is defined on most flags as it was within G_i, with the exceptions

$$\iota_G\left(i_j^{G_1}\right) = o_j^{G_2} \quad \text{and} \quad \iota_G\left(o_j^{G_2}\right) = i_j^{G_1}$$

for $1 \le j \le |\underline{c}|$. The coloring, direction, and listing of G are then defined for each flag as they were within G_i.

Note that

$$\mathrm{in}(G_1 \boxtimes G_2) = \mathrm{in}(G_2) \quad \text{and} \quad \mathrm{out}(G_1 \boxtimes G_2) = \mathrm{out}(G_1)$$

as profiles, since the former output legs of G_2 are now forming internal edges in combination with the former input legs of G_1. For ease of reference later, we now point out that grafting is an associative operation. In addition, we note that grafting reduces to disjoint union when the intermediate profile is empty, so the intuitive picture of connecting the outputs of G_2 to the inputs of G_1 should not be taken too literally.

LEMMA 3.10. (1) *Grafting of ordinary wheeled graphs is an associative operation.*

(2) *If G_1 has profiles $(\varnothing; \underline{d})$ and G_2 has profiles $(\underline{b}; \varnothing)$, then we have*

$$G_1 \boxtimes G_2 = G_1 \coprod G_2 = G_2 \coprod G_1.$$

Grafting in the general case is, in fact, also associative as well as unital (up to canonical isomorphism in the strictest sense). However, the units are exceptional wheeled graphs, while grafting involving exceptional flags has not yet been defined for technical reasons.

As a first example, special trees can be described more efficiently using grafting.

EXAMPLE 3.11. The special tree $T(\{\underline{b}^i\}; \underline{c}; d)$ in (2.1) can be constructed from corollas, via disjoint union and grafting as

$$T(\{\underline{b}^i\}; \underline{c}; d) = C_{(\underline{c}; d)} \boxtimes \left(\coprod_{i=1}^{m} C_{(\underline{b}^i; c_i)} \right).$$

3.3.2. Grafted Corollas.

DEFINITION 3.12. Given any profiles $\underline{b}$, $\underline{c}$, and $\underline{d}$, the associated **grafted corollas** is defined as the grafting

$$C_{(\underline{b}; \underline{c}; \underline{d})} = C_{(\underline{c}; \underline{d})} \boxtimes C_{(\underline{b}; \underline{c})}. \tag{3.3}$$

As above, the case $\underline{c} = \varnothing$ is allowed, in which case we have

$$C_{(\underline{b}; \varnothing; \underline{d})} = C_{(\varnothing; \underline{d})} \boxtimes C_{(\underline{b}; \varnothing)} = C_{(\varnothing; \underline{d})} \coprod C_{(\underline{b}; \varnothing)} = C_{(\underline{b}; \varnothing)} \coprod C_{(\varnothing; \underline{d})}.$$

The grafted corollas is pictorially depicted as:

(3.4)

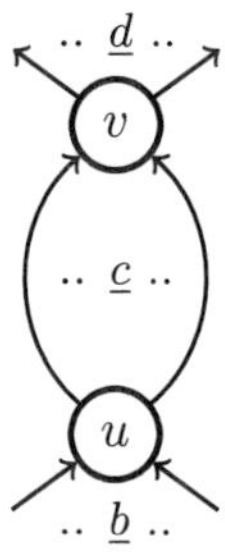

EXAMPLE 3.13. Let $\underline{b} = (b_1, b_2, b_3)$, $\underline{c} = (c_1, c_2)$, and $\underline{d} = (d_1, d_2)$ be profiles. The **rocket graph** is defined as the grafted corollas G below.

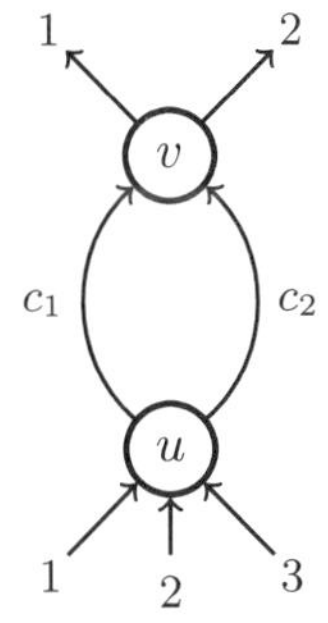

As a partitioned set, it is given by $\{\{i_1, i_2, i_3, e_{-1}, f_{-1}\}, \{e_1, f_1, o_1, o_2\}, \varnothing\}$ with empty exceptional cell and involution

$$\iota(e_j) = e_{-j}, \quad \iota(f_j) = f_{-j}, \quad \iota(i_j) = i_j, \quad \text{and} \quad \iota(o_j) = o_j,$$

with direction

$$\delta(i_j) = 1, \quad \delta(o_j) = -1, \quad \delta(e_j) = j, \quad \text{and} \quad \delta(f_j) = j,$$

coloring

$$\kappa(i_j) = b_j, \quad \kappa(o_j) = d_j, \quad \kappa(e_{\pm 1}) = c_1, \quad \text{and} \quad \kappa(f_{\pm 1}) = c_2,$$

and listing

$$\begin{gathered} \ell_G(i_j) = j = \ell_u(i_j), \quad \ell_G(o_j) = j = \ell_v(o_j), \\ \ell_u(e_{-1}) = 1 = \ell_v(e_1) \quad \text{and} \quad \ell_u(f_{-1}) = 2 = \ell_v(f_1). \end{gathered}$$

As we will discuss in section 6.1.4, general grafting is *defined* as graph substitution into a grafted corollas. When the graphs involved are ordinary, the general definition agrees with Definition 3.9.

3.3.3. Partial Grafting. In some cases, one might want to graft only a subset of the inputs of G_1 onto a subset of the outputs of G_2. The reader may want to review the description of k-segments of a profile at this point, which are intended to track these subsets. We will refer to this kind of operation as partial grafting, defined when $\underline{c}' \subset \underline{c}$ is a k-segment and $\underline{b}' \subset \underline{b}$ is a matching k-segment. Some care must now be taken to keep track of the listings for the full graph. In the notation of Definition 1.6, the inputs of the partial grafting will be of the form $\underline{c} \circ_{\underline{c}'} \underline{a}$ and the outputs will be of the form $\underline{b} \circ_{\underline{b}'} \underline{d}$. The profiles of the partial grafting are thus of the form

$$\begin{gathered} (c_1, \ldots, c_{l_c - 1}, a_1, \ldots, a_{|\underline{a}|}, c_{l_c + k}, \ldots, c_{|\underline{c}|}) \quad \text{and} \\ (b_1, \ldots, b_{l_b - 1}, d_1, \ldots, d_{|\underline{d}|}, b_{l_b + k}, \ldots, b_{|\underline{b}|}). \end{gathered}$$

DEFINITION 3.14. Given two ordinary wheeled graphs G_1 with profiles $(\underline{c}; \underline{d})$ and G_2 with profiles $(\underline{a}; \underline{b})$, together with two k-segments $\underline{c}' = (c_{l_c}, \ldots, c_{l_c + k - 1})$ and $\underline{b}' = (b_{l_b}, \ldots, b_{l_b + k - 1})$ satisfying $\underline{c}' = \underline{b}'$, define the **partial grafting**

$$G = G_1 \boxtimes_{\underline{b}'}^{\underline{c}'} G_2$$

as the disjoint union of the partitioned sets, leaving ι defined as in the component graphs with the exception that for $0 \le j \le k-1$

$$\iota_G\left(i^{G_1}_{l_c+j}\right) = o^{G_2}_{l_b+j}, \quad \text{and} \quad \iota_G\left(o^{G_2}_{l_b+j}\right) = i^{G_1}_{l_c+j},$$

with input listing for $j < l_c$, $1 \le j'' \le |\underline{a}|$, and $l_c + k - 1 < j' \le |\underline{c}|$

$$\ell_G(i^{G_1}_j) = j, \quad \ell_G(i^{G_2}_{j''}) = l_c + j'' - 1, \quad \text{and} \quad \ell_G(i^{G_1}_{j'}) = |\underline{a}| + j' - k$$

as well as output listing for $i < l_b$, $1 \le i'' \le |\underline{d}|$, and $l_b + k - 1 < i' \le |\underline{b}|$

$$\ell_G(o^{G_2}_i) = i, \quad \ell_G(o^{G_1}_{i''}) = l_b + i'' - 1, \quad \text{and} \quad \ell_G(o^{G_2}_{i'}) = |\underline{d}| + i' - k.$$

As discussed previously, there is a representing type of graph for this type of operation and we have discussed instances of it already.

Definition 3.15. Given profiles $\underline{a}$, $\underline{b}$, $\underline{c}$, and $\underline{d}$ together with $k > 0$ and k-segments $\underline{b}' \subset \underline{b}$ starting with l_b and $\underline{c}' \subset \underline{c}$ starting with l_c, define the **partially grafted corollas** $C^{l_c,l_b,k}_{\underline{a},\underline{b},\underline{c},\underline{d}}$ as $C_{(\underline{c};\underline{d})} \boxtimes^{\underline{c}'}_{\underline{b}'} C_{(\underline{a};\underline{b})}$.

A partially grafted corollas $C_{(\underline{c};\underline{d})} \boxtimes^{\underline{c}'}_{\underline{b}'} C_{(\underline{a};\underline{b})}$ has profiles $(\underline{c} \circ_{\underline{c}'} \underline{a}; \underline{b} \circ_{\underline{b}'} \underline{d})$ and can be pictured as

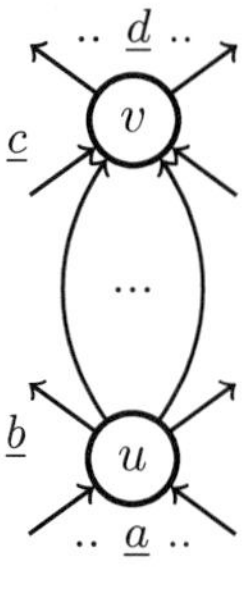

in which the internal edges have colors $\underline{c}' = \underline{b}'$.

Example 3.16. A simple tree $C_{(\underline{c};d)} \circ_i C_{(\underline{b};c_i)}$ of Definition 2.25

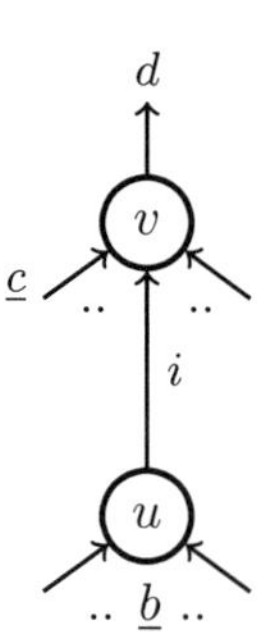

is a partially grafted corollas, over a 1-segment $\underline{c}' = (c_i)$. A basic dioperadic graph $C^{j,i}_{(\underline{a};\underline{b};\underline{c};\underline{d})}$ of Definition 2.23,

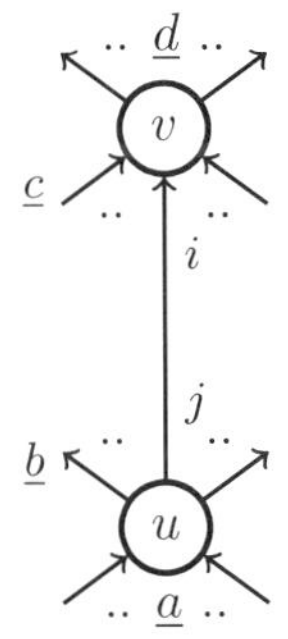

is also a partially grafted corollas over 1-segments $\underline{c}' = (c_i) = (b_j) = \underline{b}'$.

EXAMPLE 3.17. The following representation includes only the listings of the partially grafted corollas with $|\underline{a}| = 3$, $|\underline{b}| = 4$, $|\underline{c}| = 4$, and $|\underline{d}| = 4$ while $\underline{b}' = \underline{c}'$ are the middle two entries.

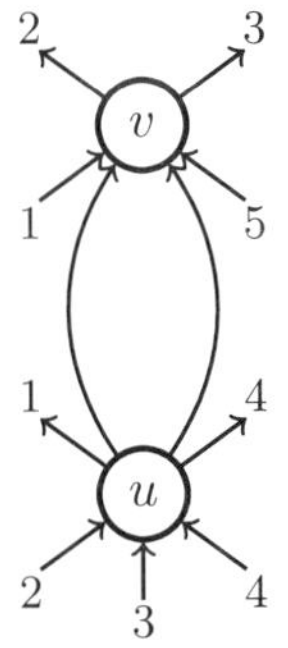

There are two special cases of partial grafting for ordinary graphs which are important, generally known as composition operations.

3.3.4. The Comp-i Operation. This operation is particularly convenient for decomposing trees systematically, and was first introduced by Gerstenhaber [**Ger63**] to describe the structure of the Hochschild cohomology of an associative algebra. Later, the operations were adapted to the operad setting by Markl [**Mar96a**], where they satisfy the same relations. In this case, the operation applied to a pair of corollas has already been introduced as the simple tree $C_{(\underline{c};d)} \circ_i C_{(\underline{b};c_i)}$ of Definition 2.25 and Ex. 3.16.

DEFINITION 3.18. Given trees G_1 with profiles $(\underline{c}; d)$ and G_2 with profiles $(\underline{b}; c_i)$, define $G = G_1 \circ_i G_2$ as the partial grafting with $k = 1$ and $\underline{c}' = (c_i)$.

The natural domain of this operation is pairs from the set of trees subject to the condition that the output color of the second matches the ith input color of the first.

EXAMPLE 3.19. The following represents $C_{(\underline{c};d)} \circ_3 C_{(\underline{b};c_3)}$ where $|\underline{c}| = 3$ and $|\underline{b}| = 2$.

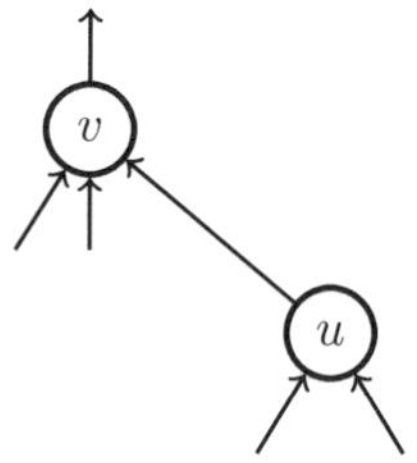

3.3.5. The j-comp-i Operation. The $\circ_i$ operation has a natural extension to dioperadic graphs as well. In this case the generating graph of the operation ${}_j\circ_i$ is a basic dioperadic graph $C^{j,i}_{(\underline{a};\underline{b};\underline{c};\underline{d})}$ of Definition 2.23.

DEFINITION 3.20. Suppose G_1 and G_2 are ordinary wheeled graphs with profiles $(\underline{c};\underline{d})$ and $(\underline{a};\underline{b})$ respectively, with $c_i = b_j$. Then define $(G_1)_j\circ_i(G_2)$ as the partial grafting with $k = 1$ where $\underline{c}' = (c_i)$ and $\underline{b}' = (b_j)$.

EXAMPLE 3.21. A basic dioperadic graph $C^{j,i}_{(\underline{a};\underline{b};\underline{c};\underline{d})}$ is $C_{(\underline{c};\underline{d})}\,{}_j\circ_i C_{(\underline{a};\underline{b})}$. Even more, the operation $\circ_i$ is just ${}_1\circ_i$ for trees.

3.4. Contraction

Suppose G is an ordinary wheeled graph with profiles $(\underline{c};\underline{d})$ such that $c_j = d_i$ for some i and j. Since the ith output leg and the jth input leg have the same color, one can imagine connecting the former to the latter, thus creating a new internal edge in the process. The resulting operation is described precisely below. Once again we restrict our attention to ordinary wheeled graphs to keep the definition technically simpler, while the general case is discussed in section 6.1.5.

3.4.1. Contraction of Ordinary Wheeled Graphs.

DEFINITION 3.22. For an ordinary $(\underline{c};\underline{d})$-wheeled graph G with $c_j = d_i$, define the $(j;i)$**-contraction**, or simply **contraction**, as a wheeled graph

$$K = \xi^i_j G$$

that is equal to G as a partitioned set and has the same coloring and direction as G, while the involution ι_K is the same as ι_G except that

$$\iota_K(o_i) = i_j \quad \text{and} \quad \iota_K(i_j) = o_i,$$

and the listing of K is the same as that of G except for decreasing the listings after the two former legs which are now part of the new internal edge

$$\ell_K(i_l) = \begin{cases} \ell_G(i_l) & \text{if } l < j, \\ \ell_G(i_l) - 1 & \text{if } l > j \end{cases}$$

and

$$\ell_K(o_k) = \begin{cases} \ell_G(o_k) & \text{if } k < i, \\ \ell_G(o_k) - 1 & \text{if } k > i. \end{cases}$$

Note that the profiles of the contraction $\xi^i_j G$ are $(\underline{c} \smallsetminus c_j; \underline{d} \smallsetminus d_i)$. The domain of ξ^i_j is the subset of wheeled graphs where $c_j = d_i$. In the contracted graph $\xi^i_j G$, it may still be the case that an output has the same color as an input, in which case a second contraction can be applied. As a consequence, this type of operation can, in some cases, be applied repeatedly.

EXAMPLE 3.23. The contraction of the two-vertex linear graph is the two-vertex cycle. In addition, the contraction ξ^1_1 of

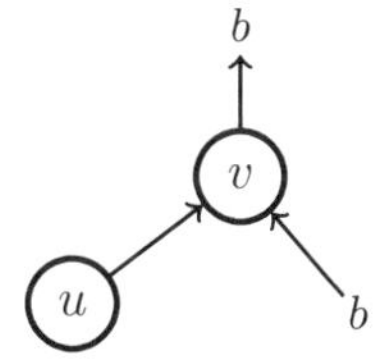

is the butterfly net graph of Example 1.38.

For the next result, keep in mind that disjoint unions of trees are ordinary graphs, allowing us to apply this restricted contraction operation we have only defined for ordinary graphs.

LEMMA 3.24. *Given ordinary wheeled graphs G_1 with profiles $(\underline{c}; \underline{d})$ and G_2 with profiles $(\underline{a}; \underline{b})$ such that $b_j = c_i$, there is an input and output relabeling such that*

$$\sigma\left((G_1)_j \circ_i (G_2)\right) \tau = \xi_i^{|\underline{d}|+j} (G_1 \sqcup G_2)$$

PROOF. In fact, all but the listing of the inputs and outputs of the full graph is identical for the composition operation and contracted union constructions. For the listing on the full graph, it is necessary to move the inputs of G_2, as a block, before the inputs from $i+1$ through $|\underline{c}|$ of G_1, and similarly for outputs. □

In the general case, where there may be exceptional legs, the contraction may not create a new internal edge at all. For example, contracting to connect flags from a pair of exceptional edges of the same color simply produces a single exceptional edge. This type of complication is related to the subtlety of graph substitution with exceptional edges, as well as being our motivation to focus here on the ordinary case.

3.4.2. Contracted Corolla.

DEFINITION 3.25. Let $\underline{c}$ and $\underline{d}$ be non-empty profiles with $c_j = d_i$ for some $1 \le j \le |\underline{c}|$ and $1 \le i \le |\underline{d}|$. Define the **contracted corolla**, which is a $(\underline{c} \smallsetminus c_j; \underline{d} \smallsetminus d_i)$-wheeled graph, as

$$C^{j,i}_{(\underline{c};\underline{d})} = \xi^i_j C_{(\underline{c};\underline{d})}. \tag{3.5}$$

In particular, we have

$$\operatorname{Leg}\left(C^{j,i}_{(\underline{c};\underline{d})}\right) = \operatorname{Leg}\left(C_{(\underline{c};\underline{d})}\right) \smallsetminus \{o^C_i, i^C_j\}$$

and

$$\operatorname{Flag_i}\left(C^{j,i}_{(\underline{c};\underline{d})}\right) = \{o^C_i, i^C_j\}.$$

The contracted corolla $C^{j,i}_{(\underline{c};\underline{d})}$ is pictorially depicted as follows.

(3.6)

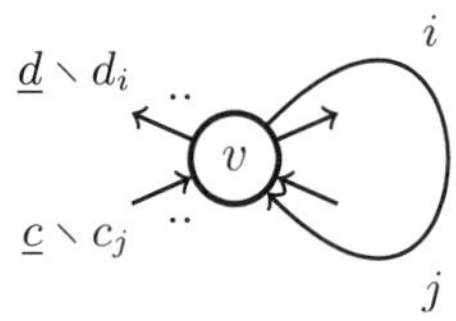

Note that $C^{j,i}_{(\underline{c};\underline{d})}$ has exactly one vertex with profiles $(\underline{c}; \underline{d})$.

As we will discuss in section 6.1.5, general contraction is *defined* as graph substitution into a contracted corolla. When the wheeled graph is ordinary, the general definition agrees with Definition 3.22.

CHAPTER 4

Graph Groupoids

In this short chapter we consider the natural groupoid structures on the collection of wheeled graphs and some of the implications. Perhaps the most obvious notion of isomorphism is what we will call strict isomorphism, which is compatible with directions, colorings, and listings. However, it is also quite useful to consider weak isomorphisms, where we ignore the question of compatibility with listings, as input and output relabeling becomes an example of an isomorphism in this way. As a consequence, the weak isomorphisms can easily be used to indicate processes which are often called bi-equivariant, or compatible with input and output relabeling.

It should come as no surprise that in considering these groupoids, we might look at automorphism groups of individual graphs. The strict automorphism group of a wheeled graph is surprisingly small, and is described generally for every wheeled graph after looking at a series of indicative examples. In particular, strict automorphisms act on the ordinary part of a graph by permuting vertices with the same profiles, if any part acts non-trivially. There is also a similar discussion of weak automorphism groups, but this is less satisfying due to the absence of the rigidity results available in the strict case. The chapter ends with a list of graph groupoids that will be considered repeatedly in later chapters, along with an indication of the common name for the associated generalized PROP structure they will be used to define.

4.1. Strict Isomorphisms

4.1.1. Strict Isomorphisms and Automorphisms. We begin with the fundamental definition of a strict isomorphism of wheeled graphs.

DEFINITION 4.1. Let G and G' be wheeled graphs. A **strict isomorphism** $\psi: G \longrightarrow G'$ is a bijection of partitioned sets that commutes with both types of involutions, and leaves invariant the colorings, the directions, and the listings.

Following our habit, we also have an alternative formulation in terms of flags and vertices.

LEMMA 4.2. *A strict isomorphism $\psi: G \longrightarrow G'$ of wheeled graphs is a bijection*

$$\psi: \mathrm{Flag}(G) \xrightarrow{\cong} \mathrm{Flag}(G')$$

such that:

(1) $\psi\iota = \iota'\psi$,
(2) *the induced map on power sets $\psi_*: 2^{\mathrm{Flag}(G)} \longrightarrow 2^{\mathrm{Flag}(G')}$ sends vertices to vertices, thereby inducing a bijection between sets of vertices,*
(3) $\psi\pi = \pi'\psi$ *on* $\mathrm{Leg}_{\mathrm{e}}(G)$,

(4) $\kappa = \kappa'\psi$,
(5) $\delta = \delta'\psi$ *and*
(6) $\ell = \ell'\psi$.

NOTATION 4.3. The **strict automorphism group** of a wheeled graph G, which consists of strict isomorphisms from G to itself, is denoted by $\mathrm{Aut}_{\mathrm{str.}}(G)$.

Later, we will be discussing a less rigid form of isomorphism, which essentially ignores the listings. Part of the point of the weaker notion is to understand constructions that are natural with respect to input and output relabeling for all pairs of profiles, called **bi-equivariant** processes. Note that, in general, the wheeled graphs G and $\sigma G\tau$ are *not* strictly isomorphic, but will be weakly isomorphic. For example, a non-trivially permuted corolla is not isomorphic to any corolla, since choosing ψ to reverse the permutation of the listing for the full graph would alter the listing at the vertex.

EXAMPLE 4.4. Here are some examples of strict automorphism groups.

(1) Suppose G is a wheeled graph such that

$$\mathrm{Flag}(G) = \mathrm{Leg}(G).$$

Then we have

$$\mathrm{Aut}_{\mathrm{str.}}(G) = \{Id\},$$

as legs must be fixed by strict automorphisms since $\ell_G = \ell_G\psi$ for a bijection ψ. In particular, this is the case for corollas, permuted corollas, and the $\underline{c}$-exceptional edge $\uparrow_{\underline{c}}$ (1.8).

(2) The strict automorphism group of either an isolated vertex or an exceptional loop is trivial.

(3) For a color $c \in \mathfrak{C}$, let $L_c \in \mathtt{Gr}^{\circlearrowright}_{\mathrm{str}}\binom{\varnothing}{\varnothing}$ denote the contracted corolla $\xi_1^1 C_{(c;c)}$ (3.5), and let G be the union

$$G = \coprod_{i=1}^{k} L_c.$$

Then we have

$$\mathrm{Aut}_{\mathrm{str.}}(G) = \Sigma_k.$$

Indeed, G has
- $2k$ flags $\{f^i_{\pm 1}\}_{i=1}^k$ and
- k vertices $\{v_i = \{f^i_{\pm 1}\}\}_{i=1}^k$

with

$$\iota(f^i_l) = f^i_{-l}.$$

The k flags f^i_1 can be permuted arbitrarily, but then the images of the other k flags f^i_{-1} are uniquely determined because a strict automorphism must preserve the involution ι.

(4) If we alter only the coloring in the previous example, now using distinct colors for each loop L_{c_i}, then there are no non-trivial strict automorphisms. This follows since a strict automorphism must preserve both the color and the direction of each flag.

(5) As in Remark 1.34, we make no distinction between different orderings of the exceptional loops in a wheeled graph. As such, for any choices of colors c_i, we would have

$$\mathrm{Aut}_{\mathrm{str.}}\left(\coprod_{i=1}^{k} \circlearrowleft_{c_i}\right) = \{Id\}.$$

(6) For G' the 2-vertex cycle (Example 1.37), we have

$$\mathrm{Aut}_{\mathrm{str.}}(G') = \mathbf{Z}/(2).$$

The only non-identity strict automorphism ψ on G' is the rotation with

$$\psi(e_{-1}) = o, \quad \psi(e_1) = i, \quad \psi(o) = e_{-1} \quad \text{and} \quad \psi(i) = e_1.$$

(7) Suppose $G \in \mathtt{Gr}^{\circlearrowleft}_{\mathrm{str}}\binom{\varnothing}{\varnothing}$ is the k-vertex cycle (Example 1.35) with a single color. Then we have

$$\mathrm{Aut}_{\mathrm{str.}}(G) = \mathbf{Z}/(k),$$

where the cyclic group is generated by the strict automorphism

$$\psi\left(e^i_{\pm 1}\right) = e^{i+1}_{\pm 1}$$

with $e^{k+1}_{\pm 1}$ denoting $e^1_{\pm 1}$ (taking indices modulo k).

If we were to choose different colors for the edges, the strict automorphism group would shrink, perhaps all the way to the trivial group. The basic idea is that one would require periodic choices of colors in order to have non-trivial strict automorphisms by permuting the cycle.

(8) Suppose G_1 and G_2 are wheeled graphs whose flags have no color in common, i.e.,

$$\kappa(\mathrm{Flag}(G_1)) \cap \kappa(\mathrm{Flag}(G_2)) = \varnothing.$$

Then we have a direct product decomposition

$$\mathrm{Aut}_{\mathrm{str.}}(G_1 \sqcup G_2) \cong \mathrm{Aut}_{\mathrm{str.}}(G_1) \times \mathrm{Aut}_{\mathrm{str.}}(G_2),$$

since a strict automorphism must preserve flag colors. For example, suppose $K = G \coprod G'$ from the last two examples. Then we have

$$\mathrm{Aut}_{\mathrm{str.}}(K) = \mathbf{Z}/(k) \times \mathbf{Z}/(2).$$

(9) Consider the ordinary 1-colored graph with no legs

(4.1)

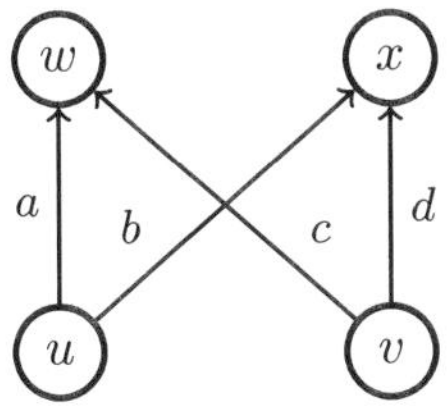

defined as the partitioned set

$$\{\{a_{-1}, b_{-1}\}, \{c_{-1}, d_{-1}\}, \{a_1, c_1\}, \{b_1, d_1\}\}$$

ι and δ are indicated by the naming convention, and the listing is

$$\ell(a_{\pm 1}) = 1 = \ell(d_{\pm 1}) \quad \text{and} \quad \ell(b_{\pm 1}) = 1 = \ell(c_{\pm 1}).$$

Define

$$\psi(a_i) = d_i, \quad \psi(b_i) = c_i, \quad \psi(c_i) = b_i \quad \text{and} \quad \psi(d_i) = a_i$$

which is compatible with ι, π, directions and colors by definition, and with listings by inspection. The only other mapping of partitioned sets here preserving directions sends a_1 to b_1, and so fails to preserve listings. As a consequence, the strict automorphism group of this graph is $\mathbf{Z}/(2)$.

4.1.2. Decomposing the Strict Automorphism Group. We begin our study of the strict automorphism group by noticing we should start by looking carefully at the strict automorphism groups of connected graphs.

LEMMA 4.5. *A strict automorphism of wheeled graphs must preserve connected components. In fact, the restriction to each connected component must remain a strict isomorphism. Finally, every* $\mathrm{Aut}_{str.}(G)$ *is finite.*

PROOF. A strict automorphism must commute with the involutions, in addition to being a map of partitioned sets, so preserves the notion of flag-connectedness. In addition, any strict isomorphism preserves isolated vertices by inspection. The fact that ψ remains a bijection of partitioned sets compatible with extra structure then implies the restriction to any component remains a strict isomorphism.

For finiteness of $\mathrm{Aut}_{\mathrm{str.}}(G)$, recall there are only finitely may flags and isolated vertices (by finiteness of the partition) in G. □

In order to understand strict automorphisms of connected graphs, we first have two results about how they interact with paths.

LEMMA 4.6. *Suppose* G *is a wheeled graph with* ψ *a strict automorphism of* G*,* v *a vertex of* G*, and* f *a flag incident to* v*. Then the following statements are equivalent:*

(1) ψ *fixes* f*,*
(2) ψ *fixes* v*,*
(3) ψ *fixes each flag incident to* v*.*

PROOF. If ψ fixes f, then as a map of partitioned sets it must not move the cell of the partition containing f, namely v. If ψ fixes the cell of the partition v, then the combination of compatibility with the direction and the listing implies it fixes each flag within that cell of the partition. □

LEMMA 4.7. *If* P *is a path in a wheeled graph* G *and* ψ *is a strict automorphism of* G*, then* ψ *fixes all elements of* P *precisely when it fixes any element of* P*.*

PROOF. If ψ fixes either a flag or a vertex of P, then Lemma 4.6 together with compatibility of ψ with the involution ι implies, inductively, that ψ fixes all edges and vertices of P. □

The next result is a rigidity statement for connected wheeled graphs with legs.

LEMMA 4.8. *If* G *is a connected graph with at least one leg, then it has only the trivial strict automorphism.*

PROOF. As in Example 4.4(1), every leg must be fixed by a strict automorphism. Since G is flag-connected, there must be at least one path connecting each flag to a leg, which must then be fixed by any strict automorphism according to Lemma 4.7. □

Now we notice that if we have two connected components which are strictly isomorphic and without legs, then we can permute them with a strict automorphism.

DEFINITION 4.9. Two *ordinary* connected components without legs G_1 and G_2 of a wheeled graph G are called **clones** if, when viewed as wheeled graphs, the two are strictly isomorphic.

Notice exceptional loops are excluded from the definition of clones. A decomposition for graphs all of whose components are clones follows.

LEMMA 4.10. *If $G = \coprod_{j=1}^{k} G_j$, where each G_j is a clone of G_1, then there is a non-canonical isomorphism of groups*

$$\mathrm{Aut}_{str.}(G) \cong (\mathrm{Aut}_{str.}(G_1))^k \times \Sigma_k.$$

PROOF. By Lemma 4.5, each component must be sent strictly isomorphically onto its image, but a permutation of the indices is also possible. □

As a consequence of Lemma 3.7, up to relabeling, one can write any wheeled graph as a union of its sequences of clones followed by the remaining (non-clone) connected components without legs, and finally the connected components with legs. That is, there exists a strict isomorphism

$$\sigma G \tau \cong \left(\coprod_n \coprod_{j=1}^{k_n} G_n \right) \sqcup \left(\coprod_m G'_m \right) \sqcup \left(\coprod_l G''_l \right)$$

where each $k_n > 1$, no G'_m is a clone, and each G''_l has legs, but each G_n or G'_m is without legs. This description leads to a decomposition of the strict automorphism group of any wheeled graph, which is surprisingly small and ignores all connected components with legs.

PROPOSITION 4.11. *For any wheeled graph G, one has $\mathrm{Aut}_{str.}(G)$ non-canonically isomorphic to a product*

$$\left(\prod_n (\mathrm{Aut}_{str.}(G_n))^{k_n} \times \Sigma_{k_n} \right) \times \left(\prod_m \mathrm{Aut}_{str.}(G'_m) \right).$$

PROOF. Combining Lemma 4.5 and the fact that all legs of G must be fixed as in Example 4.4(1) implies a strict automorphism restricts to a strict automorphism of each connected component with legs. As a consequence, the restriction of any ψ to $\coprod_l G''_l$ must be trivial by Lemma 4.8.

For the components without legs, notice each non-clone must be sent strictly automorphically to itself by assumption, so Lemma 4.10 completes the proof. □

EXAMPLE 4.12. Suppose $G = G_1 \sqcup G_1 \sqcup G_1 \sqcup G' \sqcup G''$, with G_1 a k-vertex loop as in Example 1.35, G' the graph from Example 1.35(9), and G'' is any connected graph with legs. Then $\mathrm{Aut}_{\mathrm{str.}}(G)$ is non-canonically isomorphic to

$$\mathbf{Z}/(k) \times \mathbf{Z}/(k) \times \mathbf{Z}/(k) \times \Sigma_3 \times \mathbf{Z}/(2).$$

In fact, it is not only for connected graphs with legs that we have a rigidity statement. As we will see below, all simply-connected graphs are also rigid. First we have a reduction to the question of finding a single fixed vertex in order to verify a strict isomorphism of a simply-connected graph is trivial.

LEMMA 4.13. *A strict automorphism of a simply-connected graph is uniquely determined by the image of any one vertex.*

PROOF. Suppose v is a vertex of G, which has no legs, as otherwise the result follows from Lemma 4.8. Given $\psi(v)$, the image of each flag incident to v is determined by the listings and direction. Since ψ must also preserve the involution ι, this determines the image of any full edge with a flag incident to v, and so the image of any vertex incident to such an edge. One can proceed in this manner along the unique path from v to any flag, thereby inductively determining the entire image of ψ. □

The following result explains why strict automorphisms do not appear in discussions of trees, or arbitrary simply-connected graphs.

PROPOSITION 4.14. *A simply-connected graph has only the trivial strict automorphism.*

PROOF. Suppose ψ is a strict automorphism of a simply-connected graph G without legs, as otherwise the result follows from Lemma 4.8. Consider repeated application of ψ to a fixed vertex v. By finiteness of G, there must exist natural numbers k and minimal l with $\psi^{k+l}v = \psi^k v$. Choosing $w = \psi^k v$ and, if $w \neq \psi w$, the unique path P from w to ψw, we consider the images $\psi^j P$ for $1 \leq j < l$, which must each contain the same number of flags of each direction in the same sequence. In particular, no $\psi^j P$ can be P^{op}, since that would mean P is symmetrical and ψ^j would alter the listing at the middle vertex of P, contradicting the assumption that ψ commutes with listings.

By the uniqueness of paths between any two vertices in a simply-connected graph, there must be a specific i for which $\psi^i w$ is 'farthest from' w, with the higher index images of P retracting the path of the lower index images of P to reverse the unique path from w to $\psi^i w$. This would again imply P is symmetrical, in order to avoid violating compatibility with directions, and would imply $\psi^{l-1}P = P^{op}$, which produces a contradiction as discussed above. As a consequence, we must have $\psi w = w$, in which case Lemma 4.13 implies ψ is the trivial strict automorphism. □

4.2. Weak Isomorphisms

For some purposes, it is convenient to work with a more flexible notion of isomorphism, here called weak isomorphism, of wheeled graphs. In fact, this simply removes the condition of compatibility with the listing for a strict isomorphism of wheeled graphs. As a consequence, input and output relabeling leads to weak isomorphisms that are not strict isomorphisms in general and the groupoid of graphs and weak isomorphisms becomes the natural place to index bi-equivariant processes.

DEFINITION 4.15. Let G and G' be wheeled graphs. A **weak isomorphism** $\psi: G \longrightarrow G'$ is a bijection of partitioned sets that commutes with both types of involutions and leaves invariant both the colorings and the directions.

Once again, notice there is no mention of compatibility with the listings for a weak isomorphism. Thus, choosing the same directed, colored graph but with a different listing is the canonical way to produce examples of graphs that are weakly isomorphic but not necessarily strictly isomorphic.

REMARK 4.16. By the nature of a listing, notice that for any weak isomorphism there are permutations that can be used, at each vertex and for the full graph, to alter the listing on the target to produce a strict isomorphism of graphs. Hence, the idea underlying input and output relabeling is generalized by this transition, allowing permutations to act at each vertex as well as on the legs of the full graph.

NOTATION 4.17. Let $\mathrm{Aut}_{\mathrm{w.}}(G)$ denote the group of weak automorphisms of the wheeled graph G.

Clearly $\mathrm{Aut}_{\mathrm{str.}}(G) \subset \mathrm{Aut}_{\mathrm{w.}}(G)$ but weak automorphisms form a much larger group in general. In cases where the listing is trivial, such as a $\underline{c}$-colored exceptional loop, or a linear graph, the two notions of automorphism coincide.

EXAMPLE 4.18. Here are some examples of weak isomorphisms and weak automorphism groups. Suppose $\underline{c}$ and $\underline{d}$ are profiles with $\sigma \in \Sigma_{|\underline{d}|}$, and $\tau \in \Sigma_{|\underline{c}|}$.

(1) For each wheeled graph G with profiles $(\underline{c};\underline{d})$, the identity on flags gives a weak isomorphism

$$G \longrightarrow \sigma G \tau. \tag{4.2}$$

Note that in general, G is not strictly isomorphic to $\sigma G \tau$, and the identity map need not be the only weak isomorphism between G and $\sigma G \tau$.

(2) There is a weak isomorphism of corollas

$$C_{(\underline{c};\underline{d})} \longrightarrow C_{(\underline{c}\tau;\sigma\underline{d})} \tag{4.3}$$

again given by the identity on flags. Note that these corollas are not strictly isomorphic in general.

(3) If $\underline{b}$ is another profile and $\rho \in \Sigma_{|\underline{b}|}$, then for grafted corollas the identity of partitioned sets with direction and coloring (in the presence of the evident altered listing) is a weak isomorphism

$$C_{(\underline{b};\underline{c};\underline{d})} \longrightarrow C_{(\rho\underline{b};\tau\underline{c};\sigma\underline{d})}. \tag{4.4}$$

Once again, these grafted corollas are not strictly isomorphic in general.

(4) The graphs indicated below, with listings induced by the planar presentation, are weakly isomorphic but not strictly isomorphic.

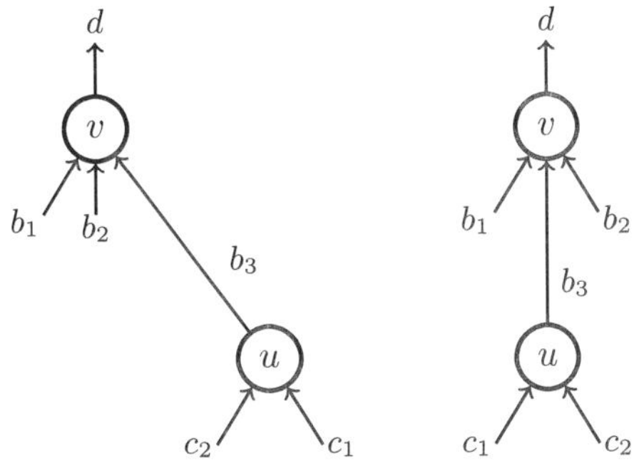

(5) Given any 1-colored $(m;n)$-corolla G, there is an isomorphism

$$\mathrm{Aut}_{\mathrm{w.}}(G) \cong \Sigma_m \times \Sigma_n,$$

while $\mathrm{Aut}_{\mathrm{str.}}(G)$ is trivial.

(6) Given $G' = \coprod_{j=1}^{k} \uparrow_c$, $\mathrm{Aut}_{\mathrm{str.}}(G')$ was trivial, while

$$\mathrm{Aut}_{\mathrm{w.}}(G') \cong \Sigma_k.$$

As with strict isomorphisms, weak isomorphisms can be studied by looking at connected components, since the proofs of Lemmas 4.5 and 4.10 make no reference to listings.

LEMMA 4.19. *A weak automorphism of wheeled graphs must preserve connected components. In fact, the restriction to each connected component must remain a weak isomorphism. Finally, every* $\mathrm{Aut}_{w.}(G)$ *is finite.*

DEFINITION 4.20. Two connected components which are not exceptional loops, possibly with legs, G_1 and G_2 of a wheeled graph G are called **weak clones** if, when viewed as wheeled graphs, the two are weakly isomorphic.

Notice, as above, that exceptional loops are explicitly excluded from being weak clones, since weak automorphisms still have no effect upon them by Remark 1.34.

LEMMA 4.21. *If* $G = \coprod_{j=1}^{k} G_j$*, where each* G_j *is a weak clone of* G_1*, then there is a non-canonical isomorphism of groups*

$$\mathrm{Aut}_{w.}(G) \cong (\mathrm{Aut}_{w.}(G_1))^k \times \Sigma_k.$$

Once again, as a consequence of Lemma 3.7, up to relabeling, one can write any wheeled graph as a union of its sequences of weak clones followed by the remaining (non-weak clone) connected components. That is, there exists a weak isomorphism

$$G \cong \left(\coprod_n \coprod_{j=1}^{k_n} G_n\right) \sqcup \left(\coprod_m G'_m\right)$$

where each $k_n > 1$ and no G'_m is a weak clone. This description leads to a decomposition of the weak automorphism group of any wheeled graph.

PROPOSITION 4.22. *For every wheeled graph G, one has* $\mathrm{Aut}_{w.}(G)$ *non-canonically isomorphic to a product*

$$\left(\prod_n (\mathrm{Aut}_{w.}(G_n))^{k_n} \times \Sigma_{k_n}\right) \times \left(\prod_m \mathrm{Aut}_{w.}(G'_m)\right).$$

The following much weaker analog of Lemma 4.6 and the absence of an analog of Lemma 4.7 are indications of the less rigid nature of weak isomorphisms. As Example 4.18(5) and (6) make clear, there is no analog here of the rigidity for strict automorphisms of connected graphs with legs.

LEMMA 4.23. *If a weak isomorphism* ψ *fixes an ordinary flag* f *of a wheeled graph G, then it must fix the vertex incident to* f.

PROOF. As a map of partitioned sets, if ψ fixes the element f of the cell v of the partition, then it must preserve that cell. □

4.3. Graph Groupoids

We now introduce a series of graph groupoids that will be considered repeatedly in what follows. In fact, our focus will be on graph groupoids that have additional structures, leading to what will be called pasting schemes for generalized PROPs.

4.3.1. Graph Groupoids Defined by Strict and Weak Isomorphisms.

DEFINITION 4.24. The groupoid of $(\underline{c};\underline{d})$-wheeled graphs and strict isomorphisms is denoted by $\mathtt{Gr}^{\circlearrowleft}_{\mathrm{str}}\binom{\underline{d}}{\underline{c}}$. The groupoid of all wheeled graphs and strict isomorphisms is denoted by $\mathtt{Gr}^{\circlearrowleft}_{\mathrm{str}}$, while that of wheeled graphs and weak isomorphisms is denoted by $\mathtt{Gr}^{\circlearrowleft}_{\mathrm{w}}$.

Notice that $\mathtt{Gr}^{\circlearrowleft}_{\mathrm{w}}$ should be decomposed in terms of orbit types of pairs of profiles, rather than specific pairs of profiles as with $\mathtt{Gr}^{\circlearrowleft}_{\mathrm{str}}$.

Since input and output relabeling alters only the listing on the entire graph, the following observation is elementary.

LEMMA 4.25. *Let $\underline{c}$ and $\underline{d}$ be profiles and $\sigma \in \Sigma_{|\underline{d}|}$ and $\tau \in \Sigma_{|\underline{c}|}$ be permutations. Then input and output relabeling yields a canonical isomorphism*

$$\mathtt{Gr}^{\circlearrowleft}_{str}\binom{\underline{d}}{\underline{c}} \xrightarrow{\cong} \mathtt{Gr}^{\circlearrowleft}_{str}\binom{\sigma\underline{d}}{\underline{c}\tau}, \quad G \mapsto \sigma G \tau \tag{4.5}$$

of groupoids.

4.3.2. Examples of Graph Groupoids. At this point we will introduce our notation for a series of different graph groupoids which will be important in what follows. These will all be examples of what we call pasting schemes, so each can be used to define a version of generalized PROPs. In order to provide some context, reference to the associated generalized PROP structure is included.

NOTATION 4.26. Here are our main examples of graph groupoids.

- **Permuted Corollas:** Let $\mathtt{Cor}$ denote the full sub-groupoid of $\mathtt{Gr}^{\circlearrowleft}_{\mathrm{w}}$ consisting of permuted corollas (3.4). This will be the smallest graph groupoid of interest to us, and the associated structures will simply be the Σ-bimodules.
- **All Wheeled Graphs:** The full graph groupoid $\mathtt{Gr}^{\circlearrowleft}_{\mathrm{w}}$ will be used to model the (one-colored) wheeled PROPs [**MMS09**] as examples of generalized PROPs. This will be the biggest graph groupoid of interest to us.
- **Connected Wheeled Graphs:** Let $\mathtt{Gr}^{\circlearrowleft}_{\mathrm{c}}$ denote the full sub-groupoid of $\mathtt{Gr}^{\circlearrowleft}_{\mathrm{w}}$ consisting of connected wheeled graphs (2.11). This will be used to model the (one-colored) wheeled properads [**MMS09**] as examples of generalized PROPs.
- **Contracted Corollas:** Let $\mathtt{Gr}^{\xi}$ denote the full sub-groupoid of $\mathtt{Gr}^{\circlearrowleft}_{\mathrm{w}}$ consisting of possibly repeatedly contracted corollas (3.25), and their input and output relabelings (3.2). In this case, all graphs have a single vertex, and the resulting structure will simply be a diagram category. This will provide a diagram category in which the vertically non-unital wheeled PROPs form a category of monoids.
- **Wheeled trees:** Let $\mathtt{Tree}^{\circlearrowleft}$ be the full sub-groupoid of $\mathtt{Gr}^{\circlearrowleft}_{\mathrm{c}}$ consisting of wheeled trees, that is, connected wheeled graphs G whose vertices each have at most one outgoing flag. This will be used later to model the (one-colored) wheeled operads introduced in [**MMS09**] as generalized PROPs.

- **Wheel-Free Graphs:** Let $\mathtt{Gr}^{\uparrow}$ denote the full sub-groupoid of $\mathtt{Gr}_{\mathrm{w}}^{\circlearrowright}$ consisting of wheel-free graphs (2.18 and 2.5). This will be used to model PROPs as examples of generalized PROPs.
- **Connected Wheel-Free Graphs:** Let $\mathtt{Gr}_{\mathrm{c}}^{\uparrow}$ be the full sub-groupoid of $\mathtt{Gr}^{\uparrow}$ consisting of connected wheel-free graphs, which will be used to model the (one-colored) properads of [**Val07**] as examples of generalized PROPs.
- **Simply-Connected Graphs:** Let $\mathtt{Gr}_{\mathrm{di}}^{\uparrow}$ be the full sub-groupoid of $\mathtt{Gr}_{\mathrm{c}}^{\uparrow}$ consisting of simply-connected graphs (2.16 and 2.5), which will be used to model the (one-colored) dioperads of [**Gan03**] as examples of generalized PROPs.
- **Half-Graphs:** Let $\mathtt{Gr}_{\frac{1}{2}}$ be the full sub-groupoid of $\mathtt{Gr}_{\mathrm{di}}^{\uparrow}$ consisting of half-graphs (2.21), which will be used to model the (one-colored) half-PROPS of [**Mar08, MV09**] as examples of generalized PROPs, originally suggested by Kontsevich in an unpublished message to Markl.
- **Level Trees:** Let $\mathtt{Tree}$ be the full sub-groupoid of $\mathtt{Gr}_{\mathrm{di}}^{\uparrow}$ consisting of level trees (2.27), which will be used to model the (one-colored) non-unital operads in the sense of Markl [**Mar08**] as examples of generalized PROPs.
- **Unital Trees:** Let $\mathtt{UTree}$ be the full sub-groupoid of $\mathtt{Gr}_{\mathrm{di}}^{\uparrow}$ consisting of level trees and every exceptional edge $\uparrow_c$, which will be used to model the (one-colored) unital operads in the original sense of May [**Mar08, May72**] as examples of generalized PROPs.
- **Horizontal Combinations:** Let $\mathtt{Gr}_{\mathrm{h}}^{\uparrow}$ denote the full sub-groupoid of $\mathtt{Gr}^{\uparrow}$ consisting of finite unions of corollas and their input and output relabelings. This will be used to model the hPROPs of [**JY09**] as examples of generalized PROPs.
- **Vertical Combinations:** Let $\mathtt{Gr}_{\mathrm{v}}^{\uparrow}$ denote the full sub-groupoid of $\mathtt{Gr}_{\mathrm{c}}^{\uparrow}$ consisting of the wheeled graphs weakly isomorphic to iterated graftings
$$C_{(\underline{c}_{n-1};\underline{c}_n)} \boxtimes C_{(\underline{c}_{n-2};\underline{c}_{n-1})} \boxtimes \cdots \boxtimes C_{(\underline{c}_1;\underline{c}_2)}$$
of corollas with each $\underline{c}_i \in \mathcal{P}(\mathfrak{C})$, $n \geq 1$, and $|\underline{c}_i| \geq 1$ for $2 \leq i \leq n-1$. This will be used to model the vPROPs of [**JY09**] as examples of generalized PROPs.
- **Linear Graphs:** Let $\mathtt{Lin}$ be the full sub-groupoid of $\mathtt{Gr}_{\mathrm{c}}^{\uparrow}$ consisting of iterated graftings
$$C_{(c_{n-1};c_n)} \boxtimes C_{(c_{n-2};c_{n-1})} \boxtimes \cdots \boxtimes C_{(c_1;c_2)}$$
of corollas with each $c_i \in \mathfrak{C}$ and $n \geq 1$, denoted **linear graphs**. This will be used to model non-unital enriched categories as generalized PROPs. Adding the single exceptional edges provides a slightly larger graph groupoid, which can be used to model enriched categories.

CHAPTER 5

Graph Substitution

In this chapter, our focus will be on the fundamental operation of graph substitution. One of our primary goals throughout this monograph is to reduce a variety of questions to consequences of graph substitution, so this requires careful study. If there are no exceptional flags involved, the construction is relatively straightforward. Thus, we begin with a discussion of the ordinary case, and then discuss a few simple examples to indicate the types of complications that arise.

The main issues come from exceptional edges in the substituting graphs. As a consequence, we treat the construction in two steps. First, we introduce a slight generalization of a graph only considered in this chapter, called a pre-graph, and having a third type of flag called ambiguous flags. It is then relatively straightforward to produce what we call the pre-substitution of pre-graphs, and we are able to show that it is nearly unital as well as being associative up to canonical strict isomorphism. In fact, in the pursuit of minimal technical requirements for later flexibility, we perform pre-subtitution without directions, colorings, or listings, and only then point out how these structures, when appropriately compatible, are inherited by the pre-substitution.

The second step is to introduce the associated graph of a pre-graph, building on the technology of paths from Chapter 2 to define ambiguous paths in pre-graphs. In essence, what we call ambiguous loops and ambiguous components are collapsed down to a single pair of flags to create exceptional loops and exceptional edges in the associated graph, while all other ambiguous flags are discarded after being used to redefine the involution ι of the associated graph.

One can now define the graph substitution as the associated graph of the pre-substition, viewing each graph in question as a very nice example of a pre-graph where there happen to be no ambiguous flags. However, the associativity of pre-substitution does not immediately imply that of the true graph substitution operation, since iterating the true graph substitution would then require two applications of the associated graph construction as well. Thus, the final technical point is to verify that choosing to perform an associated graph construction at an intermediate stage does not alter the associated graph of a pre-substitution, which requires substantial care.

After all this, the main results say graph substitution is unital and associative up to canonical strict isomorphism, as well as natural in the sense of being compatible with either strict or weak isomorphisms of the graphs involved.

5.1. Graph Substitution in the Ordinary Case

Let G be a (generalized) graph. For each vertex v in G, pick an arbitrary wheeled graph H_v and a bijection between $\mathrm{Leg}(H_v)$ and the set v, viewed as an ordinary cell in the partitioned set G. Roughly speaking, the graph substitution

$G(H_v)$ is defined by cutting a small hole around each vertex v in G and then patching in a scaled down version of the wheeled graph H_v, with the bijection saying which leg of H_v to connect to which other flag. We will say H_v **corresponds to** v or is **substituted into** v, and generally de-emphasize the choice of bijection, although in certain cases the choice is important to track. Most of the delicacy of defining graph substitution involves the exceptional legs of H_v, since this process is relatively straightforward in other cases.

Suppose we have chosen a graph G, as well as choosing for each vertex $v \in \mathrm{Vt}(G)$ a graph H_v together with a bijection

$$\psi_v : \mathrm{Leg}(H_v) \cong v.$$

Together these will be referred to as **graph substitution data**.

In order to separate the complexity of the exceptional situation from the ordinary case, and to provide a gradual introduction to the complexities of the general case, it seems reasonable to first define graph substitution when all graphs involved are ordinary graphs, with empty exceptional cell. Thus, for the moment, suppose our substitution data satisfies G and each H_v ordinary.

First we need to define a technical mechanism for using the involution ι_G to connect a leg of H_v to a leg of $H_{v'}$.

DEFINITION 5.1. Suppose f is a leg of some H_v. Then define

$$\zeta_v^G(f) = (\psi_{v'})^{-1}\iota_G\psi_v(f).$$

Notice that if the intermediate flag lies in $\mathrm{Leg}(G)$, then ζ_v^G becomes the identity, which will be used to identify the legs of the graph substitution with those of G itself. Also notice each ordinary flag of G is in the image of a unique ψ_v, which is used to find the $\psi_{v'}$ in the formula for ζ_v^G. In the ordinary case, we define the graph substitution $K = G(H_v)$ as follows, where the reader may want to revisit Remark 1.19 concerning disjoint sets of flags.

DEFINITION 5.2. As a partitioned set, the **graph substitution** for these ordinary graphs is

$$\mathrm{Flag}(K) = \coprod_{v \in \mathrm{Vt}(G)} \mathrm{Flag}(H_v)$$

with an involution

$$\iota_K(f) = \begin{cases} \iota_{H_v}(f) & \text{for } f \in \mathrm{Flag}(H_v) \setminus \mathrm{Leg}(H_v), \text{ and} \\ \zeta_v^G(f) & \text{for } f \in \mathrm{Leg}(H_v). \end{cases}$$

Keep in mind that the legs of the various H_v which are identified with legs of G become the legs of K. All other legs of the different H_v are then paired via ι_K to produce internal edges.

EXAMPLE 5.3. As a first example, consider G the two-vertex cycle of one color (see Example 1.35), with each H_j an n_j-vertex linear graph of the same color (see Example 1.36). Then $G(H_j)$ is an $(n_1 + n_2)$-vertex cycle.

EXAMPLE 5.4. Now let G denote the bat and ball graph of Example 1.40, while H_y is a copy of the walnut graph of Example 1.41, H_w is a three vertex linear graph with input removed, and H_x is a two vertex linear graph with output removed. Then $G(H_v)$ will be the union of the walnut graph with a five-vertex linear graph with both input and output removed.

If we want to work with ordinary wheeled graphs, or intermediate structures such as ordinary directed graphs, we need to require that the bijections ψ_v in the graph substitution data preserve the directions and/or colorings under consideration. Under that assumption, it is immediate to say that each flag retains the direction and/or coloring it had as a flag of H_v, thereby producing a direction and/or coloring for all of K.

For ordinary wheeled graphs, the most natural thing is to say ψ_v is essentially just the identity map between the pairs of profiles $(\text{in}(v);\text{out}(v))$ and $(\text{in}(H_v);\text{out}(H_v))$, since this is equivalent to saying it preserves both colors and listings. In most cases, we will consider this identification of pairs of profiles as the requirement for defining our graph substitutions. However, the process is slightly more robust, and does still work to produce a wheeled graph even for non-trivial choices of bijections ψ_v which do preserve the directions and colorings.

The listing for the full graph K comes from that of G, since their sets of legs are canonically identified. On the other hand, the listing at any vertex of K can simply be that of the corresponding vertex of some H_v by the definition of the partitioned set K. Thus, a non-trivial bijection ψ_v would not alter the vertices or the listings of K, but rather would alter the involution ι_K via changes to ζ^G. Some authors take advantage of this added flexibility, so we will try to provide the most general definitions throughout, but we will focus our efforts in later chapters on the stricter assumption that the profiles match.

REMARK 5.5. Without making specific choices of such isomorphisms ψ_v, it is not clear that the graph substitution is well-defined, unless one works exclusively with graphs where natural choices of such isomorphisms always exist, such as level trees. One can view any substitution with the ψ_v non-trivial as first performing input and output relabelings of the various H_v, followed by the stricter type of substitution where each ψ_v must be the identity. Many sources work with all possible choices of such relabelings and are later forced to somehow coequalize them, but we find it more economical to make the choices explicit.

5.2. Pre-Graphs and Pre-Substitution

When there are exceptional legs in the substituting graphs H_v, we must be very careful in order to deal with them properly. In fact, any exceptional flags from G, as well as any exceptional loops from any H_v will simply contribute the same to $K = G(H_v)$. The key point to understand is that exceptional legs in the H_v can serve any of three roles, which we must distinguish.

The three roles occur even in the case of linear graphs, which we now use as motivating examples.

EXAMPLE 5.6. Suppose G is a linear graph of a single color with two vertices, labeled 1 and 2 in the linear order given by the direction,

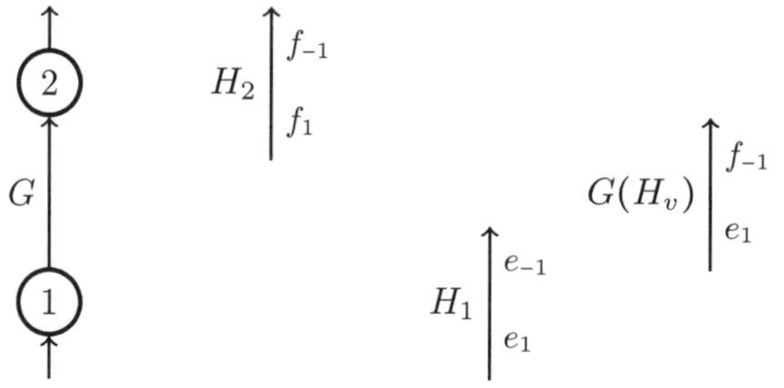

while H_1 and H_2 are exceptional edges. Then the graph substitution $G(H_v)$ should also be an exceptional edge. Thus, e_{-1} and f_1 simply serve to tell the two legs given by e_1 and f_{-1} that they are now exceptional flags paired by π.

EXAMPLE 5.7. Now suppose G is a linear graph of a single color with 4 vertices, which we can label 1 through 4 using the linear ordering given by the direction.

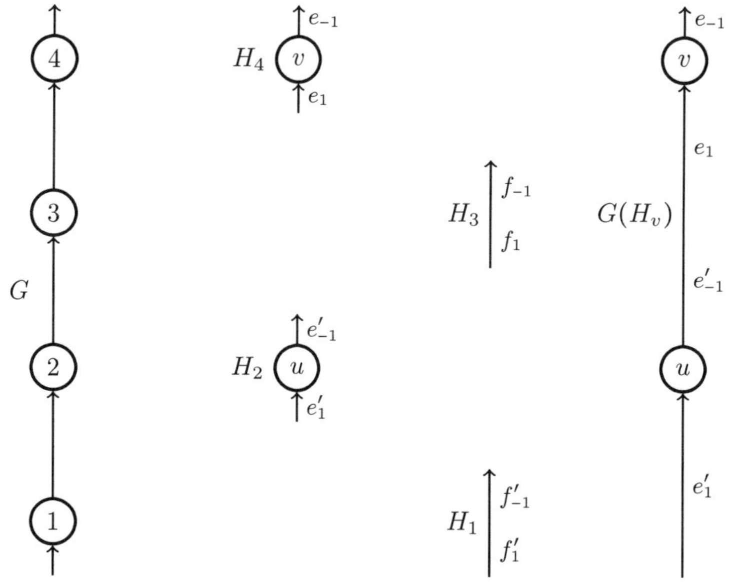

Choose H_1 and H_3 to be exceptional edges, with H_2 and H_4 corollas. In this case, the graph substitution should be a linear graph with two vertices, 2 and 4. However, that means $f'_{\pm 1}$ serve only to turn e'_1 into a leg and $f_{\pm 1}$ serve only to turn e'_{-1} and e_1 into an internal edge. Of course, the outgoing leg of $G(H_v)$ comes from the outgoing flag of H_4.

5.2.1. Pre-Graphs. In order to be explicit about our approach to treating these new roles for exceptional edges, we will introduce technical generalizations of graphs and paths used only in the remainder of this chapter.

DEFINITION 5.8. A **pre-graph** $\widehat{K}$ will consist of a finite partitioned set of flags with partitions of three types, a possibly empty collection of **ordinary cells**, a possibly empty collection of **ambiguous cells** and an **exceptional cell**, together with an involution $\iota_{\widehat{K}}$ that isolates the exceptional cell and a free involution $\pi_{\widehat{K}}$ which isolates each cell while acting on each ambiguous cell as well as on the $\iota_{\widehat{K}}$-fixed points within the exceptional cell.

Notice there is no claim that $\iota_{\widehat{K}}$ isolates the ambiguous cells from the ordinary cells, and $\pi_{\widehat{K}}$ is only defined for exceptional legs along with all ambiguous flags. Pairs of ambiguous flags connected by $\pi_{\widehat{K}}$, which must lie in the same cell by definition, will be called **ambiguous edges**, and each ambiguous flag is part of an ambiguous edge. There must also be at least one exceptional cell, although it may just be the empty set, and the ordinary and ambiguous cells are not required.

The notation $\mathrm{Ambig}(\widehat{K})$ will be used to indicate the set of ambiguous flags of $\widehat{K}$ and as with graphs the term **leg** will indicate a flag fixed by $\iota_{\widehat{K}}$. In addition, we will refer to an ambiguous flag f with $\iota_{\widehat{K}}(f)$ an ordinary flag as an **arm**, and the set of such will be indicated by $\mathrm{Arm}(\widehat{K})$, while the combination of arms and legs will be called **limbs**. Ambiguous flags which are not limbs will be called **internal ambiguous flags**, while **internal flags** will still refer to all flags not fixed by $\iota_{\widehat{K}}$.

Directions, colorings and listings for pre-graphs are defined as for graphs. Of course, a graph is just a pre-graph with no ambiguous cells.

5.2.2. Pre-Substitution. It will be useful to define pre-substitution for pre-graphs, so we note that **pre-substitution data** should consist of a pre-graph $\widehat{G}$ together with a choice for each vertex $v \in \widehat{G}$ of a pre-graph $\widehat{H}_v$ and a bijection

$$\psi_v : \mathrm{Leg}(\widehat{H}_v) \cong v.$$

As a consequence, each ordinary flag of $\widehat{G}$ is once again identified with a leg of a unique $\widehat{H}_v$ via ψ_v. However, no ambiguous flag in $\widehat{G}$ lies in the image of ψ_v, which means we must take care with arms of $\widehat{G}$.

The reader should be aware that even when each pre-graph is really a graph, there is no reason to expect the pre-substitution to be a graph. Thus, we will later be defining an associated graph functor for pre-graphs, and the graph substitution will be the associated graph of the pre-substitution.

Before defining the pre-substitution, we need a variant of the ζ_v^G used in the ordinary case to describe ι_K of the legs of the various $\widehat{H}_v$. In this case, we must define $\hat{\zeta}_v^{\widehat{G}}$ for all arms of $\widehat{G}$ and legs of any $\widehat{H}_v$.

DEFINITION 5.9. Given any leg f of $\widehat{H}_v$, define

$$\hat{\zeta}_v^{\widehat{G}}(f) = \begin{cases} \iota_{\widehat{G}}\psi_v(f) & \text{if this flag is in } \mathrm{Arm}(\widehat{G}), \text{ or} \\ (\psi_{v'})^{-1}\iota_{\widehat{G}}\psi_v(f) = \zeta_v^{\widehat{G}}(f) & \text{otherwise.} \end{cases}$$

Given an arm f of $\widehat{G}$, so $\iota_{\widehat{G}}(f)$ ordinary, we define $\hat{\zeta}_v^{\widehat{G}}(f) = (\psi_v)^{-1}\iota_{\widehat{G}}(f)$ for the unique ψ_v whose image contains this ordinary flag of $\widehat{G}$.

LEMMA 5.10. *Given pre-substitution data $\widehat{G}$ together with $\widehat{H}_v$ and ψ_v, there is an associated pre-graph $\widehat{K} = \widehat{G}(\widehat{H}_v)$ with flags*

$$\mathrm{Ambig}(\widehat{G}) \sqcup \mathrm{Flag}_{\mathrm{e}}(\widehat{G}) \sqcup \coprod_v \mathrm{Flag}(\widehat{H}_v).$$

PROOF. For the exceptional cell in our pre-substitution $\widehat{K}$, we take

$$\mathrm{Flag}_{\mathrm{e}}(\widehat{G}) \sqcup \coprod_v \mathrm{Flag}_{\mathrm{ie}}(\widehat{H}_v),$$

that is, all exceptional flags of $\widehat{G}$ together with any flags forming exceptional loops in any $\widehat{H}_v$. For the ambiguous cells of $\widehat{K}$, we take the ambiguous cells coming from either $\widehat{G}$ or any $\widehat{H}_v$, and for each $\widehat{H}_v$ with exceptional legs we form an ambiguous cell consisting of those, which are already paired by their own free involution $\pi_{\widehat{H}_v}$. As ordinary cells, we take the ordinary cells of each $\widehat{H}_v$.

Define the involution $\iota_{\widehat{K}}$ by

$$\iota_{\widehat{K}}(e) = \hat{\zeta}_v^{\widehat{G}}(e) \text{ when } e \in \mathrm{Arm}(\widehat{G}) \sqcup \coprod_v \mathrm{Leg}(\widehat{H}_v),$$

along with (for internal flags of the $\widehat{H}_v$)

$$\iota_{\widehat{K}}(f) = \iota_{\widehat{H}_v}(f) \text{ when } f \in \coprod_v \left(\mathrm{Flag}(\widehat{H}_v) \setminus \mathrm{Leg}(\widehat{H}_v)\right),$$

and (for the exceptional flags, ambiguous legs, and internal ambiguous flags of $\widehat{G}$)

$$\iota_{\widehat{K}}(g) = \iota_{\widehat{G}}(g) \text{ when } g \in \mathrm{Flag}_{\mathrm{e}}(\widehat{G}) \sqcup \left(\mathrm{Ambig}(\widehat{G}) \setminus \mathrm{Arm}(\widehat{G})\right).$$

It follows that we still need to define

$$\pi_{\widehat{K}}(f) = \pi_{\widehat{G}}(f) \text{ for } f \in \mathrm{Ambig}(\widehat{G}) \sqcup \mathrm{Leg}_{\mathrm{e}}(\widehat{G}),$$

and

$$\pi_{\widehat{K}}(e) = \pi_{\widehat{H}_v}(e) \text{ for } e \in \coprod_v \left(\mathrm{Ambig}(\widehat{H}_v) \sqcup \mathrm{Leg}_{\mathrm{e}}(\widehat{H}_v)\right).$$

□

Notice that each exceptional loop of either $\widehat{G}$ or $\widehat{H}_v$ will become an exceptional loop in the pre-substitution $\widehat{K}$, or

$$\mathrm{Flag}_{\mathrm{ie}}\left(\widehat{K}\right) = \mathrm{Flag}_{\mathrm{ie}}\left(\widehat{G}\right) \sqcup \coprod_v \mathrm{Flag}_{\mathrm{ie}}\left(\widehat{H}_v\right).$$

while each exceptional edge of $\widehat{H}_v$ will become an ambiguous edge in $\widehat{K}$. Each element of $\mathrm{Arm}(\widehat{G})$ must be connected to an ordinary flag in $\widehat{G}$, so to some element of $\coprod_v \mathrm{Leg}(\widehat{H}_v)$ in $\widehat{K}$. Moreover, $\iota_{\widehat{K}}$ is $\iota_{\widehat{H}_v}$ for internal flags in $\widehat{H}_v$ and, using a ψ_v where necessary, $\iota_{\widehat{K}}$ is given using $\iota_{\widehat{G}}$ for all other flags.

Once again, if there are directions and/or colorings chosen for the pre-graphs in such a way that they are preserved by the bijections ψ_v, then the resulting pre-substition has induced directions and colorings. Furthermore, a listing for the full graph $\widehat{G}$ together with the listings for the vertices of each $\widehat{H}_v$ combine to produce a listing for the pre-substitution as before.

5.2.3. Properties of Pre-substitution. Recall C_v denotes the corolla with the same profiles as v, so given any pre-graph $\widehat{G}$ and choosing $\widehat{H}_v = C_v$ with all ψ_v identity maps yields pre-substitution data. For any pre-graph $\widehat{G}$, we will say $\widehat{K}$ **makes the exceptional legs of** $\widehat{G}$ **ambiguous** if they share the same flags, involutions, direction, coloring, and listing, but the partition differs in that the exceptional legs of $\widehat{G}$ together form a single ambiguous cell of $\widehat{K}$. A **strict isomorphism** of pre-graphs will refer to a bijection of partitioned sets which is compatible with both involutions, as well as the directions, colorings, and listings.

We can now state our approximate unit property for pre-substitution, which is immediate from the definitions.

LEMMA 5.11. *For any pre-graph, there is a strict isomorphism* $\widehat{G}(C_v) \longrightarrow \widehat{G}$, *while* $C_{(\mathrm{in}(\widehat{G});\mathrm{out}(\widehat{G}))}(\widehat{G})$ *makes the exceptional legs of* $\widehat{G}$ *ambiguous.*

The key result in this chapter is the associativity of graph substitution, which follows with a bit more work from the following associativity statement for pre-substitution.

PROPOSITION 5.12. *Suppose* $\widehat{G}$ *together with* $\widehat{H}_v$ *and* ψ_v *are pre-substitution data, while for each* $v \in \mathrm{Vt}(\widehat{G})$, *this* $\widehat{H}_v$ *together with* $\widehat{I}_v^u$ *and isomorphisms* $\psi_{v,u}$ *are pre-substitution data. Then there is a canonical strict isomorphism between the pre-substitutions*

$$\widehat{K}_1 = \left[\widehat{G}(\widehat{H}_v)\right](\widehat{I}_v^u) \quad \textit{and} \quad \widehat{K}_2 = \widehat{G}\left[\widehat{H}_v(\widehat{I}_v^u)\right].$$

PROOF. We will construct a new pre-graph, $\widehat{K}_3$, which will be canonically strictly isomorphic to both $\widehat{K}_1$ and $\widehat{K}_2$ by definition. Define $\widehat{K}_3$ to have flags

$$\mathrm{Ambig}(\widehat{G}) \sqcup \mathrm{Flag}_{\mathrm{e}}(\widehat{G}) \sqcup \coprod_v \left(\mathrm{Ambig}(\widehat{H}_v) \sqcup \mathrm{Flag}_{\mathrm{e}}(\widehat{H}_v)\right) \sqcup \coprod_{u,v} \mathrm{Flag}(\widehat{I}_v^u),$$

with ordinary cells the union of those from the $\widehat{I}_v^u$, exceptional cell

$$\mathrm{Flag}_{\mathrm{e}}(\widehat{G}) \sqcup \coprod_v \mathrm{Flag}_{\mathrm{ie}}(\widehat{H}_v) \sqcup \coprod_{u,v} \mathrm{Flag}_{\mathrm{ie}}(\widehat{I}_v^u)$$

and ambiguous cells

$$\mathrm{Ambig}(\widehat{G}) \sqcup \coprod_v \left(\mathrm{Ambig}(\widehat{H}_v) \sqcup \mathrm{Leg}_{\mathrm{e}}(\widehat{H}_v)\right) \sqcup \coprod_{u,v} \left(\mathrm{Ambig}(\widehat{I}_v^u) \sqcup \mathrm{Leg}_{\mathrm{e}}(\widehat{I}_v^u)\right)$$

partitioned as indicated by the coproducts, with the induced definition of $\pi_{\widehat{K}_3}$. It is straightforward to verify that the involutions $\pi_{\widehat{K}_1}$ and $\pi_{\widehat{K}_2}$ correspond under the evident isomorphisms of partitioned sets coming from the universal property of coproducts. Thus, it remains to show the definitions of ι similarly correspond.

The new $\iota_{\widehat{K}_3}$ is induced by $\iota_{\widehat{G}}$ on exceptional and internal ambiguous flags of $\widehat{G}$, by $\iota_{\widehat{H}_v}$ on internal exceptional and internal ambiguous flags of $\widehat{H}_v$, and by $\iota_{\widehat{I}_v^u}$ on internal flags of $\widehat{I}_v^u$. So far, it is straightforward to see the definitions correspond, and now we must consider the arms of $\widehat{G}$ as well as the legs of each $\widehat{H}_v$, or the more complicated case of a leg of some $\widehat{I}_v^u$.

If f is a leg of some $\widehat{H}_v$, then we have

$$\iota_{\widehat{K}_3}(f) = \hat{\zeta}_v^{\widehat{G}}(f) = \begin{cases} \iota_{\widehat{G}}\psi_v(f) & \text{if this flag is in } \mathrm{Arm}(\widehat{G}), \text{ or} \\ \psi_{v'}^{-1}\iota_{\widehat{G}}\psi_v(f) = \zeta_v^{\widehat{G}}(f) & \text{otherwise.} \end{cases}$$

For e a leg of some $\widehat{I}_v^u$ with $\psi_{v,u}(e)$ an ordinary internal flag of $\widehat{H}_v$, we have

$$\iota_{\widehat{K}_3}(e) = \hat{\zeta}_u^{\widehat{H}_v}(e) = \begin{cases} \iota_{\widehat{H}_v}\psi_{v,u}(e) & \text{if this flag is in } \mathrm{Arm}(\widehat{H}_v), \text{ or} \\ (\psi_{v',u'})^{-1}\,\iota_{\widehat{H}_v}\psi_{v,u}(e) = \zeta_u^{\widehat{H}_v}(e) & \text{otherwise.} \end{cases}$$

Finally, if e' is a leg of some $\widehat{I}_v^u$ with $\psi_{v,u}(e') \in \mathrm{Leg}_\mathrm{o}(\widehat{H}_v)$, define

$$f' = \iota_{\widehat{G}}\psi_v\psi_{v,u}(e'),$$

which is either an ordinary flag or an ambiguous flag in $\widehat{G}$. Now define

$$\iota_{\widehat{K}_3}(e') = \begin{cases} f' & \text{if } f' \text{ is in } \mathrm{Arm}(\widehat{G}), \\ (\psi_{v',u'})^{-1}\,\psi_{v'}^{-1}(f') & \text{if } f' \in \mathrm{Flag}_\mathrm{o}(\widehat{G}) \text{ and } \psi_{v'}^{-1}f' \in \mathrm{Leg}_\mathrm{o}(\widehat{H}_{v'}), \\ \psi_{v'}^{-1}(f') & \text{otherwise.} \end{cases}$$

□

For later reference, we want to clarify that the third case in the definition of $\iota_{\widehat{K}_3}(e')$ above occurs when $f' \in \mathrm{Flag}_\mathrm{o}(\widehat{G})$ while $\psi_{v'}^{-1}(f')$ is either an ambiguous leg or an exceptional leg in $\widehat{H}_{v'}$.

5.3. Ambiguous Paths and Associated Graphs

Our goal at this point is to produce a graph when given a pre-graph, primarily by working with ambiguous paths consisting of ambiguous edges in a pre-graph.

5.3.1. Ambiguous Paths. Throughout this section, we will assume we have a pre-graph, which may not have any direction, coloring, or labeling. Our goal is to explain how the ambiguous edges will allow us to alter the involution ι so that we have an associated graph. Then we will point out that directions, colorings, and labelings, if present for the pre-graph, will lead to the same for the associated graph. The reader should keep in mind that ambiguous paths are an extension of the definition of a directed internal path in a graph (Definition 2.3).

DEFINITION 5.13. Two ambiguous edges are **connected by** ι if ι sends a flag of one to a flag of the other.

An **ambiguous path** P in a pre-graph $\widehat{K}$ consists of a finite sequence of distinct ambiguous edges $e^1, \ldots, e^k$, with ι connecting e^j to e^{j+1} for $1 \le j \le k-1$.

If $k > 1$, the **initial end** of P, if such exists, will refer to the flag of e^1 which is not ι of any flag of e^j with $j > 1$. Similarly, if $k > 1$, the **terminal end** of P, if such exists, will refer to the flag of e^k which is not ι of any flag of e^j with $j < k$. If $k = 1$ and ι does not pair the two flags of e^1, then we choose one of the flags to serve as the initial end, and the other becomes the terminal end.

We will refer to an ambiguous path P with $k > 1$ such that $e^k, e^1, \ldots, e^{k-1}$ is also an ambiguous path as an **ambiguous loop**. An ambiguous loop with $k = 1$ refers to a pair of ambiguous flags which are paired by both ι and π.

It should be clear that initial and terminal ends fail to exist when we consider ambiguous loops, but in all other cases both will exist.

LEMMA 5.14. *An ambiguous path is either an ambiguous loop or it has both initial and terminal ends.*

PROOF. This follows by definition if $k = 1$, so we assume $k > 1$ and there is no initial end of P. In that case, the two flags of e^1 are connected to e^2 and some e^j with $1 < j \leq k$. If $2 < j < k$, then the two flags of e^j are connected by ι to three distinct ambiguous edges, e^1, e^{j-1}, and e^{j+1}, which contradicts either the non-repeating condition on ambiguous edges in an ambiguous path, or the uniqueness of the image under the involution ι. This leaves only the option that P is an ambiguous loop, as e^1 is connected by ι to e^2 and e^k, since in the case $j = 2$ we must have $k = 2$ and both flags of e^1 are paired with flags of e^2 to avoid the two flags of e^2 again being paired with three distinct flags. □

If $\widehat{K}$ is equipped with a coloring, then compatibility of that coloring with the involutions $\iota_{\widehat{K}}$ and $\pi_{\widehat{K}}$ will imply all flags in an ambiguous path share the same color. If $\widehat{K}$ has a direction, then flags connected by ι have opposite directions, as do the two flags of each ambiguous edge (paired by π).

Our primary interest will be in maximal ambiguous paths, which cannot be ambiguous loops since both ends must exist.

DEFINITION 5.15. An ambiguous path P in a pre-graph $\widehat{K}$ is **maximal** if both ends of P are limbs. In particular, an **ambiguous component** will refer to an ambiguous path where both ends are legs, an **ambiguous connector** will refer to an ambiguous path where both ends are arms, and an **ambiguous stilt** will refer to an ambiguous path with one end an arm and one end a leg.

NOTATION 5.16. The set of all ambiguous components will be denoted by $\text{Ambig}^{\dagger}(\widehat{K})$, and the set of all ambiguous loops will be denoted by $\text{Ambig}^{\circlearrowleft}(\widehat{K})$. Be aware that ambiguous loops which are phase-shifted, that is, with the same flags but different indices, will be identified as a single element of this set (see Remark 2.4).

The basic process in forming the associated graph of a quasi-graph will be to push some ambiguous flags into the exceptional cell, and to remove other ambiguous flags after allowing them to alter the involution ι on the ordinary flags. The alterations of ι will be based upon the following concepts, where the terminology is intended to be suggestive of the role taken on in the associated graph.

DEFINITION 5.17. An ordinary flag f of a pre-graph $\widehat{K}$ will be called a **quasi-leg** if $\iota_{\widehat{K}}(f)$ is the arm end of an ambiguous stilt. Two distinct ordinary flags $e \neq f$ of $\widehat{K}$ will be called a **quasi-edge** if there is an ambiguous connector with $\iota_{\widehat{K}}(e)$ as one end and $\iota_{\widehat{K}}(f)$ as the other end. The sets of quasi-legs and quasi-edges are denoted by $\text{Leg}_{\text{q}}(\widehat{K})$ and $\text{Edge}_{\text{q}}(\widehat{K})$, respectively.

More generally, a **pre-leg** will denote either an ordinary leg or a quasi-leg. Similarly, a **pre-edge** means either an ordinary internal edge or a quasi-edge. The choice of either an exceptional edge or an ambiguous component will be called a **pre-exceptional edge**, and a **pre-exceptional loop** will indicate either an exceptional loop or an ambiguous loop.

As a consequence of the next lemma, one can sort the ambiguous flags of $\widehat{K}$ into four distinct subsets: the flags appearing in ambiguous components, ambiguous loops, ambiguous connectors, or ambiguous stilts. For $\widehat{K}$, the associated graph K will have edges the pre-edges of $\widehat{K}$, legs the pre-legs of $\widehat{K}$ and so on. Also note that pre-legs and pre-edges are made up of ordinary flags, while pre-exceptional edges and pre-exceptional loops are made up of exceptional and ambiguous flags. As a consequence, K will contain only a pair of representative flags from each ambiguous component or ambiguous loop in $\widehat{K}$, which form exceptional edges and exceptional loops of K. The flags appearing in ambiguous connectors and ambiguous stilts of $\widehat{K}$, on the other hand, will not survive to the associated graph K as flags. Of course, they will be used to alter the definition of the involution ι among the ordinary flags, since ambiguous connectors define quasi-edges and ambiguous stilts define quasi-legs.

Before producing our rectification result for pre-graphs, we need an existence and uniqueness result for maximal ambiguous paths.

LEMMA 5.18. *Every ambiguous flag in $\widehat{K}$ that is not part of an ambiguous loop is contained in a maximal ambiguous path, which is unique up to its orientation.*

PROOF. Given an ambiguous flag f, it is possible that both f and $\pi(f)$ are limbs. Then $e^1 = \{f, \pi(f)\}$ with $k = 1$ is already a maximal ambiguous path containing f. Given any ambiguous path which is not contained in an ambiguous loop and is not a maximal ambiguous path, at least one end is a flag g with $\iota(g)$ ambiguous. Thus, defining $e^{k+1} = (\iota(g), \pi\iota(g))$, or using this same ambiguous edge as e^0 and shifting indices, one has a longer ambiguous path containing the original. By finiteness of $\widehat{K}$ and the fact that flags are not repeated in an ambiguous path, this process must eventually end with a maximal ambiguous path containing the original.

For the question of uniqueness, notice that all of these choices were dictated by the structure of $\widehat{K}$ and the definition of ambiguous paths. As such, any two maximal ambiguous paths sharing an ambiguous flag must either coincide, or be reversals of one another in the sense that $e^j = \bar{e}^{k-j}$. □

5.3.2. The Associated Graph Construction. We now have enough structure to describe the associated graph of a pre-graph.

THEOREM 5.19. *Associated to any pre-graph $\widehat{K}$, there is a graph K, unique up to canonical strict isomorphism, with* $\mathrm{Flag}_{\mathrm{o}}(K) = \mathrm{Flag}_{\mathrm{o}}(\widehat{K})$ *and*

$$\mathrm{Flag}_{\mathrm{e}}(K) \cong \mathrm{Flag}_{\mathrm{e}}(\widehat{K}) \sqcup \coprod_{\mathrm{Ambig}^{\uparrow}(\widehat{K})} \{e^{\alpha}_{-1}, e^{\alpha}_{1}\} \sqcup \coprod_{\mathrm{Ambig}^{\circlearrowleft}(\widehat{K})} \{f^{\beta}_{-1}, f^{\beta}_{1}\},$$

such that ι_K pairs pre-edges and pre-exceptional loops, ι_K fixes pre-legs and pre-exceptional legs, and π_K pairs pre-exceptional edges. If $\widehat{K}$ has a direction, coloring, and listing, then K is a wheeled graph with the same profiles as $\widehat{K}$ and is unique up to strict isomorphism.

REMARK 5.20. A convenient way to think of producing the set of flags indicated in the statement above is as a quotient of the set of flags of $\widehat{K}$ by collapsing out some ambiguous edges. First, we collapse each ambiguous component or ambiguous loop to a single pair of flags by identifying to a singleton all flags in the ambiguous path that have the same direction. Notice the ends of an ambiguous connector have opposite directions, hence so do the ordinary flags they are connected to by ι. As a consequence, we can collapse each flag in an ambiguous connector to the unique ordinary flag it is eventually connected to which, in addition, has the same direction as the flag in question. Finally, each ambiguous stilt is simply collapsed to the unique ordinary flag that it is eventually connected to, which now takes on the role of a leg of K.

PROOF OF THEOREM 5.19. The statement of the theorem indicates the partitioned set and specifies the involutions ι_K and π_K. We must show ι_K isolates the exceptional cell of K, which consists of the previously isolated exceptional cell of $\widehat{K}$ along with a pair of representatives of each ambiguous component and each ambiguous loop. Since $\iota_{\widehat{K}}$ isolates each individual ambiguous component or ambiguous loop, the condition follows from the definition of ι_K. The fact that π_K is a free involution on the set of exceptional legs of K follows from the definition of the exceptional cell of K and that of both involutions. □

Notice that given a pre-graph $\widehat{K}$, the above theorem says that the ordinary legs, ordinary internal edges, exceptional edges, and exceptional loops in the associated graph K correspond to the pre-legs, pre-edges, pre-exceptional edges, and pre-exceptional loops in $\widehat{K}$.

5.3.3. Compatibility of Associated Graphs with Pre-substitution. Our goal here is to prove a technical result that will be used to conclude that the general graph substitution operation is associative.

We will continue to use hats to indicate pre-graphs and remove them to indicate the associated graph, just as we did for the associated graph of $\widehat{K}$ being K above. The reason is that we view pre-graphs as an artificial construct used only in this chapter, while the associated graphs are the actual objects of interest.

If we are given pre-substitution data $\widehat{G}$, $\widehat{H}_v$, and ψ_v, there are three possible pre-substitutions we might want to consider. We will reserve $\widehat{K}$ for the pre-substitution associated directly to this data, or $\widehat{G}(\widehat{H}_v)$. One might instead apply the associated graph construction to $\widehat{G}$ to produce G, and notice G together with $\widehat{H}_v$ and ψ_v remains pre-substitution data since the vertices and legs of G remain those of $\widehat{G}$. As a consequence, one could instead perform the pre-substitution $G(\widehat{H}_v)$ with this new data. For the third pre-substitution, one could apply the associated graph operation to each $\widehat{H}_v$, thereby producing H_v, and notice $\widehat{G}$ with H_v and ψ_v remains pre-substitution data as well, since the legs of H_v are those of $\widehat{H}_v$. Hence it also makes sense to form the pre-substitution $\widehat{G}(H_v)$. While these pre-graphs will differ in some cases, the result we need says that their associated graphs all coincide.

THEOREM 5.21. *For any pre-substitution data $\widehat{G}$, $\widehat{H}_v$ and ψ_v, the associated graphs of the pre-substitutions*

$$G(\widehat{H}_v), \quad \widehat{G}(H_v), \quad \textit{and} \quad \widehat{G}(\widehat{H}_v)$$

are strictly isomorphic.

PROOF. The ordinary flags of all three pre-graphs, hence of their associated graphs, are just

$$\coprod_v \mathrm{Flag}_{\mathrm{o}}(\widehat{H}_v).$$

Since the definitions of π, the directions, the colorings, and the listings all descend in the obvious manner from $\widehat{K}$, keeping in mind the definition of the associated graph, we must verify that the notions of pre-legs and pre-edges are identical, while pre-exceptional edges and pre-exceptional loops are canonically identified, among the three pre-graphs. These questions are addressed in Lemma 5.23 below. □

REMARK 5.22. For those working without directions, etc., notice the proof of the theorem does not rely upon their presence, other than when inducing a similar structure. This will remain the case throughout the various steps in the proof of Lemma 5.23, so keeping our presentation of graph substitution completely general.

LEMMA 5.23. *Along with having the same ordinary flags, the three pre-graphs*

$$G(\widehat{H}_v), \quad \widehat{G}(H_v), \quad \textit{and} \quad \widehat{G}(\widehat{H}_v)$$

have the same pre-legs and the same pre-edges. In addition, there are canonical bijections between the sets of pre-exceptional edges of the three pre-graphs, as well as between the sets of pre-exceptional loops of the three pre-graphs.

Our focus is now on the ambiguous paths in a pre-substitution. Since

$$\mathrm{Ambig}(\widehat{K}) = \mathrm{Ambig}(\widehat{G}) \sqcup \coprod_v \left(\mathrm{Ambig}(\widehat{H}_v) \sqcup \mathrm{Leg}_e(\widehat{H}_v)\right),$$

an ambiguous path in the pre-substitution $\widehat{K}$ consists of a sequence of ambiguous edges in $\widehat{G}$, or of either ambiguous or exceptional edges of various $\widehat{H}_v$. One can also think of looking at subsequences which remain ambiguous paths from a single $\widehat{H}_v$ or from $\widehat{G}$, and for maximal ambiguous paths in $\widehat{K}$ these subsequences will also turn out to be maximal in $\widehat{G}$ or in that $\widehat{H}_v$.

DEFINITION 5.24. Given a pre-substitution $\widehat{K} = \widehat{G}(\widehat{H}_v)$, an **ambiguous segment** p consists of an ambiguous path in $\widehat{G}$, or an ambiguous path or exceptional edge of some $\widehat{H}_v$, with initial end p_{init} and terminal end p_{term}. The flags which are not ends will be called interior flags of the segment, and choices will be made for the ends of loops.

A **decomposed ambiguous path** in $\widehat{K}$ is a finite sequence $p^1, \ldots, p^k$ of ambiguous segments with $\iota_{\widehat{K}}$ connecting p^j_{term} to p^{j+1}_{init} for $1 \le j \le k-1$.

Decomposed ambiguous loops are decomposed ambiguous paths with $\iota_{\widehat{K}}$ connecting p^k_{term} to p^1_{init}.

A decomposed ambiguous path will be called **maximal** if each end is a limb in $\widehat{K}$. In particular, **decomposed ambiguous components**, with both ends legs, **decomposed ambiguous connectors**, with both ends arms, and **decomposed ambiguous stilts**, with one end an arm and one end a leg, are the three types of maximal decomposed ambiguous paths.

First, we need to know that decomposed ambiguous paths reflect the same information as ambiguous paths. Then, the decompositions will be used to organize the proof of Lemma 5.23.

LEMMA 5.25. *Each ambiguous path in $\widehat{K}$ has a unique decomposition, which preserves the properties of being maximal, being a component, being a connector, or being a loop. In addition, no two consecutive p^j can be ambiguous components of $\widehat{G}$.*

PROOF. Given an ambiguous path P in $\widehat{K}$, the unique decomposition arises by simply cutting between each pair of limbs that appears as consecutive flags in P. Since arms cannot, by their definition, be connected to other arms, and only ambiguous stilts or ambiguous connectors of $\widehat{G}$ can occur in decomposed ambiguous paths in $\widehat{K}$ of length more than 1, the second claim follows from the observation that if $p^j \in \widehat{G}$ with $1 < j < k$, the ends must be arms of $\widehat{G}$, as legs of $\widehat{G}$ are fixed by $\iota_{\widehat{K}}$. □

Next we have a pair of technical results describing the primary correspondences that lead to the proof of Lemma 5.23.

LEMMA 5.26. (1) *If* (p^1, p^2, p^3) *is an ambiguous path in* $\widehat{K}$*, with* p^2 *an ambiguous segment in* $\widehat{G}$*, then* (p^1, p^3) *forms an ambiguous path in* $G(\widehat{H}_v)$ *and* p^2 *is the unique ambiguous connector in* $\widehat{G}$ *defining the quasi-edge consisting of* $\psi^{-1}(p^1_{term})$ *and* $\psi^{-1}(p^3_{init})$ *in* $\widehat{G}$*.*

(2) *If* (p^1, p^2) *is an ambiguous stilt in* $\widehat{K}$*, with* p^j *an ambiguous segment in* $\widehat{G}$ *for* $j = 1$ *or* 2*, then* p^{3-j} *alone is an ambiguous path in* $G(\widehat{H}_v)$ *and* p^j *is the unique ambiguous stilt in* $\widehat{G}$ *defining the quasi-leg in* $\widehat{G}$ *given by* $\psi^{-1}(p^2_{init})$ *if* $j = 1$ *or* $\psi^{-1}(p^1_{term})$ *if* $j = 2$*.*

Given p an ambiguous component in $\widehat{H}_v$, let $\bar{p}$ denote the associated exceptional edge in H_v defined by deleting the interior flags (if any) and moving it into the exceptional cell.

LEMMA 5.27. *If* (p^1, p^2, p^3) *is an ambiguous path in* $\widehat{K}$*, with both* p^1 *and* p^3 *ambiguous segments in* $\widehat{G}$*, while* p^2 *is an ambiguous component in* $\widehat{H}_v$*, then* $(p^1, \bar{p}^2, p^3)$ *forms an ambiguous path in* $\widehat{G}(H_v)$*. In fact, either* p^1 *or* p^3*, but not both, could be removed.*

We can now prove our main lemma.

PROOF OF LEMMA 5.23. We begin by describing the correspondence for pre-exceptional edges. A pre-exceptional edge in the pre-substitution $\widehat{K} = \widehat{G}(\widehat{H}_v)$ is either a pre-exceptional edge in $\widehat{G}$ or a decomposed ambiguous component in $\widehat{K}$ that is not a single ambiguous component in $\widehat{G}$. A pre-exceptional edge in $\widehat{G}$ yields an exceptional edge in the associated graph G, so a pre-exceptional edge in all three pre-substitutions. Now let P represent a decomposed ambiguous component in $\widehat{K}$ which is not a single ambiguous component in $\widehat{G}$. There is then an associated ambiguous component P_1 in $G(\widehat{H}_v)$ constructed by deletion of the ambiguous segments in $\widehat{G}$ as in Lemma 5.26. In fact, given P_1 in $G(\widehat{H}_v)$ we could also recover the original P in $\widehat{K}$ by the relevant uniqueness statement of Lemma 5.26. Likewise, there is a correspondence between P and the P_2 in $\widehat{G}(H_v)$ associated to P via relabeling as exceptional edges of H_v (and deleting the interior flags of) the ambiguous components in $\widehat{H}_v$, as in Lemma 5.27, with the uniqueness claim coming from Lemma 5.18.

The bijections for pre-exceptional loops are similar, with the main difference being that we also have to track pre-exceptional loops in $\widehat{H}_v$, which yield exceptional loops in H_v, so pre-exceptional loops in all three pre-substitutions.

Now we turn to the identifications of the pre-legs. For an ordinary flag f in $\widehat{H}_v$, define

$$\rho(f) = \begin{cases} f & \text{if } f \text{ is not a leg in } \widehat{H}_v, \\ \psi_v(f) & \text{if } f \text{ is a leg in } \widehat{H}_v. \end{cases}$$

Such an ordinary flag f is a pre-leg in $\widehat{K}$ if and only if either

- $f \in \mathrm{Leg}_o(\widehat{H}_v)$ and $\rho(f) \in \mathrm{Leg}_o(\widehat{G}) \amalg \mathrm{Leg}_q(\widehat{G}) = \mathrm{Leg}_o(G)$, or
- there is a decomposed ambiguous stilt in $\widehat{K}$ that is not a single ambiguous stilt in $\widehat{G}$ and that has internal end $\iota_{\widehat{K}}\rho(f)$.

If $\rho(f) \in \mathrm{Leg}_{\mathrm{q}}(\widehat{G})$, then the identification of pre-legs in $\widehat{K}$ with those in $G(\widehat{H}_v)$ again follows from Lemma 5.26, with a focus on the cases of ambiguous stilts. For identifying pre-legs in $\widehat{K}$ with those in $\widehat{G}(H_v)$, we use Lemma 5.27 for interior segments and an analog of the last two claims in Lemma 5.26 for ambiguous stilts from an $\widehat{H}_v$ as end segments.

For the identification of the pre-edges, suppose $e = \{f^1, f^2\}$ are two ordinary flags in $\coprod_v \mathrm{Flag}_{\mathrm{o}}(\widehat{H}_v)$. Then e is a pre-edge in $\widehat{K}$ if and only if

- e is a pre-edge in some $\widehat{H}_v$, or
- both f^j are legs in $\widehat{H}_{v_j}$ and $\{\rho(f^1), \rho(f^2)\}$ is an ordinary internal edge in $\widehat{G}$, or
- there is an ambiguous connector in $\widehat{K}$ that is not a single ambiguous connector in some $\widehat{H}_v$ with ends $\iota_{\widehat{K}}(\rho(f^1))$ and $\iota_{\widehat{K}}(\rho(f^2))$.

In the first case, e yields an ordinary internal edge in H_v and so a pre-edge in all three pre-substitions, using the same defining ambiguous connector from $\widehat{H}_v$ in the quasi-edge case. In the second case, the same conditions remain valid in all three pre-substitutions, thereby producing ordinary internal edges. In the third case, we again proceed to establish a correspondence of ambiguous connectors using Lemma 5.26 and Lemma 5.27, noting that the ends remain the relevant $\iota(\rho(f^j))$ in each case. □

5.4. General Graph Substitution

We now have all of the technical details out of the way, so we can state the fully general definition of graph substitution given actual graphs, with pre-graphs only appearing as an intermediate step in the process.

DEFINITION 5.28. Given graph substitution data G, H_v and ψ_v, define the **graph substitution** $K = G(H_v)$ by first forming, as in Lemma 5.10, the pre-substitution $\widehat{K} = \widehat{G}(\widehat{H}_v)$, with $\widehat{G} = G$ and $\widehat{H}_v = H_v$, and then taking the associated graph K as in Theorem 5.19.

EXAMPLE 5.29. Suppose G is the butterfly net graph of Example 1.38, with H_v the growth chart graph of Example 1.42, and H_u the upward ray graph of Example 1.39. Then $G(H_v)$ is the bat and ball graph of Example 1.40.

EXAMPLE 5.30. This time, choose G to be the two-vertex cycle with exceptional edge and loop of Example 1.37, where the cycle has color c, the exceptional edge has color c_1 and the exceptional loop has color c_2. Now suppose H_u consists of an exceptional loop with color a and an exceptional edge of color c, while H_v is an exceptional loop with color b and an exceptional edge of color c. Then the graph substitution $G(H_v)$ consists of one exceptional edge of color c_1 together with four exceptional loops, one each of color c_2, c, a, and b.

It will be important to note that when $\widehat{G}$ is simply a graph G viewed as a pre-graph with no ambiguous cells, the associated graph is again G as expected. On the other hand, if $\widehat{K}$ makes the exceptional legs of $\widehat{G}$ ambiguous, then their associated graphs K and G are strictly isomorphic. The reason is that these exceptional legs are transformed to ambiguous components in $\widehat{K}$ with no interior flags, hence return to exceptional legs in K without loss of flags by definition. When taken together with Lemma 5.11, we have now established that the operation of graph substitution is unital up to canonical isomorphism for graphs, or up to canonical strict isomorphism for wheeled graphs.

LEMMA 5.31. *For any wheeled graph G, there is a strict isomorphism*

$$G(C_v) \longrightarrow G$$

and a strict isomorphism

$$C_{(\mathrm{in}(G);\mathrm{out}(G))}(G) \longrightarrow G.$$

The reader may be tempted to conclude that graph substitution is now associative by combining Proposition 5.12 and Theorem 5.19. However, we need to pay attention to the fact that the associated graph operation would be applied twice in the process of forming either double substitution, hence the importance of Theorem 5.21 in proving the following associativity statement for wheeled graph substitution, which has the obvious analog even without directions, colorings, or listings.

THEOREM 5.32. *Suppose G together with H_v and ψ_v are graph substitution data, while for each $v \in \mathrm{Vt}(G)$, this H_v together with I_v^u and isomorphisms $\psi_{v,u}$ are graph substitution data. Then there is a canonical strict isomorphism between the graph substitutions*

$$K_1 = [G(H_v)](I_v^u) \quad \text{and} \quad K_2 = G[H_v(I_v^u)].$$

We also have the following naturality statement of graph substitution with respect to either weak or strict isomorphisms.

Given two choices of graph substitution data G, H_v and ψ_v, as well as G', H_v' and ψ_v', **compatible weak isomorphisms** $\theta_G : G \longrightarrow G'$ and $\theta_v : H_v \longrightarrow H_v'$ will refer to the fact that

$$\psi' \theta_v = \theta \psi.$$

PROPOSITION 5.33. *Given two choices of graph substitution data G, H_v and ψ_v, as well as G', H_v' and ψ_v', and compatible weak isomorphisms*

$$\theta_G : G \longrightarrow G' \quad \text{and} \quad \theta_v : H_v \longrightarrow H_v',$$

there is an induced weak isomorphism $\bar{\theta} : G(H_v) \longrightarrow G'(H_v')$. Furthermore, if θ_G and each θ_v is a strict isomorphism, then $\bar{\theta}$ is a strict isomorphism.

PROOF. The compatibility condition implies ambiguous paths in the pre-substitutions will continue to correspond, with the correspondence preserving both direction and coloring. Hence the induced isomorphism $\bar{\theta}$ of sets of flags for the associated graphs becomes an isomorphism of underlying graphs, which preserves the direction and coloring.

If, in addition, the listing was preserved by each θ, then $\bar{\theta}$ will also preserve the listing, thereby becoming a strict isomorphism. □

REMARK 5.34. The fact that graph substitution is only associative up to canonical strict isomorphism will sometimes require us to descend to the set of strict isomorphism classes of wheeled graphs, or subsets thereof. In fact, later we will introduce a small multicategory with strict isomorphism classes of graphs as morphisms and with the composition law coming from graph substitution. The complicated point in that context is that we think of a graph as a morphism from the profiles of its vertices to the profiles of the full graph, in order for graph substitution to function as composition.

For later consideration of graph substitution for wheel-free and simply-connected graphs, we have an additional technical result. Keep in mind that a directed cycle is also called a wheel.

LEMMA 5.35. *Every (directed) cycle in the substitution $G(H_v)$ that does not come from a (directed) cycle in some H_v implies the existence of a (directed) cycle in G.*

PROOF. In the case where all of the graphs are ordinary, consider a cycle in $G(H_v)$ which must consist of ordinary internal edges. As a consequence, the cycle decomposes as a sequence of internal edges from various H_v, which can be decomposed at transitions involving ψ_v just as we did in defining decomposed ambiguous paths. Taking just the ordinary flags in G which are identified with flags in this cycle must then produce a cycle in G by inspection. If we began with a directed cycle in $G(H_v)$, then the directions of these ordinary flags in G would be compatible as well, producing a directed wheel in G by inspection.

If there are exceptional loops in either G or some H_v, they produce exceptional loops in the substitution by definition. Exceptional legs in G simply contribute exceptional legs to the substitution, so are irrelevant to the question of existence of loops in the substitution. If there are exceptional legs in the H_v, then they become ambiguous edges in the pre-substitution, but the above argument of focusing on the ordinary flags in G to which they would attach to produce a (directed) cycle in G remains valid. □

5.4.1. Technical Variants of Graph Substitution. Since we introduced a few technical variants of graphs at the end of the first chapter, it seems appropriate to mention how they interact with graph substitution.

The first case is a simple consequence of the fact that the lexicographical ordering for substitution of ordered sets is associative, and the way ordinary flags of the graph substitution come from the ordinary flags of the innermost collection of graphs.

LEMMA 5.36. *Graph substitution is unital, associative, and natural up to canonical strict isomorphism for ordered graphs, when using the lexicographical ordering on vertices of the usual graph substitution.*

For pointed graph substitution $G(H_v)$, we must choose the basepoint vertex in $H_{\bar{v}}$, the graph which replaces the basepoint vertex $\bar{v}$ of G. As above, this selection is clearly associative for substitution of pointed sets, which proves the following.

LEMMA 5.37. *Graph substitution is unital, associative, and natural up to canonical strict isomorphism for pointed graphs, when choosing the basepoint vertex of the graph replacing the basepoint vertex in the usual graph substitution.*

One interesting point here is that the basepoints of the other substituting graphs H_v are simply forgotten. Thus, in order to produce a pointed graph using graph substitution, it is only necessary to have the graph replacing the basepoint pointed, and all others can be ordinary wheeled graphs. This idea is important when studying modules over generalized PROPs in Chapter 15.

REMARK 5.38. If interested in working with generalized PROs, rather than generalized PROPs, one simply refuses to allow choices of ψ_v which permute listings. With this in mind, the graph substitution operation just detailed is all that is necessary for studying generalized PROs.

REMARK 5.39. For the multi-stage graphs of 1.4.3, the input and output profiles of a full graph are replaced by finite sequences of profiles. As a consequence, there would also be a more technical variant of graph substitution for that context. The major complication comes from the fact that one multi-stage graph could substitute into several different vertices at once. For example, if G has two vertices with profiles $\mathrm{in}(v) = (c_1, c_2)$, $\mathrm{out}(v) = (d_2, d_1)$ and $\mathrm{in}(w) = b$, $\mathrm{out}(w) = b$, while H is the multi-stage graph of Example 1.45 then $G(H)$ could be defined.

CHAPTER 6

Properties of Graph Substitution

There are a variety of graph operations that can be phrased cleanly in terms of graph substitution using our basic graphs, as we mentioned in chapter 3. Thus, now that we have established the key properties of graph substitution, we begin this chapter with a general characterization of operations that come from graph substitution into a single graph. The characterization is surprisingly simple, namely requiring compatibility with graph substitution in the sense that operating on the outer graph G and then substituting some collection H_v determines the value of operating on the substitution $G(H_v)$. We then proceed to briefly recall a number of the most important cases, in some cases extending the definitions from chapter 3, but in others merely clarifying statements made there.

There is one operation familiar from the study of operads that does not generalize as expected, namely the shrinking of internal edges, and this will also be discussed briefly to clarify the situation. Our general viewpoint from here on is that only graph operations that arise via graph substitutions should be considered appropriate for working with graphs and graph groupoids. Thus, we will view the shrinking of internal edges instead as a method of producing 'factorizations' of graphs, or as a partial inverse of an operation, rather than as an operation of its own.

We will also establish compatibility of graph substitution with the properties of being (simply-)connected or wheel-free. Since many of our representing graphs are easily seen to have these properties, it follows immediately that the operation represented will also preserve these properties. Thus, in many cases, it is true to say an operation is compatible with a property precisely when the representing graph exhibits the property.

The last three sections of this chapter, which we call the calculus of graph substitution, is devoted to a large number of computations of relatively small graph substitutions. Each could instead be viewed as computing a certain composite of two operations, via their representing graphs, as we often make explicit, or instead as applying one operation to the representing graph of another. For the purpose of actually making calculations with generalized PROPs, algebras, or modules later, these results are very convenient to have collected somewhere, and here they can also serve as a series of examples of the complicated construction of the graph substitution. Finally, these computations set the stage for a theory of relaxed Reidemeister moves in the next chapter, which will allow us to construct large collections of graphs from strings of graph substitutions involving only simple pieces.

6.1. Operations Coming from Graph Substitution

This section begins by characterizing those graph operations given by substitution into a fixed graph, and then details a number of important examples.

6.1.1. Characterizing Representable Graph Operations. We will now require a bit more than we did of our graph operations in chapter 3, namely that they be functors from one groupoid of graphs to another (rather than just set-valued assignments). The reader should be careful that we must then keep track of which type of isomorphisms should be considered in the context of a given operation.

DEFINITION 6.1. A functor Θ from one groupoid of graphs to another will be called a **graph operation compatible with graph substitution** if there is a canonical isomorphism

$$\Theta[G^i(H_v^i)] \longrightarrow [\Theta G^i](H_v^i).$$

The basic examples will come from graph substitution itself as follows.

LEMMA 6.2. *Given any wheeled graph G, substituting into the vertices of G defines a graph operation compatible with graph substitution*

$$\prod_v \mathtt{Gr}^{\circlearrowright}_{str}(\mathrm{in}(v);\mathrm{out}(v)) \longrightarrow \mathtt{Gr}^{\circlearrowright}_{str}(\mathrm{in}(G),\mathrm{out}(G)).$$

It seems reasonable to refer to the graph operations of this lemma as **representable** operations. When the output of an operation is more than one graph, we can think of the target groupoid of graphs as a product. By the universal property of a product, a functor into such a product is uniquely determined by its projections. In other words, any graph operation can be characterized in terms of a set of graph operations with a single graph as output. Thus, a general graph operation will be called **representable** if each of its projections to an operation with a single graph as output is canonically isomorphic to one defined as in Lemma 6.2. With this decomposition property in mind, we can characterize all operations compatible with graph substitution as representable.

PROPOSITION 6.3. *A graph operation Θ is compatible with graph substitution precisely when Θ is representable.*

PROOF. Lemma 6.2 says representable operations are compatible with graph substitution if they have a single output graph. Thus, for any general representable operation, we know the projections are all compatible with graph substitution, while a canonical isomorphism in the product groupoid is equivalent to a canonical isomorphism in each projection.

Now suppose Θ is compatible with graph substitution with output a single graph. Choose C^i to be a set of corollas on the appropriate profiles to define $\Theta(C^i)$. Then for any set of inputs H^i for Θ, we have canonical isomorphisms

$$\Theta(H^i) \cong \Theta[C^i(H^i)] \cong [\Theta(C^i)](H^i)$$

by Lemma 5.31 and the assumed compatibility with substitution condition. In other words, the graph $\Theta(C^i)$ represents the operation Θ. For a general Θ compatible with graph substitution, we again consider each projection. Since each projection is a representable operation with a single output as just discussed, it follows that Θ is representable by definition. $\square$

6.1.2. Altering the Listing by Graph Substitution. Suppose G is a $(\underline{c};\underline{d})$-wheeled graph and let C_1 denote the permuted corolla $\sigma C_{(\underline{c};\underline{d})}\tau$. Then $C_1(G)$ will be the input and output relabeling of G as in Definition 3.2. Recall C_1 is the input and output relabeling of $C_{(\underline{c};\underline{d})}$ as expected from Proposition 6.3.

REMARK 6.4. One might think of the graph operations of Prop 6.3 as representable from the left, under left compatibility with graph substitution. It is possible to define graph operations from the right, where the applicability of the operation depends upon the profiles of the vertices of the graph(s) in question rather than the profiles of the full graph. Since corollas also serve as units from the right for graph substitution, a similar argument to the proof of Prop 6.3 shows that all operations from the right which are compatible with graph substitution from the right are represented by operating on a series of corollas. However, the only example we will consider of an operation from the right is the following.

Given any listing ℓ, we can alter ℓ_v simply by performing graph substitution with H_v a permuted corolla with profiles $(\text{in}(v); \text{out}(v))$ for each $v \in \text{Vt}\, G$. Thus, $G(H_v)$ will then be G with the listing at each v altered from that of ℓ_v to that of ℓ_w for w the unique vertex of H_v.

Combining these two observations, one can alter the listing of G completely by forming $C_1[G(H_v)]$.

6.1.3. Disjoint Union from Graph Substitution. Suppose $\{G_j\}_{j=1}^r$ is a finite non-empty ordered set of wheeled graphs. Choose C_j to be the corolla with the same input and output profiles as G_j and form the disjoint union of these corollas $\coprod_{j=1}^r C_j$ as in Definition 3.5. Then the graph substitution

$$\left(\coprod_{j=1}^{r} C_j\right)(G_1, \ldots, G_r),$$

where G_j is substituted into the vertex in C_j, agrees with the disjoint union $\coprod_{j=1}^r G_j$, again from Definition 3.5, on the nose.

6.1.4. Grafting from Graph Substitution. Suppose G_1 is a $(\underline{c}; \underline{d})$-wheeled graph, and G_2 is a $(\underline{b}; \underline{c})$-wheeled graph. Consider the grafted corollas $C_{(\underline{b};\underline{c};\underline{d})}$ from (3.3). Define the **grafting** of G_1 and G_2 as the graph substitution

$$G_1 \boxtimes G_2 = C_{(\underline{b};\underline{c};\underline{d})}(G_1, G_2),$$

in which G_1 is substituted into the higher vertex. Note that the inputs of $G_1 \boxtimes G_2$ are those of G_2, and its outputs are those of G_1. When G_1 and G_2 are ordinary, the above definition of grafting agrees with the one in Definition 3.9 on the nose. Also observe that the grafted corollas $C_{(\underline{b};\underline{c};\underline{d})}$ is the grafting $C_{(\underline{c};\underline{d})} \boxtimes C_{(\underline{b};\underline{c})}$ using the grafting of ordinary graphs as in Definition 3.9, again in keeping with Proposition 6.3.

It is also true that the operation of partial grafting can be represented by the partially grafted corollas, with given k-segments, of Definition 3.14, and in the case of ordinary graphs the representable operation will agree with the previous definition on the nose.

EXAMPLE 6.5. The basic dioperadic graph of (2.23) can be viewed as a partial grafting of $C_{(\underline{c};\underline{d})}$ and $C_{(\underline{a};\underline{b})}$ by choosing the 1-segments to consist of c_i and b_j.

Alternatively there is a canonical strict isomorphism from the basic dioperadic graph

$$C^{j,i}_{(\underline{a};\underline{b};\underline{c};\underline{d})} \cong \left(\uparrow_{b_{[1,j-1]}} \sqcup C_{(\underline{c};\underline{d})} \sqcup \uparrow_{b_{[j+1,l]}}\right) \boxtimes \sigma\left(\uparrow_{c_{[1,i-1]}} \sqcup C_{(\underline{a};\underline{b})} \sqcup \uparrow_{c_{[i+1,m]}}\right).$$

Here σ is the external product

$$(1\ 2)\langle i-1, j-1\rangle \times Id \times (1\ 2)\langle l-j, m-i\rangle,$$

with $(1\ 2)\langle i-1, j-1\rangle$ the block permutation induced by $(1\ 2)$ that permutes the two blocks of the indicated lengths.

EXAMPLE 6.6. The simple tree of Definition 2.25 can be viewed as a partial grafting of $C_{(\underline{c};d)}$ and $C_{(\underline{b};c_i)}$ by choosing the 1-segments to consist of c_i.

Alternatively, there is a canonical strict isomorphism from the simple tree

$$C_{(\underline{c};d)} \circ_i C_{(\underline{b};c_i)} \cong C_{(\underline{c};d)} \boxtimes \left(\uparrow_{c_{[1,i-1]}} \sqcup C_{(\underline{b};c_i)} \sqcup \uparrow_{c_{[i+1,m]}}\right).$$

6.1.5. Contraction from Graph Substitution. Suppose G is a wheeled graph with profiles $(\underline{c};\underline{d})$ and $c_j = d_i$, while $C^{j,i}_{(\underline{c};\underline{d})}$ is the contracted corolla (3.5). Define the $(j;i)$**-contraction** of G as the graph substitution

$$\xi^i_j G = \left(C^{j,i}_{(\underline{c};\underline{d})}\right)(G).$$

The profiles of $\xi^i_j G$ are $(\underline{c} \setminus c_j; \underline{d} \setminus d_i)$. We will often call $\xi^i_j G$ simply a **contraction** of G.

For example, if G is the c-colored exceptional edge $\uparrow_c$, then $\xi^1_1 G$ is the c-colored exceptional loop $\circlearrowleft_c$. Moreover, if G is an ordinary wheeled graph, then the above definition of contraction agrees with the one in Definition 3.22 on the nose.

Although apparently simple, this operation is the key to building large wheeled graphs from relatively small pieces, particularly when combined with taking unions and changing the listings. In essence, performing contraction of a union of two ordinary graphs allows us to create a new internal edge by connecting a former output to a former input.

EXAMPLE 6.7. There is a canonical strict isomorphism from the basic dioperadic graph with $b_j = c_i$, $|\underline{b}| = l$, and $|\underline{c}| = m$:

$$C^{j,i}_{(\underline{a};\underline{b};\underline{c};\underline{d})} \cong \sigma \left\{\xi^{|\underline{d}|+j}_i \left(C_{(\underline{c};\underline{d})} \sqcup C_{(\underline{a};\underline{b})}\right)\right\} \tau.$$

Here σ is the block permutation, following Notation 1.4,

$$\sigma = (1\ 2)\langle |\underline{d}|, j-1, l-j\rangle$$

induced by $(1\ 2) \in \Sigma_3$ that permutes the first two intervals of the indicated lengths. Likewise, τ is the block permutation

$$\tau = (2\ 3)\langle i-1, |\underline{a}|, m-i\rangle$$

induced by $(2\ 3) \in \Sigma_3$ that permutes the last two intervals of the indicated lengths.

6.1.6. Shrinking Internal Edges Is Not a Representable Graph Operation. Notice our contraction operation is distinct from the concept of shrinking an internal edge, familiar from a variety of induction arguments in the theory of operads.

LEMMA 6.8. *If G is a simply-connected ordinary graph and e is an internal edge, then there exists a simply-connected graph K 'shrinking e' and a basic dioperadic graph H such that $G = K(H)$.*

PROOF. To construct K, start with G as a partitioned set, remove the two flags making up e, and then take the union of the cells of the partition corresponding to the source and target of e. Leaving all of the remaining structure of G unchanged, this implies K is a wheeled graph, which will also be simply-connected (and ordinary).

For the construction of H, consider the two corollas determined by the source and target vertices of e, and partially graft them to produce the edge e. This will be a partially grafted corollas along 1-segments, so a basic dioperadic graph. Since both graphs are ordinary, the graph substitution statement follows immediately. $\square$

Unfortunately, shrinking internal edges is not, in general, compatible with graph substitution.

EXAMPLE 6.9. Suppose G is the 1-vertex cycle with e its only internal edge. Then shrinking the internal edge e would produce an isolated vertex, which has no flags. However, every graph substitution $K(G)$ has at least one flag, as does every $G(H)$, so shrinking internal edges cannot come from graph substitution (on either side) for a general G.

From the current perspective, shrinking an internal edge should not really be thought of as an operation on graphs. Instead, it is a technique for producing factorizations of graphs, since $G = (G/e)(D)$ for an appropriately chosen D with two vertices. Such factorizations are important in their own right, and are discussed more thoroughly in Section 7.1.

6.1.7. Input and Output Extensions. The following technical construction will be necessary in order to produce a common generalization of the units in operads, PROPs, and other operational structures.

DEFINITION 6.10. Given a wheeled graph G with profiles $(\underline{c};\underline{d})$ with $\underline{c} = c_{[1,m]}$ and $\underline{d} = d_{[1,n]}$, the **input extension** of G will refer to

$$G_{in} = G \boxtimes \left(\coprod_{i=1}^{m} C_{(c_i;c_i)} \right) \in \mathsf{Gr}^{\circlearrowright}_{\mathrm{str}}\binom{\underline{d}}{\underline{c}}, \tag{6.1}$$

where $C_{(c_i;c_i)}$ is the $(c_i;c_i)$-corolla.

If G is graphically represented as in (1.4), then the input extension G_{in} is represented as

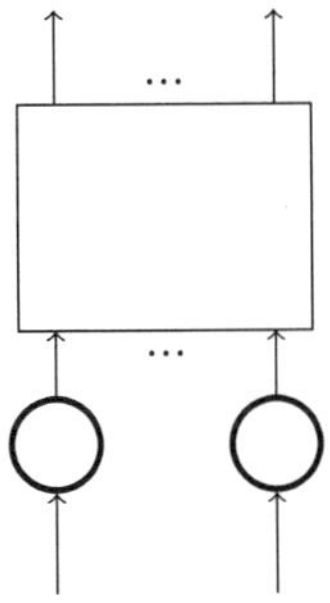

Intuitively, the input extension is obtained from G by extending every input i_j^G into an internal edge and adding a vertex v_j with an input leg of the same color and leg label as i_j^G. The representing graph in this case is the input extension of a $(\underline{c};\underline{d})$-corolla.

There is also a dual notion of output extension, once again represented by the special case of the output extension of a $(\underline{c};\underline{d})$-corolla.

DEFINITION 6.11. Given a wheeled graph G with profiles $(\underline{c};\underline{d})$ with $\underline{c} = c_{[1,m]}$ and $\underline{d} = d_{[1,n]}$, the **output extension** of G is

$$G_{out} = \left(\coprod_{j=1}^{n} C_{(d_j;d_j)}\right) \boxtimes G \in \mathtt{Gr}^{\circlearrowright}_{\text{str}}\binom{\underline{d}}{\underline{c}}. \tag{6.2}$$

If G is graphically represented as in (1.4), then the output extension G_{out} is represented as

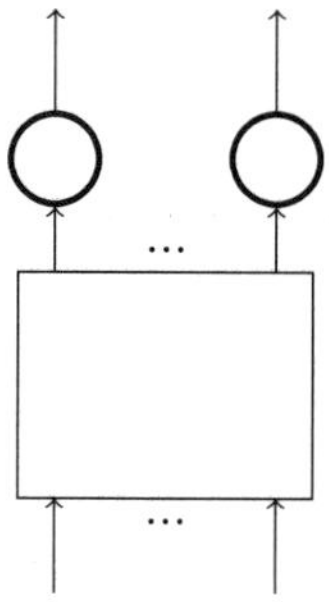

Using the graphical representation (1.5) of the $(\underline{c};\underline{d})$-corolla C, the input extension C_{in} and the output extension C_{out} are graphically represented, respectively, as:

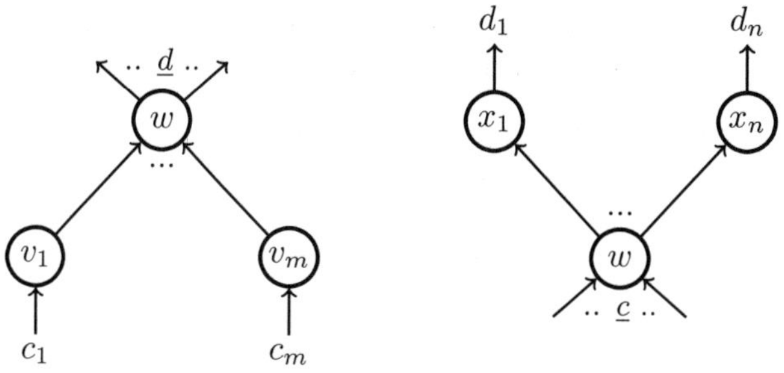

In section 10.4.4 below, we will use input and output extensions to describe the extra structure on generalized PROPs associated to unital pasting schemes.

6.2. Properties Preserved by Graph Substitution

Here we discuss how graph substitution preserves the sets of (simply-)connected and wheel-free graphs. As we have established representing graphs for many of our operations which share these properties, it follows that the operations also restrict to the sub-groupoids of $\mathtt{Gr}^{\circlearrowright}_{\text{w}}$ of interest to us.

6.2.1. Connected Graphs. First we show that graph substitution preserves connectedness. In other words, the operation of graph substitution of wheeled graphs restricts to an operation on $\mathtt{Gr}^{\circlearrowright}_{\text{c}} \subset \mathtt{Gr}^{\circlearrowright}_{\text{w}}$.

PROPOSITION 6.12. *If G is connected and H_v is connected for each $v \in \text{Vt}(G)$, then so is the graph substitution $G(H_v)$, provided it is defined.*

PROOF. If $G = \uparrow_c$, $\circlearrowleft_c$, or the isolated vertex $C_{(\varnothing;\varnothing)}$, then the assertion is clear. Otherwise, (see subsection 2.2.1) G is ordinary, flag-connected, and has no isolated vertices. In particular, if $G(H_v)$ is exceptional, then it is either $\uparrow_c$ or $\circlearrowleft_c$ by the construction of the graph substitution. On the other hand, if $G(H_v)$ is ordinary, then it cannot have isolated vertices because the H_v are connected. Moreover, $G(H_v)$ is flag-connected. Indeed, since the H_v are connected, it suffices to observe that any two legs e and e' of H_v and $H_{v'}$ respectively, are connected by a path in G since G is connected. □

Combining the previous proposition and the discussion in section 6.1, we obtain the following special cases by observing that contracted corollas and permuted corollas are connected.

COROLLARY 6.13. *Contractions and changing the listing both preserve connectedness.*

EXAMPLE 6.14. Contraction can actually create connectedness. For example, the exceptional graph $\uparrow_c \sqcup \uparrow_c$ is not connected, but

$$\xi_2^1(\uparrow_c \sqcup \uparrow_c) = \uparrow_c$$

is connected.

Likewise, because the grafted corollas $C_{(\underline{b};\underline{c};\underline{d})}$ is connected, provided $\underline{c} \neq \varnothing$, we have the following result about preservation of connectedness by grafting, and similarly for partial grafting using the partially grafted corollas

$$C^{l_c,l_b,k}_{\underline{a},\underline{b},\underline{c},\underline{d}} = C_{(\underline{c};\underline{d})} \boxtimes^{\underline{c}'}_{\underline{b}'} C_{(\underline{a};\underline{b})}.$$

COROLLARY 6.15. *Suppose G_1 and G_2 are connected wheeled graphs whose profiles are $(\underline{c};\underline{d})$ and $(\underline{b};\underline{c})$ with $\underline{c} \neq \varnothing$. Then the grafting $G_1 \boxtimes G_2$ is also a connected wheeled graph.*

COROLLARY 6.16. *Suppose G_1 with profiles $(\underline{c};\underline{d})$ and G_2 with profiles $(\underline{a};\underline{b})$ are connected wheeled graphs with a choice of matching k-segments $\underline{c}' = \underline{b}'$. Then the partial grafting $G_1 \boxtimes^{\underline{c}'}_{\underline{b}'} G_2$ is also a connected wheeled graph.*

6.2.2. Wheel-free Graphs. Next we see graph substitution preserves the property of being wheel-free (and connected). As a consequence, the operation of graph substitution restricts to the two sub-groupoids $\mathtt{Gr}^{\uparrow}_{\mathtt{c}} \subset \mathtt{Gr}^{\uparrow} \subset \mathtt{Gr}^{\circlearrowleft}_{\mathtt{w}}$.

PROPOSITION 6.17. *If G is wheel-free and H_v is wheel-free for each $v \in \mathrm{Vt}(G)$, then so is the graph substitution $G(H_v)$, provided it is defined.*

PROOF. If $G(H_v)$ contains a wheel that is not originally within any H_v, then by Lemma 5.35 there is a wheel in G. □

Combining Propositions 6.12 and 6.17, we see graph substitution also restricts to $\mathtt{Gr}^{\uparrow}_{\mathtt{c}}$.

COROLLARY 6.18. *Suppose the graph substitution $G(H_v)$ is defined. If G and the H_v are all connected and wheel-free, then so is $G(H_v)$.*

Since a permuted corolla, grafted corollas, and partially grafted corollas are connected and wheel-free, the operations they represent are also compatible with $\mathtt{Gr}^{\uparrow}$ and $\mathtt{Gr}^{\uparrow}_{\mathtt{c}}$.

COROLLARY 6.19. *Changing the listing of a (connected) wheel-free graph results in a (connected) wheel-free graph.*

COROLLARY 6.20. *Suppose G_1 and G_2 are (connected) wheel-free graphs whose profiles are $(\underline{c};\underline{d})$ and $(\underline{b};\underline{c})$ with $\underline{c} \neq \varnothing$. Then the grafting $G_1 \boxtimes G_2$ is also a (connected) wheel-free graph.*

COROLLARY 6.21. *Suppose G_1 with profiles $(\underline{c};\underline{d})$ and G_2 with profiles $(\underline{a};\underline{b})$ are (connected) wheel-free graphs with a choice of matching k-segments $\underline{c}' = \underline{b}'$. Then the partial grafting $G_1 \boxtimes_{\underline{b}'}^{\underline{c}'} G_2$ is also a (connected) wheel-free graph.*

Of course, there is no reason to expect the contraction operation to preserve the wheel-free property, since its job is to create wheels in many cases. As simple examples, contracting a linear graph of a single color produces a wheel, and contracting an exceptional edge produces an exceptional loop. As we mentioned earlier, contracting two exceptional edges together produces a single exceptional edge, so creating wheels is not the only result of the contraction operation, which can create any internal edges.

6.2.3. Simply-Connected Graphs. Finally, we see graph substitution preserves the property of being simply-connected. Thus, the operation restricts to $\mathtt{Gr}^{\uparrow}_{\mathrm{di}} \subset \mathtt{Gr}^{\circlearrowleft}_{\mathrm{w}}$.

PROPOSITION 6.22. *If G and H_v for each $v \in \mathrm{Vt}(G)$ are all simply-connected, then so is the graph substitution $G(H_v)$, provided it is defined.*

PROOF. If $G(H_v)$ contains a cycle that is not originally within any H_v, then by Lemma 5.35 there is a cycle in G. □

Since corollas and partially grafted corollas over 1-segments (but not k-segments for $k > 1$ by the example below) remain simply-connected, we have the following.

COROLLARY 6.23. *Changing the listing of a simply-connected graph results in a simply-connected graph. Partial grafting over 1-segments will also restrict to an operation on simply-connected graphs.*

A simple tree and a basic dioperadic graph are both simply-connected partial graftings of corollas over 1-segments.

EXAMPLE 6.24. Grafting and partial grafting for k-segments with $k > 1$ do *not* preserve the property of being simply-connected. For example, both corollas $C_{(\underline{c};\underline{d})}$ and $C_{(\underline{b};\underline{c})}$ are simply-connected. However, their grafting $C_{(\underline{b};\underline{c};\underline{d})}$ is not simply-connected, unless $|\underline{c}| = 1$.

EXAMPLE 6.25. Contraction does *not* preserve the property of being simply-connected, as shown by the contracted corolla $\xi^i_j C_{(\underline{c};\underline{d})} = C^{j,i}_{(\underline{c};\underline{d})}$ with the corolla $C_{(\underline{c};\underline{d})}$ simply-connected.

6.3. Substitution Properties of Basic Operations

This section begins the study of the Calculus of Graph Substitution, an extensive collection of lemmas, phrased to be examples of what will be called 'relaxed moves' when presenting a Reidemeister theory for a number of different families of graphs later. Each lemma here and in the next section is a statement about graph substitutions with very few vertices involved, and each is followed by a corollary. The corollary is stated in terms of graph operations, and may appear to be more general, but is actually equivalent when considering the representative graphs associated to operations discussed earlier in this chapter. In addition, each lemma is preceded by an informal description, usually of the corollary, since it is the corollaries which are often easiest to interpret or remember. In this section, instances of the basic operations of relabeling, union, contraction, and grafting are discussed.

6.3.1. Iterated Relabelings. Iterated input and output relabelings can always be performed in one step. One of the recurrent themes throughout this section is compatibility of various structures with input and output relabeling, or bi-equivariance, and this is just the first example.

LEMMA 6.26. *Let $\underline{c}$ and $\underline{d}$ be profiles with permutations $\sigma_i \in \Sigma_{\underline{d}}$ and $\tau_i \in \Sigma_{\underline{c}}$. Then there is an identity*

$$(\sigma_2 C_{(\underline{c}\tau_1;\sigma_1\underline{d})}\tau_2)(\sigma_1 C_{(\underline{c};\underline{d})}\tau_1) = (\sigma_2\sigma_1)C_{(\underline{c};\underline{d})}(\tau_1\tau_2).$$

In particular, choosing $\tau_2 = \tau_1^{-1}$ and $\sigma_2 = \sigma_1^{-1}$, this says it is always possible to 'untangle' the legs of a permuted corolla to recover a corolla.

Since substitution into a permuted corolla defines the general permutation operation, by associativity of graph substitution

$$(\sigma_2 C_{(\underline{c}\tau_1;\sigma_1\underline{d})})(\sigma_1 C_{(\underline{c};\underline{d})}\tau_1)[G] = (\sigma_2\sigma_1)C_{(\underline{c};\underline{d})}(\tau_1\tau_2)[G]$$

yields the following more general equation.

COROLLARY 6.27. *For any graph G, with $\tau_i \in \Sigma_{|\operatorname{in}(G)|}$ and $\sigma_i \in \Sigma_{|\operatorname{out}(G)|}$,*

$$\sigma_2(\sigma_1 G\tau_1)\tau_2 = (\sigma_2\sigma_1)G(\tau_1\tau_2).$$

6.3.2. Union. We begin with the relevant compatibility with input and output relabeling, called the bi-equivariance statement, for unions. Namely, the union of two relabelings can be viewed as a corresponding relabeling of the union of the original graphs.

LEMMA 6.28. *For any profiles $\underline{c}^i$ and $\underline{d}^i$ with $i = 1, 2$, and permutations $\sigma_i \in \Sigma_{\underline{d}^i}$ and $\tau_i \in \Sigma_{\underline{c}^i}$,*

$$\left(\sigma_1 C_{(\underline{c}^1;\underline{d}^1)}\tau_1\right) \sqcup \left(\sigma_2 C_{(\underline{c}^2;\underline{d}^2)}\tau_2\right) = (\sigma_1 \times \sigma_2)\left(C_{(\underline{c}^1;\underline{d}^1)} \sqcup C_{(\underline{c}^2;\underline{d}^2)}\right)(\tau_1 \times \tau_2).$$

Note that an input and output relabeling of a union of two corollas does *not* split as a union in general, unless each of the two permutations is an external product as above.

COROLLARY 6.29. *For any wheeled graphs G_i with $\sigma_i \in \Sigma_{|\operatorname{out}(G_i)|}$ as well as $\tau_i \in \Sigma_{|\operatorname{in}(G_i)|}$*

$$(\sigma_1 G_1\tau_1) \sqcup (\sigma_2 G_2 \tau_2) = (\sigma_1 \times \sigma_2)(G_1 \sqcup G_2)(\tau_1 \times \tau_2).$$

The next two observations say that union is unital and associative, since union for general graphs can be defined by substitution into unions of corollas.

LEMMA 6.30. *For any profiles $\underline{c}$ and $\underline{d}$,*

$$C_{(\underline{c};\underline{d})} = \left(C_{(\underline{c};\underline{d})} \sqcup C_{(\varnothing;\varnothing)}\right)\left(C_{(\underline{c};\underline{d})}, \varnothing\right) = \left(C_{(\varnothing;\varnothing)} \sqcup C_{(\underline{c};\underline{d})}\right)\left(\varnothing, C_{(\underline{c};\underline{d})}\right).$$

Again, combining the previous lemma with the associativity of graph substitution one has the following.

COROLLARY 6.31. *For any wheeled graph G,*

$$G = G \sqcup \varnothing = \varnothing \sqcup G.$$

Now we have the associativity statement for the union operation.

LEMMA 6.32. *For $\underline{c}^i$ and $\underline{d}^i$ profiles with $i = 1, 2, 3$, there is a canonical strict isomorphism*

$$\begin{aligned}&\left(C_{((\underline{c}^1,\underline{c}^2);(\underline{d}^1,\underline{d}^2))} \sqcup C_{(\underline{c}^3;\underline{d}^3)}\right)\left(C_{(\underline{c}^1;\underline{d}^1)} \sqcup C_{(\underline{c}^2;\underline{d}^2)}, C_{(\underline{c}^3;\underline{d}^3)}\right)\\ &\cong \left(C_{(\underline{c}^1;\underline{d}^1)} \sqcup C_{((\underline{c}^2,\underline{c}^3);(\underline{d}^2,\underline{d}^3))}\right)\left(C_{(\underline{c}^1;\underline{d}^1)}, C_{(\underline{c}^2;\underline{d}^2)} \sqcup C_{(\underline{c}^3;\underline{d}^3)}\right).\end{aligned}$$

COROLLARY 6.33. *For any wheeled graphs G_i with $i = 1, 2, 3$, there is a canonical strict isomorphism*

$$(G_1 \sqcup G_2) \sqcup G_3 \cong G_1 \sqcup (G_2 \sqcup G_3).$$

The next result establishes the relabeling property which is the appropriate analog of graded commutativity satisfied by unions of wheeled graphs. See Notation 1.4 for block permutations.

LEMMA 6.34. *For any pairs of profiles $(\underline{c}^i; \underline{d}^i)$ with $i = 1, 2$, if σ is the block permutation $(1\ 2)\langle |\underline{d}^1|, |\underline{d}^2|\rangle$, and τ is the block permutation $(1\ 2)\langle |\underline{c}^2|, |\underline{c}^1|\rangle$, then*

$$\sigma\left(C_{(\underline{c}^1;\underline{d}^1)} \sqcup C_{(\underline{c}^2;\underline{d}^2)}\right)\tau = C_{(\underline{c}^2;\underline{d}^2)} \sqcup C_{(\underline{c}^1;\underline{d}^1)}.$$

COROLLARY 6.35. *With the choices of permutations indicated in the previous lemma, for any wheeled graphs G_i with profiles $(\underline{c}^i; \underline{d}^i)$ for $i = 1, 2$*

$$\sigma(G_1 \sqcup G_2)\tau = G_2 \sqcup G_1.$$

6.3.3. Contraction. First we see that redundant exceptional edges can be canceled out using contraction. Elimination of redundant exceptional edges is another recurrent theme here, related to (vertical) unit properties in generalized PROPs.

LEMMA 6.36. *For any pair of profiles $(\underline{c}; \underline{d})$, there are canonical strict isomorphisms*

$$C_{(\underline{c};\underline{d})} \cong \left(\xi^i_{|\underline{c}|+1} C_{((\underline{c},d_i);(\underline{d},d_i))}\right)\left(C_{(\underline{c};\underline{d})} \sqcup \uparrow_{d_i}\right) \cong \left(\xi^{i+1}_1 C_{((d_i,\underline{c});(d_i,\underline{d}))}\right)\left(\uparrow_{d_i} \sqcup C_{(\underline{c};\underline{d})}\right)$$

and

$$C_{(\underline{c};\underline{d})} \cong \left(\xi^{|\underline{d}|+1}_j C_{((\underline{c},c_j);(\underline{d},c_j))}\right)\left(C_{(\underline{c};\underline{d})} \sqcup \uparrow_{c_j}\right) \cong \left(\xi^1_{j+1} C_{((c_j,\underline{c});(c_j,\underline{d}))}\right)\left(\uparrow_{c_j} \sqcup C_{(\underline{c};\underline{d})}\right).$$

COROLLARY 6.37. *For any wheeled graph G with profiles $(\underline{c}; \underline{d})$, there are canonical strict isomorphisms*

$$G \cong \xi^i_{|\underline{c}|+1}\left(G \sqcup \uparrow_{d_i}\right) \cong \xi^{i+1}_1\left(\uparrow_{d_i} \sqcup G\right)$$

and

$$G \cong \xi^{|\underline{d}|+1}_j\left(G \sqcup \uparrow_{c_j}\right) \cong \xi^1_{j+1}\left(\uparrow_{c_j} \sqcup G\right).$$

Next we see contraction exhibits a bi-equivariance property if, as with most operations, we are careful to identify the relevant restrictions on permutations. Given a morphism $(\tau; \sigma) \in \mathcal{P}(\mathfrak{C})^{op} \times \mathcal{P}(\mathfrak{C})$ with source $(\underline{c}; \underline{d})$ satisfying $c_j = d_i$, we define a new morphism $(\tau^{(j)}; \sigma^{(i)}) \in \mathcal{P}(\mathfrak{C})^{op} \times \mathcal{P}(\mathfrak{C})$ with source $(\underline{c} \setminus c_j; \underline{d} \setminus d_i)$ by simply omitting the indicated single entry in each permutation.

LEMMA 6.38. *Suppose the pair of profiles* $(\underline{c};\underline{d})$ *satisfies* $c_j = d_i$, *with* $(\tau^{(j)};\sigma^{(i)}) \in \mathcal{P}(\mathfrak{C})^{op} \times \mathcal{P}(\mathfrak{C})$ *as above. Then*

$$\left(\sigma^{(i)} C_{(\underline{c}\smallsetminus c_j;\underline{d}\smallsetminus d_i)}\tau^{(j)}\right)\left(\xi_j^i C_{(\underline{c};\underline{d})}\right) = \left(\xi_{\tau(j)}^{\sigma^{-1}(i)} C_{(\underline{c}\tau;\sigma\underline{d})}\right)\left(\sigma C_{(\underline{c};\underline{d})}\tau\right).$$

COROLLARY 6.39. *Suppose* G *is a wheeled graph with profiles* $(\underline{c};\underline{d})$, $c_j = d_i$, *with* $(\tau^{(j)};\sigma^{(i)}) \in \mathcal{P}(\mathfrak{C})^{op} \times \mathcal{P}(\mathfrak{C})$ *as above. Then*

$$\sigma^{(i)}\left(\xi_j^i G\right)\tau^{(j)} = \xi_{\tau(j)}^{\sigma^{-1}(i)}(\sigma G \tau).$$

Now we observe that contractions almost commute with one another.

LEMMA 6.40. *For any profiles* $(\underline{c};\underline{d})$ *such that* $c_j = d_i$ *and* $c_l = d_k$,

(1)

$$\left(\xi_j^i C_{(\underline{c}\smallsetminus c_l;\underline{d}\smallsetminus d_k)}\right)\left(\xi_l^k C_{(\underline{c};\underline{d})}\right) = \left(\xi_{l-1}^{k-1} C_{(\underline{c}\smallsetminus c_j;\underline{d}\smallsetminus d_i)}\right)\left(\xi_j^i C_{(\underline{c};\underline{d})}\right)$$

when $i < k$ *and* $j < l$, *and*

(2)

$$\left(\xi_{j-1}^i C_{(\underline{c}\smallsetminus c_l;\underline{d}\smallsetminus d_k)}\right)\left(\xi_l^k C_{(\underline{c};\underline{d})}\right) = \left(\xi_l^{k-1} C_{(\underline{c}\smallsetminus c_j;\underline{d}\smallsetminus d_i)}\right)\left(\xi_j^i C_{(\underline{c};\underline{d})}\right)$$

when $i < k$ *and* $j > l$.

The first statement above is interpreting the fact that the doubly contracted corolla $\xi_j^i \xi_l^k (C_{(\underline{c};\underline{d})}) \in \mathtt{Gr}_{\mathrm{w}}^{\circlearrowright}\binom{\underline{d}\smallsetminus\{d_i,d_k\}}{\underline{c}\smallsetminus\{c_j,c_l\}}$, pictorially depicted as:

(6.3)

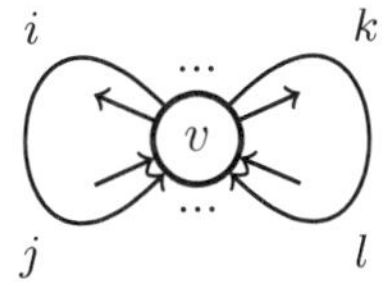

can also be constructed from the corolla $C_{(\underline{c};\underline{d})}$ by first connecting its ith output to its jth input and then connecting the output corresponding to d_k to the input corresponding to c_l. Similarly, the second statement above is interpreting the fact that there are two ways to construct the doubly contracted corolla $\xi_{j-1}^i \xi_l^k C_{(\underline{c};\underline{d})} \in \mathtt{Gr}_{\mathrm{w}}^{\circlearrowright}\binom{\underline{d}\smallsetminus\{d_i,d_k\}}{\underline{c}\smallsetminus\{c_l,c_j\}}$ depicted,

(6.4)

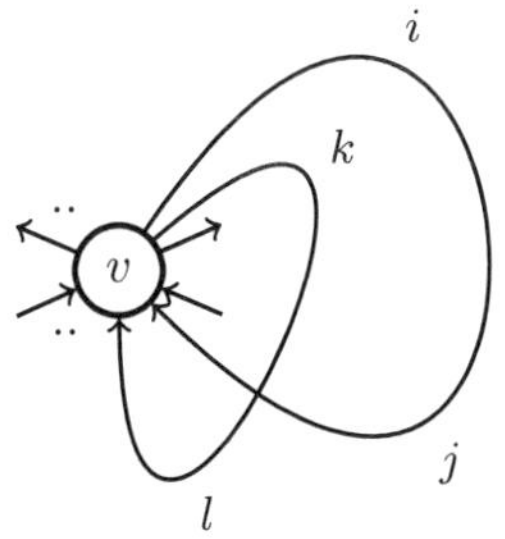

by starting with the corolla $C_{(\underline{c};\underline{d})}$.

COROLLARY 6.41. *For any wheeled graph G with profiles $(\underline{c};\underline{d})$ such that $c_j = d_i$ and $c_l = d_k$,*

(1)

$$\xi^i_j \xi^k_l G = \xi^{k-1}_{l-1} \xi^i_j G$$

when $i < k$ and $j < l$, and

(2)

$$\xi^i_{j-1} \xi^k_l G = \xi^{k-1}_l \xi^i_j G$$

when $i < k$ and $j > l$.

Finally, forming the union of a contracted graph with either a relabeled or a contracted graph produces a relabeling or contraction of the evident contracted union.

LEMMA 6.42. *Given pairs of profiles $(\underline{c}^i;\underline{d}^i)$ with $i = 1, 2$:*

(1) *If $\sigma \in \Sigma_{|\underline{d}_1|}$ and $\tau \in \Sigma_{|\underline{c}_1|}$ while $c^2_j = d^2_i$, then*

$$\left(C_{(\underline{c}^1\tau;\sigma\underline{d}^1)} \sqcup C_{(\underline{c}^2 \setminus c^2_j;\underline{d}^2 \setminus d^2_i)}\right)\left(\sigma C_{(\underline{c}^1;\underline{d}^1)}\tau, \xi^i_j C_{(\underline{c}^2;\underline{d}^2)}\right)$$
$$= \left[\left(\sigma \times Id^{|\underline{d}^2|}\right) C' \left(\tau \times Id^{|\underline{c}^2|}\right)\right]\left(\xi^{|\underline{d}^1|+i}_{|\underline{c}^1|+j} C_{((\underline{c}^1,\underline{c}^2);(\underline{d}^1,\underline{d}^2))}\right)\left(C_{(\underline{c}^1;\underline{d}^1)} \sqcup C_{(\underline{c}^2;\underline{d}^2)}\right),$$

where $C' = C_{((\underline{c}^1,\underline{c}^2 \setminus c^2_j);(\underline{d}^1,\underline{d}^2 \setminus d^2_i))}$.

(2) *If $c^1_j = d^1_i$ and $c^2_l = d^2_k$ then*

$$\left(C_{(\underline{c}^1 \setminus c^1_j;\underline{d}^1 \setminus d^1_i)} \sqcup C_{(\underline{c}^2 \setminus c^2_l;\underline{d}^2 \setminus d^2_k)}\right)\left(\xi^i_j C_{(\underline{c}^1;\underline{d}^1)}, \xi^k_l C_{(\underline{c}^2;\underline{d}^2)}\right)$$
$$= \left(\xi^{|\underline{d}^1|-1+k}_{|\underline{c}^1|-1+l} C_{((\underline{c}^1 \setminus c^1_j,\underline{c}^2);(\underline{d}^1 \setminus d^1_i,\underline{d}^2))}\right)\left(\xi^i_j C_{((\underline{c}^1,\underline{c}^2);(\underline{d}^1,\underline{d}^2))}\right)\left(C_{(\underline{c}^1;\underline{d}^1)} \sqcup C_{(\underline{c}^2;\underline{d}^2)}\right).$$

EXAMPLE 6.43. Choosing the permutations σ and τ to be the identity, and reversing the order of the union by symmetry, the first condition in Lemma 6.42 can be restated as a canonical strict isomorphism of

$$\xi^i_j C_{((\underline{c},\underline{a});(\underline{d},\underline{b}))}\left(C_{(\underline{c};\underline{d})} \sqcup C_{(\underline{a};\underline{b})}\right)$$

with, for $1 \leq j \leq |\underline{c}|$ and $1 \leq i \leq |\underline{d}|$,

$$\xi^i_j C_{(\underline{c};\underline{d})} \sqcup C_{(\underline{a};\underline{b})}$$

depicted

(6.5)

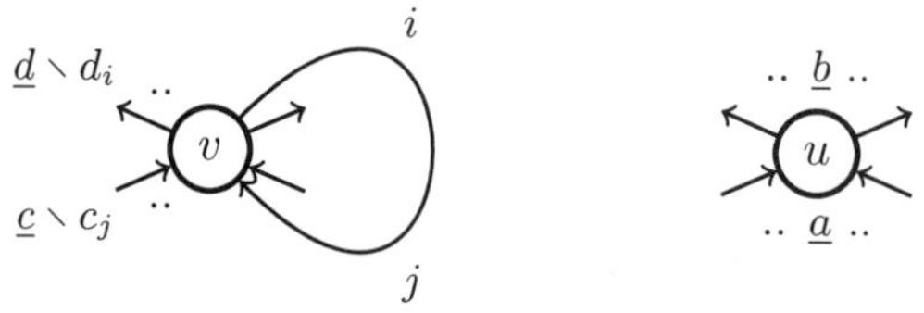

while for $j > |\underline{c}|$ and $i > |\underline{d}|$, we set $l = j - |\underline{c}|$ and $k = i - |\underline{d}|$ to instead recover

$$C_{(\underline{c};\underline{d})} \sqcup \xi^i_j C_{(\underline{a};\underline{b})}$$

depicted

(6.6)

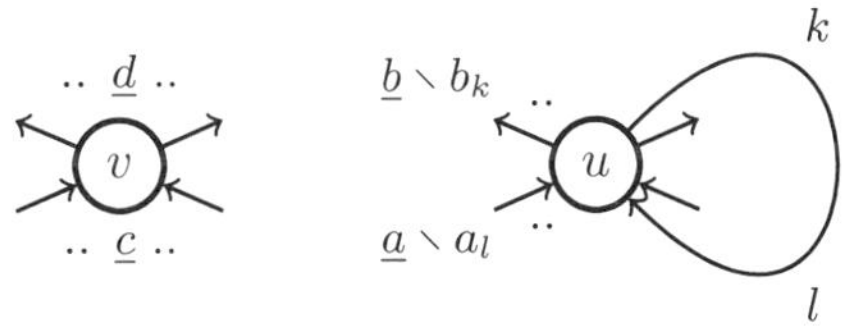

COROLLARY 6.44. *Suppose G_1 and G_2 have profiles $(\underline{c}^1;\underline{d}^1)$ and $(\underline{c}^2;\underline{d}^2)$.*

(1) *If $\sigma \in \Sigma_{|\underline{d}_1|}$ and $\tau \in \Sigma_{|\underline{c}_1|}$ while $c^2_j = d^2_i$, then*

$$(\sigma G_1 \tau) \sqcup (\xi^i_j G_2) = \left(\sigma \times Id^{|\underline{d}^2|}\right)\left[\xi^{|\underline{d}^1|+i}_{|\underline{c}^1|+j}(G_1 \sqcup G_2)\right]\left(\tau \times Id^{|\underline{c}^2|}\right).$$

(2) *If $c^1_j = d^1_i$ and $c^2_l = d^2_k$ then*

$$(\xi^i_j G_1) \sqcup (\xi^k_l G_2) = \xi^{|\underline{d}^1|-1+k}_{|\underline{c}^1|-1+l}\xi^i_j(G_1 \sqcup G_2).$$

6.3.4. Grafting. Here we start with the observation that grafting has the exceptional edges as two-sided units, using the $\boxtimes$ notation for grafting as in Definition 5.2, and $C_{(\underline{b};\underline{c};\underline{d})} = C_{(\underline{c};\underline{d})} \boxtimes C_{(\underline{b};\underline{c})}$ for the grafted corollas as in Definition 3.12.

LEMMA 6.45. *Given a pair of profiles $(\underline{c};\underline{d})$, there are canonical strict isomorphisms*

$$\begin{aligned} C_{(\underline{c};\underline{d})} &\cong C_{(\underline{c};\underline{c};\underline{d})}\left(C_{(\underline{c};\underline{d})}, \uparrow_{\underline{c}}\right) \\ &\cong C_{(\underline{c};\underline{d};\underline{d})}\left(\uparrow_{\underline{d}}, C_{(\underline{c};\underline{d})}\right). \end{aligned}$$

COROLLARY 6.46. *For any wheeled graph G with profiles $(\underline{c};\underline{d})$, there are canonical strict isomorphisms*

$$G \cong G \boxtimes \uparrow_{\underline{c}} \cong \uparrow_{\underline{d}} \boxtimes G.$$

Now we show the grafting operation is bi-equivariant, although one needs to be careful to cancel any permutation of the intermediate profile properly.

LEMMA 6.47. *Suppose $\underline{b}$, $\underline{c}$, and $\underline{d}$ are profiles, while $\sigma \in \Sigma_{|\underline{d}|}$, $\tau \in \Sigma_{|\underline{c}|}$ and $\rho \in \Sigma_{|\underline{b}|}$. Then*

$$[\sigma C_{(\underline{b};\underline{d})}\rho](C_{(\underline{b};\underline{c};\underline{d})}) = C_{(\underline{b}\rho;\underline{c}\tau;\sigma\underline{d})}(\sigma C_{(\underline{c};\underline{d})}\tau, \tau^{-1}C_{(\underline{b};\underline{c})}\rho).$$

EXAMPLE 6.48. In 7.1.2, we will introduce analogs of Reidemeister moves for manipulation of multiply-iterated graph substitutions. In that context, we can restate Lemma 6.47 as 'moving a grafted corollas outside of permuted corollas'. On the other hand, the grafting of two permuted corollas without cancellation of any intermediate permutation is not necessarily the input and output relabeling of a grafted corolla, so the reverse movement is not always possible. For example, with a single color, the wheel-free graph

$$C_{(2;0)} \boxtimes (1\ 2)C_{(0;2)} = \left(C_{(2;0)} \boxtimes C_{(0;2)}\right)\left(C_{(2;0)}, (1\ 2)C_{(0;2)}\right)$$

is not the input and output relabeling of any grafted corollas because it has no inputs or outputs, and it is not a grafted corollas.

COROLLARY 6.49. *Suppose G_1 is a wheeled graph with profiles $(\underline{c};\underline{d})$, and G_2 is a wheeled graph with profiles $(\underline{b};\underline{c})$, $\sigma \in \Sigma_{|\underline{d}|}$, $\tau \in \Sigma_{|\underline{c}|}$ and $\rho \in \Sigma_{|\underline{b}|}$. Then*

$$\sigma\left(G_1 \boxtimes G_2\right)\rho = (\sigma G_1 \tau) \boxtimes (\tau^{-1} G_2 \rho).$$

Next we notice that grafting is associative up to canonical strict isomorphism for wheeled graphs.

LEMMA 6.50. *Suppose $\underline{a}$, $\underline{b}$, $\underline{c}$, and $\underline{d}$ are profiles. Then there is a canonical strict isomorphism*

$$C_{(\underline{a};\underline{b};\underline{d})}\left(C_{(\underline{b};\underline{c};\underline{d})}, C_{(\underline{a};\underline{b})}\right) \cong C_{(\underline{a};\underline{c};\underline{d})}\left(C_{(\underline{c};\underline{d})}, C_{(\underline{a};\underline{b};\underline{c})}\right)$$

COROLLARY 6.51. *If G_1 is a wheeled graph with profiles $(\underline{c};\underline{d})$, G_2 is a wheeled graph with profiles $(\underline{b};\underline{c})$, and G_3 is a wheeled graph with profiles $(\underline{a};\underline{b})$, then there is a canonical strict isomorphism*

$$(G_1 \boxtimes G_2) \boxtimes G_3 \cong G_1 \boxtimes (G_2 \boxtimes G_3).$$

It will be useful when describing wheeled PROPs to know that grafting can be recovered from a union via repeated contractions. In fact, a similar statement holds for partial grafting with a bit more care about indices lying in the relevant segment (Lemma 6.79).

LEMMA 6.52. *Given profiles $\underline{b}$, $\underline{c}$, and $\underline{d}$, there is a canonical strict isomorphism*

$$C_{(\underline{b};\underline{c};\underline{d})} \cong (\xi_1^{|\underline{d}|+1})^{|\underline{c}|}(C_{(\underline{c};\underline{d})} \sqcup C_{(\underline{b};\underline{c})}).$$

COROLLARY 6.53. *Given wheeled graphs G_1 with profiles $(\underline{c};\underline{d})$ and G_2 with profiles $(\underline{b};\underline{c})$, there is a canonical strict isomorphism*

$$G_1 \boxtimes G_2 \cong (\xi_1^{|\underline{d}|+1})^{|\underline{c}|}(G_1 \sqcup G_2)$$

Another important observation is that the union of (single) graftings can be rewritten as a grafting of unions. In particular, the first isomorphism in the statements below induces the so-called Interchange Law for PROPs.

LEMMA 6.54. *For all choices of the evident profiles, there are canonical strict isomorphisms*

$$\begin{aligned}&\left(C_{(\underline{b}^1;\underline{d}^1)} \sqcup C_{(\underline{b}^2;\underline{d}^2)}\right)\left(C_{(\underline{b}^1;\underline{c}^1;\underline{d}^1)}, C_{(\underline{b}^2;\underline{c}^2;\underline{d}^2)}\right)\\ &\cong C_{((\underline{b}^1,\underline{b}^2);(\underline{c}^1,\underline{c}^2);(\underline{d}^1,\underline{d}^2))}\left(C_{(\underline{c}^1;\underline{d}^1)} \sqcup C_{(\underline{c}^2;\underline{d}^2)}, C_{(\underline{b}^1;\underline{c}^1)} \sqcup C_{(\underline{b}^2;\underline{c}^2)}\right),\end{aligned}$$

and

$$\begin{aligned}C_{(\underline{b}^1;\underline{c}^1;\underline{d}^1)} \sqcup C_{(\underline{c}^2;\underline{d}^2)} &\cong C_{((\underline{b}^1,\underline{c}^2);(\underline{c}^1,\underline{c}^2);(\underline{d}^1,\underline{d}^2))}\left(C_{(\underline{c}^1;\underline{d}^1)} \sqcup C_{(\underline{c}^2;\underline{d}^2)}, C_{(\underline{b}^1;\underline{c}^1)} \sqcup \uparrow_{\underline{c}^2}\right)\\ &\cong C_{((\underline{b}^1,\underline{c}^2);(\underline{c}^1,\underline{d}^2);(\underline{d}^1,\underline{d}^2))}\left(C_{(\underline{c}^1;\underline{d}^1)} \sqcup \uparrow_{\underline{d}^2}, C_{(\underline{b}^1;\underline{c}^1)} \sqcup C_{(\underline{c}^2;\underline{d}^2)}\right).\end{aligned}$$

EXAMPLE 6.55. The first claim of Lemma 6.54 can be depicted as

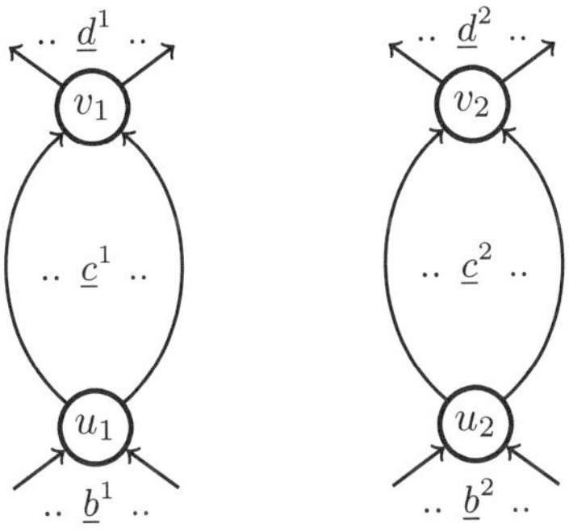

or restated as 'grafted corollas can be pulled outside of unions of corollas'. However, the result is still just a special case, so the reverse move is not always possible. For example, with a single color, the connected wheel-free graph

$$\left(C_{(2;0)} \boxtimes C_{(2;2)}\right)\left(C_{(2;0)}, C_{(1;1)} \sqcup C_{(1;1)}\right)$$

cannot be written as a non-trivial union because such unions are never connected. Another such example is the connected wheel-free graph

$$\left(C_{(3;0)} \boxtimes C_{(0;3)}\right)\left(C_{(2;0)} \sqcup C_{(1;0)}, C_{(0;1)} \sqcup C_{(0;2)}\right).$$

COROLLARY 6.56. *Suppose G_1 has profiles $(\underline{c}^1;\underline{d}^1)$, G_2 has profiles $(\underline{b}^1;\underline{c}^1)$, H_1 has profiles $(\underline{c}^2;\underline{d}^2)$, H_2 has profiles $(\underline{b}^2;\underline{c}^2)$, and K has profiles $(\underline{c}^3;\underline{d}^3)$ as wheeled graphs. Then there are canonical strict isomorphisms*

$$(G_1 \boxtimes G_2) \sqcup (H_1 \boxtimes H_2) \cong (G_1 \sqcup H_1) \boxtimes (G_2 \sqcup H_2),$$

and

$$\begin{aligned}(G_1 \boxtimes G_2) \sqcup K &\cong (G_1 \sqcup K) \boxtimes (G_2 \sqcup \uparrow_{\underline{c}^3}) \\ &\cong (G_1 \sqcup \uparrow_{\underline{d}^3}) \boxtimes (G_2 \sqcup K).\end{aligned}$$

6.4. Substitution Properties of Partial Grafting

This section continues the development of the Calculus of Graph Substitution, with the focus here being on partial grafting and how it interacts with the basic operations. The next section provides the reader interested only in one special case of partial grafting, such as trees or basic dioperadic graphs, with statements more appropriate to that particular context. However, as we will need the general formulations as well, it seems appropriate to begin with the general formulations provided here. As in the previous section, each lemma is a graph substitution statement followed by an equivalent corollary stated in terms of general graph operations, and preceded by an informal description.

6.4.1. Partial Grafting. In the remainder of this chapter, $\underline{a}$, $\underline{b}$, etc. will denote arbitrary profiles, while $\underline{b}' \subset \underline{b}$, $\underline{c}' \subset \underline{c}$, etc. will denote segments. A partially grafted corollas

$$C_{(\underline{c};\underline{d})} \boxtimes^{\underline{c}'}_{\underline{b}'} C_{(\underline{a};\underline{b})}$$

may be pictured as

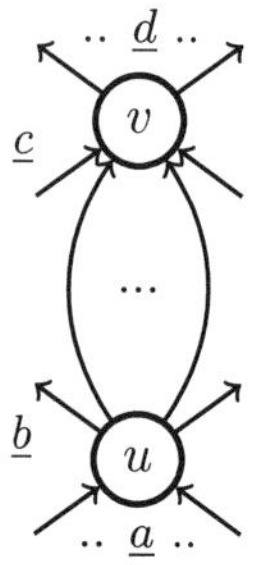

in which the internal edges have colors $\underline{c}' = \underline{b}'$. The reader may also want to revisit Definition 1.6 and Lemma 1.7 related to the notation for profiles of a partial grafting.

First, we have a unit property from graph substitution of exceptional edges into a partially grafted corollas.

LEMMA 6.57. *There are canonical strict isomorphisms*

$$\begin{aligned} C_{(\underline{c};\underline{d})} &\cong \left(C_{(\underline{c};\underline{d})} \boxtimes_{\underline{c}'}^{\underline{c}'} C_{(\underline{c}';\underline{c}')}\right)\left(C_{(\underline{c};\underline{d})}, \uparrow_{\underline{c}'}\right) \\ &\cong \left(C_{(\underline{d}';\underline{d}')} \boxtimes_{\underline{d}'}^{\underline{d}'} C_{(\underline{c};\underline{d})}\right)\left(\uparrow_{\underline{d}'}, C_{(\underline{c};\underline{d})}\right). \end{aligned}$$

COROLLARY 6.58. *For any wheeled graph G with profiles $(\underline{c};\underline{d})$, there are canonical strict isomorphisms*

$$G \cong G \boxtimes_{\underline{c}'}^{\underline{c}'} \uparrow_{\underline{c}'} \cong \uparrow_{\underline{d}'} \boxtimes_{\underline{d}'}^{\underline{d}'} G.$$

Next we have two bi-equivariance lemmas, one for permuting within a segment and another for permuting the segment as a block. As we have seen already, the notation for the types of permutations preserved by certain graph substitutions can become tedious. Thus, we introduce a general notation for the type of permutation most relevant here.

DEFINITION 6.59. Suppose $\underline{b}' = b_{[l,l+k-1]} \subset \underline{b}$ is a k-segment, $\lambda' \in \Sigma_{|\underline{b}|-k+1}$, and $\sigma \in \Sigma_n$. Suppose λ is the block permutation

$$\lambda = \lambda'\langle \underbrace{1, \ldots, 1}_{l-1}, k, 1, \ldots, 1\rangle \in \Sigma_{|\underline{b}|}$$

induced by λ'. Define the permutation

$$\lambda \circ_{\underline{b}'} \sigma = \lambda'\langle \underbrace{1, \ldots, 1}_{l-1}, n, 1, \ldots, 1\rangle \circ \left(Id^{\times l-1} \times \sigma \times Id^{\times |\underline{b}|-k-l+1}\right)$$

in $\Sigma_{|\underline{b}|-k+n}$.

If $\underline{b}' = (b_l)$ is a 1-segment (i.e., $k = 1$), then $\lambda = \lambda'$ and we may write

$$\lambda \circ_l \sigma = \lambda \circ_{b_l} \sigma$$

The following lemma describes what happens when the vertex listings of a partially grafted corollas are permuted in such a way that the segments $\underline{b}'$ and $\underline{c}'$ are permuted as blocks. We will refer to this lemma as an outer bi-equivariance condition.

LEMMA 6.60. *Suppose* $\underline{b}' = \underline{c}'$, *while* $\sigma \in \Sigma_{|\underline{d}|}$, $\tau \in \Sigma_{|\underline{c}|}$, $\lambda \in \Sigma_{|\underline{b}|}$, *and* $\rho \in \Sigma_{|\underline{a}|}$ *are permutations such that* τ *(resp.,* λ*) permutes* $\underline{c}'$ *(resp.,* $\underline{b}'$*) as a block. Then there is a canonical strict isomorphism*

$$\begin{aligned}&\Big(C_{(\underline{c}\tau;\sigma\underline{d})} \boxtimes_{\underline{b}'}^{\underline{c}'} C_{(\underline{a}\rho;\lambda\underline{b})}\Big)\big(\sigma C_{(\underline{c};\underline{d})}\tau, \lambda C_{(\underline{a};\underline{b})}\rho\big)\\ &\cong \Big((\lambda \circ_{\underline{b}'} \sigma) C_{(\underline{c}\circ_{\underline{c}'}\underline{a};\underline{b}\circ_{\underline{b}'}\underline{d})}(\tau \circ_{\underline{c}'} \rho)\Big)\Big(C_{(\underline{c};\underline{d})} \boxtimes_{\underline{b}'}^{\underline{c}'} C_{(\underline{a};\underline{b})}\Big).\end{aligned}$$

REMARK 6.61. In Lemma 6.60, if $\underline{b}' = \underline{c}'$ are 1-segments, then the partially grafted corollas are basic dioperadic graphs. In this case, the assumptions about τ and λ hold for trivial reasons, which gives us Lemmas 6.89, 6.95, and 6.103 in the next section.

COROLLARY 6.62. *Suppose* G_1 *has profiles* $(\underline{c};\underline{d})$ *and* G_2 *has profiles* $(\underline{a};\underline{b})$ *as wheeled graphs such that* $\underline{b}' = \underline{c}'$, *along with the permutations* σ, τ, λ, *and* ρ *as in Lemma* 6.60. *Then there is a canonical strict isomorphism*

$$(\sigma G_1 \tau) \boxtimes_{\underline{b}'}^{\underline{c}'} (\lambda G_2 \rho) \cong (\lambda \circ_{\underline{b}'} \sigma)\Big(G_1 \boxtimes_{\underline{b}'}^{\underline{c}'} G_2\Big)(\tau \circ_{\underline{c}'} \rho).$$

The following lemma describes what happens when the segments in a partially grafted corolla are internally permuted. We will refer to this lemma as an inner bi-equivariance condition.

LEMMA 6.63. *Suppose* $\underline{b}' = b_{[l_b, l_b+k-1]} = c_{[l_c, l_c+k-1]} = \underline{c}'$ *are* k*-segments,* $\theta \in \Sigma_k$,

$$\tau = Id^{\times l_c - 1} \times \theta \times Id^{\times |\underline{c}| - l_c - k + 1} \in \Sigma_{|\underline{c}|}, \quad \text{and} \quad \lambda = Id^{\times l_b - 1} \times \theta \times Id^{\times |\underline{b}| - l_b - k + 1} \in \Sigma_{|\underline{b}|}.$$

Then there is a canonical strict isomorphism

$$C_{(\underline{c};\underline{d})} \boxtimes_{\underline{b}'}^{\underline{c}'} C_{(\underline{a};\underline{b})} \cong \big(C_{(\underline{c};\underline{d})}\tau\big) \boxtimes_{(\theta^{-1}\underline{b}')}^{(\underline{c}'\theta)} \big(\lambda^{-1} C_{(\underline{a};\underline{b})}\big).$$

REMARK 6.64. Lemma 6.63 becomes vacuous if $k = 1$ (i.e., if we use a tree or a basic dioperadic graph), so we will not have versions of it in the discussions of these special cases of partial grafting in the next section.

COROLLARY 6.65. *Suppose* G_1 *has profiles* $(\underline{c};\underline{d})$ *and* G_2 *has profiles* $(\underline{a};\underline{b})$ *as wheeled graphs such that* $\underline{b}' = \underline{c}'$, *along with permutations* θ, τ, *and* λ *as in Lemma* 6.63. *Then there is a canonical strict isomorphism*

$$G_1 \boxtimes_{\underline{b}'}^{\underline{c}'} G_2 \cong (G_1 \tau) \boxtimes_{(\theta^{-1}\underline{b}')}^{(\underline{c}'\theta)} \big(\lambda^{-1} G_2\big).$$

Next we have four associativity lemmas. The first of these describes two ways of making the **caterpillar graph**

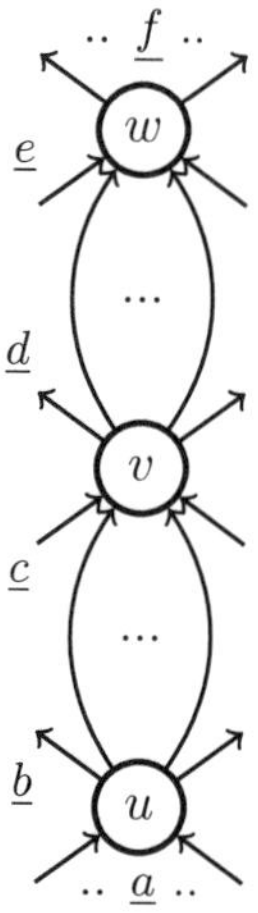

using two partially grafted corollas, and the corollary is the standard associativity property for partial grafting.

LEMMA 6.66. *Suppose $\underline{b}' = \underline{c}'$ and $\underline{d}' = \underline{e}'$. Then there is a canonical strict isomorphism*

$$\begin{aligned}&\left(C_{(\underline{e}\circ_{\underline{e}'}\underline{c};\underline{d}\circ_{\underline{d}'}\underline{f})} \boxtimes^{\underline{c}'}_{\underline{b}'} C_{(\underline{a};\underline{b})}\right)\left(C_{(\underline{e};\underline{f})} \boxtimes^{\underline{e}'}_{\underline{d}'} C_{(\underline{c};\underline{d})}, C_{(\underline{a};\underline{b})}\right)\\ &\cong \left(C_{(\underline{e};\underline{f})} \boxtimes^{\underline{e}'}_{\underline{d}'} C_{(\underline{c}\circ_{\underline{c}'}\underline{a};\underline{b}\circ_{\underline{b}'}\underline{d})}\right)\left(C_{(\underline{e};\underline{f})}, C_{(\underline{c};\underline{d})} \boxtimes^{\underline{c}'}_{\underline{b}'} C_{(\underline{a};\underline{b})}\right).\end{aligned}$$

COROLLARY 6.67. *Suppose G_1 has profiles $(\underline{e};\underline{f})$, G_2 has profiles $(\underline{c};\underline{d})$, and G_3 has profiles $(\underline{a};\underline{b})$ as wheeled graphs such that $\underline{b}' = \underline{c}'$ and $\underline{d}' = \underline{e}'$. Then there is a canonical strict isomorphism*

$$\left(G_1 \boxtimes^{\underline{e}'}_{\underline{d}'} G_2\right) \boxtimes^{\underline{c}'}_{\underline{b}'} G_3 \cong G_1 \boxtimes^{\underline{e}'}_{\underline{d}'} \left(G_2 \boxtimes^{\underline{c}'}_{\underline{b}'} G_3\right).$$

The next lemma describes two ways of making the **lighthouse graph**

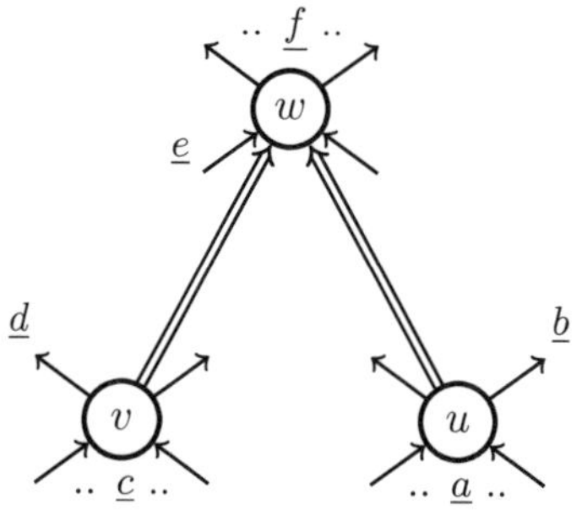

using two partially grafted corollas. The corollary involves the first of three variants of associativity when two partial graftings are possible, with the upper graph G_1 remaining fixed in this case. Here the arrow $\Longrightarrow$ means there is at least one internal edge with that direction between the indicated vertices. To denote two segments within $\underline{e}$, we will consistently assume $\underline{e}'$ occurs before $\underline{e}''$, so $\underline{e} = (\ldots, \underline{e}', \ldots, \underline{e}'', \ldots)$ where any of the three dotted segments could be empty.

LEMMA 6.68. *Suppose $\underline{d}' = \underline{e}'$ and $\underline{b}' = \underline{e}''$ are segments such that $\underline{e}'$ occurs before $\underline{e}''$. Then there is a canonical strict isomorphism*

$$\begin{aligned}&\left(C_{(\underline{e}\circ_{\underline{e}'}\underline{c};\underline{d}\circ_{\underline{d}'}\underline{f})} \boxtimes_{\underline{b}'}^{\underline{e}''} C_{(\underline{a};\underline{b})}\right)\left(C_{(\underline{e};\underline{f})} \boxtimes_{\underline{d}'}^{\underline{e}'} C_{(\underline{c};\underline{d})}, C_{(\underline{a};\underline{b})}\right)\\ &\cong \left(\sigma C_{((\underline{e}\circ_{\underline{e}''}\underline{a})\circ_{\underline{e}'}\underline{c};\underline{d}\circ_{\underline{d}'}(\underline{b}\circ_{\underline{b}'}\underline{f}))}\right)\left(C_{(\underline{e}\circ_{\underline{e}''}\underline{a};\underline{b}\circ_{\underline{b}'}\underline{f})} \boxtimes_{\underline{d}'}^{\underline{e}'} C_{(\underline{c};\underline{d})}\right)\left(C_{(\underline{e};\underline{f})} \boxtimes_{\underline{b}'}^{\underline{e}''} C_{(\underline{a};\underline{b})}, C_{(\underline{c};\underline{d})}\right),\end{aligned}$$

where σ is the unique block permutation of Lemma 1.7.

COROLLARY 6.69. *Suppose G_1 has profiles $(\underline{e};\underline{f})$, G_2 has profiles $(\underline{c};\underline{d})$, and G_3 has profiles $(\underline{a};\underline{b})$ as wheeled graphs, while $\underline{d}' = \underline{e}'$ and $\underline{b}' = \underline{e}''$ are segments such that $\underline{e}'$ occurs before $\underline{e}''$. Then there is a canonical strict isomorphism*

$$\left(G_1 \boxtimes_{\underline{d}'}^{\underline{e}'} G_2\right) \boxtimes_{\underline{b}'}^{\underline{e}''} G_3 \cong \sigma\left[\left(G_1 \boxtimes_{\underline{b}'}^{\underline{e}''} G_3\right) \boxtimes_{\underline{d}'}^{\underline{e}'} G_2\right],$$

where σ is the unique block permutation of Lemma 1.7.

The next lemma describes two ways of making the **fireworks graph**

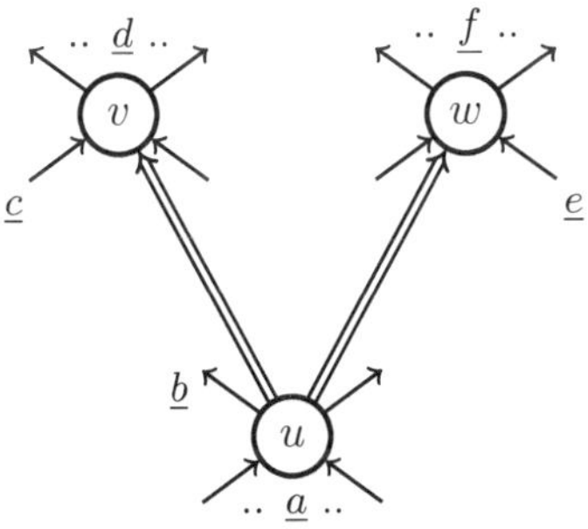

using two partially grafted corollas. This time the corollary is a variant of associativity when two types of partial graftings are possible with the lower graph G_3 remaining fixed.

LEMMA 6.70. *Suppose $\underline{b}' = \underline{c}'$ and $\underline{b}'' = \underline{e}'$ are segments such that $\underline{b}'$ occurs before $\underline{b}''$. Then there is a canonical strict isomorphism*

$$\begin{aligned}&\left(C_{(\underline{e};\underline{f})} \boxtimes_{\underline{b}''}^{\underline{e}'} C_{(\underline{c}\circ_{\underline{c}'}\underline{a};\underline{b}\circ_{\underline{b}'}\underline{d})}\right)\left(C_{(\underline{e};\underline{f})}, C_{(\underline{c};\underline{d})} \boxtimes_{\underline{b}'}^{\underline{c}'} C_{(\underline{a};\underline{b})}\right)\\ &\cong \left(C_{\underline{c}\circ_{\underline{c}'}(\underline{e}\circ_{\underline{e}'}\underline{a});(\underline{b}\circ_{\underline{b}''}\underline{f})\circ_{\underline{b}'}\underline{d})}\tau\right)\left(C_{(\underline{c};\underline{d})} \boxtimes_{\underline{b}'}^{\underline{c}'} C_{(\underline{e}\circ_{\underline{e}'}\underline{a};\underline{b}\circ_{\underline{b}''}\underline{f})}\right)\left(C_{(\underline{c};\underline{d})}, C_{(\underline{e};\underline{f})} \boxtimes_{\underline{b}''}^{\underline{e}'} C_{(\underline{a};\underline{b})}\right),\end{aligned}$$

where τ is the unique block permutation of Lemma 1.7.

COROLLARY 6.71. *Suppose G_1 has profiles $(\underline{e};\underline{f})$, G_2 has profiles $(\underline{c};\underline{d})$, and G_3 has profiles $(\underline{a};\underline{b})$ as wheeled graphs, while $\underline{b}' = \underline{c}'$ and $\underline{b}'' = \underline{e}'$ are segments such that $\underline{b}'$ occurs before $\underline{b}''$. Then there is a canonical strict isomorphism*

$$G_1 \boxtimes_{\underline{b}''}^{\underline{e}'} \left(G_2 \boxtimes_{\underline{b}'}^{\underline{c}'} G_3\right) \cong \left[G_2 \boxtimes_{\underline{b}'}^{\underline{c}'} \left(G_1 \boxtimes_{\underline{b}''}^{\underline{e}'} G_3\right)\right]\tau,$$

where τ is the unique block permutation of Lemma 1.7.

The final associativity lemma describes two ways of making the **bow graph**

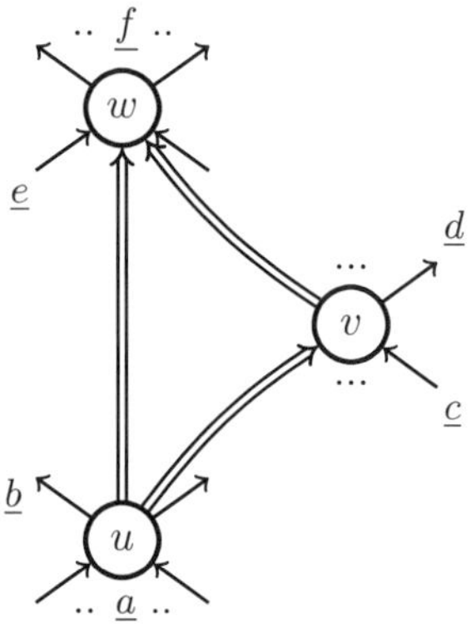

using two partially grafted corollas. Here the corollary is the symmetrical case of associativity with two partial graftings involved, although stated in a restrictive way to avoid involving several permutations.

LEMMA 6.72. *Suppose* $\underline{b}' = \underline{e}'$, $\underline{b}'' = \underline{c}'$, *and* $\underline{d}' = \underline{e}''$ *are segments such that*

$$\underline{b} = (\ldots, \underline{b}', \underline{b}'', \ldots), \quad \underline{e} = (\ldots, \underline{e}', \underline{e}'', \ldots), \quad \underline{c} = (\underline{c}', \ldots), \quad \text{and} \quad \underline{d} = (\underline{d}', \ldots).$$

Then there is a canonical strict isomorphism

$$\left(C_{(\underline{e}\circ_{\underline{e}''}\underline{c};\underline{d}\circ_{\underline{d}'}\underline{f})} \boxtimes^{(\underline{e}',\underline{c}')}_{(\underline{b}',\underline{b}'')} C_{(\underline{a};\underline{b})}\right)\left(C_{(\underline{e};\underline{f})} \boxtimes^{\underline{e}''}_{\underline{d}'} C_{(\underline{c};\underline{d})}, C_{(\underline{a};\underline{b})}\right)$$
$$\cong \left(C_{(\underline{e};\underline{f})} \boxtimes^{(\underline{e}',\underline{e}'')}_{(\underline{b}',\underline{d}')} C_{(\underline{c}\circ_{\underline{c}'}\underline{a};\underline{b}\circ_{\underline{b}''}\underline{d})}\right)\left(C_{(\underline{e};\underline{f})}, C_{(\underline{c};\underline{d})} \boxtimes^{\underline{c}'}_{\underline{b}''} C_{(\underline{a};\underline{b})}\right).$$

COROLLARY 6.73. *Suppose* G_1 *has profiles* $(\underline{e}; \underline{f})$, G_2 *has profiles* $(\underline{c}; \underline{d})$, *and* G_3 *has profiles* $(\underline{a}; \underline{b})$ *as wheeled graphs, while* $\underline{b}' = \underline{e}'$, $\underline{b}'' = \underline{c}'$, *and* $\underline{d}' = \underline{e}''$ *are segments such that*

$$\underline{b} = (\ldots, \underline{b}', \underline{b}'', \ldots), \quad \underline{e} = (\ldots, \underline{e}', \underline{e}'', \ldots), \quad \underline{c} = (\underline{c}', \ldots), \quad \text{and} \quad \underline{d} = (\underline{d}', \ldots).$$

Then there is a canonical strict isomorphism

$$\left(G_1 \boxtimes^{\underline{e}''}_{\underline{d}'} G_2\right) \boxtimes^{(\underline{e}',\underline{c}')}_{(\underline{b}',\underline{b}'')} G_3 \cong G_1 \boxtimes^{(\underline{e}',\underline{e}'')}_{(\underline{b}',\underline{d}')} \left(G_2 \boxtimes^{\underline{c}'}_{\underline{b}''} G_3\right).$$

Now we have a very technical combination of these results which will be necessary later for understanding how to decompose arbitrary connected wheel-free graphs in terms of partially grafted corollas.

LEMMA 6.74. *Suppose* $\underline{d}' = \underline{e}'$ *are* k*-segments and* $\underline{g} = (\underline{e} \circ_{\underline{e}'} \underline{c})\tau$ *with* $\underline{g}' = \underline{b}'$ *and* σ *acts on* $\underline{d} \circ_{\underline{d}'} \underline{f}$. *Then there are pairs of permutations* $(\tau_i; \sigma_i)$ *for* $i = 1, \ldots 4$ *and a canonical strict isomorphism*

$$\left(C_{(\underline{g};\underline{h})} \boxtimes^{\underline{g}'}_{\underline{b}'} C_{(\underline{a};\underline{b})}\right)\left(\sigma C_{(\underline{e}\circ_{\underline{e}'}\underline{c};\underline{d}\circ_{\underline{d}'}\underline{f})}\tau, C_{(\underline{a};\underline{b})}\right)\left(C_{(\underline{e};\underline{f})} \boxtimes^{\underline{e}'}_{\underline{d}'} C_{(\underline{c};\underline{d})}, C_{(\underline{a};\underline{b})}\right) \cong$$
$$\sigma_1\left[K\left(\sigma_4 C_{(\underline{e};\underline{f})}\tau_4, \sigma_3 C_{(\underline{c};\underline{d})}\tau_3, \sigma_2 C_{(\underline{a};\underline{b})}\tau_2\right)\right]\tau_1$$

where K *represents*

$$\left(C_{(\underline{e}\tau_4\circ_{\underline{e}'\tau_4}\underline{c}\tau_3;\sigma_3\underline{d}\circ_{\sigma_3\underline{d}'}\sigma_4\underline{f})} \boxtimes^{?}_{\sigma_2\underline{b}'} C_{(\underline{a}\tau_2;\sigma_2\underline{b})}\right)\left(C_{(\underline{e}\tau_4;\sigma_4\underline{f})} \boxtimes^{\underline{e}'\tau_4}_{\sigma_3\underline{d}'} C_{(\underline{c}\tau_3;\sigma_3\underline{d})}, C_{(\underline{a}\tau_2;\sigma_2\underline{b})}\right).$$

PROOF. First, we notice that the presented graph G is connected wheel-free with 3 vertices, so it must be *weakly* isomorphic to one of the four types of graphs considered above, which we label K. In addition, G cannot be a fireworks graph, since that would require the inner partially grafted corollas to be substituted into the lower, rather than the upper, vertex of the outer partially grafted corollas. Notice each remaining choice of K has a presentation as indicated, by the combination of the last four Lemmas. Now by relabeling each of the input corollas and the entire graph, we can convert the weak isomorphism between G and K to a strict isomorphism as indicated. □

REMARK 6.75. In the statement above, τ_2 and σ_4 can be chosen to be identities, with σ_1 and τ_1 in a particularly nice form. See Lemma 6.77 below for comparison.

There is also a variant of this lemma for when the permuted corolla is substituted into the lower vertex of the outer partially grafted corollas, with the presentation of K also employing a substitution of a partially grafted corollas into the lower vertex of the outer partially grafted corollas. In that case, it is the lighthouse graph which is the excluded possibility for K.

The results above will provide one way of building connected wheel-free graphs, but we will also consider a second. This second approach depends upon instead working with basic pieces of the form

$$C_{(\underline{c};\underline{d})}\tau \boxtimes_{\underline{b}'}^{\underline{c}'} \sigma C_{(\underline{a};\underline{b})}$$

where the convention here is that $\underline{c}' \subset \underline{c}\tau$ and $\underline{b}' \subset \sigma\underline{b}$ are matching segments. The reason for this change is that there is no reason to expect all flags from one vertex connected to another chosen vertex to be sequential among the output flags, and similarly on the input side. Thus, the current building blocks are closer to the arbitrary piece of a connected wheel-free graph, and as a consequence will have a stronger bi-equivariance statement.

LEMMA 6.76. *Suppose $\underline{b}' \subset \nu\lambda\underline{b}$ and $\underline{c}' \subset \underline{c}\tau\delta$ are segments with $\underline{b}' = \underline{c}'$. Then there is a canonical strict isomorphism*

$$\begin{gathered}\left(C_{(\underline{c}\tau;\sigma\underline{d})}\delta \boxtimes_{\underline{b}'}^{\underline{c}'} \nu C_{(\underline{a}\rho;\lambda\underline{b})}\right)\left(\sigma C_{(\underline{c};\underline{d})}\tau, \lambda C_{(\underline{a};\underline{b})}\rho\right) \cong \\ \left((1 \circ_{\underline{b}'} \sigma) C_{(\underline{c}\tau\delta\circ_{\underline{c}'}\underline{a};\nu\lambda\underline{b}\circ_{\underline{b}'}\underline{d})}(1 \circ_{\underline{c}'} \rho)\right)\left(C_{(\underline{c};\underline{d})}\tau\delta \boxtimes_{\underline{b}'}^{\underline{c}'} \nu\lambda C_{(\underline{a};\underline{b})}\right)\end{gathered}$$

Now we have the single associativity statement, which combines all four of those above, although to get the lighthouse and firework graphs we must allow the degenerate case where one of the partial graftings is over a 0-segment, which just gives a union. This involvement of disconnected pieces is the primary reason this approach to describing connected wheel-free graphs is a secondary effort.

LEMMA 6.77. *Suppose $\underline{d}' \subset \sigma\underline{d}$ and $\underline{e}'' \subset \underline{e}\rho$ are matching segments, while $\underline{b}' \subset \lambda\underline{b}$ and $\underline{g}' \subset \underline{g} = (\underline{e}\rho \circ_{\underline{e}''} \underline{c})\delta$ are also matching segments. Then there exist ν, τ, π, $\underline{b}''$, $\underline{e}' = \underline{b}' \circ_{\underline{b}''} \underline{d}' = \underline{h}'$ with $\underline{h}' \subset \underline{h} = \nu(\lambda\underline{b} \circ_{\underline{b}''} \underline{d})$, and a strict isomorphism*

$$\begin{gathered}\left(C_{(\underline{e}\rho\circ_{\underline{e}''}\underline{c};\sigma\underline{d}\circ_{\underline{d}'}\underline{f})}\delta \boxtimes_{\underline{b}'}^{\underline{g}'} \lambda C_{(\underline{a};\underline{b})}\right)\left(C_{(\underline{e};\underline{f})}\rho \boxtimes_{\underline{d}'}^{\underline{e}''} \sigma C_{(\underline{c};\underline{d})}, C_{(\underline{a};\underline{b})}\right) \cong \\ \left[\left(C_{(\underline{e};\underline{f})}\rho \boxtimes_{\underline{h}'}^{\underline{e}'} \nu C_{(\underline{c}\tau\circ_{\underline{c}'}\underline{a};\lambda\underline{b}\circ_{\underline{b}''}\underline{d})}\right)\left(C_{(\underline{e};\underline{f})}, C_{(\underline{c};\underline{d})}\tau \boxtimes_{\underline{b}''}^{\underline{c}'} \lambda C_{(\underline{a};\underline{b})}\right)\right]\pi\,.\end{gathered}$$

PROOF. In this case, choose $\underline{b}''$ to consist of those entries of $\underline{b}'$ which are to be grafted to flags from $\underline{c}$, with $\underline{c}'$ the corresponding entries from $\underline{g}'$. The remaining entries of $\underline{b}'$ must then be connected to flags from $\underline{e}$, which is accomplished by the indicated expansion of $\underline{e}''$ to $\underline{e}'$. Form $\underline{c}\tau$ by simply taking the entries from $\underline{c}$ in the order they appear in $\underline{g}$, so τ is a form of restriction of δ. To define ν, it must switch $\lambda\underline{b} \circ_{\underline{b}''} \underline{d}$ to $\lambda\underline{b} \circ_{\underline{b}'} \left(\sigma\underline{d} \circ_{\underline{d}'} (\underline{b}' \circ_{\underline{b}''} \underline{d}')\right)$ which depends solely on the lengths of the segments and profiles involved. Finally, π must act from the right to send $\underline{e}\rho \circ_{\underline{e}''} (\underline{c}\tau \circ_{\underline{c}'} \underline{a})$ to $(\underline{e}\rho \circ_{\underline{e}''} \underline{c})\delta \circ_{\underline{c}'} \underline{a}$ by a variant of Lemma 1.7. □

REMARK 6.78. There is a dual statement to Lemma 6.77 as well, allowing one to start with an arbitrary case of the partially grafted permuted corollas substituted into the lower vertex. The general case is a combination of the caterpillar and bow graphs, since when $\underline{b}' = \underline{b}''$ we have a relabeled caterpillar graph and when $\underline{e}'' = \underline{e}'$ we have a relabeled bow graph. We also recover a relabeled fireworks graph when $\underline{e}'' = \varnothing = \underline{d}'$ and a relabeled lighthouse graph when $\underline{b}'' = \varnothing = \underline{c}'$, although these involve intermediate graphs which are not connected.

6.4.2. Partial Grafting and Contraction. Now we have a series of results concerning combinations of partial grafting and contraction. The first is the generalization of Lemma 6.52 for partial graftings, where contractions are used on either a union of two corollas or a basic dioperadic graph to produce more general partially grafted corollas.

LEMMA 6.79. *Suppose $\underline{b}' = b_{[l_b, l_b+k-1]} = c_{[l_c, l_c+k-1]} = \underline{c}'$ are k-segments. Then there are canonical strict isomorphisms*

$$\begin{aligned} &C_{(\underline{c};\underline{d})} \boxtimes_{\underline{b}'}^{\underline{c}'} C_{(\underline{a};\underline{b})} \\ &\cong \left(\xi_{l_c+|\underline{a}|}^{l_b+|\underline{d}|}\right)^{k-1} \left(C_{(\underline{c};\underline{d})} \boxtimes_{b_{l_b}}^{c_{l_c}} C_{(\underline{a};\underline{b})}\right) \\ &\cong \sigma\left[\left(\xi_{l_c}^{l_b+|\underline{d}|}\right)^{k} \left(C_{(\underline{c};\underline{d})} \sqcup C_{(\underline{a};\underline{b})}\right)\right]\tau, \end{aligned}$$

where σ is the block permutation of the first two blocks

$$\sigma = (1\ 2)\langle |\underline{d}|, l_b - 1, |\underline{b}| - (l_b + k) + 1\rangle$$

induced by the permutation $(1\ 2) \in \Sigma_3$, and similarly τ is the block permutation of the last two blocks

$$\tau = (2\ 3)\langle l_c - 1, |\underline{a}|, |\underline{c}| - (l_c + k) + 1\rangle$$

induced by the permutation $(2\ 3) \in \Sigma_3$.

REMARK 6.80. The geometric meaning of the previous lemma is that, in order to construct a partially grafted corolla G, we can start with a basic dioperadic graph whose only internal edge is the left-most internal edge of G. Then we proceed from left to right and construct the other $k-1$ internal edges by connecting, in each step, an input and an output via a contraction. Of course, this is only one of $k!$ such decompositions of a partially grafted corolla, but starting from the left keeps the notation simpler.

Graphically, the previous lemma suggests two ways of making the **little green man graph**

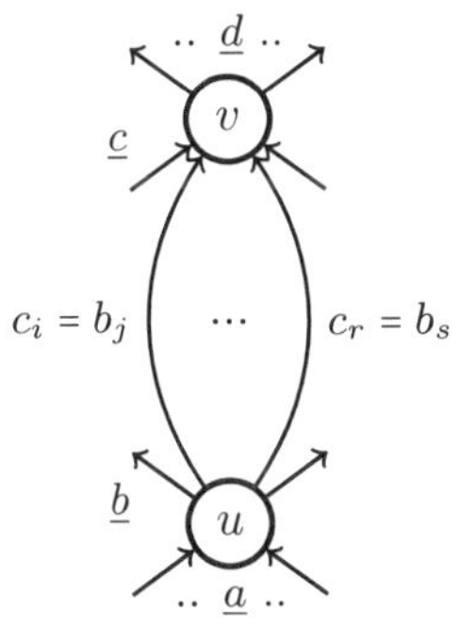

using a basic dioperadic graph and a contracted corolla. We can either start with the left internal edge and construct the right one by contraction, or start with the right internal edge and construct the left one by contraction.

COROLLARY 6.81. *Suppose G_1 has profiles $(\underline{c};\underline{d})$ and G_2 has profiles $(\underline{a};\underline{b})$ as wheeled graphs such that $\underline{b}' = \underline{c}'$. Then there are canonical strict isomorphisms*

$$G_1 \boxtimes_{\underline{b}'}^{\underline{c}'} G_2 \cong \left(\xi_{l_c+|\underline{a}|}^{l_b+|\underline{d}|}\right)^{k-1} \left(G_1 \boxtimes_{b_{l_b}}^{c_{l_c}} G_2\right) \cong \sigma\left[\left(\xi_{l_c}^{l_b+|\underline{d}|}\right)^{k} (G_1 \sqcup G_2)\right]\tau,$$

with σ and τ as in Lemma 6.79.

The next lemma describes two ways of making the **snowboard graph**

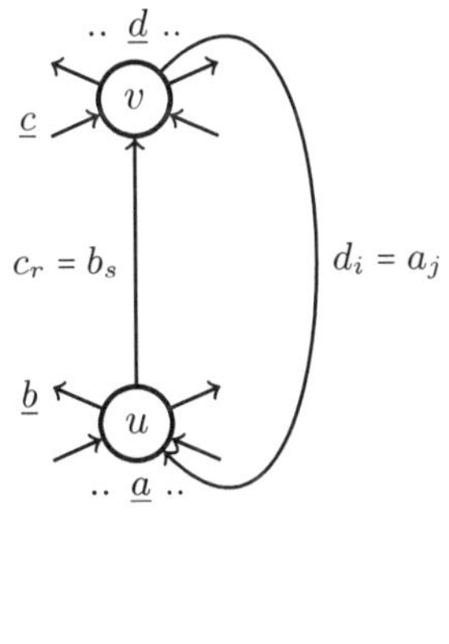

using a basic dioperadic graph and a contracted corolla. One way views the vertex with profiles $(\underline{c};\underline{d})$ as the top vertex, while the other views it as the bottom vertex. Considering the previous lemma, we know each represents a double contraction of the union of the two corollas up to relabeling, so their agreement up to relabeling is not surprising.

LEMMA 6.82. *Supose $c_r = b_s$ and $d_i = a_j$. Then there is a canonical strict isomorphism*

$$\begin{aligned}&\left(\xi^{s-1+i}_{r-1+j} C_{(\underline{c}\circ_r \underline{a};\underline{b}\circ_s \underline{d})}\right)\left(C_{(\underline{c};\underline{d})} \boxtimes^{c_r}_{b_s} C_{(\underline{a};\underline{b})}\right)\\ &\cong \left(\sigma C_{(\underline{a}\circ_j(\underline{c}\circ_r \varnothing);\underline{d}\circ_i(\underline{b}\circ_s\varnothing))}\tau\right)\left(\xi^{s-1+i}_{r-1+j} C_{(\underline{a}\circ_j \underline{c};\underline{d}\circ_i \underline{b})}\right)\left(C_{(\underline{a};\underline{b})} \boxtimes^{a_j}_{d_i} C_{(\underline{c};\underline{d})}\right),\end{aligned}$$

where σ and τ are the unique block permutations for output and input, respectively, of Lemma 1.7.

When applying Lemma 1.7 above, the output profiles of the two contracted graphs should be viewed as

$$\underline{b}\circ_{b_s}(\underline{d}\circ_{d_i}\varnothing) \quad \text{and} \quad \underline{d}\circ_{d_i}(\underline{b}\circ_{b_s}\varnothing)$$

and similarly for the input profiles. This ability to view removing an entry as a partial composite of a profile with an empty profile over a 1-segment is useful in other situations as well.

COROLLARY 6.83. *Suppose G_1 has profiles $(\underline{c};\underline{d})$ and G_2 has profiles $(\underline{a};\underline{b})$ as wheeled graphs such that $c_r = b_s$ and $d_i = a_j$. Then there is a canonical strict isomorphism*

$$\xi^{s-1+i}_{r-1+j}\left(G_1 \boxtimes^{c_r}_{b_s} G_2\right) \cong \sigma\left[\xi^{s-1+i}_{r-1+j}\left(G_2 \boxtimes^{a_j}_{d_i} G_1\right)\right]\tau,$$

where σ and τ are the unique block permutations for output and input, respectively, of Lemma 1.7.

Among other things, the previous two lemmas detail situations when basic dioperadic graphs are substituted into contracted corollas. Together, the following two lemmas allow one to always rewrite a contracted corolla substituted into a partially grafted corollas as another instance of substituting a partially grafted corollas into a contracted corolla. The first corollary below is the case where the original contraction is applied to the upper graph in a partial grafting, and the second involves a contraction of the lower graph.

Now we describe two ways of making a partially grafted corollas with a loop at the top:

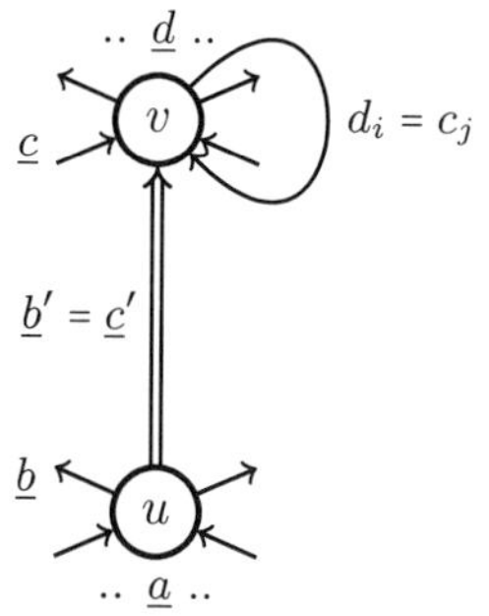

LEMMA 6.84. *Suppose $\underline{b}' = \underline{c}'$ and $c_j = d_i$ such that $\underline{c}'$ occurs before c_j. Then there is a canonical strict isomorphism*

$$\begin{aligned}&\left(C_{(\underline{c}\smallsetminus c_j;\underline{d}\smallsetminus d_i)} \boxtimes^{\underline{c}'}_{\underline{b}'} C_{(\underline{a};\underline{b})}\right)\left(\xi^i_j C_{(\underline{c};\underline{d})}, C_{(\underline{a};\underline{b})}\right)\\ &\qquad\cong \left(\xi^{l_b-1+i}_{j-|\underline{c}'|+|\underline{a}|} C_{(\underline{c}\circ_{\underline{c}'}\underline{a};\underline{b}\circ_{\underline{b}'}\underline{d})}\right)\left(C_{(\underline{c};\underline{d})} \boxtimes^{\underline{c}'}_{\underline{b}'} C_{(\underline{a};\underline{b})}\right).\end{aligned}$$

Notice, if c_j occurs before $\underline{c}'$, just replace $j - |\underline{c}'| + |\underline{a}|$ with j as the lower index of the second contraction both in the lemma above and the corollary below.

COROLLARY 6.85. *Suppose G_1 has profiles $(\underline{c};\underline{d})$ and G_2 has profiles $(\underline{a};\underline{b})$ as wheeled graphs, while $\underline{b}' = \underline{c}'$ and $c_j = d_i$ such that $\underline{c}'$ occurs before c_j. Then there is a canonical strict isomorphism*

$$\left(\xi^i_j G_1\right) \boxtimes^{\underline{c}'}_{\underline{b}'} G_2 \cong \xi^{l_b - 1 + i}_{j - |\underline{c}'| + |\underline{a}|}\left(G_1 \boxtimes^{\underline{c}'}_{\underline{b}'} G_2\right).$$

Finally, we describe two ways of making a partially grafted corollas with a loop at the bottom:

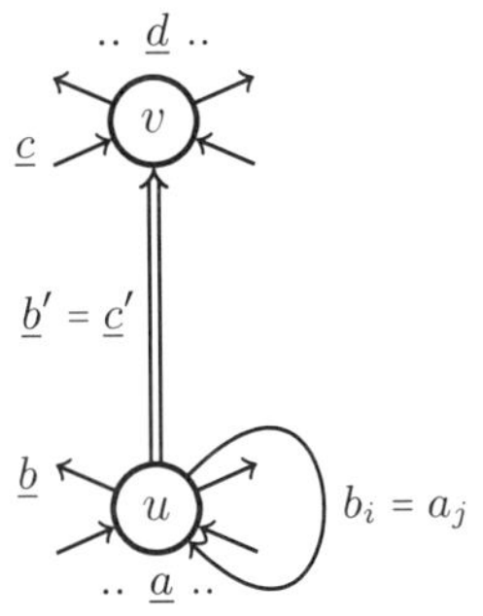

LEMMA 6.86. *Suppose $\underline{b}' = \underline{c}'$ and $b_i = a_j$ such that $\underline{b}'$ occurs before b_i. Then there is a canonical strict isomorphism*

$$\left(C_{(\underline{c};\underline{d})} \boxtimes^{\underline{c}'}_{\underline{b}'} C_{(\underline{a}\smallsetminus a_j;\underline{b}\smallsetminus b_i)}\right)\left(C_{(\underline{c};\underline{d})}, \xi^i_j C_{(\underline{a};\underline{b})}\right)$$
$$\cong \left(\xi^{i-|\underline{b}'|+|\underline{d}|}_{l_c - 1 + j} C_{(\underline{c}\circ_{\underline{c}'}\underline{a};\underline{b}\circ_{\underline{b}'}\underline{d})}\right)\left(C_{(\underline{c};\underline{d})} \boxtimes^{\underline{c}'}_{\underline{b}'} C_{(\underline{a};\underline{b})}\right).$$

Once again, if b_i occurs before $\underline{b}'$, then in both the lemma above and the corollary below replace the upper index $i - |\underline{b}'| + |\underline{d}|$ in the second contraction with i.

COROLLARY 6.87. *Suppose G_1 has profiles $(\underline{c};\underline{d})$ and G_2 has profiles $(\underline{a};\underline{b})$ as wheeled graphs, while $\underline{b}' = \underline{c}'$ and $b_i = a_j$ such that $\underline{b}'$ occurs before b_i. Then there is a canonical strict isomorphism*

$$G_1 \boxtimes^{\underline{c}'}_{\underline{b}'} \left(\xi^i_j G_2\right) \cong \xi^{i-|\underline{b}'|+|\underline{d}|}_{l_c - 1 + j}\left(G_1 \boxtimes^{\underline{c}'}_{\underline{b}'} G_2\right).$$

6.5. Substitution Properties for Trees and Basic Dioperadic Graphs

Almost all of the results in this final Calculus of Graph Substitution section are special cases of results in the previous section on partial graftings (using 1-segments). However, for the convenience of the reader only interested in trees, or in basic dioperadic graphs, we provide the simpler statements appropriate to that context here. As a consequence, we do not provide a reformulation of each lemma as a general corollary, instead providing a reference to the lemma of which it is a special case, and whose corollary was stated earlier.

6.5.1. Simple Trees. The first result here is an instance of Lemma 6.57, allowing cancellation of exceptional edges with simple trees, which will lead to a unit property later.

LEMMA 6.88. *There are canonical strict isomorphisms*

$$\begin{aligned} C_{(\underline{c};d)} &\cong \big(C_{(\underline{c};d)} \circ_i C_{(c_i;c_i)}\big)\big(C_{(\underline{c};d)}, \uparrow_{c_i}\big) \\ &\cong \big(C_{(d;d)} \circ_1 C_{(\underline{c};d)}\big)\big(\uparrow_d, C_{(\underline{c};d)}\big). \end{aligned}$$

Now we will see that permuting within the two types of inputs of a simple tree is equivalent to a special type of input relabeling of the full simple tree, a special case of the outer bi-equivariance property of Lemma 6.60.

LEMMA 6.89. *There is a canonical strict isomorphism*

$$\begin{aligned} &\big(C_{(\underline{c}\circ_{c_i}\underline{b};d)}(\tau \circ_i \rho)\big)\big(C_{(\underline{c};d)} \circ_i C_{(\underline{b};c_i)}\big) \\ &\qquad\cong \big(C_{(\underline{c}\tau;d)} \circ_{\tau(i)} C_{(\underline{b}\rho;c_i)}\big)\big(C_{(\underline{c};d)}\tau, C_{(\underline{b};c_i)}\rho\big). \end{aligned}$$

EXAMPLE 6.90. It is *not* always possible to write the input relabeling of a simple tree by substituting permuted corollas into a simple tree when the input permutation is not of the form $\tau \circ_i \rho$. One such example is obtained from the simple tree $C_{(\underline{c};d)} \circ_i C_{(\underline{b};c_i)}$ by permuting an input b_r with an input c_s $(s \neq i)$.

The next observation is a special case of Lemma 6.66, here saying that when building a 3-level tree with two internal edges via graph substitution from simple trees, the order in which the internal edges are constructed is irrelevant.

LEMMA 6.91. *There is a canonical strict isomorphism*

$$\begin{aligned} &\big(C_{(\underline{c};d)} \circ_i C_{(b_{[1,j-1]},\underline{a},b_{[j+1,l]};c_i)}\big)\big(C_{(\underline{c};d)}, C_{(\underline{b};c_i)} \circ_j C_{(\underline{a};b_j)}\big) \\ &\cong \big(C_{(c_{[1,i-1]},\underline{b},c_{[i+1,m]};d)} \circ_{i-1+j} C_{(\underline{a};b_j)}\big)\big(C_{(\underline{c};d)} \circ_i C_{(\underline{b};c_i)}, C_{(\underline{a};b_j)}\big). \end{aligned}$$

Here a representation of the result is the following.

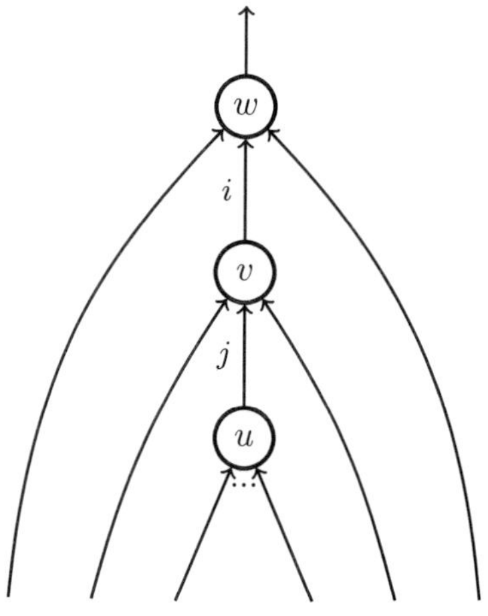

Likewise, the next observation is a special case of Lemma 6.68, which here says that when building a 2-level tree with two internal edges via graph substitution from simple trees, the order in which the internal edges are constructed is irrelevant.

LEMMA 6.92. *If $|\underline{b}| = l$ and $i < j$, then there is a canonical strict isomorphism*

$$\left(C_{(\underline{c}_{[1,i-1]},\underline{b},\underline{c}_{[i+1,m]};d)} \circ_{j-1+l} C_{(\underline{a};c_j)}\right)\left(C_{(\underline{c};d)} \circ_i C_{(\underline{b};c_i)}, C_{(\underline{a};c_j)}\right)$$
$$\cong \left(C_{(\underline{c}_{[1,j-1]},\underline{a},\underline{c}_{[i+1,m]};d)} \circ_i C_{(\underline{b};c_i)}\right)\left(C_{(\underline{c};d)} \circ_j C_{(\underline{a};c_j)}, C_{(\underline{b};c_i)}\right).$$

Either approach above results in the following tree.

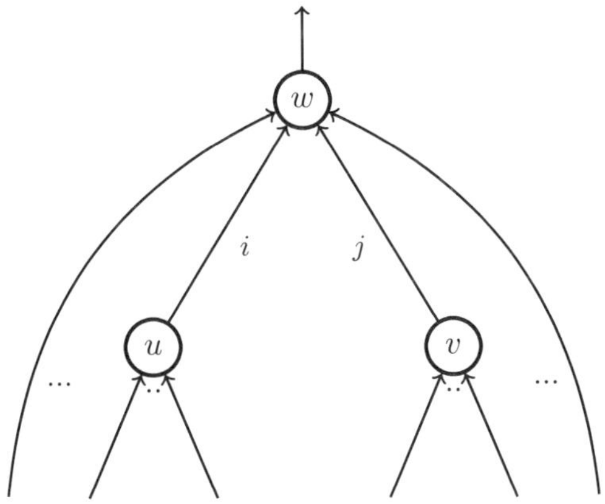

Finally, one can apply Lemma 6.89, substituting the pair consisting of a corolla and a simple tree, to produce 3-level trees which occur fairly often in practice.

COROLLARY 6.93. *There are canonical strict isomorphisms*

$$\left(C_{(\underline{c}\circ_i(\underline{b}\circ_j\underline{a});d)}(\tau \circ_i id)\right)\left(C_{(\underline{c};d)} \circ_i C_{(\underline{b}\circ_j\underline{a};c_i)}\right)\left(C_{(\underline{c};d)}, C_{(\underline{b};c_i)} \circ_j C_{(\underline{a};b_j)}\right)$$
$$\cong \left(C_{(\underline{c}\tau;d)} \circ_{\tau(i)} C_{(\underline{b}\circ_j\underline{a};c_i)}\right)\left(C_{(\underline{c};d)}\tau, C_{(\underline{b};c_i)} \circ_j C_{(\underline{a};b_j)}\right)$$

and

$$\left(C_{((\underline{c}\circ_i\underline{b})\circ_j\underline{a};d)}(id \circ_j \rho)\right)\left(C_{(\underline{c}\circ_i\underline{b};d)} \circ_j C_{(\underline{a};(\underline{c}\circ_i\underline{b})_j)}\right)\left(C_{(\underline{c};d)} \circ_i C_{(\underline{b};c_i)}, C_{(\underline{a};(\underline{c}\circ_i\underline{b})_j)}\right)$$
$$\cong \left(C_{(\underline{c}\circ_i\underline{b};d)} \circ_j C_{(\underline{a}\rho;(\underline{c}\circ_i\underline{b})_j)}\right)\left(C_{(\underline{c};d)} \circ_i C_{(\underline{b};c_i)}, C_{(\underline{a};(\underline{c}\circ_i\underline{b})_j)}\rho\right).$$

PROOF. For the first claim, simply choose $\rho = id$, and replace $\underline{b}$ by $\underline{b} \circ_{b_j} \underline{a}$ in Lemma 6.89, then substitute into both graphs the pair

$$\left(C_{(\underline{c};d)}, C_{(\underline{b};c_i)} \circ_j C_{(\underline{a};b_j)}\right),$$

with the second claim proven similarly. □

6.5.2. Special Trees. As in 2.5.3, a special tree $T\left(\{\underline{b}^i\}; \underline{c}; d\right)$ can be pictorially represented as:

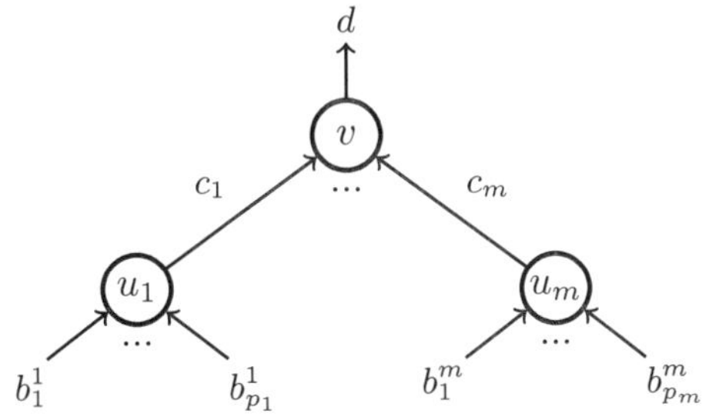

or viewed as a variant of a lighthouse graph.

The first observation here is that redundant exceptional edges can be canceled using a special tree, as in Lemma 6.57, a property which is later used to produce a unit property for operads.

LEMMA 6.94. *There are canonical strict isomorphisms*

$$\begin{aligned} C_{(\underline{c};d)} &\cong T(\{c_i\};\underline{c};d)(C_{(\underline{c};d)},\uparrow_{c_i}) \\ &\cong T(\{\underline{c}\};d;d)(\uparrow_d, C_{(\underline{c};d)}). \end{aligned}$$

Now we notice that permuting within the various input types of a special tree is just a special case of permuting all of the inputs, as an instance of Lemma 6.60.

LEMMA 6.95. *There is a canonical strict isomorphism*

$$\begin{aligned} &T\left(\left\{\underline{b}^{\tau^{-1}(i)}\rho_{\tau^{-1}(i)}\right\}_{i=1}^{m};\underline{c}\tau;d\right)\left(C_{(\underline{c};d)}\tau,\left\{C_{(\underline{b}^i;c_i)}\rho_i\right\}_{i=1}^{m}\right) \\ &\cong \left(C_{((\underline{b}^1,\dots,\underline{b}^m);d)}\left(\tau\langle p_1,\dots,p_m\rangle\circ(\rho_1\times\cdots\times\rho_m)\right)\right)\left(T(\{\underline{b}^i\}_{i=1}^m;\underline{c};d)\right), \end{aligned}$$

where the external product of the ρ_j is denoted $\rho_1\times\cdots\times\rho_m$, and $\tau\langle p_1,\dots,p_m\rangle$ is the block permutation induced by τ that permutes the m blocks of the indicated lengths.

The next observation says that when building a fully 3-level tree via graph substitution from special trees, the order in which the different levels are constructed is irrelevant. We have not stated the corresponding iterated associativity result for partial grafting, but it follows as an application of Lemma 6.68 and the associativity of graph substitution, upon viewing a special tree as a variant of a lighthouse graph.

LEMMA 6.96. *Suppose $|\underline{b}_i| = p_i$, $P_i = p_1+\cdots+p_i$, $P_0 = 0$, and $P = P_m$, while*

$$\underline{b} = (\underline{b}_1,\dots,\underline{b}_m) = (b_1,\dots,b_P).$$

Then there is a canonical strict isomorphism

$$\begin{aligned} &T\left(\left\{\underline{a}_{[P_{i-1}+1,P_i]}\right\}_{i=1}^{m};\underline{c};d\right)\left(C_{(\underline{c};d)},\left\{T\left(\left\{\underline{a}_{P_{i-1}+j}\right\}_{j=1}^{p_i};\underline{b}^i;c_i\right)\right\}_{i=1}^{m}\right) \\ &\cong T\left(\left\{\underline{a}^j\right\}_{j=1}^{P};\underline{b};d\right)\left(T(\{\underline{b}^i\};\underline{c};d),\left\{C_{(\underline{a}^j;b_j)}\right\}_{j=1}^{P}\right). \end{aligned}$$

These both yield presentations of a graph K of the form:

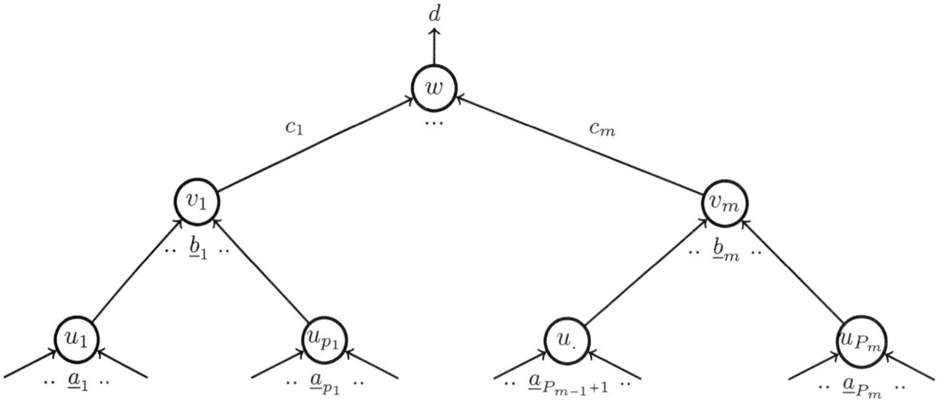

6.5.3. Truncated and Contracted Trees. A Truncated tree is a tree with its output leg removed and a contracted tree is simply the contraction of a tree, just as the names suggest.

EXAMPLE 6.97. • Using the notation in (2.1) for special trees, one 2-level truncated tree is

$$T\left(\{\underline{b}^i\}_{i=1}^m; \underline{c}; \varnothing\right) = C_{(\underline{c};\varnothing)} \boxtimes \left(C_{(\underline{b}^1;c_1)} \cup \cdots \cup C_{(\underline{b}^m;c_m)}\right) \in \mathtt{Tree}^{\mathfrak{C}}\binom{\varnothing}{\underline{b}},$$

which can be pictorially depicted as:

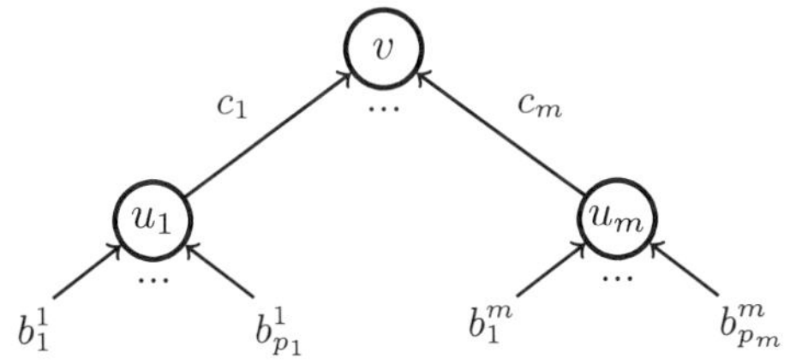

• Another example of a truncated tree is the (permuted) corolla

$$C_{(\underline{c};\varnothing)}\tau = \left\{C_{(\underline{c};\varnothing)} \boxtimes \left(\uparrow_{c_1} \cup \cdots \cup \uparrow_{c_m}\right)\right\}\tau \in \mathtt{Tree}^{\mathfrak{C}}\binom{\varnothing}{\underline{c}\tau}$$

with empty output.

We will require truncated variations of the results on special trees from 6.5.2. First we observe that redundant exceptional edges can be canceled using a truncated special tree, a consequence of Lemma 6.57, viewed as a truncated version of Lemma 6.94.

LEMMA 6.98. *There is a canonical strict isomorphism*

$$C_{(\underline{c};\varnothing)} \cong T(\{c_i\}; \underline{c}; \varnothing)(C_{(\underline{c};\varnothing)}, \uparrow_{c_i})$$

Next we notice that permuting within the various input types of a truncated (special) tree is just a special case of permuting all of the inputs, as in Lemma 6.60.

LEMMA 6.99. *Suppose C is the corolla with profiles $(\underline{b}^1, \ldots, \underline{b}^m; \varnothing)$, the external product of the ρ_j is denoted $\rho_1 \times \cdots \times \rho_m$, and $\tau\langle p_1, \ldots, p_m\rangle$ is the block permutation induced by τ that permutes the m blocks of the indicated lengths. Then there is a canonical strict isomorphism*

$$\begin{aligned}&T\left(\left\{\underline{b}^{\tau^{-1}(i)}\rho_{\tau^{-1}(i)}\right\}_{i=1}^m; \underline{c}\tau; \varnothing\right)\left(C_{(\underline{c};\varnothing)}\tau, \left\{C_{(\underline{b}^i;c_i)}\rho_i\right\}_{i=1}^m\right)\\ &\cong \left(C\left(\tau\langle p_1, \ldots, p_m\rangle \circ (\rho_1 \times \cdots \times \rho_m)\right)\right)\left(T(\{\underline{b}^i\}_{i=1}^m; \underline{c}; \varnothing)\right).\end{aligned}$$

We also need the truncated version of Lemma 6.96.

LEMMA 6.100. *Suppose $|\underline{b}_i| = p_i$, $P_i = p_1 + \cdots + p_i$, $P_0 = 0$, and $P = P_m$, while*

$$\underline{b} = (\underline{b}_1, \ldots, \underline{b}_m) = (b_1, \ldots, b_P).$$

Then there is a canonical strict isomorphism

$$\begin{aligned}&T\left(\left\{\underline{a}_{[P_{i-1}+1,P_i]}\right\}_{i=1}^m; \underline{c}; \varnothing\right)\left(C_{(\underline{c};\varnothing)}, \left\{T\left(\left\{\underline{a}_{P_{i-1}+j}\right\}_{j=1}^{p_i}; \underline{b}^i; c_i\right)\right\}_{i=1}^m\right)\\ &\cong T\left(\left\{\underline{a}^j\right\}_{j=1}^P; \underline{b}; \varnothing\right)\left(T(\{\underline{b}^i\}; \underline{c}; \varnothing), \left\{C_{(\underline{a}^j;b_j)}\right\}_{j=1}^P\right).\end{aligned}$$

Here the picture of K becomes:

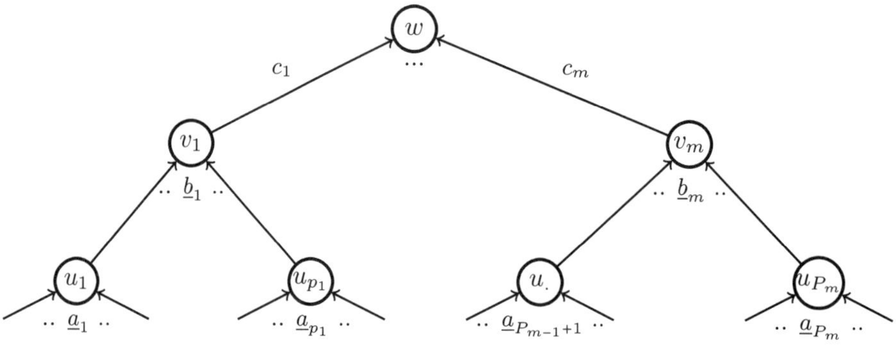

Finally, we will need a result about contracted trees as well.

LEMMA 6.101. *If* $\underline{b} = (\underline{b}^1, \dots, \underline{b}^m)$ *and* $b_j = c_j$, *there is a canonical strict isomorphism*

$$\begin{aligned} &T\left(\{\underline{b}^i\}_{i \neq j}; \underline{c} \smallsetminus c_j; \varnothing\right)(\xi_j^1 C_{(\underline{c};c_j)}, C_{(\underline{b}^1,c_1)}, \dots, C_{(\underline{b}^{j-1},c_{j-1})}, C_{(\underline{b}^{j+1},c_{j+1})}, \dots, C_{(\underline{b}^m,c_m)}) \\ &\quad \cong \xi_j^1 C_{\underline{b};c_j}\left(T\left(\{\underline{b}^i\}; \underline{c}; c_j\right)\right) \\ &\qquad \times \left(C_{(\underline{c};c_j)}, C_{(\underline{b}^1,c_1)}, \dots, C_{(\underline{b}^{j-1},c_{j-1})}, \uparrow_{c_j}, C_{(\underline{b}^{j+1},c_{j+1})}, \dots, C_{(\underline{b}^m,c_m)})\right) \end{aligned}$$

In other words, we have two ways of building a wheeled graph of the following form:

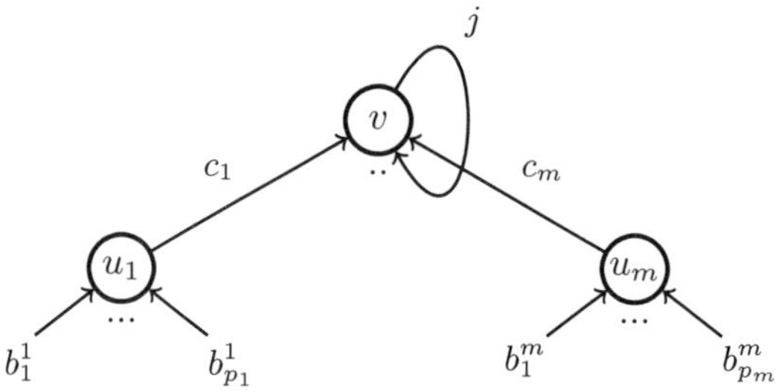

6.5.4. Basic Dioperadic Graphs. Recall from 2.4.2, a basic dioperadic graph is a partially grafted corollas over 1-segments $C^{j,i}_{(\underline{a};\underline{b};\underline{c};\underline{d})} = C_{(\underline{c};\underline{d})} \boxtimes^{c_i}_{b_j} C_{(\underline{a};\underline{b})}$ and it can be depicted as follows.

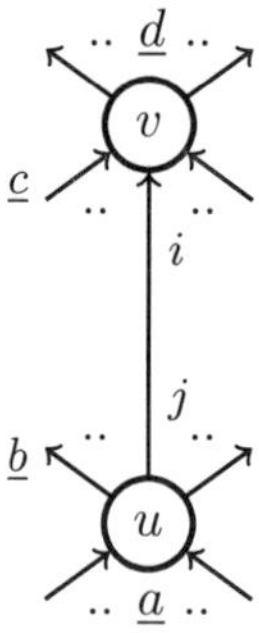

The first observation here says that exceptional edges can be canceled using a basic dioperadic graph, as in Lemma 6.57.

LEMMA 6.102. *There are canonical strict isomorphisms*

$$\begin{aligned} C_{(\underline{c};\underline{d})} &\cong C^{1,i}_{(c_i;c_i;\underline{c};\underline{d})}\left(C_{(\underline{c};\underline{d})}, \uparrow_{c_i}\right) \\ &\cong C^{j,1}_{(\underline{c};\underline{d};d_j;d_j)}\left(\uparrow_{d_j}, C_{(\underline{c};\underline{d})}\right). \end{aligned}$$

Next we notice that permuting within the various types of inputs to a basic dioperadic graph is a special case of permuting the inputs of the full graph, another instance of Lemma 6.60.

LEMMA 6.103. *If $b_j = c_i$, then there is a canonical strict isomorphism*

$$C^{\lambda^{-1}(j),\tau(i)}_{(\underline{a}\pi;\lambda\underline{b};\underline{c}\tau;\sigma\underline{d})}\left(\sigma C_{(\underline{c};\underline{d})}\tau, \lambda C_{(\underline{a};\underline{b})}\pi\right) \cong \left[(\lambda \circ_j \sigma)\, C_{(\underline{c}\circ_i\underline{a};\underline{b}\circ_j\underline{d})}\,(\tau \circ_i \pi)\right]\left(C^{j,i}_{(\underline{a};\underline{b};\underline{c};\underline{d})}\right).$$

EXAMPLE 6.104. It is *not* always possible to write the input and output relabeling of a basic dioperadic graph as the grafting of the input and output relabeling of the component corollas when the permutations are not of the form $\lambda \circ_j \sigma$ and $\tau \circ_i \pi$. One such example is obtained from the dioperadic graph $C^{j,i}_{(\underline{a};\underline{b};\underline{c};\underline{d})}$ by permuting an input a_r with an input c_s $(s \neq i)$.

The following three lemmas correspond to the three distinct ways one can substitute one basic dioperadic graph into another (up to weak isomorphism). These are all instances of the associativity lemmas for partial grafting, but there is no analog of the fourth here since the bow graph is not simply connected.

First, we have a standard associativity statement, an application of Lemma 6.66, producing an instance of a caterpillar graph.

LEMMA 6.105. *If $b_j = c_i$ and $d_s = e_r$, then there is a canonical strict isomorphism*

$$\begin{aligned} &C^{j,i+r-1}_{(\underline{a};\underline{b};\underline{e}\circ_r\underline{c};\underline{d}\circ_s\underline{f})}\left(C^{s,r}_{(\underline{c};\underline{d};\underline{e};\underline{f})}, C_{(\underline{a};\underline{b})}\right) \\ &\qquad \cong C^{j+s-1,r}_{(\underline{c}\circ_i\underline{a};\underline{b}\circ_j\underline{d};\underline{e};\underline{f})}\left(C_{(\underline{e};\underline{f})}, C^{j,i}_{(\underline{a};\underline{b};\underline{c};\underline{d})}\right). \end{aligned}$$

Either presentation yields a simply-connected graph of the form:

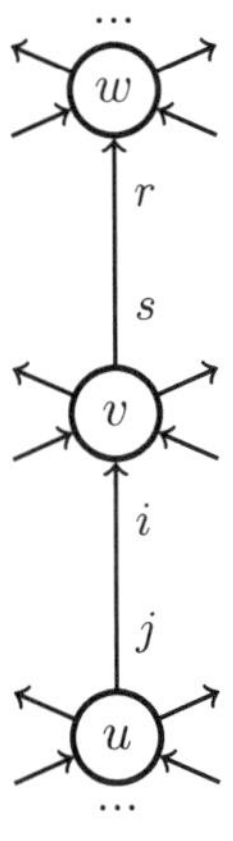

Next we have an instance of a lighthouse graph, by Lemma 6.68.

LEMMA 6.106. *If $d_s = e_r$ and $b_j = e_i$, with $r < i$, then there is a canonical strict isomorphism*

$$C^{j,i+m-1}_{(\underline{a};\underline{b};\underline{e}\circ_r\underline{c};\underline{d}\circ_s\underline{f})}\left(C^{s,r}_{(\underline{c};\underline{d};\underline{e};\underline{f})}, C_{(\underline{a};\underline{b})}\right)$$
$$\cong \left[\sigma C_{((\underline{e}\circ_i\underline{a})\circ_r\underline{c};\underline{d}\circ_s(\underline{b}\circ_j\underline{f}))}\right]\left(C^{s,r}_{(\underline{c};\underline{d};\underline{e}\circ_i\underline{a};\underline{b}\circ_j\underline{f})}\right)\left(C^{j,i}_{(\underline{a};\underline{b};\underline{e};\underline{f})}, C_{(\underline{c};\underline{d})}\right),$$

where σ is the unique block permutation of Lemma 1.7.

Here both approaches give a simply-connected graph of the form:

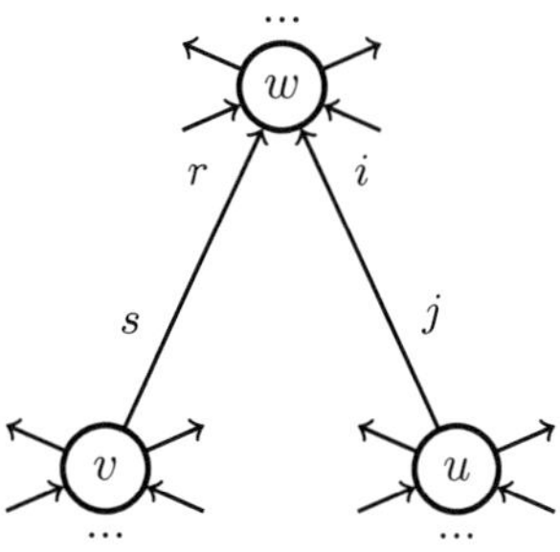

Finally, we have a special case of Lemma 6.70, or a fireworks graph.

LEMMA 6.107. *If $b_j = c_i$ and $b_s = e_r$, with $j < s$, then there is a canonical strict isomorphism*

$$C^{|\underline{d}|-1+s,r}_{(\underline{c}\circ_i\underline{a};\underline{b}\circ_j\underline{d};\underline{e};\underline{f})}(C_{(\underline{e};\underline{f})}, C^{j,i}_{(\underline{a};\underline{b};\underline{c};\underline{d})})$$
$$\cong \left[C_{(\underline{c}\circ_i(\underline{e}\circ_r\underline{a});(\underline{b}\circ_s\underline{f})\circ_j\underline{d})}\tau\right](C^{j,i}_{(\underline{e}\circ_r\underline{a};\underline{b}\circ_s\underline{f};\underline{c};\underline{d})})(C_{(\underline{c};\underline{d})}, C^{s,r}_{(\underline{a};\underline{b};\underline{e};\underline{f})}),$$

where τ is the unique block permutation of Lemma 1.7.

These are both presentations of a simply-connected graph of the form:

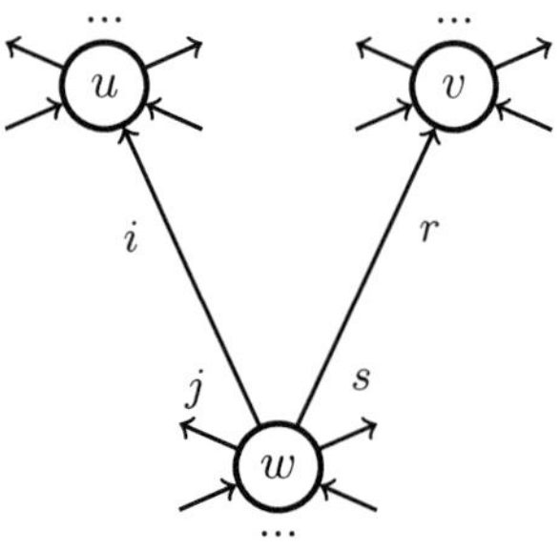

CHAPTER 7

Generators for Graphs

Our generalized PROPs, as well as their modules, will be defined as algebras over a monad, whereas we would like to be able to characterize them in terms of collections of objects with structure maps that must be related in certain ways. Thus, the primary purpose of this chapter is to provide a method of translating results from the language of monads into more concrete descriptions, for all of our pasting schemes of interest. The basic idea is to create a theory of generators and relations for a pasting scheme, or more generally a graph groupoid, but we must go farther to establish the correspondence we seek. The underlying notion of a generating set, as one would expect, is that all graphs in the groupoid can be produced by graph substitution from the graphs in the generating set. However, the notion of relations becomes tenuous unless we have what we call a strong generating set below.

In each case, our monads are built from coproducts indexed over long strings of possible graph substitutions. It then seems clear that we should start the chapter by introducing the notion of graph simplices, which we use to organize these strings. However, in taking this idea seriously, we realize that graph simplices have an analog of the familiar relaxed Reidemeister moves from knot theory, with technical details provided by the three calculus of graph substitution sections at the end of the previous chapter. As a consequence, we define equivalent graph simplices as those whose eventual composites agree, and ask for an analog of Reidemeister's Theorem, that all equivalent graph simplices be connected by a finite number of relaxed moves. A generating set then has the property that all graphs can be written as the composite of a graph simplex with entries in the generating set, called a presentation of the graph, but a strong generating set goes farther to insist any two such presentations be connected by a finite string of relaxed moves. It is this imposition of finite ways from any presentation to any other that allows us to characterize generalized PROPs, algebras and modules in terms of structure maps associated to any graph in the generating set, requiring commutative diagrams corresponding to all of the presentations of such graphs. These are the so-called biased presentations of Markl [**Mar08**] for the cases he considered and beyond, where the term biased refers to a bias towards operations associated to the graphs in the generating set. The most efficient way of describing the required commutative diagrams is to provide an explicit list of relaxed moves which is sufficient to connect any equivalent graph simplices. In each case of interest, we provide such a list based upon a notion of stratified presentation of that type of graph.

The stratified presentations, thought of as standard approaches to building graphs working within the graph groupoid in question, can be of interest in their own right. It is also of some interest to compare the techniques in the different cases. The basic moves are always provided by the calculus of graph substitution sections at the end of the last chapter, but the real ideas come from thinking about graphs as being built from very small pieces.

For the full collection of wheeled graphs, one begins to describe some G by first forming the union of the corollas associated to the vertices and an exceptional edge for each of either the exceptional edges or the exceptional loops of G. Then the contraction operation, embodied by the contracted corollas, suffices to create internal edges (including exceptional loops). As a later consequence, wheeled PROPs can be described in terms of a horizontal composition corresponding to the unions and a contraction operation.

For the wheel-free graphs, the contraction operation is no longer representable. Thus, we instead turn to grafting to create internal edges. A notion called the depth of a vertex allows us to think of the graph G as an iterated grafting of a sequence of horizontal strips, where each strip is a union of corollas and exceptional edges. This embodies the fact that MacLane's PROPs can be thought of as having horizontal compositions, coming from the unions as above, and vertical compositions coming from the graftings, with vertical units coming from the exceptional edges.

Next the discussion turns to various forms of trees. The level trees are built inductively from iterated substitutions of simple trees, which are the embodiment of Gerstenhaber's $\circ_i$ operation, and this means Markl's non-unital operads are described thereby. The unital trees could be presented in terms of combining simple trees and an exceptional edge of each color, so giving Markl's approach to operads. However, we instead pursue a presentation using iterated substitutions of special trees, hence following May's approach to operads. The case of wheeled trees is the most involved in this chapter, since we essentially split the groupoid into three pieces consisting of the unital trees, their truncated versions where the output is removed, and their contracted versions where the output is connected to an input of the same color. As a consequence, our decomposition proceeds by iterated substitutions of special trees, their truncated versions, or a single contracted version. As a consequence, wheeled operads have an operadic composition map as well as a possibly surprising truncated version and the expected contraction operation.

For simply-connected graphs, the decomposition comes from inserting one internal edge at a time, by looking at iterated substitutions of basic dioperadic graphs, which are partially grafted corollas with a single internal edge. Perhaps the biggest surprise here is the need for three different associativity conditions along with the ${}_j\circ_i$ operation for dioperads.

The case of half-graphs is closely analogous to that of level trees. The analogs of the tree with one vertex becomes the permuted corollas where the product of the cardinalities of the profiles is at least 2. Then the role of the simple trees is played by those examples of basic dioperadic graphs which are themselves half-graphs. This leads to strong connections between half-PROPs, operads, and dioperads.

For the case of the connected wheel-free graphs, one would still like to insert the edges between two vertices at a time, but this is not as straightforward as in the simply-connected case. Here we must introduce the notion of two vertices which are called closest neighbors, so that when thinking of building G in a sequence of substitutions we are sure both pieces remain wheel-free. The basic building blocks this time are relabelings of partially grafted corollas, and there are actually four versions of associativity conditions for properads when viewed in terms of these basic structure maps. One might be tempted to think the decomposition for wheel-free graphs, by grafting slices together, would be appropriate here. The problem is that unions are not representable in the world of connected graphs, so the slices would not remain in the graph groupoid under investigation. Similar comments comparing different cases are possible throughout.

Finally, we discuss building connected wheeled graphs by first building a simply-connected graph from basic dioperadic graphs, and then imposing the necessary cycles and wheels by contraction. Thus, wheeled properads may be view in terms of an underlying dioperad structure and a contraction operation which is appropriately compatible.

7.1. Graph Simplices and Generating Sets

We will be repeatedly discussing iterated graph substitutions, which become notationally complex very quickly. Thus, it is worthwhile to introduce some notation for this situation.

7.1.1. Equivalent Graph Simplices.

DEFINITION 7.1. A **graph n-simplex** will refer to a finite string

$$\mathcal{H} = (\mathcal{H}^1, \ldots, \mathcal{H}^n),$$

where $\mathcal{H}^j$ for each $1 \le j \le n$ denotes a finite family of wheeled graphs indexed on the vertices which appear in $\mathcal{H}^{j+1}$ such that the full graph $\mathcal{H}^j_v$ and the vertex $v \in \mathcal{H}^{j+1}$ have the same input/output profiles.

In keeping with the abbreviations for profiles, a graph n-simplex as above will be abbreviated to $\mathcal{H}^{[1,n]}$, and the sub-simplex $(\mathcal{H}^i, \ldots, \mathcal{H}^j)$ will be abbreviated to $\mathcal{H}^{[i,j]}$.

REMARK 7.2. There is no claim in general that $\mathcal{H}^j_v$ is a connected graph, so the reader should be aware that it is dangerous to think of $\mathcal{H}^j$ as a union without taking care with the assignments labeled by v.

In most cases, we will assume $\mathcal{H}^n$ is a single graph, so $\mathcal{H}^{n-1}$ will be a family of graphs indexed on the vertices of $\mathcal{H}^n$ and so forth. Thus, we are really setting the stage for keeping track of strings of possible graph substitutions. If one would like to be more flexible and allow choices of bijections between the vertex profiles, as in our most general definition of graph substitution, it becomes cumbersome at best and misleading at worst.

Also notice that it is possible to concatenate an n-simplex and an m-simplex in order to produce an $(n+m)$-simplex by choosing an assignment of the graphs in the last stage of the first string to the vertices which appear in the first stage of the second string. Of course, there could be more than one choice of such an assignment, since two vertices can have the same profiles, so one should be careful to refer to a choice of concatenation and so forth.

DEFINITION 7.3. Given a graph n-simplex $\mathcal{H}$ for $n > 1$, the **substitution** $\operatorname{sub}(\mathcal{H})$ will denote the associated graph 1-simplex formed by performing all of the possible graph substitutions indicated. Call $\mathcal{H}$ a **presentation of** $\operatorname{sub}(\mathcal{H})$.

Of course, if $\mathcal{H}^n$ is a single graph, then $\operatorname{sub}(\mathcal{H})$ is also a single graph, but in general $\operatorname{sub}(\mathcal{H})$ is a finite family of graphs.

DEFINITION 7.4. Two graph simplices $\mathcal{H}$ and $\mathcal{H}'$ are said to be **equivalent**, if they are presentations of strictly isomorphic families, i.e., $\mathrm{sub}(\mathcal{H}) \cong \mathrm{sub}(\mathcal{H}')$.

7.1.2. Relaxed Moves. Our goal here is to introduce an analog of Reidemeister moves from knot theory, with the intention of presenting an analog of Reidemeister's Theorem, allowing us to connect any two equivalent graph simplices by a finite string of relaxed moves.

DEFINITION 7.5. Given a graph n-simplex $\mathcal{H}$ and $k < n$, a **k-move** of $\mathcal{H}$ is a graph n-simplex $\mathcal{H}'$ and the choice of $1 \leq i \leq n - k$ such that

- $\mathcal{H}^j = (\mathcal{H}')^j$ except when $i \leq j < i + k$, and
- there is a strict isomorphism
$$\mathrm{sub}\left(\mathcal{H}^{[i,i+k-1]}\right) \cong \mathrm{sub}\left((\mathcal{H}')^{[i,i+k-1]}\right)$$
which is compatible with the indexing in terms of vertices of $\mathcal{H}^{i+k} = (\mathcal{H}')^{i+k}$.

In other words, we are changing only a graph k-simplex within our graph n-simplex, keeping the length k fixed along with the equivalence class of the k-simplex. Along the way, we need to make sure the graphs are not only the same, but are being substituted into the same vertices at the next stage, hence the assumption of compatibility with indexing in terms of vertices of the next family of graphs. For technical reasons, we must generalize a bit more by allowing the internal simplices to have different lengths, both bounded by k as follows.

DEFINITION 7.6. Given a graph n-simplex $\mathcal{H}$ and $k < n$, a **relaxed k-move** of $\mathcal{H}$ is the choice of $1 \leq k' \leq k$ and $1 \leq k'' \leq k$ together with a graph $n - k' + k''$-simplex $\mathcal{H}'$ and the choice of $1 \leq i \leq n - k'$ such that

- $\mathcal{H}^j = (\mathcal{H}')^j$ for $1 \leq j < i$,
- $\mathcal{H}^j = (\mathcal{H}')^{j-k'+k''}$ for $i + k' \leq j \leq n$, and
- there is a strict isomorphism
$$\mathrm{sub}\left(\mathcal{H}^{[i,i+k'-1]}\right) \cong \mathrm{sub}\left((\mathcal{H}')^{[i,i+k''-1]}\right)$$
which is compatible with the indexing in terms of vertices of $\mathcal{H}^{i+k'} = (\mathcal{H}')^{i+k''}$.

In other words, we are changing a graph simplex of length $k' \leq k$ for a graph k''-simplex with $k'' < k$ but keeping fixed both the equivalence class of these shorter graph simplices and the vertices into which they are substituted. Relaxed moves are analogous to the first two Reidemeister moves, which relate knot diagrams with different numbers of crossings.

EXAMPLE 7.7. Depending on the choices of the graphs G, the calculus of graph substitution Lemmas 6.26 through 6.107 yield relaxed 2-moves and relaxed 3-moves.

Our technique to establish our analogs of Reidemeister's Theorem will be to provide stratified presentations of graphs in various graph groupoids. We will also need to have the option of working relative to a choice of graph groupoid as below.

DEFINITION 7.8. Given a graph groupoid $\mathtt{G}$ (or a collection of graphs $\mathcal{T}$), a graph n-simplex $\mathcal{H}$ **lies in** $\mathtt{G}$ (or **lies in** $\mathcal{T}$) provided for each j, the family $\mathcal{H}^j$ consists of graphs in $\mathtt{G}$ (or graphs in $\mathcal{T}$) indexed by the vertices of the graphs in

$\mathcal{H}^{j+1}$. Then a **(relaxed) k-move within $\mathtt{G}$** will simply impose the condition that all graph simplices under discussion in the definition of (relaxed) k-move lie in $\mathtt{G}$. The definition of **(relaxed) k-move within $\mathcal{T}$** is similar to that of (relaxed) k-move with all graph simplices lying in $\mathcal{T}$.

7.1.3. Pointed Graph Simplices. Let $\mathcal{T}$ be a collection of wheeled graphs. For our discussion of modules over generalized PROPs, we will need the pointed versions of graph simplices, relaxed moves, equivalence, and so forth. The basic idea is that every time we see a pointed wheeled graph, we simply recognize it as such and keep track of the distinguished vertices.

DEFINITION 7.9. Denote by $\mathcal{T}_*$ the collection of pointed wheeled graphs associated to wheeled graphs in $\mathcal{T}$. In other words, $\mathcal{T}_*$ contains a pointed wheeled graph (G, v) whenever $G \in \mathcal{T}$ and v is a vertex in G.

DEFINITION 7.10. A **pointed graph n-simplex** is a graph n-simplex $\mathcal{H}^{[1,n]}$ such that

- each $H^n \in \mathcal{H}^n$ is a pointed wheeled graph, and
- if $H_v^j \in \mathcal{H}^j$ is substituted into $v \in H^{j+1} \in \mathcal{H}^{j+1}$, then H^{j+1} is a pointed wheeled graph with distinguished vertex v if and only if H_v^j is a pointed wheeled graph.

A **pointed graph n-simplex in $\mathcal{T}$** is a pointed graph n-simplex in which each (pointed) wheeled graph is in $\mathcal{T}$ (or $\mathcal{T}_*$). The collection of pointed wheeled graphs in a pointed graph simplex will be called the **pointed part** of the pointed graph simplex.

Be careful that if H_v^j is a pointed wheeled graph, every other H_u^j may still have a non-empty set of vertices, although it does not have a distinguished vertex. A pointed graph simplex in $\mathcal{T}$ may involve exceptional wheeled graphs in $\mathcal{T}$. In particular, in forming various pointed graph simplices in $\mathcal{T}$, we usually have to use all of $\mathcal{T}$, not just those wheeled graphs with a non-empty set of vertices.

EXAMPLE 7.11.
(1) A pointed graph 1-simplex is a finite set of pointed wheeled graphs. In particular, the substitution of a pointed graph simplex with a single element in $\mathcal{H}^n$ is a pointed wheeled graph.
(2) A pointed graph 2-simplex of the form
$$(\{H_u\}, (G, v))$$
means that $(\{H_u\}, G)$ is a graph 2-simplex, (G, v) and H_v are pointed, and other H_u are not pointed.

DEFINITION 7.12. Two pointed graph simplices are **equivalent** if their substitutions are strictly isomorphic families of pointed wheeled graphs.

DEFINITION 7.13. A **(relaxed) k-move** connecting two pointed graph simplices is defined as in Definitions 7.5 and 7.6, except that the equalities and strict isomorphisms involving the pointed parts are to be interpreted in the sense of pointed wheeled graphs.

7.1.4. Generating Set for a Graph Groupoid.

DEFINITION 7.14. A set $\mathcal{T}$ of graphs is a **generating set for a graph groupoid** $\mathtt{G}$ if $\mathcal{T} \subset \mathtt{G}$ and all graphs in $\mathtt{G}$ have a presentation in $\mathcal{T}$.

In other words, up to strict isomorphism, every $G \in \mathtt{G}$ can be written as an iterated graph substitution where the component graphs all lie in $\mathcal{T}$.

EXAMPLE 7.15.
- The set consisting of all permuted corollas and all unions of two permuted corollas is a generating set for the Horizontal Combinations graph groupoid $\mathtt{Gr}^{\uparrow}_{\mathrm{h}}$.
- The set of permuted corollas and grafted corollas is a generating set for the Vertical Combinations graph groupoid $\mathtt{Gr}^{\updownarrow}_{\mathrm{v}}$.
- The set of corollas with only one input and one output together with the set of grafted corollas with all three profiles a single color form a generating set for the Linear Graphs groupoid $\mathtt{Lin}$.

The following really demands an analog of Reidemeister's Theorem to apply in the context of $\mathtt{G}$, and will be the key to comparing generalized PROPs defined by monads to their so-called biased definitions (following Markl [**Mar08**]).

DEFINITION 7.16. A set $\mathcal{W}$ of relaxed n-moves in a generating set $\mathcal{T}$ for a graph groupoid $\mathtt{G}$ is **n-strong in $\mathcal{T}$** if any two equivalent graph simplices in $\mathcal{T}$ are connected by a finite sequence of relaxed n-moves in $\mathcal{W}$.

A generating set $\mathcal{T}$ for a graph groupoid $\mathtt{G}$ is **n-strong** if there exists a set $\mathcal{W}$ of relaxed n-moves which is n-strong in $\mathcal{T}$.

For all of the graph groupoids of general interest, we will present a strong generating set. In fact, in most cases we will be able to go further and give stratified presentations, which are even more structured, and will lead to explicit descriptions of $\mathcal{W}$ and $\mathcal{T}$.

It is clear that choosing $\mathcal{W}$ as the set of all relaxed n-moves in $\mathcal{T}$ simply says that $\mathcal{T}$ satisfies the analog of Reidemeister's Theorem, but the ability to choose a smaller set $\mathcal{W}$ will be important when describing generalized PROPs in more concrete terms. As is common in such cases, when the precise choice of $\mathcal{W}$ is irrelevant, we will simply speak of $\mathcal{T}$ as n-strong, and when the choice is important we will be more careful.

7.2. Strong Generating Set for Wheeled Graphs

In this section we use stratified presentations to prove we have a 2-strong generating set for the full collection of wheeled graphs.

DEFINITION 7.17. Given a finite set of non-empty wheeled graphs $\{G_\alpha\}$, a **stratified presentation** is the choice of a presentation of $\{G_\alpha\}$ by a graph n-simplex $\mathcal{H}$ such that:

- $\mathcal{H}^1$ consists of corollas and exceptional edges,
- there exists i such that each layer in $\mathcal{H}^{[2,i]}$ consists of corollas and unions of two corollas,
- each layer in $\mathcal{H}^{[i+1,n-1]}$ consists of contracted corollas or corollas, and
- $\mathcal{H}^n$ consists of permuted corollas.

where any of the last three types might be missing.

Notice that for a single graph, in order for $\mathrm{sub}\,(\mathcal{H}) = G$ to hold, the profiles of the corollas in $\mathcal{H}^1$ must be those of the vertices of G, while the profiles of the full graph of the single element of $\mathcal{H}^n$ must be those of the full graph G.

PROPOSITION 7.18. *Every finite set of non-empty wheeled graphs has a stratified presentation.*

PROOF. Since one can combine presentations for individual graphs to build presentations of finite families, it suffices to consider a single graph G. Begin by choosing an ordering of the vertices and of the set of exceptional edges and exceptional loops. Now let

$$\mathcal{H}^1 = \left\{C_{v_j}, E_k\colon \text{ all } j,k\right\},$$

where C_{v_j} is the $(\text{in}(v_j);\text{out}(v_j))$-corolla and E_k is an exceptional edge with the same color as the kth exceptional edge or exceptional loop.

Next choose $\mathcal{H}^{[2,i]}$ such that each layer consists of corollas and unions of two corollas and that

$$\text{sub}\left(\mathcal{H}^{[1,i]}\right) = \bigcup_j C_{v_j} \cup \bigcup_k E_k.$$

Of course, there are a number of such choices, but there is no claim to uniqueness of a stratified presentation. Now choose an ordering of the internal edges of G, which include the exceptional loops. Define $\mathcal{H}^{[i+1,n-1]}$ such that each layer is a single contracted corolla, which creates one of the internal edges in G. Then the graph simplex $\mathcal{H}^{[1,n-1]}$ provides a presentation which may differ from G only by the labeling of the inputs and outputs of the full graph. As a consequence, choosing $\mathcal{H}^n$ to consist of the required single permuted corolla to perform this input and output relabeling will yield a presentation of G, completing the proof. □

Intuitively, the previous proof gives a constructive definition of wheeled graphs. One could easily envision a child's toy where the basic pieces are 'hubs' with a series of flexible colored tubes coming in and going out from the hub. Then one could use connectors to attach a tube from one hub to a tube from another. Here the connectors are created by the contraction operation, and the basic hubs are the corollas. For either type of exceptional component, one begins with an exceptional edge, analogous to a single strand without a hub, again applying connectors to both ends of the same strand as necessary to produce exceptional loops.

The following corollary says that all non-empty wheeled graphs are generated by some basic wheeled graphs via three basic graph operations.

COROLLARY 7.19. *The c-colored exceptional edges* $\uparrow_c$ *(for* $c \in \mathfrak{C}$*) and the corollas* $C_{(\underline{c};\underline{d})}$ *(for* $(\underline{c};\underline{d}) \in \mathcal{P}(\mathfrak{C})^{op} \times \mathcal{P}(\mathfrak{C})$*) generate all non-empty wheeled graphs via the operations of input and output relabeling* 6.1.2*, union* 6.1.3*, and contraction* 6.1.5.

Not only will we be able to use this corollary to produce a strong generating set, but we will give an explicit n-strong set of moves, setting the stage for a number of similar results below.

DEFINITION 7.20.
- Let $\mathcal{T}^{\circlearrowleft}$ consist of all contracted corollas, permuted corollas, unions of 2 corollas, an exceptional edge of each color, and the empty graph.
- Let $W^{\circlearrowleft}$ consist of the moves referred to in Lemmas 6.26, 6.28, 6.30, 6.32, 6.34, 6.36, 6.38, 6.40, and Example 6.43.

The following is now a simple application of the presentation of the operations in terms of graph substitutions. The reader should keep in mind that trivially permuted corollas are just corollas, so corollas are included in $\mathcal{T}^{\circlearrowleft}$.

COROLLARY 7.21. *The set* $\mathcal{T}^{\circlearrowleft}$ *forms a generating set for wheeled graphs.*

Our analog of Reidemeister's Theorem is really a corollary of the following Proposition.

PROPOSITION 7.22. *The set of moves* $W^{\circlearrowleft}$ *is 2-strong in* $\mathcal{T}^{\circlearrowleft}$*, so* $\mathcal{T}^{\circlearrowleft}$ *is a 2-strong generating set for wheeled graphs.*

The proof of this proposition will be broken into the two Lemmas which follow, as will be our habit in other cases as well.

LEMMA 7.23. *Every presentation in* $\mathcal{T}^{\circlearrowleft}$ *of a finite set of non-empty wheeled graphs* $\{G_\alpha\}$ *is connected to a stratified presentation of* $\{G_\alpha\}$ *by a finite sequence of relaxed 2-moves in* $W^{\circlearrowleft}$.

PROOF. By finiteness, it again suffices to consider presentations of a single wheeled graph. First, we observe that for any presentation in $\mathcal{T}^{\circlearrowleft}$, a finite string of relaxed 2-moves will allow us to assume that there is a single permuted corolla, which appears in the last term $\mathcal{H}^n$. To see this, one uses Lemmas 6.26, 6.28, 6.34, and 6.38, which all involve 2-moves. Also, if any empty graphs appear in the presentation, use the relaxed 2-move of Lemma 6.30 to remove them.

Next, we observe that once there are no (non-trivially) permuted corollas involved, then contracted corollas can always be moved to have higher indices than unions of corollas, so to form $\mathcal{H}^{[i+1,n-1]}$, by Example 6.43, which is also a 2-move. The remaining layers in the presentation, which will become $\mathcal{H}^{[1,i]}$, must then consist of corollas, unions of two corollas, and exceptional edges. Given the unit property of corollas with respect to graph substitution, the last stage is then to observe that one can always move exceptional edges to $\mathcal{H}^1$ using a series of 2-moves, because exceptional edges do not have vertices, so no wheeled graphs can be substituted into them. □

Now we have the more delicate question of trying to align two specific presentations. This type of argument relies heavily upon uniqueness up to demonstrably unimportant choices at various stages in most cases that follow.

LEMMA 7.24. *Any two stratified presentations of the same finite set of non-empty wheeled graphs* $\{G_\alpha\}$ *can be connected by a finite sequence of relaxed 2-moves in* $W^{\circlearrowleft}$.

PROOF. By finiteness, it suffices to show any presentation $\mathcal{H}$ of a single wheeled graph G may be connected to a choice of presentation $\mathcal{K}$ built as in Proposition 7.18. Each contracted corolla in $\mathcal{H}$ either creates an internal edge of G or eliminates an extraneous exceptional edge as in Lemma 6.36. Thus, if $\mathcal{H}$ has more contracted corollas than the number of internal edges in G, one should use relaxed 2-moves to move the relevant exceptional edges to $\mathcal{H}^i$ and then use the relaxed 2-move of Lemma 6.36 to eliminate them.

As noted above, the profiles of the corollas in $\mathcal{H}^1$ will be the vertex profiles of the wheeled graph G. Thus $\mathcal{H}^1$ is fully determined by G itself, now that we have removed any extraneous exceptional edges, hence agrees with $\mathcal{K}^1$.

Similarly, the union of corollas and exceptional edges used to construct G, that is $\operatorname{sub}\left(\mathcal{H}^{[1,i]}\right)$, is uniquely determined by G except for the order of the unions now that extraneous exceptional edges have been removed. As a consequence, Lemma

6.32 and Lemma 6.34 provide a series of relaxed 2-moves which will allow the series of unions in $\mathcal{H}^{[2,i]}$ to be aligned with those from $\mathcal{K}$. Notice that applying Lemma 6.34 will require the subsequent use of a 2-move from Lemma 6.38 to get the intermediate presentation of G back into stratified form.

At this point, there are no internal edges yet constructed. There must also be a bijection between the number of internal edges of G and the number of contracted corollas in our intermediate presentation. Even more, the profiles of the contracted corollas are also determined, again up to a choice of order, by G itself. Thus, Lemma 6.40 provides 2-moves allowing us to align the contractions in the (already altered) presentation $\mathcal{H}^{[i+1,n-1]}$ with those from $\mathcal{K}$.

Finally, the graph G, including its vertex profiles, is now recovered except for the listing of the full graph by $\mathrm{sub}\left(\mathcal{H}^{[1,n-1]}\right)$. Thus, there remains a single permuted corolla in $\mathcal{H}^n$ which must be applied to relabel the inputs and outputs properly. Since the intermediate presentation agrees up to this point, the permutations will also agree since they are determined as the filler between two bijections (the current and final listings of the legs). In other words, $\mathcal{H}^n = \mathcal{K}^n$, and this completes the process of connecting these two presentations by relaxed 2-moves. □

We can now state our analog of Reidemeister's Theorem, while the proof works both for Proposition 7.22 and the Theorem itself.

THEOREM 7.25. *Any two equivalent graph simplices in $W^{\circlearrowleft}$ can be connected via a finite string of relaxed 2-moves.*

PROOF. Since $\mathrm{sub}(\mathcal{H}) \cong \mathrm{sub}(\mathcal{H}')$ by definition whenever $\mathcal{H}$ and $\mathcal{H}'$ are equivalent, we can view both as presentations of the finite set of wheeled graphs $\mathrm{sub}(\mathcal{H})$. Now apply Lemma 7.23 twice to connect $\mathcal{H}$ and $\mathcal{H}'$ to stratified presentations of $\mathrm{sub}(\mathcal{H})$. Then apply Lemma 7.24 to these stratified presentations. □

7.3. Strong Generating Set for Wheel-Free Graphs

We now move on to a similar discussion of wheel-free graphs $\mathtt{Gr}^{\uparrow}$, which are used to model PROPs.

As above, the result follows from exploiting a stratified presentation. In fact, we will be able to give an explicit n-strong set of moves, as we will in all cases of interest.

DEFINITION 7.26.
- Let $\mathcal{T}^{\uparrow}$ consist of all unions of two corollas, all graftings of two corollas, all permuted corollas, all $\underline{c}$-exceptional edges for $\varnothing \neq \underline{c} \in \mathcal{P}(\mathfrak{C})$, and the empty graph.
- Let $\mathcal{W}^{\uparrow}$ denote the set of moves which appear in Lemmas 6.26, 6.28, 6.30, 6.32, 6.34, 6.45, 6.47, 6.50, and 6.54.

We can now state the main theorem, as well as the key proposition from which it follows.

THEOREM 7.27. *The set $\mathcal{T}^{\uparrow}$ is a 2-strong generating set for the graph groupoid of wheel-free graphs $\mathtt{Gr}^{\uparrow}$.*

PROPOSITION 7.28. *The set of moves $\mathcal{W}^{\uparrow}$ is 2-strong in $\mathcal{T}^{\uparrow}$.*

Here we detail the appropriate notion of stratified presentation within this groupoid of wheel-free graphs.

DEFINITION 7.29. Given a finite set of wheel-free graphs $\{G_\alpha\}$, a **stratified presentation** is the choice of a presentation of $\{G_\alpha\}$ by a graph n-simplex $\mathcal{H}$ such that:

- $\mathcal{H}^1$ consists of corollas and $\underline{c}$-exceptional edges,
- there exists i such that each layer of $\mathcal{H}^{[2,i-1]}$ consists of corollas and unions of two corollas,
- $\mathcal{H}^i$ consists of permuted corollas, and
- each layer of $\mathcal{H}^{[i+1,n]}$ consists of corollas and grafted corollas,

where any of the last three types could be missing.

Note that unlike the case of all wheeled graphs, this stratified presentation has a layer of permuted corollas somewhere in the middle of a graph simplex. This discrepancy between the wheeled and the wheel-free cases can be explained as follows. In the former case, internal edges are created by contraction, i.e., graph substitution into contracted corollas. As we pointed out in Lemma 6.38, permuted corollas can be moved past a contracted corolla in either direction using a 2-move. On the other hand, as we will explain further below, in a wheel-free graph, internal edges are created by grafting, i.e., graph substitution into grafted corollas. However, permuted corollas can only have their indices reduced relative to a grafted corolla in general by Lemma 6.47 and Example 6.48.

The following definition will be used as an inductive framework for verifying the existence of stratified presentations below.

DEFINITION 7.30. The **depth** of a vertex in a wheel-free graph G is the largest integer n such that there exists a directed internal path of length n whose terminal vertex is the given vertex. A vertex without any incoming internal edges has depth 0.

The real value in the concept of depth, which only makes sense in a context without wheels, lies in the resulting decomposition of a wheel-free graph. By writing all of the vertices of a given depth at the same height, we can present a picture of a wheel-free G so that all directed edges increase depth and cutting horizontally between two consecutive choices of depth produces a relabeled union of corollas and exceptional edges. Even more, the grafting of these 'slices' then recovers the original G, provided we view exceptional edges of G as extending through every slice. Notice, a directed edge can easily increase depth by more than one in G, which would lead to additional exceptional edges in that slice, not coming from the exceptional part of G. This presentation is unique for G except for the order of the vertices within each slice, and also for the presentation of each slice as a union.

LEMMA 7.31. *Every finite set of wheel-free graphs has a stratified presentation.*

PROOF. As discussed previously, we need only consider a single wheel-free graph G. Suppose N is the largest integer among the depths of the vertices of G. We can use the depth to decompose G into a grafting of $N+1$ horizonal slices, where only the vertices of a fixed depth are in any slice. For $0 \leq k \leq N$, the kth slice is an input and output relabeling of a union of corollas on the vertices of depth k and exceptional edges.

We begin by taking $\mathcal{H}^1$ to be the set of all of the corollas on the vertices of G and all of the exceptional edges which appear in these slices. Then define $\mathcal{H}^{[2,i-1]}$ such that each of these $\mathcal{H}^j$ layers consists of corollas and unions of two corollas and so that $\mathrm{sub}(\mathcal{H}^{[1,i-1]})$ is the set of the $N+1$ horizontal slices, but without any input and output relabelings that might be required. The permuted corollas in $\mathcal{H}^i$ should then be chosen to produce the required input and output relabelings of the slices. As a consequence, $\mathrm{sub}(\mathcal{H}^{[1,i]})$ is precisely the set of the $N+1$ horizontal slices. Finally, define $\mathcal{H}^{[i+1,n]}$ such that each layer consists of corollas and grafted corollas and that $\mathrm{sub}(\mathcal{H}^{[1,n]})$ is the iterated grafting of these layers, i.e., G itself. □

Given any presentation, we can use relaxed 2-moves in $\mathcal{W}^{\uparrow}$ to gradually push it to this more inductive style and then compare those stratified presentations. To verify this fact, we have the analogs of two previous results, which complete the proof of the theorem by proving the proposition.

LEMMA 7.32. *Every presentation in $\mathcal{T}^{\uparrow}$ of a finite set of wheel-free graphs $\{G_\alpha\}$ is connected to a stratified presentation of $\{G_\alpha\}$ by a finite sequence of relaxed 2-moves in $\mathcal{W}^{\uparrow}$.*

PROOF. Given any presentation of a single graph, which suffices by finiteness, we can increase the indices of all grafted corollas beyond those of permuted corollas by Lemma 6.47, and beyond those of unions of two corollas by Lemma 6.54, which involve 2-moves. Now use relaxed 2-moves to move the exceptional edges to $\mathcal{H}^1$, and 2-moves as in Lemma 6.28 to increase the indices of permuted corollas beyond those of unions of two corollas. If any empty graphs appear, simply remove them using the relaxed 2-move of Lemma 6.30. □

LEMMA 7.33. *Any two stratified presentations of the same finite set of wheel-free graphs $\{G_\alpha\}$ can be connected by a finite sequence of relaxed 2-moves in $\mathcal{W}^{\uparrow}$.*

PROOF. As before, we will consider a stratified presentation $\mathcal{H}$ of a single graph G and compare it to a presentation $\mathcal{K}$ constructed as in Lemma 7.31. First, notice the ordinary part of $\mathcal{H}^1$ is completely determined by the vertex profiles of G, while there is some ambiguity about the exceptional edges in a stratified presentation, as in Lemma 6.45. Since exceptional edges are fully compatible with relabeling via 2-moves, the relaxed 2-moves from Lemma 6.45 and 6.54 will allow us to alter $\mathcal{H}^1$ to match $\mathcal{K}^1$.

Now the unions of corollas and exceptional edges which will form (each slice in) $\mathrm{sub}\left(\mathcal{H}^{[1,i-1]}\right)$ is uniquely determined by G, using its depth information, except for the order of the unions. Thus, Lemma 6.32 and Lemma 6.34 provide a series of relaxed 2-moves which will allow the unions of two corollas in $\mathcal{H}^{[2,i]}$ to be aligned with those of $\mathcal{K}$. Notice that applying Lemma 6.34 will require the subsequent use of a relaxed 2-move from Lemma 6.26 to get the intermediate presentation of G back into stratified form. The permuted corollas forming $\mathcal{H}^i$ are then completely determined by the full graph listings, so those must now agree with $\mathcal{K}^i$. Finally, the collection of graftings, but not their order, is uniquely determined by G. Thus, one can exploit Lemma 6.50 to provide 2-moves to align the collection of graftings of two corollas of the intermediate presentation with that of $\mathcal{K}$. □

7.4. Strong Generating Set for Level Trees

Now we move on to considering the graph groupoid of Level Trees, which we denote `Tree`, used to model Markl's non-unital operads. Notice these are *not* the same as May's non-unital operads [**Mar08**] (section 2). Markl's non-unital operads are generated by Gerstenhaber's $\circ_i$ operations [**Ger63**], which are the algebraic manifestation of the simple trees in Definition 2.25 (see 3.3.4).

DEFINITION 7.34.
- Let $\mathcal{T}^{Tree}$ consist of all permuted corollas with one output and all simple trees.
- Let $\mathcal{W}^{Tree}$ consist of the moves defined by Lemmas 6.89, 6.91, and 6.92.

Note that a simple tree has exactly one internal edge, and every level tree with exactly one internal edge is the input relabeling of a simple tree. Iterated graph substitution of simple trees allows us to build any level tree up to input relabeling, one internal edge at a time.

THEOREM 7.35. *The set $\mathcal{T}^{Tree}$ is a* 3*-strong generating set for the graph groupoid* `Tree`.

PROPOSITION 7.36. *The set of moves $\mathcal{W}^{Tree}$ is* 3*-strong in $\mathcal{T}^{Tree}$.*

PROOF. While Corollary 6.93 and a variation are used in the proofs below, they both arise by substitution from the result of Lemma 6.89, so are simply included for convenience rather than being formally required. □

As above, the result follows from exploiting a stratified presentation and comparing other presentations in $\mathcal{T}^{Tree}$ to stratified presentations.

DEFINITION 7.37. Given a finite set of Level Trees $\{G_\alpha\}$, a **stratified presentation** is the choice of a presentation of $\{G_\alpha\}$ by a graph n-simplex $\mathcal{H}$ such that:

- each layer in $\mathcal{H}^{[1,n-1]}$ consists of corollas with one output and simple trees, and
- $\mathcal{H}^n$ consists of permuted corollas with one output.

The induction step of the next lemma is a standard example of a shrink-and-expand argument.

LEMMA 7.38. *Every finite set of level trees has a stratified presentation.*

PROOF. Once again, we need only consider a single level tree G, and we proceed by induction on the number of internal edges in G. If the number of internal edges in G is either 0 or 1, then G must be a permuted corolla with one output or an input relabeling of a simple tree, so the claim follows.

Suppose G has $n+1$ internal edges with $n \geq 1$ and is a k-level tree, i.e., k is the largest integer for which there is a vertex of level k, so $2 \leq k \leq n+2$. Pick a level k vertex v. All the incoming flags of v, if there are any, are input legs of G. There is also a unique internal edge e from v to a level $k-1$ vertex u. Suppose (the $\delta = 1$ flag of) e is the ith incoming flag of u.

Construct a new level tree G' from G by removing the internal edge e, and deleting v, after reassigning all the incoming flags of v to u and, denoting the image of u in G' by u', redefining $\operatorname{in} u' = \operatorname{in} u \circ_i \operatorname{in} v$. Note that G' has the same profiles as G and has n internal edges. In addition, there is a unique simple tree T whose profiles are the same as those of u', whose only internal edge e' has the same color as e and is the ith incoming flag of its terminal vertex. In fact, T can be defined using the vertices u and v and flags adjacent to them in G. This yields a strict isomorphism

$$G \cong G'(T),$$

where the graph substitution on the right means that T is substituted into u' and a corolla is substituted into every other vertex. Since the induction hypothesis applies to G', the induction is complete. □

It follows immediately from the previous lemma that $\mathcal{T}^{Tree}$ is a generating set for `Tree`.

We can now complete the proof of the proposition and theorem with the following two, by now expected, results.

LEMMA 7.39. *Every presentation in $\mathcal{T}^{Tree}$ of a finite set of level trees $\{G_\alpha\}$ is connected to a stratified presentation of $\{G_\alpha\}$ by a finite sequence of relaxed 3-moves in $\mathcal{W}^{Tree}$.*

PROOF. This follows from Lemma 6.89, particularly the two portions of its immediate Corollary 6.93. □

LEMMA 7.40. *Any two stratified presentations of the same finite set of level trees $\{G_\alpha\}$ can be connected by a finite sequence of 2-moves in $\mathcal{W}^{Tree}$.*

PROOF. Since corollas are the units of graph substitution, the number of simple trees in $\mathcal{H}^{[1,n-1]}$ is uniquely determined by G, although their order is not determined. Similarly, the permuted corollas in $\mathcal{H}^n$ are determined by G itself. Thus, we simply apply the 2-moves of Lemmas 6.91 and 6.92 to connect the two. □

7.5. Strong Generating Set for Unital Trees

To obtain a 3-strong generating set for `UTree`, we could simply add to $\mathcal{T}^{Tree}$ the exceptional edges involving a single color. However, that would make comparing `UTree`-PROPs with May's (unital) operads a bit indirect. Therefore, we will work with a generating set corresponding to May's original definition of a unital operad. Recall that wheeled graphs in `UTree` are called unital trees, so a unital tree is either a level tree or an exceptional edge.

Note that every fully 2-level tree is an input relabeling of a special tree (2.1) (or Example 3.11).

THEOREM 7.41. *The set $\mathcal{T}^{UTree}$ consisting of all permuted corollas with one output, all special trees, and the exceptional edge of each color is a 3-strong generating set for the graph groupoid* `UTree`.

PROPOSITION 7.42. *The set $\mathcal{W}^{UTree}$ consisting of the moves in Lemmas 6.94, 6.95, and 6.96 is 3-strong in $\mathcal{T}^{UTree}$.*

Once again, we compare via stratified presentations.

DEFINITION 7.43. Given a finite set of unital trees $\{G_\alpha\}$, a **stratified presentation** is the choice of a presentation of $\{G_\alpha\}$ by a graph n-simplex $\mathcal{H}$ such that:

- $\mathcal{H}^1$ consists of corollas with one output and exceptional edges involving a single color,
- each layer of $\mathcal{H}^{[2,n-1]}$ consists of special trees and corollas with one output, and
- $\mathcal{H}^n$ consists of permuted corollas with one output.

As above, the theorem follows from the next three lemmas.

LEMMA 7.44. *Every finite set of unital trees has a stratified presentation.*

PROOF. It suffices to consider a single unital tree. Each exceptional edge forms a graph 1-simplex, which is a stratified presentation. For level trees, first note that every simple tree is a graph 2-simplex in $\mathcal{T}^{UTree}$:

$$C_{(\underline{c};d)} \circ_i C_{(\underline{b};c_i)} = T\left(C_{(\underline{c};d)}, C_{(\underline{b};c_i)}, \uparrow_{c_j}\right),$$

where T is the special tree

$$T = T(\{\underline{b}^j\}_{j=1}^m; \underline{c}; d)$$

with $\underline{b}^j = c_j$ if $j \neq i$ and $\underline{b}^i = \underline{b}$. The exceptional edge $\uparrow_{c_j}$ is substituted into the level 2 vertex connected to the input leg c_j for $j \neq i$. Now use Lemma 7.38, the fact that corollas are two-sided units for graph substitution, and that exceptional edges have no vertices, to conclude that every unital tree has a stratified presentation. □

LEMMA 7.45. *Every presentation in $\mathcal{T}^{UTree}$ of a finite set of unital trees $\{G_\alpha\}$ is connected to a stratified presentation of $\{G_\alpha\}$ by a finite sequence of relaxed 3-moves in $\mathcal{W}^{UTree}$.*

PROOF. This follows from Lemmas 6.94 and 6.95 and the consequences of the latter in which some of the permuted corollas are replaced by special trees. □

LEMMA 7.46. *Any two stratified presentations of the same finite set of unital trees $\{G_\alpha\}$ can be connected by a finite sequence of relaxed 2-moves in $\mathcal{W}^{UTree}$.*

PROOF. Apply Lemmas 6.94 and 6.96, considering the uniqueness of such a presentation except for the order. □

7.6. Strong Generating Set for Wheeled Trees

7.6.1. Building from Level Trees to Wheeled Trees. We must first discuss the transition from (unital) level trees to wheeled trees before presenting our strong generating set for wheeled trees.

Recall that wheeled trees are defined as connected graphs with at most one output from any vertex. On the other hand, level trees are characterized in Lemma 2.29 as simply-connected ordinary graphs with exactly one output from each vertex. It may be a bit surprising how simply the former can be characterized in terms of a slight generalization of the latter.

We saw in section 2.5.4 that truncated trees are examples of wheeled trees that are not level trees. Here are some more examples of wheeled trees that are not themselves level trees.

EXAMPLE 7.47. Given any level tree G, if the output color matches one of the input colors for the full graph, we can connect the output leg of the graph to define a wheeled tree. To be more precise, suppose G is a level tree with output leg the same color, say c_j, as the jth input leg. Then $\xi_j^1(G)$ uses graph contraction to connect these two flags and leave the rest of the graph alone. The result is then a wheeled tree, which will be referred to as a **contracted tree**

Recall from 4.26 that unital trees consist of level trees together with all exceptional edges $\uparrow_c$, so together with Lemma 2.29, the graph groupoid `UTree` can be characterized as consisting of all simply-connected graphs with exactly one output from each vertex. Notice the restriction that they be ordinary has been removed in order to include the various $\uparrow_c$, but the simply-connected condition will exclude exceptional loops. Notice it also makes sense to contract $\uparrow_c$ to produce an exceptional loop, but it does not make sense to truncate $\uparrow_c$ since there is no level 1 vertex (and the result would be a free-standing leg rather than an edge).

NOTATION 7.48.

- Let $\mathtt{Tr.Tree}$ denote the full sub-groupoid of $\mathtt{Gr}^{\mathfrak{C}}_{\mathrm{w}}$ consisting of all truncated trees as in Definition 2.35.
- Let $\xi(\mathtt{UTree})$ denote the full sub-groupoid of $\mathtt{Gr}^{\mathfrak{C}}_{\mathrm{w}}$ consisting of all contracted trees as in 7.47 and all exceptional loops.

Notice $\xi(\mathtt{UTree})$ consists of all results of contractions applied to elements of $\mathtt{UTree}$, while $\mathtt{Tr.Tree}$ consists of all results of truncations of elements of $\mathtt{UTree}$.

The main result here is the following decomposition.

PROPOSITION 7.49. *The graph groupoid of wheeled trees* $\mathtt{Tree}^{\mathfrak{C}}$ *decomposes as a disjoint union of three smaller graph groupoids,*

$$\mathtt{Tree}^{\mathfrak{C}} = \mathtt{UTree} \sqcup \mathtt{Tr.Tree} \sqcup \xi(\mathtt{UTree}).$$

As a consequence, we will produce the three pieces of our decomposition based on the three possibilities:

(1) G contains neither a wheel nor a vertex without outgoing flags (for $\mathtt{UTree}$).
(2) G contains exactly one vertex without outgoing flags and no wheels (for $\mathtt{Tr.Tree}$).
(3) G contains exactly one wheel and no vertex without outgoing flags (for $\xi(\mathtt{UTree})$).

Recall two of these characterizations were already established, the first from combining Lemma 2.29 and Lemma 2.34, and the second case instead using 2.37. The remaining case is proved in the next lemma by a 'cut and reconnect' argument.

LEMMA 7.50. *Suppose G is a wheeled tree. Then G is a contracted tree if and only if it contains a wheel.*

PROOF. Clearly, each contracted tree contains a wheel by starting at the level 1 vertex, following the edge created by the contraction to a different level, and then returning to the level 1 vertex.

Conversely, suppose G is a wheeled tree containing a wheel, and we would like to produce a $G' \in \mathtt{UTree}$ together with a contraction of G' to G. If G is an exceptional loop, we simply choose G' to be $\uparrow_c$ for the same color. Otherwise, G must be an ordinary graph, since it is connected, so G' should also be chosen as ordinary. In the wheel (unique by Lemma 2.34) in G, choose an internal edge $e = \{e_{-1}, e_1\}$, so we have $\iota_G(e_i) = e_{-i}$. Define a new graph G' to be the same as G except for the involution ι and the listing ℓ. In G' we redefine

$$\iota_{G'}(e_i) = e_i,$$

and $\iota_{G'}$ is the same as ι_G on other flags. In particular, in G' the flag e_{-1} (resp., e_1) is an output (resp., input). We must also choose $\ell_{G'}$, which we do by declaring the new input to be the first input, and otherwise $\ell_{G'}(f) = \ell_G(f) + 1$ (and similarly for the outputs of G'). Note that G' is still a wheeled tree, which has neither a wheel nor a vertex with no outgoing flags, so $G' \in \mathtt{UTree}$ as a consequence of Lemma 2.29 and Lemma 2.34. Of course, we can also reconstruct G as

$$G = \xi_1^1(G')$$

since the contraction ξ_1^1 connects the (unique) output e_{-1} to the input e_1. $\square$

7.6.2. Strong Generating Set for Wheeled Trees. Given our comparison to level trees, we are now able to offer a generating set for wheeled trees.

DEFINITION 7.51. • Let $\mathcal{T}^{\mathrm{Q}Tree}$ denote the set of wheeled trees

$$\mathcal{T}^{\mathrm{Q}Tree} = \{T(\{\underline{b}^i\};\underline{c};d), C_{(\underline{c};d)}\mathcal{T}, \uparrow_c\} \sqcup \{\xi_j^1 C_{(\underline{c};c_j)}\} \sqcup \{C_{(\underline{c};\varnothing)}\mathcal{T}, T(\{\underline{b}^i\};\underline{c};\varnothing)\}.$$

• Let $\mathcal{W}^{\mathrm{Q}Tree}$ denote the set of moves in Lemmas 6.38 (when $m = i = 1$), 6.82 (when $|\underline{b}| = |\underline{d}| = s = i = 1$) 6.94, 6.95, 6.96, 6.98, 6.99 6.100, and 6.101.

LEMMA 7.52. *The set $\mathcal{T}^{\mathrm{Q}Tree}$ is a generating set of* $\mathtt{Tree}^{\mathrm{Q}}$.

PROOF. Using Proposition 7.49 it suffices to show that $\mathcal{T}^{\mathrm{Q}Tree}$ generates each of the three (disjoint) sub-groupoids of $\mathtt{Tree}^{\mathrm{Q}}$. First observe that the first subset of $\mathcal{T}^{\mathrm{Q}Tree}$ is a (3-strong) generating set of $\mathtt{UTree}$. Since the graph contraction operation ξ_j^1 is generated via graph substitution by the contracted corolla $\xi_j^1 C_{(\underline{c};c_j)}$, the set $\mathcal{T}^{\mathrm{Q}Tree}$ also generates all the contracted trees. Finally, every truncated tree can be written as a substitution

$$\left(T(\{\underline{b}^i\};\underline{c};\varnothing)\right)\left(C_{(\underline{c};\varnothing)}, T_1, \ldots, T_m\right) \tag{7.1}$$

with each $T_i \in \mathtt{UTree}$. Since $\mathtt{UTree}$ is generated by the first subset of $\mathcal{T}^{\mathrm{Q}Tree}$, we conclude that $\mathcal{T}^{\mathrm{Q}Tree}$ generates $\mathtt{Tr.Tree}$ as well. □

Observe that, although $\mathtt{Tree}^{\mathrm{Q}}$ is about three times as large as $\mathtt{UTree}$, graph substitution in $\mathtt{Tree}^{\mathrm{Q}}$ follows a very rigid pattern. Indeed, to build a graph in $\mathtt{UTree}$ using graph substitutions involving only the generators, one can only use generators in $\mathtt{UTree}$. To build a contracted tree, one must use a contracted corolla $\xi_j^1 C$ exactly once in the graph simplex. To build a truncated tree, one cannot use any contracted tree. Moreover, the graph simplex must satisfy $\mathcal{H}^n$ a 2-level truncated tree $T(\{\underline{b}^i\};\underline{c};\varnothing)$, and only a truncated tree can then be substituted into the top vertex. The next proof is a minor modification and extension of the proof that the first subset of $\mathcal{T}^{UTree}$ is a 3-strong generating set for $\mathtt{UTree}$.

THEOREM 7.53. *The set $\mathcal{T}^{\mathrm{Q}Tree}$ is a 3-strong generating set for* $\mathtt{Tree}^{\mathrm{Q}}$.

PROPOSITION 7.54. *The set of moves $\mathcal{W}^{\mathrm{Q}Tree}$ is 3-strong in $\mathcal{T}^{\mathrm{Q}Tree}$.*

PROOF. Consider graphs in the three disjoint sub-groupoids in Proposition 7.49 separately. As remarked above, graphs in $\mathtt{UTree}$ can only be built by the generators in $\mathtt{UTree}$, so if the presented graph $G \in \mathtt{UTree}$ we refer to Proposition 7.42. The proof is similar if $G \in \mathtt{Tr.Tree}$ by replacing the results of subsection 6.5.2 with those of 6.5.3.

Given any presentation $\mathcal{H}$ of a contracted tree G, we can form a presentation of an associated 'pre-contraction' tree G' of Lemma 7.50 as follows. Replace the single contracted tree which appears in $\mathcal{H}$ with a pre-contraction version, and then add an output of the color of the contracted edge to any truncated tree into which this is subsequently substituted. This gives $\mathcal{H}'$ which is a presentation of G' at the expense of trading truncated 2-level trees or contracted 2-level trees for ordinary 2-level trees. The only possible issue here comes from the fact that there could be different choices of pre-contracted versions, since the wheel is unique but there is a required choice of edge within that wheel in Lemma 7.50. However, any two such choices can be connected using iterates of the 3-move which is the case of Lemma 6.82 when $|\underline{b}| = |\underline{d}| = s = i = 1$.

For any two presentations $\mathcal{H}_1$ and $\mathcal{H}_2$ of G, we can thereby produce two presentations $\mathcal{H}_1'$ and $\mathcal{H}_2'$ of G'. Furthermore, these presentations of G' lie in $\mathcal{T}^{UTree}$ of Proposition 7.42, which produces a finite string of relaxed 3-moves in $\mathcal{W}^{UTree}$ connecting the two presentations of G'. Applying the contraction operator to the copy of the 'pre-contraction' tree and the truncation operator to every tree into which it is subsequently substituted throughout the graph simplex then produces a finite string of relaxed 3-moves in $\mathcal{W}^{QTree}$ connecting the original two presentations of G, again replacing the results of subsection 6.5.2 with those of 6.5.3. ☐

7.7. Strong Generating Set for Simply-Connected Graphs

Here we provide a strong generating set for the graph groupoid of simply-connected graphs, denoted $\mathtt{Gr}^{\uparrow}_{\mathrm{di}}$.

The following observations concern simply-connected graphs with few vertices.

LEMMA 7.55. *Suppose G is a simply-connected graph with exactly n vertices.*

(1) *If $n = 0$, then G is the exceptional edge $\uparrow_c$ for some $c \in \mathfrak{C}$.*
(2) *If $n = 1$, then G is a permuted corolla.*
(3) *If $n = 2$, then G is weakly isomorphic to a graph of the form*

$$\xi_i^{|\underline{d}|+j}\left(C_{(\underline{c};\underline{d})} \cup C_{(\underline{a};\underline{b})}\right) \in \mathtt{Gr}^{\uparrow}_{di}\binom{\underline{d}, \underline{b} \setminus b_j}{\underline{c} \setminus c_i, \underline{a}} \tag{7.2}$$

with $b_j = c_i$.

PROOF. The first two cases are immediate from the definition. Suppose G has two vertices. Being simply-connected implies that there is a unique internal edge connecting the two vertices, and there are no other internal edges. Therefore, G must have the form (7.2). ☐

Note that the wheeled graph in (7.2) can be obtained, up to strict isomorphism, from a basic dioperadic graph by input and output relabeling.

DEFINITION 7.56.
- Let $\mathcal{T}^{di}$ consist of all permuted corollas, the exceptional edge of each color, and the basic dioperadic graphs $C^{j,i}_{(\underline{a};\underline{b};\underline{c};\underline{d})}$.
- Let $\mathcal{W}^{di}$ consist of the moves from Lemmas 6.102, 6.103, 6.105, 6.106, and 6.107.

THEOREM 7.57. *The set $\mathcal{T}^{di}$ is a 3-strong generating set for the graph groupoid $\mathtt{Gr}^{\uparrow}_{di}$.*

PROPOSITION 7.58. *The set of moves $\mathcal{W}^{di}$ is 3-strong in $\mathcal{T}^{di}$.*

The theorem again follows by exploiting a stratified presentation.

DEFINITION 7.59. Given a finite set of simply-connected graphs $\{G_\alpha\}$, a **stratified presentation** is the choice of a presentation of $\{G_\alpha\}$ by a graph n-simplex $\mathcal{H}$ such that:

- $\mathcal{H}^1$ consists of corollas and exceptional edges,
- each layer of $\mathcal{H}^{[2,n-1]}$ consists of basic dioperadic graphs and corollas, and
- $\mathcal{H}^n$ consists of permuted corollas.

Each exceptional edge clearly has a stratified presentation. For ordinary simply-connected graphs, the construction is very similar to that of Lemma 7.38, since an essentially identical shrink-and-expand argument allows us to build the internal edges one at a time.

LEMMA 7.60. *Every finite set of simply-connected graphs has a stratified presentation.*

PROOF. Induct upon the number of internal edges using Lemma 6.8. □

LEMMA 7.61. *Every presentation in $\mathcal{T}^{di}$ of a finite set of simply-connected graphs $\{G_\alpha\}$ is connected to a stratified presentation of $\{G_\alpha\}$ by a finite sequence of relaxed 3-moves in $\mathcal{W}^{di}$.*

PROOF. Apply Lemmas 6.102 and 6.103, in addition to the consequences of Lemma 6.103 where either a basic dioperadic graph or an exceptional edge is substituted into one of the permuted corollas. □

LEMMA 7.62. *Any two stratified presentations of the same finite set of simply-connected graphs $\{G_\alpha\}$ can be connected by a finite sequence of relaxed 3-moves in $\mathcal{W}^{di}$.*

PROOF. If there are extraneous exceptional edges in either resolution, they can be removed by the relaxed 2-move of Lemma 6.102. Otherwise, $\mathcal{H}^1$ is determined by the graph G being presented. As we have seen, the outermost layer $\mathcal{H}^n$ is also determined by G once the intermediate layers of the two presentations have been connected. Thus, considering that the collection of internal edges in G determines the set of basic dioperadic graphs necessary to construct it other than their order, the result follows from the three types of associativity statements possible among basic dioperadic graphs, namely the relaxed 3-moves of Lemmas 6.105, 6.106, and 6.107. □

7.8. Strong Generating Set for Half-Graphs

Recall from Definition 2.21 and Example 2.24 that some dioperadic graphs are half-graphs. Let $S_{1/2}$ denote the full sub-groupoid of $\mathcal{P}(\mathfrak{C})^{op} \times \mathcal{P}(\mathfrak{C})$ consisting of pairs of profiles $(\underline{c};\underline{d})$ with $|\underline{c}||\underline{d}| \geq 2$ and a **(permuted) $S_{1/2}$-corolla** similarly must satisfy $|\underline{c}||\underline{d}| \geq 2$.

DEFINITION 7.63.

- Let $\mathcal{T}^{1/2}$ consist of all permuted $S_{1/2}$-corollas and the half graphs $C^{1,i}_{(\underline{a};c_i;\underline{c};\underline{d})}$ and $C^{j,1}_{(\underline{a};\underline{b};b_j;\underline{d})}$ from Example 2.24.
- Let $\mathcal{W}^{1/2}$ consist of moves coming from the special cases of
  - Lemma 6.103 with $l = 1 = j$ or with $m = 1 = i$,
  - Lemma 6.105 with $l = n = 1 = j = s$ or $m = p = 1 = i = r$ or $l = p = 1 = j = r$,
  - Lemma 6.106 with $l = n = 1 = j = s$, and
  - Lemma 6.107 with $m = p = 1 = i = r$.

The results here follow as in the previous section, where the strong generating condition is proved by a minor modification of the proof of Proposition 7.36, which dealt with $\mathcal{T}^{Tree}$. Instead of permuted corollas with one output, here we have permuted $S_{1/2}$-corollas. The roles of the simple trees are now played by the half-graphs in $\mathcal{T}^{1/2}$, which build internal edges one at a time.

THEOREM 7.64. *The set $\mathcal{T}^{1/2}$ is a 3-strong generating set for the graph groupoid $\mathtt{Gr}_{\frac{1}{2}}$ of half-graphs.*

PROPOSITION 7.65. *The set of moves $\mathcal{W}^{1/2}$ is 3-strong in $\mathcal{T}^{1/2}$.*

7.9. Strong Generating Sets for Connected Wheel-Free Graphs

Here we provide two strong generating sets for the graph groupoid $\mathtt{Gr}^{\uparrow}_{\mathtt{c}}$ of connected wheel-free graphs. As in several of the previous cases, we would like to work inductively to insert internal edges. However, here we do not have a contraction operation from graph substitution, so need to insert all internal edges between a fixed pair of vertices at once. As a consequence, we work with some version of a partially grafted corollas as our basic building block in both cases.

DEFINITION 7.66.
- Let $\mathcal{T}^{\uparrow}_{\mathtt{c}}$ consist of all permuted corollas, the exceptional edge of each color, and the partially grafted corollas.
- Let $\mathcal{W}^{\uparrow}_{\mathtt{c}}$ consist of the relaxed 4-moves from Lemmas 6.26, 6.102, 6.60, 6.63, 6.66, 6.68, 6.70, 6.72, and 6.74, as well as Remark 6.75.

THEOREM 7.67. *The set $\mathcal{T}^{\uparrow}_{c}$ is a 4-strong generating set for the graph groupoid $\mathtt{Gr}^{\uparrow}_{c}$.*

PROPOSITION 7.68. *The set of moves $\mathcal{W}^{\uparrow}_{c}$ is 4-strong in $\mathcal{T}^{\uparrow}_{c}$.*

Once again we exploit a stratified presentation.

DEFINITION 7.69. Given a finite set of connected wheel-free graphs $\{G_\alpha\}$, a **stratified presentation** is the choice of a presentation of $\{G_\alpha\}$ by a graph n-simplex $\mathcal{H}$ such that:
- $\mathcal{H}^1$ consists of permuted corollas and exceptional edges,
- each layer of $\mathcal{H}^{[2,n]}$ consists of permuted corollas and partially grafted corollas.

To see that each connected wheel-free graph has a stratified presentation, we will work inductively via the following concept.

DEFINITION 7.70. Suppose u and v are two different vertices in a wheel-free graph G. Then u and v are called **closest neighbors** if the following two conditions hold:
- There is at least one internal edge connecting them.
- There are no directed paths beginning at one of them and ending at the other that involve a third vertex.

In this case, we also call v a closest neighbor of u.

EXAMPLE 7.71.
- In a simply-connected graph, two vertices are closest neighbors precisely when they are connected by an internal edge.
- In the connected wheel-free graph

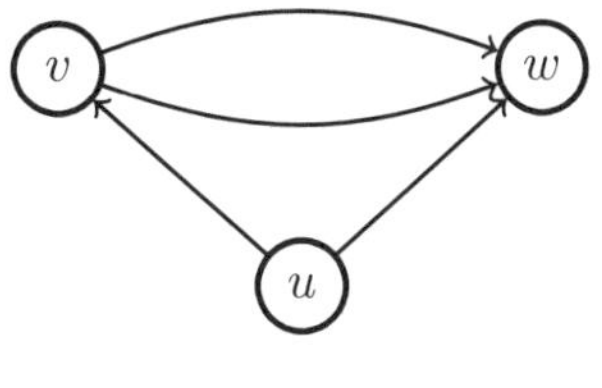

the vertices u and v are closest neighbors, as are v and w. On the other hand, the vertices u and w are not closest neightbors.

REMARK 7.72. In [**Val07**, Sec. 4], the author introduced a very similar, possibly equivalent, notion he referred to as "a pair of adjacent vertices".

It will be helpful to introduce the following generalization of the corolla determined by a vertex.

EXAMPLE 7.73. Given a pair of closest neighbors u and v in a connected wheel-free graph G, there is an associated relabeling of a partially grafted corollas. First, observe that there cannot be edges between u and v with both orientations, due to the wheel-free assumption. Thus, without loss of generality, say $u \rightrightarrows v$ and notice this defines a partially grafted corollas with internal edges corresponding to the edges from u to v. Unfortunately, this may have the wrong vertex listings, since there is no reason to expect that all of the flags to be grafted will be in a single sequential grouping. In order to understand the full graph listings of the graph we are constructing, we simply view it as $C_v\tau \boxtimes \sigma C_u$. Denote this vertex relabeling of a partially grafted corollas by $C_{u,v}$.

The real usefulness of this closest neighbors concept lies with the following observation, which will serve as the induction step in our proof below. This really shows that the process of building a simply-connected graph up from a corolla by each time substituting a basic dioperadic graph to insert a single internal edge has an analogue for connected wheel-free graphs.

LEMMA 7.74. *Suppose u and v are closest neighbors in $G \in \mathtt{Gr}^{\uparrow}_{c}$. Then there is a decomposition $G \cong K(C_{u,v})$ where $C_{u,v}$ is the vertex relabeling of a partially grafted corollas of Example 7.73 and $K \in \mathtt{Gr}^{\uparrow}_{c}$ has one fewer vertex than G.*

PROOF. To define K, we start with the definition of G, remove the flags corresponding to edges from u to v, and then take the union of the remaining flags in the two cells corresponding to u and v to define a single vertex $u' \in K$. The fact that K is a wheeled graph with one fewer vertex than G is immediate from the definitions, and the graph substitution statement follows easily since no exceptional pieces are involved. Thus, it suffices to observe that $K \in \mathtt{Gr}^{\uparrow}_{c}$, since up to choices of edges from u to v, there is a correspondence between (directed) paths in K and those of the same type in G. □

We will need the following observation to show that each connected wheel-free graph has a stratified presentation.

LEMMA 7.75. *Suppose G is a connected wheel-free graph with at least two vertices. Then each vertex has a closest neighbor.*

PROOF. Pick a vertex v in G. Since G is connected wheel-free and has at least one other vertex, there exists a vertex u_0 and an internal edge between v and u_0. By the ability to decompose paths into directed segments and symmetry of the proof, we may assume that there exists a directed path from u_0 to v. If they are not closest neighbors, then among all the directed paths from u_0 to v involving at least one other vertex, choose P to be one with the greatest length (i.e., the greatest number of internal edges), which must be finite. Let u_1 be the next-to-last vertex in P, so by construction there is an internal edge from u_1 to v.

If u_1 and v are still not closest neighbors, then there exists a directed path Q from u_1 to v that involves a third vertex w. If Q were to also include any of the earlier vertices in P, there would be a wheel in G. That implies we can form the concatenation of the portion of P from u_0 to u_1 with Q to produce a new directed path from u_0 to v which is strictly longer than P, a contradiction. As a consequence, u_1 and v are closest neighbors. □

REMARK 7.76. In the previous lemma, the proof actually provides an algorithm for finding a closest neighbor of a chosen vertex.

LEMMA 7.77. *Every finite set of connected wheel-free graphs has a stratified presentation.*

PROOF. It suffices to consider a single connected wheel-free graph G with n vertices. If $n = 0$, then G is an exceptional edge of a single color. If $n = 1$, then G is a permuted corolla. If $n = 2$, then up to relabeling G is a partially grafted corolla. For $n > 2$, we proceed by induction, with the induction step just the combination of Lemma 7.75 and Lemma 7.74, keeping in mind that $C_{u,v} = H(C_v\tau, \sigma C_u)$ for a partially grafted corollas H (and an exceptional edge has no vertices). □

LEMMA 7.78. *Every presentation in $\mathcal{T}_c^\uparrow$ of a finite set of connected wheel-free graphs $\{G_\alpha\}$ is connected to a stratified presentation of $\{G_\alpha\}$ by a finite sequence of relaxed 2-moves in $\mathcal{W}_c^\uparrow$.*

PROOF. As ever, we push exceptional edges to $\mathcal{H}^1$ with relaxed 2-moves. If necessary, corollas can be inserted to keep the partially grafted corollas out of $\mathcal{H}^1$ by the unit property, which is a relaxed 2-move. □

For the final lemma, a few words are in order about the uniqueness which plays an important role in these proofs. First, the collection of $C_{u,v}$ which could appear in Lemma 7.74 are determined by G, and taken together they determine G up to relabeling and the order in which they are inserted. Notice that all of the edges between a pair of vertices must be inserted in a single step, since contraction, generated by contracted corollas which are excluded from this pasting scheme, is the only operation to connect edges within a single graph. Of course, trying to use pasting through several steps involving different vertices is also not allowed, since then there would be an intermediate vertex 'in the way'. As a consequence, the number of partially grafted corollas, along with the lengths of the segments used for pasting, in any stratified presentation are determined by G itself, at least once extraneous exceptional edges are eliminated.

Thus, the fact that in this case we cannot use simple moves to get all permuted corollas 'out of the way' is the only real impediment to proceeding as we did when building simply-connected graphs from basic dioperadic graphs, only this time using partially grafted corollas. The real concern is that a segment to be grafted can be spread into several different components by a permutation, but since the grafting must all be performed at once, the relevant pieces of such a segment will all then be brought back together before that grafting is applied in any presentation of a fixed G.

LEMMA 7.79. *Any two stratified presentations of the same finite set of connected wheel-free graphs $\{G_\alpha\}$ can be connected by a finite sequence of relaxed 4-moves in $\mathcal{W}_c^\uparrow$.*

PROOF. As usual, we begin by eliminating extraneous exceptional edges with relaxed 2-moves by Lemma 6.102. We can once again compare an arbitrary stratified presentation (with extraneous exceptional edges removed) $\mathcal{H}$ of a graph G to another presentation $\mathcal{K}$ of the form presented in the proof of Lemma 7.77.

The vertex profiles and exceptional edges in $\mathcal{H}^1$ are determined by G, so also agree with those of $\mathcal{K}^1$, although there may be non-trivially permuted corollas in

$\mathcal{H}^1$. As noted above, any such permutations which separate later segments into non-sequential pieces will be re-collected at some point before the relevant grafting occurs. Otherwise, these permutations are either inner or outer permutations when viewed by the graph into which they are substituted, so can be manipulated using 2-moves by Lemmas 6.60 and 6.63.

We would like to use the relaxed 3-moves of Lemmas 6.66, 6.68, 6.70, and 6.72 to alter the order with which the required partial graftings are performed. However, in this case, we may require the relaxed 4-moves of Lemma 6.74 and Remark 6.75 as well, since there may be a permuted corolla between the two partially grafted corollas whose order we would like to interchange. Once again, as necessary we can use Lemmas 6.26, 6.60, and 6.63 to also push the permuted corollas into the expected positions as well. □

7.9.1. An Alternative Strong Generating Set for Connected Wheel-Free Graphs. If we instead work with basic pieces of the form

$$C_{(\underline{c};\underline{d})}\tau \boxtimes^{\underline{c}'}_{\underline{b}'} \sigma C_{(\underline{a};\underline{b})}$$

then we instead have a 3-strong generating set with a single type of generating graph beyond the permuted corollas.

DEFINITION 7.80.
- A **basic properadic graph** will refer to a graph of the form
$$C_{(\underline{c};\underline{d})}\tau \boxtimes^{\underline{c}'}_{\underline{b}'} \sigma C_{(\underline{a};\underline{b})}$$
with $\underline{b}' \subset \sigma\underline{b}$ and $\underline{c}' \subset \underline{c}\tau$ matching non-empty segments.
- Let $\mathcal{T}^{\uparrow}_{c,2}$ consist of all permuted corollas, the exceptional edge of each color, and all basic properadic graphs.
- Let $\mathcal{W}^{\uparrow}_{c,2}$ consist of the relaxed 3-moves from Lemmas 6.26, 6.68, 6.70, 6.76, 6.77, and Remark 6.78, along with the relaxed 2-moves given by applying Cor. 6.58 with $G = C_{(\underline{c};\underline{d})}\tau$ or $G = \sigma C_{(\underline{c};\underline{d})}$.

THEOREM 7.81. *The set $\mathcal{T}^{\uparrow}_{c,2}$ is a 3-strong generating set for the graph groupoid $\mathtt{Gr}^{\uparrow}_{c}$.*

PROPOSITION 7.82. *The set of moves $\mathcal{W}^{\uparrow}_{c,2}$ is 3-strong in $\mathcal{T}^{\uparrow}_{c,2}$.*

We now have the associated notion of a stratified presentation.

DEFINITION 7.83. Given a finite set of connected wheel-free graphs $\{G_\alpha\}$, an **alternative stratified presentation** is the choice of a presentation of $\{G_\alpha\}$ by a graph n-simplex $\mathcal{H}$ such that:

- $\mathcal{H}^1$ consists of corollas and exceptional edges,
- each layer of $\mathcal{H}^{[2,n-1]}$ consists of basic properadic graphs, and
- $\mathcal{H}^n$ consists of permuted corollas.

LEMMA 7.84. *Every finite set of connected wheel-free graphs has an alternative stratified presentation.*

PROOF. Proceed by induction using Lemmas 7.74 and 7.75. □

LEMMA 7.85. *Every presentation in $\mathcal{T}^{\uparrow}_{c,2}$ of a finite set of connected wheel-free graphs $\{G_\alpha\}$ is connected to an alternative stratified presentation of $\{G_\alpha\}$ by a finite sequence of relaxed 2-moves in $\mathcal{W}^{\uparrow}_{c,2}$.*

PROOF. If necessary, insert a layer of corollas as $\mathcal{H}^1$ using the relaxed 2-move from the identity property of corollas. Also, use relaxed 2-moves to push exceptional edges to $\mathcal{H}^1$. Finally, using Lemmas 6.26 and 6.76 allows one to push the permuted corollas to $\mathcal{H}^n$. □

LEMMA 7.86. *Any two alternative stratified presentations of the same finite set of connected wheel-free graphs $\{G_\alpha\}$ can be connected by a finite sequence of relaxed 3-moves in $\mathcal{W}^{\uparrow}_{c,2}$.*

PROOF. First, using Lemma 6.57 we eliminate any extraneous exceptional edges. Then the collection of basic properadic graphs involved is determined by the pairs of closest neighbors in the connected wheel-free graph G being presented, except for the order of inserting them. However, that order is not an obstruction to connecting the two presentations by using the relaxed 3-moves of Lemma 6.76. Keep in mind that when either of the lower index partial graftings is over 0-segments, we have a relabeled lighthouse or fireworks graph, so we can instead use the relaxed 3-moves of Lemmas 6.68 or 6.70 to rewrite the presentation in either order without using (disjoint) unions. □

7.10. Strong Generating Set for Connected Wheeled Graphs

Here we provide a strong generating set for the graph groupoid $\mathtt{Gr}_c^{\circlearrowright}$ of connected wheeled graphs.

DEFINITION 7.87.
- Let $\mathcal{T}_c^{\circlearrowright}$ consist of all permuted corollas, the exceptional edge of each color, the contracted corollas, and the basic dioperadic graphs.
- Let $\mathcal{W}_c^{\circlearrowright}$ consist of the relaxed moves from Lemmas 6.26, 6.38, 6.40, 6.79, 6.82, 6.84, 6.86, 6.102, 6.103, 6.105, 6.106, and 6.107.

THEOREM 7.88. *The set $\mathcal{T}_c^{\circlearrowright}$ is a 3-strong generating set for the graph groupoid $\mathtt{Gr}_c^{\circlearrowright}$ of connected wheeled graphs.*

PROPOSITION 7.89. *The set of moves $\mathcal{W}_c^{\circlearrowright}$ is 3-strong in $\mathcal{T}_c^{\circlearrowright}$.*

As before the theorem follows by exploiting a stratified presentation.

DEFINITION 7.90. Given a finite set of connected wheeled graphs $\{G_\alpha\}$, a **stratified presentation** is the choice of a presentation of $\{G_\alpha\}$ by a graph n-simplex $\mathcal{H}$ such that:
- $\mathcal{H}^1$ consists of corollas and exceptional edges,
- each layer of $\mathcal{H}^{[2,i]}$ consists of basic dioperadic graphs and corollas for some i,
- each layer of $\mathcal{H}^{[i+1,n-1]}$ consists of contracted corollas and corollas, and
- $\mathcal{H}^n$ consists of permuted corollas.

Let $\mathrm{sc}\mathcal{H} = \mathrm{sub}\left(\mathcal{H}^{[1,i]}\right)$ denote the **simply-connected portion** of this presentation.

The reader should keep in mind that $\mathrm{sc}\mathcal{H}$ must be simply-connected by Proposition 6.22. In addition, the contracted corollas and permuted corollas in the later layers will not alter the set of vertices. As a consequence, for a single $G \neq \uparrow_c$, $\mathrm{sc}\mathcal{H}$ must contain all of the vertices of G and so have $|\mathrm{Vt}(G)| - 1$ internal edges as all simply-connected graphs with this set of vertices do. In addition, $\mathcal{H}^i$ must consist of a single basic dioperadic graph as a consequence.

LEMMA 7.91. *Every finite set of connected wheeled graphs has a stratified presentation.*

PROOF. As usual, it suffices to consider a single connected wheeled graph G. It is clear that both $\uparrow_c$ and $\circlearrowleft_c = \xi_1^1 \uparrow_c$ have stratified presentations, so we focus on ordinary G and assume G has m internal edges. If G is simply-connected, then we already know that it has such a stratified presentation, which does not involve contracted corollas, by Lemma 7.60. Next we consider the cases with few internal edges, noticing if $m = 0$, then G is a permuted corolla. If $m = 1$, then G is either a contracted corolla or a basic dioperadic graph up to input and output relabeling. We now proceed by induction, so we assume $m > 1$ and G is not simply connected.

Since G is not simply connected, it must contain a cycle, so we can pick a single internal edge e in a cycle of G. Thus, by altering G to create a new connected wheeled graph H with the edge e disconnected, as in the proof of Lemma 7.50, we can write

$$G = \xi_j^i H.$$

However, this implies H has $m - 1$ internal edges, so has a stratified presentation by the induction hypothesis (or the simply connected case). Now, by the 2-move in Lemma 6.38, it follows that G also has a stratified presentation. □

REMARK 7.92. Let us explain the geometric meaning of the stratified presentation above. Suppose G is an ordinary connected wheeled graph. If it has a cycle P, then disconnect one of the internal edges in P. The resulting graph G_1 is still connected, and if G_1 has no cycles, then it is simply-connected, so we apply Lemma 7.60. Otherwise, we choose a cycle in G_1 and disconnect one of its internal edges to obtain a connected graph G_2, and so on. Regardless, at some finite stage we have written (a relabeling of) G as an iterated contraction of a simply-connected graph, to which we can apply Lemma 7.60, where we built the internal edges one at a time using substitutions of basic dioperadic graphs.

LEMMA 7.93. *Every presentation in $\mathcal{T}_c^{\circlearrowleft}$ of a finite set of connected wheeled graphs $\{G_\alpha\}$ is connected to a stratified presentation of $\{G_\alpha\}$ by a finite sequence of relaxed 2-moves in $\mathcal{W}_c^{\circlearrowleft}$.*

PROOF. As usual, we can use relaxed 2-moves to push all exceptional edges to $\mathcal{H}^1$. Now Lemmas 6.38 and 6.103 allow us to use 2-moves to push the permuted corollas to $\mathcal{H}^n$, where they can be combined as a single layer using the relaxed 2-moves of Lemma 6.26.

At this point, it suffices to apply Lemmas 6.84 and 6.86, which provide the relaxed 2-moves necessary to increase the indices of the contracted corollas beyond those of the basic dioperadic graphs. □

In this case, the now familiar question of connecting two specific presentations is particularly subtle, and requires a bit of preparation. First, we have a definition and result about simply-connected graphs which will be applied to the simply-connected portions of stratified presentations below. Basically, any simply-connected ordinary graph splits into two disjoint pieces by cutting any edge.

DEFINITION 7.94. Given a simply-connected graph H and an edge e from u to v in H, say a vertex x of H is e-closer to u if the unique internal path from x to u does not include e.

LEMMA 7.95. *Given a simply-connected graph H and an edge e from u to v in H, there are two disjoint, simply-connected graphs determined by the vertices e-closer to u or to v together with the remaining structure inherited from H.*

PROOF. The uniqueness of internal paths connecting two vertices in a simply-connected graph implies the two collections of vertices are disjoint. Even more, the edge e is the unique connection between the two types of vertices, again by simple-connectivity of H. In order to specify two graphs, we must declare the listing of the new input or output to be the first, with all other related input or output listings coming from H incremented by one. The fact that each subset produces a simply-connected graph follows from the inclusion of paths and this unique connecting edge property. □

DEFINITION 7.96. If $\mathcal{H}$ and $\mathcal{H}'$ are both stratified presentations of the same connected G, say they have **presentation distance** k if $\mathrm{sc}\mathcal{H}$ and $\mathrm{sc}\mathcal{H}'$ share $|\mathrm{Vt}(G)| - k - 1$ internal edges.

With the number of internal edges fixed, presentation distance k can instead be described by $\mathrm{sc}\mathcal{H}'$ contains k internal edges which are not in $\mathrm{sc}\mathcal{H}$ (and vice-versa). Then presentation distance 0 means $\mathrm{sc}\mathcal{H}$ and $\mathrm{sc}\mathcal{H}'$ are weakly isomorphic, and we will be focused on decreasing presentation distance. The basic approach is to carefully choose a pair of edges, and remove the basic dioperadic graph which creates one in $\mathcal{H}'$ while replacing it with the basic dioperadic graph which creates the other in $\mathcal{H}$.

LEMMA 7.97. *Suppose $\mathcal{H}$ and $\mathcal{H}'$ have presentation distance $k > 1$, while e' is an edge of G which is in $\mathcal{H}'$ but not in $\mathcal{H}$. Then there exists a third stratified presentation $\mathcal{H}''$ of G with presentation distance 1 from $\mathcal{H}'$ and presentation distance $k - 1$ from $\mathcal{H}$.*

PROOF. First, decompose $\mathrm{sc}\mathcal{H}'$ into the two pieces of Lemma 7.95 based on the edge e'. Since $\mathrm{sc}\mathcal{H}$ is connected, with the same set of vertices, there exists an edge e in $\mathcal{H}$ between the two types of vertices, which then cannot be an edge of $\mathrm{sc}\mathcal{H}'$. Applying the relaxed 3-moves of Lemmas 6.105, 6.106, and 6.107, we can assume e' is constructed by the unique basic dioperadic graph H' in $(\mathcal{H}')^i$ and e is constructed by the unique basic dioperadic graph H in $\mathcal{H}^i$. Replacing H' with H in $\mathcal{H}'$ then leads to a new stratified presentation of G, since the resulting graph remains simply-connected by e being the unique edge connecting the two pieces of Lemma 7.95. Even more, the new presentation has presentation distance 1 from $\mathcal{H}'$ and presentation distance $k - 1$ from $\mathcal{H}$ by construction. □

LEMMA 7.98. *Any two stratified presentations of the same finite set of connected wheeled graphs $\{G_\alpha\}$ can be connected by a finite sequence of relaxed 3-moves in $\mathcal{W}_c^{\mathbb{Q}}$.*

PROOF. First notice, without loss of generality, we may assume G is ordinary. Now if the simply-connected portions of $\mathcal{H}$ and $\mathcal{H}'$ agree, then we may apply Lemma 7.62 to connect the presentations up to layer i using relaxed 3-moves in $\mathcal{W}^{di} \subset \mathcal{W}_c^Q$, and the graph G being presented now determines the set of contractions and relabeling to be applied, although not their order. Hence, since contractions are nearly commutative and bi-equivariant, by the relaxed 2-moves of Lemmas 6.38 and 6.40, the original presentations were connected by relaxed 3-moves in $\mathcal{W}_c^Q$. This completes the case of presentation distance 0.

Once we handle the case of presentation distance 1, the result follows by induction using Lemma 7.97. Notice, we cannot simply start our induction with the case $k = 0$ because one piece of this Lemma is always $k = 1$.

For the presentation distance 1 case, choose the unique internal edge e' in $\mathrm{sc}\mathcal{H}'$ which is not in $\mathrm{sc}\mathcal{H}$. Then this edge is created by a contraction in $\mathcal{H}$, but via a partial grafting through a basic dioperadic graph in $\mathcal{H}'$. Similarly, there is a unique internal edge e in $\mathrm{sc}\mathcal{H}$ which is not in $\mathrm{sc}\mathcal{H}'$, so e must be constructed by a contraction in $\mathcal{H}'$. Applying relaxed 3-moves both above and below i, we may assume the basic dioperadic graphs constructing e and e' in $\mathrm{sc}\mathcal{H}$ and $\mathrm{sc}\mathcal{H}'$ respectively are the unique basic dioperadic graphs in layer i, while the contractions creating the two in the opposite context are the only non-corollas in layer $i+1$. Now let $J = \mathrm{sub}\left(\mathcal{H}^{[i,i+1]}\right)$ and define J' similarly using $\mathcal{H}'$.

We now know J and J' each consist of a single contraction of a basic dioperadic graph. Notice that neither e nor e' can be a loop at a single vertex, as edges in simply-connected graphs. As a consequence, J and J' cannot be basic dioperadic graphs with a loop at either vertex, hence must be either snowboard graphs or little green man graphs, depending on whether the edges e and e' have different or the same orientations. If e and e' share an orientation, then both are little green man graphs, and the two ways of constructing such are connected by the relaxed 3-move of Lemma 6.79. Similarly, if the orientations differ, both are snowboard graphs, so they are connected by the relaxed 3-move of Lemma 6.82. Hence, we have completed a direct connection between the two presentations up to layer $i+1$, at which point we proceed as in the presentation distance 0 case to compare the remaining contractions and final relabeling. □

CHAPTER 8

Pasting Schemes

In this chapter, we finally define pasting schemes, which one uses to parametrize the various operations in generalized PROPs. Each pasting scheme consists of a graph groupoid that is closed under graph substitution and contains the units of graph substitution. Slightly more is required to define a (vertically) unital pasting scheme, which must contain input and output extensions as well as certain exceptional edges, keeping in mind that exceptional edges are the units of grafting. Once again, a large list of examples is included and a diagram of their containments is included for the convenience of the reader.

The second section in this chapter is devoted to the study of free products of pasting schemes. It is not difficult to see that the union of generating sets for two pasting schemes gives a generating set for the free product pasting scheme. However, it is technically difficult to verify a similar result for strong generating sets, and the technical details are all included. The goal is to set the stage for later free product decompostion results. Two special examples are considered throughout. One is a decomposition of $\mathtt{Gr}^{\uparrow}$ using the vertical and horizontal combinations pasting schemes, while the second is a decomposition of $\mathtt{Gr}_{\mathrm{w}}^{\circlearrowleft}$ using the contractions and horizontal combinations pasting schemes.

The chapter ends by setting the stage for a later result, that when the pasting scheme is "monogenic", i.e. each graph has a single vertex, then the associated generalized PROPs will be simply a diagram category. Here the focus is on producing the small category associated to a monogenic pasting scheme, essentially by viewing any monogenic graph as built from our basic operations on a corolla. There is also a brief discussion of what we call a virtual pasting scheme, which generalizes this small category to an appropriate multicategory, a concept which is discussed further in Chapter 14.

8.1. Definitions and First Examples

Here we will present the long-awaited definition of a pasting scheme, together with an extensive list of examples and a diagram of the relationships between them. Together with an appropriate graph groupoid, we will consider a sub-groupoid $S \subset \mathcal{P}(\mathfrak{C})^{op} \times \mathcal{P}(\mathfrak{C})$ of the pairs of profiles. The point is to allow us to be explicit about considering different underlying notions of bimodules as part of the structure, but for most of the examples S is simply all of $\mathcal{P}(\mathfrak{C})^{op} \times \mathcal{P}(\mathfrak{C})$. For example, when considering operads, one only wants to consider pairs of profiles where there is a unique output, while PROPs are defined on arbitrary pairs of profiles. As a consequence, the ability to compare the two, in order to see that a PROP has an underlying operad structure, depends on restricting both the profiles under consideration and the groupoids of graphs involved.

8.1.1. Definition of a Pasting Scheme. Here we define the pasting schemes for our generalized PROPs. Recall a subcategory is called **replete** in a larger category if it contains any object isomorphic (in the larger category) to an object of the subcategory. In what follows, our replete subcategory $S \subset \mathcal{P}(\mathfrak{C})^{op} \times \mathcal{P}(\mathfrak{C})$ will often be the entire category. Recall $\mathtt{Gr}^{\circlearrowleft}_{\mathrm{w}}$ denotes the groupoid of wheeled graphs and weak isomorphisms.

If $\mathtt{G} \subseteq \mathtt{Gr}^{\circlearrowleft}_{\mathrm{w}}$ is a replete subgroupoid, let $\mathtt{G}\binom{\underline{d}}{\underline{c}}$ denote the groupoid of $(\underline{c};\underline{d})$-wheeled graphs contained in $\mathtt{G}$ and strict isomorphisms, as weak isomorphisms could alter the pair of profiles for the full graph. Notice that for any morphism $(\tau;\sigma)$ of $\mathcal{P}(\mathfrak{C})^{op} \times \mathcal{P}(\mathfrak{C})$, the induced weak isomorphism given by input and output relabeling will also be called

$$(\tau;\sigma) \colon \mathtt{G}\binom{\underline{d}}{\underline{c}} \xrightarrow{\cong} \mathtt{G}\binom{\sigma\underline{d}}{\underline{c}\tau}, \quad G \longrightarrow \sigma G \tau.$$

Next we have a bit of notation for the property of a graph groupoid to be closed under graph substitution.

DEFINITION 8.1. Let $\mathtt{G}_S$ denote the replete and full sub-groupoid of $\mathtt{G}$ consisting of wheeled graphs whose vertices v all satisfy $\binom{\mathrm{out}(v)}{\mathrm{in}(v)} \in S$. We call $(S, \mathtt{G})$ **hereditary** if given any $G \in \mathtt{G}_S$ and collection $H_v \in \mathtt{G}\binom{\mathrm{out}(v)}{\mathrm{in}(v)}$ for $v \in \mathrm{Vt}(G)$, the graph substitution $G(H_v)$ also lies in $\mathtt{G}$.

DEFINITION 8.2. A **$\mathfrak{C}$-colored pasting scheme**, or simply a **pasting scheme**, is a pair

$$\mathcal{G} = (S, \mathtt{G})$$

in which

(1) S is a replete and full subcategory of $\mathcal{P}(\mathfrak{C})^{op} \times \mathcal{P}(\mathfrak{C})$, and
(2) $\mathtt{G}$ is a replete and full sub-groupoid of $\mathtt{Gr}^{\circlearrowleft}_{\mathrm{w}}$

such that:

(1) $\mathtt{G}$ is hereditary, and
(2) $\mathtt{G}$ contains all the $(\underline{c};\underline{d})$-corollas for all pairs of profiles $(\underline{c};\underline{d}) \in S$.

Notice that being closed under weak isomorphism and containing all corollas, $\mathtt{G}$ must also contain all permuted corollas. Since graph substitution with permuted corollas produces input and output relabeling, $\mathtt{G}$ must then be closed under this operation. In fact, whenever the representing graphs are contained in $\mathtt{G}$, it must be closed under all of the operations of section 6.1. In essence, this definition is the usual model of defining an algebraic construction as a collection with both an operation and a unit, taken fairly close to its logical extreme.

Our convention will be that whenever S is not specified, it will be the entire category $\mathcal{P}(\mathfrak{C})^{op} \times \mathcal{P}(\mathfrak{C})$. If S is clear from the context, we will sometimes denote a pasting scheme $(S, \mathtt{G})$ by $\mathtt{G}$.

8.1.2. Partial Ordering on Pasting Schemes.

DEFINITION 8.3. Define a partial ordering on the pasting schemes by declaring

$$\mathcal{G} \leq \mathcal{G}'$$

whenever there is a containment of subcategories in each variable.

Later we will be using pasting schemes to define our operational structures, and this containment will lead to a forgetful functor, for which we will also produce a left adjoint in general.

It will also be useful later that there are only a set of pasting schemes, which follows from the fact that the isomorphism classes of graphs form a set and that the larger category where S resides is a small category. Thus, we have what is commonly referred to as a poset of pasting schemes using this partial ordering.

8.1.3. Restriction on S. If $S' \subseteq S$, and $(S, \mathtt{G})$ is a pasting scheme, then $(S', \mathtt{G})$ is also a pasting scheme. Here are some examples of pasting schemes that can be obtained this way.

(1) Given any subset $\mathfrak{C}'$ of the set $\mathfrak{C}$ of colors, one can choose S' to consist of the pairs of profiles involving only those colors, which will lead to restriction of colors.
(2) Let $S_{\leq n}$ and S_n be the subcategories of pairs of profiles $\binom{\underline{d}}{\underline{c}}$ with $|\underline{d}| \leq n$ and $|\underline{d}| = n$, respectively. The usual notion of operads lands in S_1. Wheeled operads use $S_{\leq 1}$.

8.1.4. First Examples of Pasting Schemes.

NOTATION 8.4. Here we give a series of simple examples of pasting schemes from among the graph groupoids of Notation 4.26. All but the last are compatible with any choice of S.

- **Minimal Pasting Scheme:** For each replete and full subcategory S of $\mathcal{P}(\mathfrak{C})^{op} \times \mathcal{P}(\mathfrak{C})$, let
$$Min(S) = (S, \mathtt{Cor})$$
be the pasting scheme in which $\mathtt{Cor}$ contains only the permuted corollas (see 4.26). With respect to the partial ordering $\leq$ on pasting schemes (Definition 8.3), $Min(S)$ is the smallest pasting scheme corresponding to S.
- **Maximal Pasting Scheme:** The groupoid $\mathtt{Gr}_{\mathrm{w}}^{\circlearrowright}$ of all $\mathfrak{C}$-colored wheeled graphs, together with any choice of S, also forms a (unital) pasting scheme, and will correspond to the maximal element of our partial ordering with $\mathfrak{C}$ and S fixed. (In Section 1.4, a few more general constructions are mentioned which might lead to different maximal elements of interest in modified technical conditions.)
- **Contractions Pasting Scheme:** The pasting scheme $\mathtt{Gr}^{\xi}$ consists of corollas and (repeatedly) contracted corollas (see 4.26), together with any choice of S. Thus, all objects have a single vertex, and may be constructed by the graph contraction operation applied to a corolla some finite number of times. As such, it has the hereditary property, since the graph contraction operation is represented by performing a graph substitution into a contracted corolla.
- **The Horizontal Pasting Scheme:** Recall the graph groupoid of finite unions of corollas $\mathtt{Gr}_{\mathrm{h}}^{\uparrow}$ (see 4.26), which is a pasting scheme with any choice of S since the union operation is generated by graph substitution into finite unions of corollas.

- **The Vertical Pasting Scheme:** Recall (see 4.26) $\mathtt{Gr}^{\uparrow}_{\mathrm{v}}$ is the full sub-groupoid of $\mathtt{Gr}^{\uparrow}_{\mathrm{c}}$ (the groupoid of connected, wheel-free graphs and weak isomorphisms) consisting of the iterated graftings of corollas. Then $\mathtt{Gr}^{\uparrow}_{\mathrm{v}}$ with any choice of S is a pasting scheme, since the grafting operation is generated by graph substitution into grafted corollas.
- **The Linear Pasting Scheme:** Recall (see 4.26) $\mathtt{Lin}$ is the full sub-groupoid of $\mathtt{Gr}^{\uparrow}_{\mathrm{c}}$ consisting of iterated grafting of corollas with one input and one output. Let $S_{1,1}$ be the full-subcategory of $\mathcal{P}(\mathfrak{C})^{op} \times \mathcal{P}(\mathfrak{C})$ of pairs $(c; d)$, where each entry is a profile of length 1. Then $(S_{1,1}, \mathtt{Lin})$ is a pasting scheme since graph substitution into grafted corollas of $S_{1,1}$-type generates grafting in this context.

8.1.5. Unital Pasting Schemes.

DEFINITION 8.5. Suppose S is a sub-category of $\mathcal{P}(\mathfrak{C})^{op} \times \mathcal{P}(\mathfrak{C})$ and $c \in \mathfrak{C}$. With a slight abuse of notation, we define $c \in S$ to mean that there exist profiles $\underline{d} = (d_1, \ldots, d_n)$ and $\underline{c} = (c_1, \ldots, c_m)$ such that

(1) $c = c_i$ for some $i \in \{1, \ldots, m\}$ and
(2) either $(\underline{c}; \underline{d}) \in S$ or $(\underline{d}; \underline{c}) \in S$.

In other words, $c \in S$ if c is an entry in some profile that appears in S. If $S = \mathcal{P}(\mathfrak{C})^{op} \times \mathcal{P}(\mathfrak{C})$, then $c \in S$ for all $c \in \mathfrak{C}$.

DEFINITION 8.6. A **unital pasting scheme** is defined as a pasting scheme $\mathcal{G} = (S, \mathtt{G})$ that satisfies the following two additional conditions:

(1) If $(\underline{c}; \underline{d}) \in S$ and $C = C_{(\underline{c};\underline{d})}$ is the $(\underline{c}; \underline{d})$-corolla (see 1.3.1), then the input extension C_{in} and the output extension C_{out} (see 6.1.7) are both in $\mathtt{G}$.
(2) If $c \in S$ for some $c \in \mathfrak{C}$, then $(c; c) \in S$ and $\uparrow_c \in \mathtt{G}$.

The maximal pasting scheme listed above is unital, but the minimal, contractions, horizontal, vertical, and linear pasting schemes are non-unital. However, the following provides a unital version of the linear pasting scheme and similar extensions are always possible (see Lemma 8.14).

EXAMPLE 8.7. Let $\mathtt{ULin}$ be the full sub-groupoid of $\mathtt{Gr}^{\uparrow}_{\mathrm{c}}$ consisting of linear graphs and the exceptional graphs $\uparrow_c$ for $c \in \mathfrak{C}$. Then $(S_{1,1}, \mathtt{ULin})$ is a unital pasting scheme which contains $\mathtt{Lin}$. As a consequence, we see that being contained in a unital pasting scheme is *not* sufficient to imply the smaller pasting scheme is unital.

The fact that restriction of colors preserves unitality is immediate from the definitions.

LEMMA 8.8. *Suppose that $\mathfrak{C}' \subseteq \mathfrak{C}$ and that S' and $\mathtt{G}'$ are the restrictions of S and $\mathtt{G}$, respectively, to the colors in $\mathfrak{C}'$ as in* 8.1.3(1). *Then $\mathcal{G}'$ is also a unital pasting scheme.*

8.1.6. Key Examples of Pasting Schemes. Expanding upon the collection of graph groupoids from 4.26, we can give more interesting examples of pasting schemes. In many cases, they are independent of the choice of S, so we will separate these from the cases where a specific choice of S (or smaller) is vital.

NOTATION 8.9. In each of these cases, the choice of S is arbitrary to produce a pasting scheme

- **The Connected Wheeled Pasting Scheme:** Recall $\mathtt{Gr}_{\mathrm{c}}^{\circlearrowleft}$ denotes the full sub-groupoid of $\mathtt{Gr}_{\mathrm{w}}^{\circlearrowleft}$ consisting of connected wheeled graphs. When paired with any choice of S, this produces a unital pasting scheme, later used to model wheeled properads.
- **The Wheel-Free Pasting Scheme:** Recall $\mathtt{Gr}^{\uparrow}$ denotes the full sub-groupoid of $\mathtt{Gr}_{\mathrm{w}}^{\circlearrowleft}$ consisting of wheel-free graphs. When paired with any choice of S, this produces a unital pasting scheme, later used to model PROPs.
- **The Connected Wheel-Free Pasting Scheme:** Recall $\mathtt{Gr}_{\mathrm{c}}^{\uparrow}$ denotes the full sub-groupoid of $\mathtt{Gr}_{\mathrm{w}}^{\circlearrowleft}$ consisting of connected, wheel-free graphs. When paired with any choice of S, this produces a unital pasting scheme, later used to model properads.
- **The Simply-Connected Pasting Scheme:** Recall $\mathtt{Gr}_{\mathrm{di}}^{\uparrow}$ denotes the full sub-groupoid of $\mathtt{Gr}_{\mathrm{w}}^{\circlearrowleft}$ consisting of simply-connected, wheel-free graphs. When paired with any choice of S, this produces a unital pasting scheme, later used to model dioperads.
- **The Half-Graphs Pasting Scheme:** Recall $\mathtt{Gr}_{\frac{1}{2}}$ (section 4.26) denotes the full sub-groupoid of $\mathtt{Gr}_{\mathrm{di}}^{\uparrow}$ consisting of half-graphs (Definition 2.21). When paired with any choice of S, this produces a non-unital pasting scheme, later used to model half-PROPs.

The remaining few examples of pasting schemes are dependent upon a specific choice of S, although restricting to a smaller S would also produce a pasting scheme. As such, we will exhibit them in terms of their maximal possible choice of S, thinking of the previous cases as those having maximal case S being the whole of $\mathcal{P}(\mathfrak{C})^{op} \times \mathcal{P}(\mathfrak{C})$.

NOTATION 8.10. In the case of variations of trees, it is important to consider only pairs of profiles whose outputs have length 1 or at most 1, labeled S_1 and $S_{\leq 1}$ in 8.1.3(2).

- **The Wheeled Trees Pasting Scheme:** Recall $\mathtt{Tree}^{\circlearrowleft}$ denotes the full sub-groupoid of $\mathtt{Gr}_{\mathrm{c}}^{\circlearrowleft}$ consisting of connected wheeled graphs G such that $|\mathrm{out}(v)| \leq 1$ for each $v \in \mathrm{Vt}(G)$. When paired with $S_{\leq 1}$, this produces a unital pasting scheme, later used to model wheeled operads.
- **The (Unital) Level Trees Pasting Scheme:** Recall $\mathtt{Tree}$ denotes the full sub-groupoid of $\mathtt{Gr}_{\mathrm{c}}^{\uparrow}$ consisting of level trees 2.27. When paired with S_1, this produces a non-unital pasting scheme, later used to model Markl's non-unital operads. For operads with units, we need to use the slightly larger pasting scheme $\mathtt{UTree}$ 4.26 which also includes the exceptional graphs $\uparrow_c$ for all $c \in \mathfrak{C}$. Then $(S_1, \mathtt{UTree})$ is a unital pasting scheme, which will be used to model $\mathfrak{C}$-colored May's operads.

8.1.7. Summary of Relationships Between Pasting Schemes. The above examples give the following inclusions of pasting schemes:

(8.1)

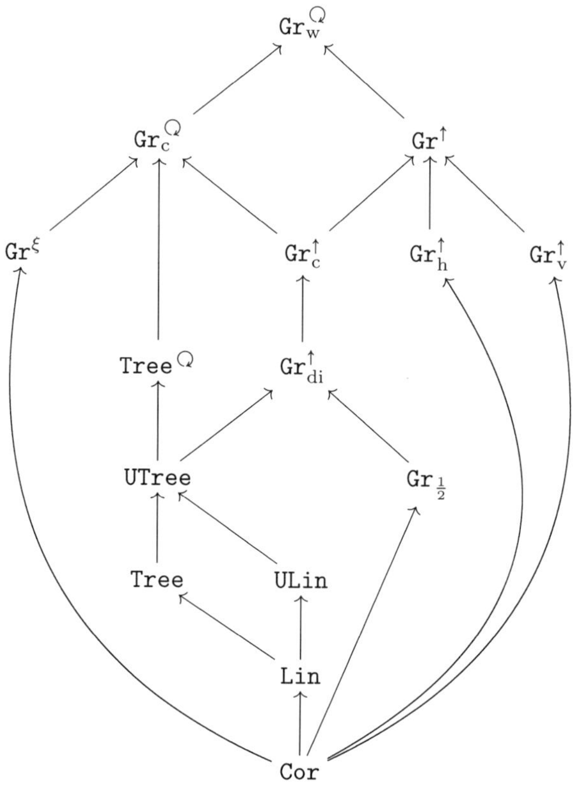

8.1.8. Virtual Pasting Schemes. The reader may notice that a pasting scheme looks a bit like a multicategory whose objects are the pairs of profiles in S, with a graph as a morphism from its pairs of vertex profiles to the pair of profiles for the full graph. Even more, one might try to use the graph substitution to define a composition law, but this operation is only associative up to strict isomorphism. However, by passing to strict isomorphism classes of graphs as morphisms, one can define such a multicategory $\mathsf{U}_{\mathcal{G}}$ as we will show in Chapter 14. Even more, the multifunctors from this $\mathsf{U}_{\mathcal{G}}$ to a symmetric monoidal category $\mathcal{E}$ gives an alternative description of the $\mathcal{G}$-PROPs.

DEFINITION 8.11. A **virtual pasting scheme** will be any multicategory $\mathfrak{M}$ whose objects are the pairs of profiles in $\mathfrak{C}$.

Given the description above, it is reasonable to define the generalized PROPs over the virtual pasting scheme $\mathfrak{M}$ as the multifunctors $\mathfrak{M} \longrightarrow \mathcal{E}$, and substantial portions of the theory of generalized PROPs can be phrased in these terms.

8.1.9. Non-Σ Pasting Schemes. One might want to restrict attention to situations where the symmetric group actions would be distracting. This works as with the symmetric group actions, but looking at pasting schemes that need not contain the permuted corollas. This will lead to what might be called generalized PROs rather than the generalized PROPs in our main approach. However, there are very few other changes required in order to deal with this variation.

8.2. Free Product Decompositions of Pasting Schemes

In [**JY09**], we were able to show that colored PROPs are monoids in a category of monoids, using two nicely related monoidal structures on the category of bimodules. The more general analog of such a result comes from considering generalized PROPs indexed on what we call a free product of pasting schemes. Thus, we must first detail the notion of free product of pasting schemes, and then investigate the appropriate compatibility statements for strong generating sets to produce a strong generating set of the free product. This leads us to consider conditions we call separating and balanced generating sets.

8.2.1. Free Product. We begin with the definition of our free product, which is forced by the need for the hereditary condition.

DEFINITION 8.12. The **free product** $\mathcal{G}_1 * \mathcal{G}_2$ of two pasting schemes $\mathcal{G}_1$ and $\mathcal{G}_2$ is the full graph sub-groupoid $\mathtt{G}_1 * \mathtt{G}_2 \leq \mathtt{Gr}_{\mathrm{w}}^{\circlearrowright}$ with objects those graphs having a presentation $\mathcal{H}$ with each $\mathcal{H}^i$ in $\mathtt{G}_1 \cup \mathtt{G}_2$, together with $S = S_1 \cup S_2$.

In other words, the free product of two pasting schemes consists of those graphs that can be written using iterated graph substitution involving entries from either. This collection is hereditary, and the presentation of a corolla by considering it as a graph 1-simplex implies the corolla containment condition with this choice of S is also immediate.

Our key examples of decompositions of pasting schemes as free products require us to first introduce a few more pasting schemes.

DEFINITION 8.13. The **relabeled unions of corollas** will refer to all finite, possibly empty, unions of corollas and their input and output relabelings, which together form the pasting scheme $\mathtt{Gr}_{\mathrm{h}}^{\uparrow}$.

Let $\overline{\mathtt{Gr}_{\mathrm{h}}^{\uparrow}}$ be the pasting scheme consisting of relabeled unions of corollas and $\underline{c}$-exceptional edges.

Let $\mathtt{Gr}_{\mathrm{ord}}^{\circlearrowright}$ denote the pasting scheme consisting of all ordinary wheeled graphs.

Given S, let $\mathtt{Cor}_S^{\mathrm{in/out}}$ denote the minimal pasting scheme restricted to S, denoted $\mathtt{Cor}_S$, together with the exceptional edge of each color occuring in S and the finite iterations of the input and/or output extensions of each corolla in $\mathtt{Cor}_S$.

Every non-unital pasting scheme has a unital extension by taking the free product with the minimal unital pasting scheme by the following result. In particular, $\mathtt{Cor}_S^{\mathrm{in/out}}$ will be the minimal unital pasting scheme for the given S.

LEMMA 8.14. *A pasting scheme* $\mathcal{G} = (S, \mathtt{G})$ *is unital precisely when*

$$\mathcal{G} * \mathtt{Cor}_S^{in/out} = \mathcal{G}.$$

PROOF. Once a pasting scheme contains the input and output extensions of corollas, the hereditary condition implies it must also contain the iterated input and/or output extensions of all corollas, since e.g. $C_{in}(C_{out})$ is the input and output extension while $C_{in}(C_{in})$ is an iterated input extension. Thus, the containment

$$\mathcal{G} * \mathtt{Cor}_S^{\mathrm{in/out}} \leq \mathcal{G}$$

with $\mathcal{G}$ a pasting scheme is equivalent to the definition of a unital pasting scheme, while the opposite containment always holds. □

We now have several examples of free product decompositions.

PROPOSITION 8.15. *The free product of* $\overline{\mathtt{Gr}^{\uparrow}_{h}}$ *and* $\mathtt{Gr}^{\uparrow}_{v}$ *is* $\mathtt{Gr}^{\uparrow}$, *or*

$$\overline{\mathtt{Gr}^{\uparrow}_{h}} * \mathtt{Gr}^{\uparrow}_{v} = \mathtt{Gr}^{\uparrow}.$$

PROOF. Since both $\overline{\mathtt{Gr}^{\uparrow}_{\mathrm{h}}} \leq \mathtt{Gr}^{\uparrow}$ and $\mathtt{Gr}^{\uparrow}_{\mathrm{v}} \leq \mathtt{Gr}^{\uparrow}$, it follows that

$$\overline{\mathtt{Gr}^{\uparrow}_{\mathrm{h}}} * \mathtt{Gr}^{\uparrow}_{\mathrm{v}} \leq \mathtt{Gr}^{\uparrow}.$$

On the other hand, the generating set $\mathcal{T}^{\uparrow}$ in Theorem 7.27 is a generating set for $\mathtt{Gr}^{\uparrow}$ consisting of elements that lie either in $\overline{\mathtt{Gr}^{\uparrow}_{\mathrm{h}}}$ or in $\mathtt{Gr}^{\uparrow}_{\mathrm{v}}$. Thus, it follows that

$$\mathtt{Gr}^{\uparrow} \leq \overline{\mathtt{Gr}^{\uparrow}_{\mathrm{h}}} * \mathtt{Gr}^{\uparrow}_{\mathrm{v}},$$

completing the proof. □

As mentioned earlier, there will be a characterization of the generalized PROPs determined by certain free products (Theorem 11.9).

This time using the generating set $\mathcal{T}^{\circlearrowleft}$ of Corollary 7.21 for $\mathtt{Gr}^{\circlearrowleft}_{\mathrm{w}}$, the same proof also yields the following.

PROPOSITION 8.16. *We can identify the free products*

$$\overline{\mathtt{Gr}^{\uparrow}_{h}} * \mathtt{Gr}^{\xi} = \mathtt{Gr}^{\circlearrowleft}_{w} \quad \textit{and} \quad \mathtt{Gr}^{\uparrow}_{h} * \mathtt{Gr}^{\xi} = \mathtt{Gr}^{\circlearrowleft}_{ord}.$$

Additional examples of free products include the following.

PROPOSITION 8.17. *One can describe* $\mathtt{Gr}^{\circlearrowleft}_{c}$ *in two ways as a free product.*

$$\mathtt{Gr}^{\uparrow}_{di} * \mathtt{Gr}^{\xi} = \mathtt{Gr}^{\circlearrowleft}_{c} = \mathtt{Gr}^{\uparrow}_{c} * \mathtt{Gr}^{\xi}.$$

PROOF. The decomposition $\mathtt{Gr}^{\uparrow}_{\mathrm{di}} * \mathtt{Gr}^{\xi} = \mathtt{Gr}^{\circlearrowleft}_{\mathrm{c}}$ follows as above from the strong generating set already established.

For the second decomposition, notice any generating set coming from the factors is sufficient for this same argument to apply. Thus, it suffices to show that basic properadic graphs together with the exceptional edge of each color, permuted corollas, and contracted corollas form a generating set for $\mathtt{Gr}^{\circlearrowleft}_{\mathrm{c}}$, without claiming it is *strong*.

The argument follows as in the proof of Lemma 7.91, but only breaking an edge in each wheel (or *directed* cycle) of G to construct a connected wheel-free graph G', rather than a simply-connected graph sc$\mathcal{H}$. This connected wheel-free graph then has a presentation in terms of basic properadic graphs, the exceptional edge of each color, and the permuted corollas by Theorem 7.81. Now use contractions to start with G' and reproduce the original G. □

REMARK 8.18. Unfortunately for the consistency of our notation, the groupoid of wheeled trees is *not* the free product of $\mathtt{Gr}^{\xi}$ and $\mathtt{UTree}$, as wheeled trees also contain the truncated trees.

8.2.2. Generating Sets for Free Products.

PROPOSITION 8.19. *Suppose* $S_1 = S_2$. *Given a generating set* $\mathcal{T}_1$ *for* $\mathcal{G}_1$ *and a generating set* $\mathcal{T}_2$ *for* $\mathcal{G}_2$*, their union* $\mathcal{T}_1 \cup \mathcal{T}_2$ *forms a generating set for* $\mathcal{G}_1 * \mathcal{G}_2$.

PROOF. Choose $G \in \mathcal{G}_1 * \mathcal{G}_2$ and so without a loss of generality a presentation $\mathcal{H}$ with each $\mathcal{H}^j$ consisting of graphs in just one of the G_k. Now form a presentation in $\mathcal{T}_k$ of each $H \in \mathcal{H}^j$. Combining the layers then provides a presentation in $\mathcal{T}_1 \cup \mathcal{T}_2$ of G, completing the proof. □

We would like to be able to understand when the process of the previous result produces a strong generating set. There is an unfortunate amount of technical machinery necessary for this discussion, which we view as justified by subsequent results on the levels of modules and of generalized PROPs.

For the rest of this section, suppose $\mathcal{T}_1$ and $\mathcal{T}_2$ are strong generating sets for two pasting schemes $\mathcal{G}_1$ and $\mathcal{G}_2$, respectively, with $S_1 = S_2$.

8.2.3. Separating Generating Sets for Free Products. In order to detect when the union of strong generating sets remains a strong generating set for the free product of pasting schemes, we will need some sense of how long the relaxed moves must be in order to regroup so that all of the terms in one generating set come first. As we will show below, this requires some care, to avoid a combinatorial explosion making the appropriate regrouping impossible to verify.

DEFINITION 8.20. Call a graph n-simplex $\mathcal{H}$ with each layer $\mathcal{H}^i$ wholly within one of the two $\mathcal{T}_j$ **weakly separating**. Given a weakly separating $\mathcal{H}$, the **1-length** will denote the number of layers wholly within $\mathcal{T}_1$.

For a weakly separating graph n-simplex $\mathcal{H}$ with 1-length i and $\mathcal{T}_1$-layers having indices $k_1 < \cdots < k_i$, define its **deviation** as

$$D(\mathcal{H}) = \sum_{j=0}^{i-1} \left(1 - \delta_{n-j,k_{i-j}}\right),$$

where $\delta_{x,y}$ is the Kronecker delta.

Call a graph n-simplex $\mathcal{H}$ in $\mathcal{T}_1 \cup \mathcal{T}_2$ **separating** if it is weakly separating with deviation zero. In particular, call $\mathcal{H}$ **upper separating** if it is separating and has 1 as its 1-length, or **lower separating** if it is separating and has $(n-1)$ as its 1-length.

Thus, a graph simplex is:

- separating precisely when there exists j with $\mathcal{H}^i$ in $\mathcal{T}_2$ for $i \leq j$ and $\mathcal{H}^i$ in $\mathcal{T}_1$ for $i > j$;
- lower separating if $\mathcal{H}^1$ is in $\mathcal{T}_2$ and $\mathcal{H}^i$ is in $\mathcal{T}_1$ for $i > 1$;
- upper separating if $\mathcal{H}^i$ is in $\mathcal{T}_2$ for $i < n$ and $\mathcal{H}^n$ is in $\mathcal{T}_1$.

The deviation is in the interval $[0, i]$, where i is the 1-length, and the sum $D(\mathcal{H})$ always has the form

$$0 + \cdots + 0 + 1 + \cdots + 1,$$

where the string of 0's or of 1's may be empty. If the first number is 0, that indicates $k_i = n$, so there is no term from $\mathcal{T}_2$ with higher index than the last term from $\mathcal{T}_1$. If the second number is 0, that indicates $k_{i-1} = n - 1$, so again there can be no term from $\mathcal{T}_2$ with higher index than this penultimate term from $\mathcal{T}_1$ and so on. The purpose of the deviation is to measure how far $\mathcal{H}$ is from being separating, or equivalently, how many terms from $\mathcal{T}_1$ have some term from $\mathcal{T}_2$ with higher index, thereby implying the $\mathcal{T}_1$ term is out of position.

EXAMPLE 8.21. If we use $(\mathcal{H}')^j$ to indicate a generic layer wholly within $\mathcal{T}_j$, consider a graph simplex of the form

$$((\mathcal{H}')^1, (\mathcal{H}')^1, (\mathcal{H}')^2, (\mathcal{H}')^1, (\mathcal{H}')^2, (\mathcal{H}')^1)$$

which will have 1-length 4 and deviation $0 + 1 + 1 + 1 = 3$, while a graph simplex of the form

$$((\mathcal{H}')^2, (\mathcal{H}')^2, (\mathcal{H}')^2, (\mathcal{H}')^1, (\mathcal{H}')^1)$$

has 1-length 2 and deviation $0 + 0 = 0$ so is separating.

We now need to extend our terminology for the two extreme cases of saying every graph 2-simplex has a presentation that is separating.

DEFINITION 8.22. Call $\mathcal{T}_1$ N-**lower separating over** $\mathcal{T}_2$ if every graph 2-simplex $\mathcal{H}'$ of the form $((\mathcal{H}')^1, (\mathcal{H}')^2)$ with $(\mathcal{H}')^1$ in $\mathcal{T}_1$ and $(\mathcal{H}')^2$ in $\mathcal{T}_2$ has a presentation of length at most N that is lower separating. Similarly, $\mathcal{T}_1$ is N-**upper separating over** $\mathcal{T}_2$ if each such $\mathcal{H}'$ has a presentation of length at most N that is upper separating.

Our examples rely upon the calculus of graph substitution computations of Section 6.3.

EXAMPLE 8.23. Suppose $\mathcal{T}_1$ is the 2-strong generating set for $\mathtt{Gr}^{\xi}$ consisting of contracted corollas and permuted corollas. Suppose $\mathcal{T}_2$ is the 3-strong generating set for $\mathtt{Gr}^{\uparrow}_{\mathrm{h}}$ consisting of unions of two corollas and permuted corollas. Then $\mathcal{T}_1$ is 3-lower separating over $\mathcal{T}_2$ (Section 6.3).

EXAMPLE 8.24. Suppose $\mathcal{T}_1$ is the 3-strong generating set for $\mathtt{Gr}^{\uparrow}_{\mathrm{v}}$ consisting of grafted corollas and permuted corollas. Suppose $\mathcal{T}_2$ is the 3-strong generating set for $\overline{\mathtt{Gr}^{\uparrow}_{\mathrm{h}}}$ consisting of unions of two corollas, permuted corollas, all exceptional edges, and the empty graph. Then $\mathcal{T}_1$ is 3-upper separating over $\mathcal{T}_2$ (Section 6.3).

We now have our technical result avoiding the combinatorial explosion problem via the separating condition.

LEMMA 8.25. *Suppose $\mathcal{T}_1$ is N-lower or N-upper separating over $\mathcal{T}_2$ with $N \geq 2$. Then every graph simplex in $\mathcal{T}_1 \cup \mathcal{T}_2$ is connected to a separating graph simplex by a finite number of relaxed N-moves.*

PROOF. Using relaxed 2-moves and corollas, which lie in any pasting scheme, we can rewrite our $\mathcal{H}$ to be weakly separating. Suppose $\mathcal{T}_1$ is N-upper separating over $\mathcal{T}_2$, and induct upon the deviation of $\mathcal{H}$ assuming a fixed 1-length. When the deviation is zero, $\mathcal{H}$ is already separating, so the claim is immediate. For the induction step, suppose $\mathcal{H}^{k_j}$ is the $\mathcal{T}_1$ layer with the highest index that satisfies $k_j < n - (i - j)$, while

$$k_i = n, k_{i-1} = n - 1, \ldots, k_{j+1} = n - (i - j - 1).$$

Then we apply upper separating relaxed N-moves to this $\mathcal{T}_1$ layer a total of $[n - (i - j) - k_j]$ times. The resulting graph simplex is still weakly separating, has the same 1-length, and is 1 lower in deviation, which suffices for the induction step.

If $\mathcal{T}_1$ is N-lower separating over $\mathcal{T}_2$, the proof is dual after introducing the dual definitions of 2-length and 2-deviation. □

One might be tempted to weaken the definitions of upper or lower separating to consider variations of separating presentations where the transition occurs other than at the two ends. However, the following example should demonstrate the resulting potential for combinatorial explosion.

EXAMPLE 8.26. Suppose $\mathcal{T}_1$ is neither upper nor lower separating over $\mathcal{T}_2$, but it is true that any graph 2-simplex of the form $((\mathcal{H}')^1,(\mathcal{H}')^2)$ in the notation of Example 8.21 has a presentation of the form

$$((\mathcal{H}')^2,(\mathcal{H}')^2,(\mathcal{H}')^1,(\mathcal{H}')^1).$$

Then a graph 3-simplex of the form $((\mathcal{H}')^1,(\mathcal{H}')^2,(\mathcal{H}')^2)$ need not have any presentation that is separating. The naive replacement when trying to increase the index of the $\mathcal{T}_1$ term would be of the form

$$((\mathcal{H}')^2,(\mathcal{H}')^2,(\mathcal{H}')^1,(\mathcal{H}')^1,(\mathcal{H}')^2).$$

Still trying to increase the index of the last $\mathcal{T}_1$-term term would then yield a simplex of the form

$$((\mathcal{H}')^2,(\mathcal{H}')^2,(\mathcal{H}')^1,(\mathcal{H}')^2,(\mathcal{H}')^2,(\mathcal{H}')^1,(\mathcal{H}')^1),$$

but notice the original 3-simplex occurs again in the middle. As a consequence, the same two steps would yield something of the form

$$((\mathcal{H}')^2,(\mathcal{H}')^2,(\mathcal{H}')^2,(\mathcal{H}')^2,(\mathcal{H}')^1,(\mathcal{H}')^2,(\mathcal{H}')^2,(\mathcal{H}')^1,(\mathcal{H}')^1,(\mathcal{H}')^1,(\mathcal{H}')^1),$$

and so always keeping two $\mathcal{T}_2$ terms with higher index than one of the $\mathcal{T}_1$-terms.

NOTATION 8.27. Given two N-strong generating sets $\mathcal{T}_1$ and $\mathcal{T}_2$, certain relaxed N-moves for graph n-simplices in $\mathcal{T}_1 \cup \mathcal{T}_2$ will be singled out:

- relaxed N-moves within a single $\mathcal{T}_i$ will be called **insular moves**,
- relaxed N-moves of the form considered in the definition of upper separating will be called **separating moves**, and
- relaxed N-moves replacing a sub-N-simplex which is separating by a graph 1-simplex in $\mathtt{G}_1 \cap \mathtt{G}_2$ will be called **balanced moves**.

DEFINITION 8.28. Given two N-strong generating sets $\mathcal{T}_1$ and $\mathcal{T}_2$ with $\mathcal{T}_1$ N-upper separating over $\mathcal{T}_2$, they will be called **N-upper balanced** if any two equivalent separating presentations can be connected by a finite number of relaxed N-moves of the three types considered above. The notion of **N-lower balanced** is defined similarly with lower separating replacing upper separating.

THEOREM 8.29. *If $\mathcal{T}_1$ and $\mathcal{T}_2$ are N-strong generating sets that are also N-upper balanced or N-lower balanced with $N \geq 2$, then $\mathcal{T}_1 \cup \mathcal{T}_2$ is an N-strong generating set for the free product of $\mathcal{G}_1$ and $\mathcal{G}_2$.*

PROOF. The generating set condition comes from Lemma 8.19. Now suppose $\mathcal{H}$ and $\mathcal{H}'$ are equivalent graph n-simplices in $\mathcal{T}_1 \cup \mathcal{T}_2$. Then each is connected to a separating simplex by a finite string of relaxed N-moves by Lemma 8.25. These equivalent separating simplices are then connected by a finite string of more specialized relaxed N-moves by the definition of balanced. □

EXAMPLE 8.30. The $\mathcal{T}_1$ and $\mathcal{T}_2$ of Example 8.23 satisfy the hypotheses of Theorem 8.29 with $N = 3$ (Section 6.3), as do the generating sets in Example 8.24. Taken together with Propositions 8.16 and 8.15 this gives a 3-strong generating set for $\mathtt{Gr}^{\circlearrowleft}_{\mathrm{ord}}$ and another for $\mathtt{Gr}^{\uparrow}$.

8.3. Monogenic Pasting Schemes

As discussed along with the description of a virtual pasting scheme, it is possible to form a multicategory with strict isomorphism classes of ordered wheeled graphs as the morphisms, pairs of profiles as the objects, and with composition coming from the graph substitution operation. If one restricts attention to the case of graphs with exactly one vertex, the restriction is an ordinary small category, which we would like to understand better. In fact, for a pasting scheme where each graph has a single vertex, the generalized PROPs will simply become the diagrams indexed on this small category.

DEFINITION 8.31. A collection of wheeled graphs is **monogenic** if each graph it contains has exactly one vertex, and similarly for a pasting scheme.

Our two primary examples throughout this discussion are the minimal pasting scheme `Cor`, and the contractions pasting scheme $\mathtt{Gr}^{\xi}$, which are both monogenic.

The following lemma characterizes a monogenic graph groupoid as a graph groupoid having a monogenic generating set.

LEMMA 8.32. *Let* `G` *be a graph groupoid. The following statements are equivalent.*

(1) `G` *is monogenic.*
(2) *Every generating set for* `G` *is monogenic.*
(3) *There exists a monogenic generating set for* `G`.

PROOF. A generating set for `G` is contained in `G`, so the first condition implies the second. Next, `G` itself is a generating set for `G`, so the second condition implies the third. Finally, suppose $\mathcal{T}$ is a monogenic generating set for `G`. To see that `G` is monogenic, simply observe that if G and H are both one-vertex wheeled graphs, then so is the graph substitution $G(H)$. Therefore, the substitution of any $\mathcal{T}$-simplex is a finite set of one-vertex wheeled graphs. □

By considering the proof of Proposition 7.18, we can characterize the graphs with exactly one vertex.

LEMMA 8.33. *A wheeled graph has a unique vertex v precisely when it can be written as*

$$\sigma\left(\xi^k\left(C_v \sqcup \coprod_j E_j\right)\right)\tau$$

where ξ^k denotes a series of k contractions, C_v denotes a corolla, and each E_j is an exceptional edge of a single color.

Keep in mind that a contraction applied to a single exceptional edge yields an exceptional loop, while a contraction that pairs two different exceptional edges of the same color simply reduces the pair to a single exceptional edge. As a consequence of the proof of 7.18, there are no extraneous exceptional edges included, so we can assume each contraction in the exceptional portion forms a single exceptional loop.

DEFINITION 8.34. Suppose $(\underline{c};\underline{d})$ is a pair of profiles with $d_i = c_j$. The $(j;i)$-**contraction** of $(\underline{c};\underline{d})$ is the pair of profiles

$$\xi_j^i(\underline{c};\underline{d}) = (\underline{c} \setminus c_j; \underline{d} \setminus d_i).$$

We will often call $\xi^i_j(\underline{c};\underline{d})$ a **contraction** of $(\underline{c};\underline{d})$ and ξ^i_j a **contraction operator**.

As a consequence of Theorem 5.32 and surrounding results, we can build a category from a monogenic pasting scheme and the previous lemma can be used to index its morphisms.

DEFINITION 8.35. Let $\mathcal{G} = (S, \mathtt{G})$ be a monogenic pasting scheme. Define the **associated category** $\mathcal{C}(\mathcal{G})$ with

- objects the profile pairs in S,
- $\mathcal{C}(\mathcal{G})(s,t)$ the set of strict isomorphism classes $[G]$ in $\mathcal{G}$ in which the unique vertex has profiles s and the full graph G has profiles t,
- composition given by graph substitution, so $[G] \circ [H] = [G(H)]$, and
- identity morphisms the strict isomorphism classes of corollas.

LEMMA 8.36. *The morphisms in $\mathcal{C}(\mathcal{G})$ with source $s = (\underline{c};\underline{d})$ can be indexed by*

$$(\xi^k, \underline{e}, U, \sigma, \tau)$$

where ξ^k is an ordered sequence of k contractions of s, $\underline{e}$ a single profile, U a subset of $\{1, 2, \ldots, |\underline{e}|\}$, σ a permutation of $|\underline{d}| - k + |\underline{e}| - |U|$ letters, and τ a permutation of $|\underline{c}| - k + |\underline{e}| - |U|$ letters.

PROOF. The profile $\underline{e}$ simply keeps track of the union of exceptional edges added to C_s in Lemma 8.33, ξ^k the contractions of the ordinary part, U the contractions of the exceptional part, σ the output relabeling, and τ the input relabeling. □

In fact, one could go farther and give a completely combinatorial description of $\mathcal{C}(\mathcal{G})$ and its morphisms, exploiting the calculus of graph substitution lemmas of Section 6.3 to give analogs of the simplicial identities for presenting the contraction operations and so forth. However, for now, we will simply indicate that our consideration of $\mathcal{C}(\mathcal{G})$ will be justified by identifying the $\mathcal{G}$-PROPs as the $\mathcal{C}(\mathcal{G})$ diagrams in $\mathcal{E}$ later.

CHAPTER 9

Well-Matched Pasting Schemes

This chapter deals with two technical issues, with the goal of understanding when the relationship between two pasting schemes is particularly tight. Our primary goal is to understand the notion of a well-matched pair of pasting schemes which, among other things, gives a simple way of building from a generalized PROP or module on the smaller pasting scheme to a corresponding generalized PROP or module on the larger pasting scheme. This will make a variety of constructions simpler in this context, eventually giving stronger homotopy-theoretic results as a consequence.

Along the way, we introduce the Kontsevich groupoid for the pair, building on the observation that led Kontsevich to originally propose half-PROPs. The Kontsevich groupoid is described as maximal with the property of being both orthogonal and prime to the smaller groupoid. Orthogonality is fairly simple, meaning the intersection of the two pasting schemes is really just some permuted corollas (and possibly the empty graph). Being a prime subgroupoid is more complicated, but essentially means in each reasonable decomposition of a graph not in the smaller groupoid $G = K(H_v)$, if the H_v lie in the smaller groupoid then they must be corollas. A list of calculations of Kontsevich groupoids of various important examples is included.

The well-matched condition requires that each non-empty graph in the larger groupoid has an essentially unique presentation $G = K(H_v)$ with K in the Kontsevich groupoid and each H_v in the smaller groupoid. There is always the option of performing relabelings on the vertices of K and a series of input and output relabelings on the H_v, so the uniqueness condition is really up to weak isomorphism. As a consequence, the larger groupoid is the free product of the smaller groupoid and the Kontsevich groupoid in a very specific way, due to this uniqueness property. The chapter closes with a series of examples of pairs that are, or just as importantly pairs that are not, well-matched.

9.1. Kontsevich Groupoid

In this section we discuss what we call the Kontsevich groupoid. In nice cases, the Kontsevich groupoid tells us how a bigger pasting scheme can be generated in a simple way from a smaller pasting scheme. Later, this Kontsevich groupoid will be used to generalize a particularly useful relationship first observed by Kontsevich involving half-PROPs.

9.1.1. Intersection.

DEFINITION 9.1. The intersection of two pasting schemes $\mathcal{G}_1$ and $\mathcal{G}_2$ will be formed by intersecting the two components, so

$$\mathcal{G}_1 \cap \mathcal{G}_2 = (S_1 \cap S_2, \mathtt{G}_1 \cap \mathtt{G}_2).$$

It is immediate from the definitions that the intersection is a pasting scheme, as well as forming an infimum in the partially ordered set of pasting schemes. Furthermore, if both pasting schemes are unital, then the intersection is also unital by inspection. One could work with the intersection of a set of (unital) pasting schemes as well, and the intersection of graph groupoids is also set-theoretically defined.

EXAMPLE 9.2.
- $\mathtt{Gr}^{\uparrow}_{\mathrm{c}} = \mathtt{Gr}^{\uparrow} \cap \mathtt{Gr}^{\circlearrowleft}_{\mathrm{c}}$.
- $Min(S) = \mathtt{Gr}^{\uparrow}_{\mathrm{h}} \cap \mathtt{Gr}^{\uparrow}_{\mathrm{v}}$.
- $Min(S) = \mathtt{Gr}^{\uparrow}_{\mathrm{h}} \cap \mathtt{Gr}^{\xi}$.
- $\mathtt{Lin} = \mathtt{Tree} \cap \mathtt{ULin}$.
- $\mathtt{UTree} \sqcup \mathtt{Tr.Tree} = \mathtt{Tree}^{\circlearrowleft} \cap \mathtt{Gr}^{\uparrow}_{\mathrm{di}}$.

It may be difficult to identify generating sets for an intersection in general, as there is no reason an intersection of generating sets must produce a generating set for the intersection of pasting schemes.

9.1.2. Orthogonality. The notion of orthogonal pasting schemes is intended to function a bit like the familiar notion of linear independence. The following notation will be convenient below.

NOTATION 9.3. Given a graph groupoid $\mathtt{G}$, denote by $\mathtt{G}_{cor}$ the subgroupoid of permuted corollas in $\mathtt{G}$, and if the empty graph $\varnothing$ lies in $\mathtt{G}$, we include it to form the **extended corollas** $\mathtt{G}_{ecor}$. Let the **non-corollas** denote the complement $\mathtt{G}_{ncor} = \mathtt{G} \setminus \mathtt{G}_{ecor}$, and let $\mathtt{G}_{ntriv}$ denote the subgroupoid of $\mathtt{G}$ consisting of wheeled graphs that do not contain exceptional legs.

Loosely speaking, two pasting schemes are orthogonal if their intersection is the minimal pasting scheme, possibly with the empty graph added.

DEFINITION 9.4. Let $\mathtt{G}$ and $\mathtt{G}'$ be two groupoids of graphs.

(1) The groupoids are **orthogonal** if

$$\mathtt{G} \cap \mathtt{G}' = \mathtt{G}_{ecor} \cap \mathtt{G}'_{ecor}.$$

(2) Two pasting schemes are **orthogonal** if their component graph groupoids are orthogonal.

The following result says that the horizontal combinations pasting scheme $\mathtt{Gr}^{\uparrow}_{\mathrm{h}}$ is orthogonal to any pasting scheme consisting solely of connected graphs.

PROPOSITION 9.5. *If either*
- $\mathcal{G} \le \mathtt{Gr}^{\circlearrowleft}_{c}$ *and* $\mathcal{G}' \le \mathtt{Gr}^{\uparrow}_{h}$, *or*
- $\mathcal{G} \le \mathtt{Gr}^{\circlearrowleft}_{c} \cap \mathtt{Gr}^{\circlearrowleft}_{ord}$ *and* $\mathcal{G}' \le \overline{\mathtt{Gr}^{\uparrow}_{h}}$,

then $\mathcal{G}$ *and* $\mathcal{G}'$ *are orthogonal.*

PROOF. If $G \in \mathtt{Gr}^{\uparrow}_{\mathrm{h}}$ is connected, then it is by definition a permuted corolla, which also lies in $\mathtt{Gr}^{\circlearrowleft}_{\mathrm{c}}$. If $G \in \overline{\mathtt{Gr}^{\uparrow}_{\mathrm{h}}}$ is connected and non-empty, then it is either a permuted corolla or some c-exceptional edge. If, moreover, G is ordinary, then it must be a permuted corolla. □

COROLLARY 9.6. *The pasting schemes* $\overline{\mathtt{Gr}^{\uparrow}_{h}}$ *and* $\mathtt{Gr}^{\uparrow}_{v}$ *are orthogonal, as are the pasting schemes* $\mathtt{Gr}^{\uparrow}_{h}$ *and* $\mathtt{Gr}^{\xi}$*, as well as the pasting schemes* $\overline{\mathtt{Gr}^{\uparrow}_{h}}$ *and* $\mathtt{Gr}^{\xi}$.

For the importance of these pairs of pasting schemes, recall from Propositions 8.15 and 8.16 that their free products are $\mathtt{Gr}^{\uparrow}$, $\mathtt{Gr}^{\circlearrowleft}_{\mathrm{ord}}$, and $\mathtt{Gr}^{\circlearrowleft}_{\mathrm{w}}$, respectively, while they have 3-strong generating sets which produce the same for the free products, as discussed in Example 8.30.

9.1.3. Prime Graph Groupoids. As usual, the following definitions are complicated by the need to deal with exceptional flags carefully. The basic idea is that a graph G is prime to a graph groupoid $\mathtt{G}$ if the only H_v from $\mathtt{G}$ which can occur in a presentation as $G = K(H_v)$ are corollas.

DEFINITION 9.7. Given a graph subgroupoid $\mathtt{G}_1$ of $\mathtt{G}'$, we say a wheeled graph $G \in \mathtt{G}' \setminus \mathtt{G}_1$ is **prime to** $\mathtt{G}_1$ if, when writing

$$G = K(H_v) \quad \text{with } K \in \mathtt{G}' \text{ and each } H_v \in (\mathtt{G}_1)_{ntriv}, \tag{9.1}$$

it must follow that each $H_v \in (\mathtt{G}_1)_{cor}$, and whenever G contains any exceptional flags, it must have an exceptional connected component that is not itself in $\mathtt{G}_1$.

The subgroupoid of all wheeled graphs $G \in \mathtt{G}' \setminus \mathtt{G}_1$ that are prime to $\mathtt{G}_1$ is denoted by $Pr(\mathtt{G}' \setminus \mathtt{G}_1)$. A subgroupoid $\mathtt{G}_2$ of $\mathtt{G}'$ is **prime to** $\mathtt{G}_1$ in $\mathtt{G}'$ if every $G \in \mathtt{G}_2 \setminus \mathtt{G}_1$ is prime to $\mathtt{G}_1$, i.e.,

$$\mathtt{G}_2 \setminus \mathtt{G}_1 \subseteq Pr(\mathtt{G}' \setminus \mathtt{G}_1).$$

EXAMPLE 9.8. The empty graph is prime to $\mathtt{Gr}^{\circlearrowleft}_{\mathrm{c}}$ in $\mathtt{Gr}^{\circlearrowleft}_{\mathrm{w}}$, but no graph containing exceptional flags can be prime to $\mathtt{Gr}^{\circlearrowleft}_{\mathrm{c}}$ since every exceptional connected component is a connected wheeled graph.

9.1.4. Kontsevich Groupoid. In defining the Kontsevich groupoid, it is important to look at the restriction of the larger graph groupoid to the smaller collection of profile pairs.

DEFINITION 9.9. Given two pasting schemes

$$\mathcal{G} = (S, \mathtt{G}) \leq (S', \mathtt{G}') = \mathcal{G}',$$

the **Kontsevich groupoid** is the maximal subgroupoid of $\mathtt{G}'_S$ that is both orthogonal to $\mathtt{G}_S$ and prime to $\mathtt{G}_S$ in $\mathtt{G}'_S$. It is denoted by $\mathtt{Kont}(\mathcal{G}, \mathcal{G}')$.

The next proposition gives an alternative presentation of the Kontsevich groupoid, while also dealing with any questions about the existence of such a maximal subgroupoid.

PROPOSITION 9.10. *The Kontsevich groupoid is the disjoint union*

$$Kont(\mathcal{G}, \mathcal{G}') = (\mathtt{G}_S)_{ecor} \coprod Pr(\mathtt{G}'_S \setminus \mathtt{G}_S).$$

PROOF. The disjoint union is both orthogonal and prime to $\mathtt{G}_S$ in $\mathtt{G}'_S$ by construction. Even more, no larger groupoid than the disjoint union can remain both prime to $\mathtt{G}_S$ in $\mathtt{G}'_S$ by the definition of Pr and orthogonal to $\mathtt{G}_S$. As a consequence, the disjoint union is the required maximal subgroupoid. □

In Chapter 12, we will use the Kontsevich groupoid to generalize a result due to Kontsevich about the relationship between half-PROPs and dioperads or PROPs. That is, the left adjoint to the forgetful functor will be particularly well-behaved when the pasting schemes are what we call well-matched below.

An indicative example of the computation of a Kontsevich groupoid is given below.

LEMMA 9.11. *The Kontsevich groupoid*

$$\mathtt{Kont}\left(\mathtt{Gr}_c^{\circlearrowleft}, \mathtt{Gr}_w^{\circlearrowleft}\right)$$

consists of the relabeled unions of corollas.

PROOF. Suppose G lies in the Kontsevich groupoid, and we would like to show it is a relabeled union of corollas. However, by Proposition 9.10, it suffices to consider the case $G \notin \mathtt{Gr}_c^{\circlearrowleft}$ is prime to $\mathtt{Gr}_c^{\circlearrowleft}$. Furthermore, each exceptional connected component lies in $\mathtt{Gr}_c^{\circlearrowleft}$ by definition, so no graph containing exceptional flags can be prime to $\mathtt{Gr}_c^{\circlearrowleft}$. Thus, it suffices to assume G is ordinary and show it is a relabeled union of corollas. To this end, decompose such an ordinary G in terms of its connected components as $K(H_v)$, with K a relabeled union of corollas, and each $H_v \in \mathtt{Gr}_c^{\circlearrowleft}$ a connected component of G. Since G is ordinary and prime to $\mathtt{Gr}_c^{\circlearrowleft}$, each H_v is then a permuted corolla. Therefore, G itself is a relabeled union of corollas.

Conversely, suppose G is a non-empty relabeled union of corollas, so G is ordinary and contains no internal edges, as well as lying in $\mathtt{Gr}_w^{\circlearrowleft} \setminus \mathtt{Gr}_c^{\circlearrowleft}$. We will show that it is either a permuted corolla or prime to $\mathtt{Gr}_c^{\circlearrowleft}$, hence belongs to the Kontsevich groupoid as described in Proposition 9.10. If G is connected, then it must be a permuted corolla. On the other hand, suppose G is not connected and write $G = K(H_v)$ with each H_v connected and containing no exceptional legs. If some H_v contains an internal edge, either ordinary or exceptional, then so does $G = K(H_v)$, which is a contradiction. Therefore, each H_v is connected and contains neither exceptional flag nor internal edge, so it must be a permuted corolla. In other words, we have verified that G is prime to $\mathtt{Gr}_c^{\circlearrowleft}$ by definition, which completes the proof. □

The examples below can all be proved with similar arguments based on Proposition 9.10.

EXAMPLE 9.12. (1) For any pasting scheme $\mathcal{G} = (S, \mathtt{G})$, we have the Kontsevich groupoid

$$\mathtt{Kont}\,(Min(S), \mathcal{G}) = \mathtt{G}_S,$$

where $Min(S) = (S, \mathtt{Cor})$.

(2) The Kontsevich groupoid

$$\mathtt{Kont}\left(\mathtt{Tree}^{\circlearrowleft}, \mathtt{Gr}_w^{\circlearrowleft}\right)$$

consists of relabeled unions of corollas, each with at most one output.

(3) The Kontsevich groupoid

$$\mathtt{Kont}\left(\mathtt{Tree}^{\circlearrowleft}, \mathtt{Gr}_{\mathrm{c}}^{\circlearrowleft}\right)$$

consists of corollas with at most one output and their input relabelings.

(4) The Kontsevich groupoid

$$\mathtt{Kont}\left(\mathtt{Gr}_{\mathrm{c}}^{\uparrow}, \mathtt{Gr}^{\uparrow}\right)$$

consists of relabeled unions of corollas.

(5) The Kontsevich groupoid

$$\mathtt{Kont}\left(\mathtt{UTree}, \mathtt{Gr}^{\uparrow}\right)$$

consists of relabeled unions of corollas, each with exactly one output.

(6) The Kontsevich groupoid

$$\mathtt{Kont}\left(\mathtt{UTree}, \mathtt{Gr}_{\mathrm{c}}^{\uparrow}\right)$$

consists of corollas with exactly one output and their input relabelings.

(7) The Kontsevich groupoid

$$\mathtt{Kont}\left(\mathtt{Gr}_{\frac{1}{2}}, \mathtt{Gr}_{\mathrm{di}}^{\uparrow}\right)$$

consists of simply-connected wheeled graphs satisfying the following conditions:

- Each vertex v satisfies $|\mathrm{in}(v)||\mathrm{out}(v)| \geq 2$.
- Each directed internal edge (e_{-1}, e_1) has the property that if $e_{-1} \in u$ and $e_1 \in v$, then e_{-1} is not the only outgoing flag of u, and e_1 is not the only incoming flag of v.

(8) The Kontsevich groupoid

$$\mathtt{Kont}\left(\mathtt{Gr}_{\frac{1}{2}}, \mathtt{Gr}^{\uparrow}\right)$$

admits the same description as $\mathtt{Kont}\left(\mathtt{Gr}_{\frac{1}{2}}, \mathtt{Gr}_{\mathrm{di}}^{\uparrow}\right)$ with simply-connected wheeled graphs replaced by wheel-free graphs.

(9) The Kontsevich groupoid

$$\mathtt{Kont}\left(\mathtt{Gr}_{\mathrm{di}}^{\uparrow}, \mathtt{Gr}^{\uparrow}\right)$$

contains the grafted corollas (3.3) $C_{(\underline{b};\underline{c};\underline{d})}$ with $|\underline{c}| \geq 2$. Finite unions of such grafted corollas, as well as their input and output relabelings, are also in this Kontsevich groupoid.

(10) The Kontsevich groupoid

$$\mathtt{Kont}\left(\mathtt{Gr}_{\mathrm{di}}^{\uparrow}, \mathtt{Gr}_{\mathrm{c}}^{\uparrow}\right)$$

also includes the grafted corollas $C_{(\underline{b};\underline{c};\underline{d})}$ with $|\underline{c}| \geq 2$.

(11) The Kontsevich groupoid

$$\mathtt{Kont}\left(\mathtt{Gr}^{\uparrow}, \mathtt{Gr}_{\mathrm{w}}^{\circlearrowleft}\right)$$

contains the finite Hawaiian earring

$$\left(\xi_1^1\right)^{|\underline{c}|} C_{(\underline{c};\underline{c})} \in \mathtt{Gr}_{\mathrm{w}}^{\circlearrowleft}\binom{\varnothing}{\varnothing}$$

with $|\underline{c}| \geq 1$, pictorially depicted as:

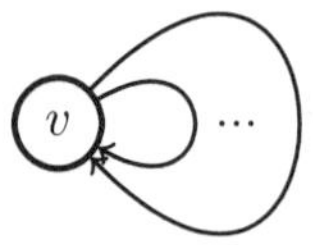

Finite unions of such Hawaiian earrings are also in this Kontsevich groupoid.

(12) The Kontsevich groupoid

$$\texttt{Kont}\left(\texttt{Gr}_{\texttt{c}}^{\uparrow}, \texttt{Gr}_{\texttt{c}}^{\circlearrowright}\right)$$

also includes the Hawaiian earring $\left(\xi_1^1\right)^{|\underline{c}|} C_{(\underline{c};\underline{c})}$ with $|\underline{c}| \geq 1$ as well.

9.2. Well-Matched Pasting Schemes

A well-matched pair of pasting schemes will yield a particularly nice form of the so-called free functor or induction functor, defined as the left adjoint to the forgetful functor, between the two associated categories of generalized PROPs. As such, it will be important to know which of our key examples do, or do not, satisfy this technical condition.

Definition 9.13. Call a pair of pasting schemes

$$\mathcal{G} = (S, \texttt{G}) \leq (S', \texttt{G}') = \mathcal{G}'$$

well-matched if each non-empty graph $G \in \texttt{G}'_S \setminus \texttt{G}_S$ has a presentation in $\mathcal{G}'$ that is unique up to weak isomorphism, by $\mathcal{H} = (\mathcal{H}^1, \mathcal{H}^2)$, where

- $\mathcal{H}^2$ is in the Kontsevich groupoid $\texttt{Kont}(\mathcal{G}, \mathcal{G}')$, and
- each $H \in \mathcal{H}^1$ belongs to $\texttt{G}_S$.

Proposition 9.14. *If the pair $\mathcal{G} \leq \mathcal{G}'$ is well-matched, then*

$$\texttt{G}'_S = \texttt{G}_S * \mathit{Kont}(\mathcal{G}, \mathcal{G}').$$

Proof. By well-matchedness, every $G \in \texttt{G}'_S$ that is not already in $\texttt{G}_S$ has a presentation consisting of wheeled graphs in $\texttt{G}_S$ and the Kontsevich groupoid. □

Let us emphasize that being well-matched is a much stronger condition than saying $\texttt{G}'_S$ is the free product of $\texttt{G}_S$ and the Kontsevich groupoid. In fact, being well-matched says that every non-empty wheeled graph in $\texttt{G}'_S \setminus \texttt{G}_S$ has a weakly unique 2-simplex presentation whose higher index entry is in the Kontsevich groupoid and whose lower index entries are in $\texttt{G}_S$.

9.2.1. Well-Matched Examples.

EXAMPLE 9.15. For any pasting scheme over S,

$$Min(S) \le \mathcal{G}$$

is a well-matched pair.

EXAMPLE 9.16. The pairs of pasting schemes

$$\mathtt{Gr}_{\frac{1}{2}} \le \mathtt{Gr}^{\uparrow}_{\mathrm{di}} \quad \text{and} \quad \mathtt{Gr}_{\frac{1}{2}} \le \mathtt{Gr}^{\uparrow}$$

are both well-matched. This is proven in [**MV09**], although they did not use the same terminology as we do here.

In cases where $\mathtt{G}'_S \smallsetminus \mathtt{G}_S$ is the empty set, the well-matched condition is vacuously satisfied, giving the following simple condition.

LEMMA 9.17. *If* $\mathtt{G}'_S = \mathtt{G}_S$, *then the pair* $\mathcal{G} \le \mathcal{G}'$ *is well-matched.*

EXAMPLE 9.18. The pairs of pasting schemes

$$\mathtt{Tree}^{\circlearrowleft} \le \mathtt{Gr}^{\circlearrowleft}_{\mathrm{c}} \quad \text{and} \quad \mathtt{UTree} \le \mathtt{Gr}^{\uparrow}_{\mathrm{c}}$$

are both well-matched by Lemma 9.17.

Next we have an extension of Lemma 9.11 to show another pair is well-matched.

LEMMA 9.19. *The pair of pasting schemes*

$$\mathtt{Gr}^{\circlearrowleft}_{c} \le \mathtt{Gr}^{\circlearrowleft}_{w}$$

is well-matched.

PROOF. Indeed, suppose we have a non-empty wheeled graph G that is not connected. There is a connected component decomposition

$$G = \sigma\left(\coprod_{i=1}^{p} G_i \sqcup \uparrow_{\underline{b}} \sqcup \circlearrowleft_{\underline{a}}\right)\tau, \tag{9.2}$$

in which each $G_i \in \mathtt{Gr}^{\circlearrowleft}_{\mathrm{c}}\binom{\underline{d}_i}{\underline{c}_i}$ is an ordinary connected component, $\underline{b} = b_{1,r}$, $\underline{a} = a_{1,s}$, and $p + r + s \ge 2$. Define

$$K = \sigma\left(\coprod_{i=1}^{p} C_{(\underline{c}_i;\underline{d}_i)} \sqcup \coprod_{k=1}^{r} C_{(b_k;b_k)} \sqcup \coprod_{l=1}^{s} C_{(\varnothing;\varnothing)}\right)\tau \in \mathtt{Kont}(\mathtt{Gr}^{\circlearrowleft}_{\mathrm{c}}, \mathtt{Gr}^{\circlearrowleft}_{\mathrm{str}}),$$

$$\mathrm{Vt}(K) = \{v_i\}_{i=1}^{p} \sqcup \{u_k\}_{k=1}^{r} \sqcup \{w_l\}_{l=1}^{s},$$

$$H_v = \begin{cases} G_i & \text{if } v = v_i \text{ for } 1 \le i \le p, \\ \uparrow_{b_k} & \text{if } v = u_k \text{ for } 1 \le k \le r, \\ \circlearrowleft_{a_l} & \text{if } v = w_l \text{ for } 1 \le l \le s. \end{cases}$$

Here v_i (resp., u_k and w_l) is the unique vertex in the corolla $C_{(\underline{c}_i;\underline{d}_i)}$ (resp., $C_{(b_k;b_k)}$ and $C_{(\varnothing;\varnothing)}$). Then $(\{H_v\}, K)$ is the unique, up to weak isomorphism, presentation of G that shows that the pair $\mathtt{Gr}^{\circlearrowleft}_{\mathrm{c}} \le \mathtt{Gr}^{\circlearrowleft}_{\mathrm{w}}$ is well-matched. □

A modification of the previous proof can also be applied to verify the following.

EXAMPLE 9.20. The pairs of pasting schemes

$$\mathtt{Tree}^{\circlearrowleft} \le \mathtt{Gr}^{\circlearrowleft}_{\mathrm{w}}, \quad \mathtt{Gr}^{\uparrow}_{\mathrm{c}} \le \mathtt{Gr}^{\uparrow}, \quad \text{and} \quad \mathtt{UTree} \le \mathtt{Gr}^{\uparrow}$$

are well-matched.

9.2.2. Examples That Are Not Well-Matched.

EXAMPLE 9.21. The pairs of pasting schemes

$$\mathtt{Gr}^{\uparrow}_{\mathrm{di}} \leq \mathtt{Gr}^{\uparrow} \quad \text{and} \quad \mathtt{Gr}^{\uparrow}_{\mathrm{di}} \leq \mathtt{Gr}^{\uparrow}_{\mathrm{c}}$$

in Example 9.12 ((9) and (10)) are *not* well-matched because the uniqueness part of well-matchedness fails. An example illustrating this non-uniqueness can be found in [**MV09**] (Remark 7). In plain language, this means that in general there is no unique way, up to weak isomorphism, to compose away all the simply-connected graphs in a (connected) wheel-free graph.

LEMMA 9.22. *The pairs of pasting schemes*

$$\mathtt{Gr}^{\uparrow} \leq \mathtt{Gr}^{\circlearrowleft}_{w} \quad \textit{and} \quad \mathtt{Gr}^{\uparrow}_{c} \leq \mathtt{Gr}^{\circlearrowleft}_{c}$$

are not *well-matched.*

PROOF. We will show that the uniqueness part of well-matchedness fails in both cases by demonstrating a single graph G with two such presentations that are not weakly isomorphic. The colors are not important here, so we fix an arbitrary color $c \in \mathfrak{C}$ and assume all flags involved are this color. The wheeled graph in question is $G \in \mathtt{Gr}^{\circlearrowleft}_{\mathrm{c}}\binom{\varnothing}{\varnothing}$

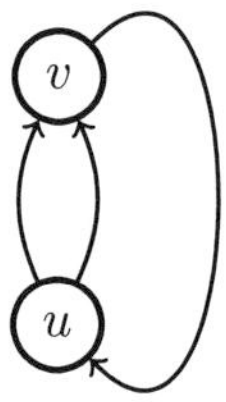

defined as

$$G = \xi_1^1\left(C_{(c;c,c;c)}\right),$$

where $C_{(c;c,c;c)}$ is a grafted corollas (3.3).

Let

$$K_1 = \xi_1^1\left(C_{(c;c)}\right) = C^{1,1}_{(c;c)} \in \mathtt{Gr}^{\circlearrowleft}_{\mathrm{c}}\binom{\varnothing}{\varnothing},$$

which is a contracted corolla (3.5) and is pictorially depicted as:

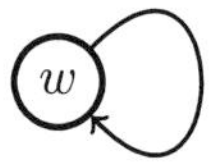

Note that K_1 belongs to the Kontsevich groupoid $\mathtt{Kont}(\mathtt{Gr}^{\uparrow}_{\mathrm{c}}, \mathtt{Gr}^{\circlearrowleft}_{\mathrm{c}})$. Choose

$$H_1 = C_{(c;c,c;c)} \in \mathtt{Gr}^{\uparrow}_{\mathrm{c}}\binom{c}{c},$$

which is a grafted corollas and is pictorially depicted as follows.

Then we have the presentation (H_1, K_1) of $G = K_1(H_1)$.

Let K_2 denote the doubly contracted corolla

$$K_2 = \xi_1^1 \xi_1^1 C_{(c,c;c,c)} \in \mathtt{Gr}_{\mathtt{c}}^{\circlearrowleft}\binom{\varnothing}{\varnothing},$$

which is pictorially depicted as:

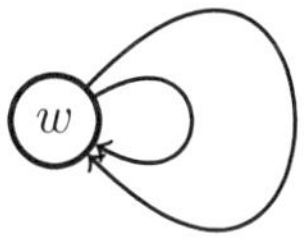

and belongs to the Kontsevich groupoid $\mathtt{Kont}(\mathtt{Gr}_{\mathtt{c}}^{\uparrow}, \mathtt{Gr}_{\mathtt{c}}^{\circlearrowleft})$. Choose

$$H_2 = C_{(c,c;c,c)} \in \mathtt{Gr}_{\mathtt{c}}^{\uparrow}\binom{c, c}{c, c},$$

which is a grafted corollas and is pictorially depicted as:

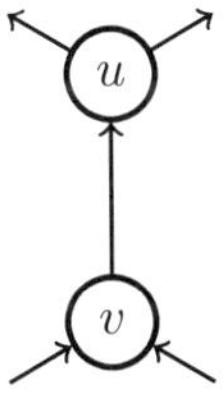

Then we have another presentation (H_2, K_2) of $G = K_2(H_2)$.

Notice K_1 is *not* weakly isomorphic to K_2 because K_1 has 2 flags, while K_2 has 4 flags. Therefore, these two presentations of G are not weakly isomorphic either, which show that the pair $\mathtt{Gr}_{\mathtt{c}}^{\uparrow} \le \mathtt{Gr}_{\mathtt{c}}^{\circlearrowleft}$ is not well-matched. The same presentations of the wheeled graph G also show that the pair $\mathtt{Gr}^{\uparrow} \le \mathtt{Gr}_{\mathtt{w}}^{\circlearrowleft}$ is not well-matched. $\square$

Part 2

Generalized PROPs, Algebras, and Modules

CHAPTER 10

Generalized PROPs

The main purposes of this chapter are to define generalized PROPs corresponding to pasting schemes and to give some initial examples of generalized PROPs, while the more complicated examples are addressed in careful detail in Chapter 11. We begin with some categorical background material, about symmetric monoidal categories, unordered tensor products, and algebras over a monad. The reason is that our definition of generalized PROPs will be as the algebras over a monad built as coproducts of unordered tensor products in a symmetric monoidal category.

The second section is devoted to a variant of the notion of a monad which can be used to describe constructions like the modules over a ring. As you will recall, even for the notion of a left X-set for a monoid X, the obvious associativity condition of the action map does not follow the usual paradigm of an algebra over a monad. However, the category of left G-sets is monadic, and in general it is possible to associate a monad to recover the category we would consider the modules. This construction can be treated as a black box for our purposes, since we will only use it to deduce general properties of our modules. The intent is to set the stage for the category of modules over a generalized PROP, including the existence of (co)limits of modules and the existence of a free module functor.

The next section is devoted to understanding Σ-bimodules and decorated wheeled graphs. The bimodules will be the basic home for the entries of generalized PROPs, and decorated graphs will consist of unordered tensor products that serve as the source of the structure maps of generalized PROPs.

Section 10.4 details the monad $F_{\mathcal{G}}$ associated to a pasting scheme, and then defines the associated $\mathcal{G}$-PROPs both as the algebras over this monad and in more concrete terms using structure maps out of decorated graphs. The basic premise is that structure maps of any algebraic type should consist of maps out of tensor products, so each graph in the pasting scheme induces a "multiplication" from a tensor product associated to the vertices of the graph into an entry associated to the full graph profiles. The formal associativity property of the monad then translates into the statement that

$$\gamma_{[G(H_v)]} = \gamma_{[G]} \circ \bigotimes_v \gamma_{[H_v]} \,.$$

In particular, any two presentations of the same graph will produce the same structure map, which is a more familiar sort of structure statement. These are what Markl would call unbiased definitions, since we have made no effort to focus attention on any subset of the graphs in question. Consideration of the so-called biased definitions of the main examples, where one's attention is biased toward the graphs in a chosen strong generating set and so produces more finitary descriptions, is postponed until Chapter 11.

In addition, Section 10.5 provides a few of the simpler examples, including a characterization of generalized PROPs over a monogenic pasting scheme as a category of diagrams. This last result applies to both the minimal pasting scheme `Cor`, where the diagram category is just the bimodules, and the contractions pasting scheme $\mathtt{Gr}^{\xi}$.

10.1. Categorical Preliminaries

In this section, we briefly recall some basic definitions regarding (symmetric) monoidal categories, (un)ordered tensor product, and monads. Some standard references are [**Bor94, Kel82, Mac65, Mac98**].

If A and B are objects in a category $\mathcal{C}$, then the set of morphisms from A to B is denoted by $\mathcal{C}(A,B)$.

10.1.1. Monoidal Categories.

DEFINITION 10.1. A **monoidal category**

$$(\mathcal{E}, \otimes, I, \alpha, \lambda, \rho)$$

consists of:

- a category $\mathcal{E}$;
- a **unit** object $I \in \mathcal{E}$;
- a bifunctor
  $$\otimes : \mathcal{E} \times \mathcal{E} \longrightarrow \mathcal{E},$$
  called the **monoidal product** or the **tensor product**;
- an **associativity isomorphism**
  $$\alpha_{ABC} : (A \otimes B) \otimes C \longrightarrow A \otimes (B \otimes C)$$
  that is natural in A, B, and C;
- a **left unit isomorphism**
  $$\lambda_A : I \otimes A \longrightarrow A$$
  that is natural in A;
- a **right unit isomorphism**
  $$\rho_A : A \otimes I \longrightarrow A$$
  that is natural in A;

such that $\lambda_I = \rho_I$ and the following two diagrams are commutative.

Pentagon axiom:

$$\begin{array}{ccc}
((A \otimes B) \otimes C) \otimes D & \xrightarrow{\quad\alpha\quad} & (A \otimes B) \otimes (C \otimes D) \\
{\scriptstyle \alpha \otimes Id}\downarrow & & \\
(A \otimes (B \otimes C)) \otimes D & & \downarrow{\scriptstyle \alpha} \\
{\scriptstyle \alpha}\downarrow & & \\
A \otimes ((B \otimes C) \otimes D) & \xrightarrow{\quad Id \otimes \alpha\quad} & A \otimes (B \otimes (C \otimes D))
\end{array}$$

Unit coherence:

$$\begin{array}{ccc} (A \otimes I) \otimes B & \xrightarrow{\alpha} & A \otimes (I \otimes B) \\ & \underset{\rho \otimes Id}{\searrow} & \downarrow{\scriptstyle Id \otimes \lambda} \\ & & A \otimes B \end{array}$$

A monoidal category is **strict** if the associativity and unit isomorphisms are identity maps.

We will often denote a monoidal category $(\mathcal{E}, \otimes, I, \alpha, \lambda, \rho)$ by $(\mathcal{E}, \otimes, I)$ or even just $\mathcal{E}$.

REMARK 10.2. We follow the convention in [**Bor94, Kel82**] that the associativity isomorphism α moves the parentheses to the right. The reader should be aware that in [**Mac98**] the associativity isomorphism moves the parentheses to the left.

It is not surprising that the correct notion of functor between monoidal categories now requires a compatibility with the monoidal products and another with the units. A similar statement also holds for the correct notion of natural transformation which follows.

DEFINITION 10.3. A **monoidal functor**

$$(F, F_2, F_0) : \mathcal{D} \longrightarrow \mathcal{E}$$

between two monoidal categories consists of:

- a functor $F : \mathcal{D} \longrightarrow \mathcal{E}$;
- a morphism
$$F_2 : F(A) \otimes F(B) \longrightarrow F(A \otimes B) \in \mathcal{E}$$
that is natural in $A, B \in \mathcal{D}$;
- a morphism
$$F_0 : I \longrightarrow F(I) \in \mathcal{E}.$$

such that the following three diagrams are commutative.

Associativity coherence:

$$\begin{array}{ccc} (F(A) \otimes F(B)) \otimes F(C) & \xrightarrow{\alpha} & F(A) \otimes (F(B) \otimes F(C)) \\ {\scriptstyle F_2 \otimes Id}\downarrow & & \downarrow{\scriptstyle Id \otimes F_2} \\ F(A \otimes B) \otimes F(C) & & F(A) \otimes F(B \otimes C) \\ {\scriptstyle F_2}\downarrow & & \downarrow{\scriptstyle F_2} \\ F((A \otimes B) \otimes C) & \xrightarrow{F(\alpha)} & F(A \otimes (B \otimes C)) \end{array}$$

Left unit coherence:

$$\begin{array}{ccc} I \otimes F(A) & \xrightarrow{\lambda} & F(A) \\ {\scriptstyle F_0 \otimes Id}\downarrow & & \uparrow{\scriptstyle F(\lambda)} \\ F(I) \otimes F(A) & \xrightarrow{F_2} & F(I \otimes A) \end{array}$$

Right unit coherence:

$$\begin{array}{ccc} F(A)\otimes I & \xrightarrow{\rho} & F(A) \\ {\scriptstyle Id\otimes F_0}\downarrow & & \uparrow{\scriptstyle F(\rho)} \\ F(A)\otimes F(I) & \xrightarrow{F_2} & F(A\otimes I) \end{array}$$

A monoidal functor is **strong** (resp., **strict**) if the morphisms F_0 and F_2 are isomorphisms (resp., identity maps).

As before, a monoidal functor (F, F_2, F_0) is often abbreviated to F.

DEFINITION 10.4. Let $\mathcal{D}$ and $\mathcal{E}$ be monoidal categories, and let both F and G be monoidal functors from $\mathcal{D}$ to $\mathcal{E}$. A **monoidal natural transformation**

$$\theta\colon F \longrightarrow G$$

is a natural transformation of the underlying functors such that the following two diagrams are commutative.

$$\begin{array}{ccc} F(A)\otimes F(B) & \xrightarrow{F_2} & F(A\otimes B) \\ {\scriptstyle \theta\otimes\theta}\downarrow & & \downarrow{\scriptstyle \theta} \\ G(A)\otimes G(B) & \xrightarrow{G_2} & G(A\otimes B) \end{array}$$

$$\begin{array}{ccc} I & \xrightarrow{F_0} & F(I) \\ & {\scriptstyle G_0}\searrow & \downarrow{\scriptstyle \theta} \\ & & G(I) \end{array}$$

10.1.2. Symmetric Monoidal Categories. As in studying rings, certain operations become difficult or even impossible without imposing some form of commutativity condition on the product, as we do here for monoidal products.

DEFINITION 10.5. A **symmetric monoidal category** is a monoidal category $\mathcal{E}$ equipped with a natural isomorphism

$$\tau\colon A\otimes B \longrightarrow B\otimes A$$

for $A, B \in \mathcal{E}$, called the **symmetry**, such that the following three diagrams are commutative.

Associativity coherence:

$$\begin{array}{ccc} (A\otimes B)\otimes C & \xrightarrow{\tau\otimes Id} & (B\otimes A)\otimes C \\ {\scriptstyle \alpha}\downarrow & & \downarrow{\scriptstyle \alpha} \\ A\otimes(B\otimes C) & & B\otimes(A\otimes C) \\ {\scriptstyle \tau}\downarrow & & \downarrow{\scriptstyle Id\otimes\tau} \\ (B\otimes C)\otimes A & \xrightarrow{\alpha} & B\otimes(C\otimes A) \end{array}$$

Unit coherence:

$$\begin{array}{ccc} A \otimes I & \xrightarrow{\tau} & I \otimes A \\ & \searrow^{\rho} & \downarrow{\scriptstyle\lambda} \\ & & A \end{array}$$

Symmetry axiom:

$$\begin{array}{ccc} A \otimes B & \xrightarrow{\tau} & B \otimes A \\ & \searrow^{Id} & \downarrow{\scriptstyle\tau} \\ & & A \otimes B \end{array}$$

The following condition may appear unusual at first glance, but the adjoint relationship provides a strong tool to translate a number of relationships between the two types of functors.

DEFINITION 10.6. A **symmetric monoidal closed category** is a symmetric monoidal category $\mathcal{E}$ in which for each object $A \in \mathcal{E}$, the functor

$$- \otimes A : \mathcal{E} \longrightarrow \mathcal{E}$$

has a right adjoint

$$\mathrm{Hom}_{\mathcal{E}}(A, -) : \mathcal{E} \longrightarrow \mathcal{E},$$

called the **internal hom functor**.

When desperate to save space, e.g. in commutative diagrams, we will sometimes write $[A, ?]_{\mathcal{E}}$ or even just $[A, ?]$ as shorthand for $\mathrm{Hom}_{\mathcal{E}}(A, -)$.

EXAMPLE 10.7. Here are some standard examples of symmetric monoidal closed categories. The monoidal product is the usual (tensor/smash) product in each case.

(1) **Set** - the category of sets.
(2) **Mod** - the category of (graded) modules over an associative commutative ring with unit.
(3) **Ch** - the category of chain complexes over an associative commutative ring with unit.
(4) **SSet** - the category of simplicial sets.
(5) **SMod** - the category of simplicial modules over an associative commutative ring with unit.
(6) **SGrp** - the category of simplicial groups.
(7) **Top** - the category of compactly generated Hausdorff spaces with the Kelly functor applied to the Cartesian product.
(8) $\mathbf{Top}_*$ the pointed analog of **Top**, using the Kelly smash product.
(9) S-**Mod** - the category of S-modules [**EKMM97**], where S is the sphere spectrum.
(10) $\mathbf{Sp}^{\Sigma}$ - the category of symmetric spectra [**HSS00**].

All the symmetric monoidal closed categories above have all small limits and colimits.

DEFINITION 10.8. A **symmetric monoidal functor** between two symmetric monoidal categories is a monoidal functor (F, F_2, F_0) such that the square

$$\begin{array}{ccc} F(A)\otimes F(B) & \xrightarrow{\tau} & F(B)\otimes F(A) \\ {\scriptstyle F_2}\downarrow & & \downarrow{\scriptstyle F_2} \\ F(A\otimes B) & \xrightarrow{F(\tau)} & F(B\otimes A) \end{array}$$

commutes.

Notice that a monoidal natural transformation remains the appropriate way of mapping between symmetric monoidal functors, since the diagram

$$\begin{array}{ccc} F(A\otimes B) & \xrightarrow{F(\tau)} & F(B\otimes A) \\ {\scriptstyle \theta}\downarrow & & \downarrow{\scriptstyle \theta} \\ G(A\otimes B) & \xrightarrow{G(\tau)} & G(B\otimes A) \end{array}$$

automatically commutes by naturality of θ and the fact that τ is a morphism in $\mathcal{D}$.

We now include a standard fact which provides a simple criterion for verifying that both functors in an adjunction are symmetric monoidal.

LEMMA 10.9. *Given an adjoint pair* $L : \mathcal{E} \rightleftarrows \mathcal{D} : R$ *between symmetric monoidal categories with* L *strong symmetric monoidal, it follows that* R *is also a symmetric monoidal functor*

PROOF. We must produce natural maps $R_2 : R(A)\otimes R(B) \longrightarrow R(A\otimes B)$, which we describe as adjoint to the natural composite

$$L\left(R(A)\otimes R(B)\right) \cong LR(A)\otimes LR(B) \longrightarrow A\otimes B$$

where the isomorphism comes from the strong assumption and the second map is the tensor product of two counit maps. Similarly, we must produce $R_0 : I \longrightarrow R(I)$ which is adjoint to the isomorphism $L(I) \cong I$, and in both cases the required diagrams follow from adjoint diagrams working with L. □

CONVENTION 10.10. From this point forward, unless otherwise specified, all the symmetric monoidal categories under consideration, which will generally be denoted $\mathcal{E}$, are assumed to have all small colimits.

10.1.3. Ordered and Unordered Tensor Products. The following presentation of (un)ordered tensor products in a symmetric monoidal category $\mathcal{E}$ is similar to that in [**MSS02**] (I.1.7).

Let X be a set of cardinality n. An **ordering** of X is a bijection

$$\sigma : X \xrightarrow{\cong} \{1, \ldots, n\}.$$

The set of all orderings of X is denoted by $\mathrm{Ord}(X)$.

Let $A_x \in \mathcal{E}$ be an object for each $x \in X$. For each ordering σ of X, define the **ordered tensor product**

$$\bigotimes_{\sigma} A_x = A_{\sigma^{-1}(1)} \otimes \cdots \otimes A_{\sigma^{-1}(n)} \in \mathcal{E}.$$

If $X = \varnothing$, then the ordered tensor product is defined as the unit object I. If $\rho \in \Sigma_n$ is a permutation, then the symmetry in $\mathcal{E}$ determines an isomorphism

$$\rho : \bigotimes_{\sigma} A \xrightarrow{\cong} \bigotimes_{\rho\sigma} A,$$

which defines a Σ_n-action on the coproduct $\coprod_{\sigma \in \mathrm{Ord}(X)} \bigotimes_{\sigma} A_x$. Define the **unordered tensor product** as the resulting Σ_n-coinvariants:

$$\begin{aligned} \bigotimes_{x \in X} A_x &= \operatorname*{colim}_{\rho \in \Sigma_n} \left(\coprod_{\sigma \in \mathrm{Ord}(X)} \bigotimes_{\sigma} A_x \xrightarrow{\rho} \coprod_{\sigma \in \mathrm{Ord}(X)} \bigotimes_{\sigma} A_x \right) \\ &= \left(\coprod_{\sigma \in \mathrm{Ord}(X)} \bigotimes_{\sigma} A_x \right)_{\Sigma_n} . \end{aligned} \tag{10.1}$$

By construction, for each ordering σ of X, there is a natural map

$$\bigotimes_{\sigma} A_x \longrightarrow \bigotimes_{x \in X} A_x$$

given by including one summand followed with the natural quotient to the colimit. In particular, each morphism from the unordered tensor product restricts to a morphism from each ordered tensor product of the same factors. Conversely, suppose

$$f : \bigotimes_{\sigma} A_x \longrightarrow B$$

is a morphism from an ordered tensor product. Then by symmetry there is a unique induced morphism

$$\overline{f} : \bigotimes_{x \in X} A_x \longrightarrow B \tag{10.2}$$

from the unordered tensor product, whose restriction to the ordered tensor product $\bigotimes_{\sigma} A_x$ is f.

Suppose X is the set $\{1, \ldots, n\}$. Then we also write

$$\bigotimes_{i=1}^{n} A_i = A_1 \otimes \cdots \otimes A_n$$

for the ordered tensor product and

$$\bigodot_{i=1}^{n} A_i = A_1 \odot \cdots \odot A_n = \bigotimes_{i \in X} A_i$$

for the unordered tensor product.

10.1.4. Monads and Their Algebras.

DEFINITION 10.11. A **monad** (T, μ, ν) on a category $\mathcal{C}$ consists of

- a functor $T : \mathcal{C} \longrightarrow \mathcal{C}$, and
- natural transformations

$$\mu : T^2 \longrightarrow T \quad \text{and} \quad \nu : Id \longrightarrow T,$$

called the **multiplication** and the **unit**, respectively, such that the following associativity and unity diagrams are commutative.

$$\begin{array}{ccc} T^3 & \xrightarrow{T\mu} & T^2 \\ {\scriptstyle \mu T}\downarrow & & \downarrow{\scriptstyle \mu} \\ T^2 & \xrightarrow[\mu]{} & T \end{array} \qquad \begin{array}{ccccc} T & \xrightarrow{T\nu} & T^2 & \xleftarrow{\nu T} & T \\ & {\scriptstyle Id}\searrow & \downarrow{\scriptstyle \mu} & \swarrow{\scriptstyle Id} & \\ & & T & & \end{array} \tag{10.3}$$

A monad (T,μ,ν) is often abbreviated to T. In fact, any adjoint pair $L:\mathcal{C}\longrightarrow\mathcal{D}$ and $R:\mathcal{D}\longrightarrow\mathcal{C}$ defines a monad on $\mathcal{C}$, by considering RL.

EXAMPLE 10.12. The standard example of a monad in **Set** comes from the word construction, which we describe formally as $TX = \coprod_{n\geq 0} X^n$, where X^n is the Cartesian product of n copies of X, and this represents a singleton when $n = 0$. Then ν is the inclusion of X as X^1. To describe μ, it is first important to identify

$$T^2X = \coprod_{m\geq 0}(\coprod_{n\geq 0} X^n)^m \cong \coprod_{m,n_1,\ldots,n_m} X^{n_1}\times\cdots\times X^{n_m}\,.$$

Now define

$$\mu_{m,n_1,\ldots,n_m} : X^{n_1}\times\cdots\times X^{n_m} \longrightarrow X^{\Sigma_i n_i}$$

simply by concatenating Cartesian products. It follows that including the target into the coproduct TX and taking these composites as the summands defines the required map μ. In other words, this is just the concatenation operation in the usual description of words in X.

To verify the unit diagrams, notice ν_{TX} has summands including X^n into T^2X as the copy indexed on the pair $(1,n)$, so thinking of a word in X as a word of words which just happens to have a single letter in the outer word. On the other hand $T\nu_X$ includes X^n as the copy of $X^1\times\cdots\times X^1$ indexed on $(n,1,\ldots,1)$, so thinking of a word in X as a word of words by parenthesizing each letter.

For describing T^3X, we use a similar isomorphism

$$T^3X \cong \coprod_{m,n_1,\ldots,n_m,k_1^1,\ldots,k_{n_1}^1,k_1^2,\ldots,k_{n_m}^m} X^{k_1^1}\times\cdots\times X^{k_{n_m}^{n_m}}\,.$$

Then on a given summand, $T\mu_X$ tells us to concatenate each $X^{k_1^i}\times\cdots\times X^{k_{n_i}^i}$ to a single $X^{\Sigma_j k_j^i}$, and the subsequent μ_X then tells us to concatenate further to a single $X^{\Sigma_i\Sigma_j k_j^i}$. On the other hand, on the same summand μ_{TX} tells us to first view this string of copies as indexed on $\Sigma_i n_i$ terms, and then μ_X tells us to concatenate these long strings into a single product $X^{\Sigma_j\Sigma_i k_j^i}$. As these two iterated concatenations agree, we have verified commutativity of the diagram one summand at a time.

With this formal description, the same approach suffices to describe a monad on any symmetric monoidal *closed* category, where the closed condition is used in our decomposition of T^2 and T^3 via distributivity of the $\otimes$ over $\coprod$.

Every monad has an associated category of algebras, which are a natural generalization of the notion of sets acted upon by a monoid.

DEFINITION 10.13. Given (T,μ,ν) a monad on $\mathcal{C}$, a T-**algebra** (X,γ) consists of

- an object $X\in\mathcal{C}$ and

- a morphism

$$\gamma : TX \longrightarrow X \in \mathcal{C},$$

  called the **structure map**

such that the following associativity and unity diagrams are commutative.

$$\begin{array}{ccc} T^2X & \xrightarrow{T\gamma} & TX \\ {\scriptstyle \mu_X}\downarrow & & \downarrow{\scriptstyle \gamma} \\ TX & \xrightarrow[\gamma]{} & X \end{array} \qquad\qquad \begin{array}{ccc} X & \xrightarrow{\nu_X} & TX \\ & {\scriptstyle Id}\searrow & \downarrow{\scriptstyle \gamma} \\ & & X \end{array}$$

If (Y, γ^Y) is another T-algebra, then a **morphism** of T-algebras

$$f : (X, \gamma^X) \longrightarrow (Y, \gamma^Y)$$

is a morphism $f : X \longrightarrow Y$ in $\mathcal{C}$ such that the square

$$\begin{array}{ccc} TX & \xrightarrow{T(f)} & TY \\ {\scriptstyle \gamma^X}\downarrow & & \downarrow{\scriptstyle \gamma^Y} \\ X & \xrightarrow[f]{} & Y \end{array}$$

commutes. The category of algebras over T will be denoted $\mathbf{Alg}(T)$.

EXAMPLE 10.14. When T comes from the word construction as in Example 10.12, a T-algebra is simply an associative monoid (in **Set**), and a morphism of T-algebras is just a morphism of monoids. To see this, notice that the associativity condition for a monoid says that there is a unique iterated multiplication $\gamma_n : X^n \longrightarrow X$ for $n > 1$, while the map $\gamma_1 : X^1 \longrightarrow X$ must be the identity by the unit condition for a T-algebra, and the map $\gamma_0 : X_0 = \{*\} \longrightarrow X$ will simply pick out the monoidal unit element. This defines the map $\gamma_X : TX \longrightarrow X$ when X is an associative monoid, and verifies the unit condition for a T-algebra structure.

Considering the associativity condition for a T-algebra structure, restricted to a summand indexed by $(m, n_1, \ldots, n_m)$, the map $T\gamma_X$ sends $X^{n_1} \times \cdots \times X^{n_m}$ to $X^1 \times \ldots \times X^1$ indexed on $(m, 1, \ldots, 1)$, and the subsequent γ_X multiplies once again to end up in X. In other words, this direction takes a word of words and first multiplies within the parentheses, followed by multiplying those answers together to produce a single element. In the other direction, μ_X simply concatenates to a single long string in $X^{\Sigma_i n_i}$ and then the subsequent γ_X multiplies all of these terms at once to produce a single element of X.

Comparing the special cases of $(2, 2, 1)$ and $(2, 1, 2)$ verify that γ_3 in fact corresponds to the associative multiplication of a monoid in the usual sense,

$$(10.4)\qquad \begin{array}{ccccc} X^2 \times X^1 & \xrightarrow{\cong} & X^1 \times X^2 & \xrightarrow{\gamma^1 \times \gamma^2} & X^1 \times X^1 \\ & {\scriptstyle \mu_{2,2,1}}\searrow & \downarrow{\scriptstyle \mu_{2,1,2}} & & \\ {\scriptstyle \gamma^2 \times \gamma^1}\downarrow & & X^3 & & \downarrow{\scriptstyle \gamma^2} \\ & & & {\scriptstyle \gamma^3}\searrow & \\ X^1 \times X^1 & & \xrightarrow[\gamma^2]{} & & X^1 \end{array}$$

which relies upon compatibility of the summand μ maps with the associativity isomorphism for Cartesian product. In fact, all higher instances can then be established by induction, but the T-algebra approach deals with all cases at once.

Similarly, a morphism of T-algebras must be compatible with all of the multiplications γ_n, which is equivalent by induction to the usual notion of a morphism of monoids. As alluded to above, this approach will suffice to describe monoids in any symmetric monoidal closed category, which is the primary reason for the formal phrasing.

The category of T-algebras has all (co)limits which exist in the underlying category, $\mathcal{C}$. Given any diagram of T-algebras, the limit of the underlying objects gets a structure map from composing with the natural map into the limit

$$T(\lim_\alpha X_\alpha) \longrightarrow \lim_\alpha TX_\alpha \longrightarrow \lim_\alpha X_\alpha \ .$$

For colimits, one instead takes the free algebra associated to the colimit of the underlying diagrams. (See e.g. [**Bor94**, 4.3.1-4.3.4].)

It will be important later to determine which functors preserve monads and their algebras, which is the point of the following.

LEMMA 10.15. *Suppose $K : \mathcal{D} \longrightarrow \mathcal{E}$ is a functor, while R is a monad on $\mathcal{D}$ and T is a monad on $\mathcal{E}$. Assume $\psi : TK \longrightarrow KR$ is a natural transformation that is multiplicative in the sense that the diagram*

$$\begin{array}{ccccc} TTK & \xrightarrow{T\psi} & TKR & \xrightarrow{\psi R} & KRR \\ \Big\downarrow{\scriptstyle \mu_T K} & & & & \Big\downarrow{\scriptstyle K\mu_R} \\ TK & & \xrightarrow{\quad\psi\quad} & & KR \end{array} \tag{10.5}$$

commutes, and unital in the sense that the diagram

$$\begin{array}{ccc} & \overset{\nu_T}{\nearrow} & TK \\ & & \Big\downarrow{\scriptstyle \psi} \\ K & \xrightarrow{K\nu_R} & KR \end{array}$$

commutes. Then K induces a functor from R-algebras to T-algebras.

PROOF. Suppose $X \in \mathcal{D}$ is an R-algebra, so equipped with a choice of structure map $\gamma : RX \longrightarrow X$. Then choose $K(\gamma) \circ \psi_X : TKX \longrightarrow KRX \longrightarrow KX$ as the structure map for KX. The unity diagram follows from applying K to the unity diagram for X and the unity assumption on ψ

$$\begin{array}{ccc} & \overset{\nu_T}{\nearrow} & TKX \\ & & \Big\downarrow{\scriptstyle \psi} \\ KX & \xrightarrow{K\nu_R} & KRX \\ & \underset{Id}{\searrow} & \Big\downarrow{\scriptstyle K\gamma} \\ & & KX. \end{array}$$

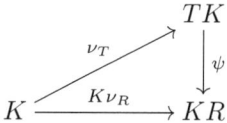

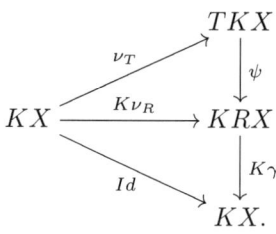

Similarly, the associativity diagram follows as in the diagram

$$\begin{array}{ccccc}
TTKX & \xrightarrow{T\psi_X} & TKRX & \xrightarrow{TK\gamma} & TKX \\
 & & \downarrow{\scriptstyle \psi_{RX}} & & \downarrow{\scriptstyle \psi_X} \\
{\scriptstyle \mu_T K(X)}\downarrow & & KRRX & \xrightarrow{KR\gamma} & KRX \\
 & & {\scriptstyle K\mu_R(X)}\downarrow & & \downarrow{\scriptstyle K\gamma} \\
TKX & \xrightarrow{\psi_X} & KRX & \xrightarrow{K\gamma} & KX
\end{array}$$

where the commutativity of the rectangle on the left is the multiplicativity assumption on ψ, the top square commutes by naturality of ψ together with the fact that γ is a map in $\mathcal{D}$, and the bottom square is K applied to the associativity diagram of X.

To see that the construction above preserves morphisms of algebras, consider the commutative diagram

$$\begin{array}{ccc}
TKX & \xrightarrow{TKf} & TKY \\
{\scriptstyle \psi_X}\downarrow & & \downarrow{\scriptstyle \psi_Y} \\
KRX & \xrightarrow{KRf} & KRY \\
{\scriptstyle K\gamma_X}\downarrow & & \downarrow{\scriptstyle K\gamma_Y} \\
KX & \xrightarrow{Kf} & KY
\end{array}$$

where the top square commutes by naturality of ψ and the bottom square is K applied to the square defining $f : X \longrightarrow Y$ a morphism of algebras. Composition and identities are constructed as in the underlying category, so compatibility with them follows from K an underlying functor. □

10.2. Pointed Extensions of Monads

Suppose we have a multiplicative structure defined solely in terms of an algebra over a monad, and we would like to understand the corresponding notion of module over such a structure. A natural approach would be to ask if there is a 2-variable functor related to the functor defining our monad, so that $T(X, X)$ is closely related to TX, but $T(X, M) \longrightarrow M$ would represent the action of X on M.

DEFINITION 10.16. Given a monad $T : \mathcal{C} \longrightarrow \mathcal{C}$, a **pointed extension of** T is a functor of two variables $T(?, ?) : \mathcal{C} \times \mathcal{C} \longrightarrow \mathcal{C}$, together with

- a natural unit map $\nu_{X,M} : M \longrightarrow T(X, M)$ and
- a natural multiplication map $\mu_{X,M} : T(TX, T(X, M)) \longrightarrow T(X, M)$,

such that defining $\overline{T}_X M = T(TX, T(X, M))$, $T_X M = T(X, M)$, and

$$\lambda_{X,M} : T_X^2 M \xrightarrow{T(\nu_X, Id)} \overline{T}_X M \xrightarrow{\mu_{X,M}} T_X M \; ,$$

for any T-algebra (X, γ_X), it is required that $(T_X, \lambda_{X,?}, \nu_{X,?})$ is a monad on $\mathcal{C}$ and the following unit and associativity diagrams commute.

(10.6)
$$\begin{array}{ccc} T_X M & \xrightarrow{\nu_{X,T_X M}} & T_X^2 M \\ & \searrow^{Id} & \downarrow \lambda_{X,M} \\ & & T_X M \end{array}$$

(10.7)
$$\overline{T}_{TX}(T_X^2 M) \underset{T(T\gamma_X, \mu_{X,T_X M})}{\overset{\mu_{TX,T_X^2 M}}{\rightrightarrows}} \overline{T}_X(T_X M) \xrightarrow{\mu_{X,T_X M}} T_X^2 M$$

EXAMPLE 10.17. Given the word monad T of Example 10.12, there is a pointed extension for left X-sets with $T(X,M) = \coprod_{n>0} X^{n-1} \times M$. The unit map still just identifies M with the entry $n = 1$, while the multiplication map is still just concatenation of words. Notice that if a string of words in X is followed by a (possibly empty) word in X with a letter from M added at the end, then the concatenation is a single word in X with a letter from M at the end. Also notice the empty word is missing, so TX is then $T(X,X) \coprod \{*\}$.

Now we would like to use a pointed extension of T to define when an object has an action by an algebra over the monad T.

DEFINITION 10.18. Suppose T is a monad on $\mathcal{C}$ and (X, γ_X) is an algebra over T, while $T(?,?)$ is a pointed extension of T. Then **an X-module** with respect to T will consist of a pair (M, λ), where $\lambda : T(X,M) \longrightarrow M$ in $\mathcal{C}$, making the following associativity and unit diagrams commute.

$$\begin{array}{ccc} T(TX, T(X,M)) & \xrightarrow{\mu_{X,M}} & T(X,M) \\ {\scriptstyle T(\gamma_X,\lambda)}\downarrow & & \downarrow{\scriptstyle \lambda} \\ T(X,M) & \xrightarrow{\lambda} & M \end{array} \qquad \begin{array}{ccc} M & \xrightarrow{\nu_{X,M}} & T(X,M) \\ & \searrow^{Id} & \downarrow{\scriptstyle \lambda} \\ & & M \end{array}$$

Given two modules (M, λ^M) and (N, λ^N) over X, a **morphism of modules** is an $f : M \longrightarrow N$ in $\mathcal{C}$ which makes the following diagram commute.

$$\begin{array}{ccc} T(X,M) & \xrightarrow{\lambda^M} & M \\ {\scriptstyle T(Id,f)}\downarrow & & \downarrow{\scriptstyle f} \\ T(X,N) & \xrightarrow{\lambda^N} & N \end{array}$$

The category of X-modules will be denoted $\mathrm{Module}_T(X)$ or $\mathrm{Module}(X)$.

REMARK 10.19. Being an X-module is a stronger condition than being an algebra over the monad $(T_X, \lambda_{X,?}, \nu_{X,?})$ by considering the following diagram,

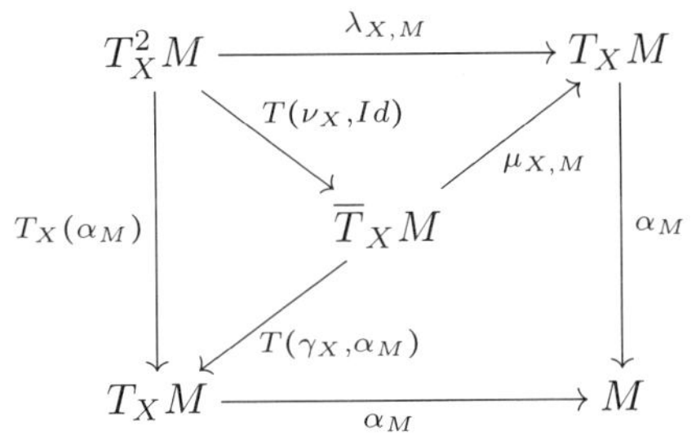

where the top right triangle is the definition of $\lambda_{X,M}$, and the left triangle commutes by the unit condition for γ_X. This should not be surprising, given that we did not need to assume X was an algebra over T in order to define the structure maps of the monad $(T_X, \lambda_{X,?}, \nu_{X,?})$. The point is that defining traditional modules requires an interaction between the multiplication in the ring and the module action map, which is more closely related to the lower right distorted square in the diagram above, or equivalently, our module associativity diagram.

It is also comforting to note that the associativity diagram for a pointed extension of a monad is an instance of that for a module, just as we saw with algebras over a monad.

EXAMPLE 10.20. With our running example of the word construction, a module over the pointed extension consisting of words in X with last letter in M, an X-module for a monoid is just a left X-**Set**, and a morphism of X-modules is a map of left X-**Set**s.

There is a slight variation which suffices to define bimodules over an associative ring. The point is that the associativity diagram above becomes the extension of the familiar condition for a bimodule that

$$\lambda(x_1, \lambda(x_2, m)) = \lambda(x_1 x_2, m)$$

as well as, following the comparison of two such structure maps as in the previous section,

$$\lambda(\lambda(x_1, m)x_2) = \lambda(x_1, \lambda(m, x_2)).$$

It will be important later, when defining modules over a generalized PROP, that the modules in the sense above remain a category of algebras over a monad. However, the free X-module on M is not just $T_X M$, since that need not satisfy the module associativity condition. One must instead force that associativity condition by looking at a quotient of $T_X M$ via a coequalizer.

10.2.1. Monad Replacement for Pointed Extensions of a Monad. The remainder of this section is devoted to the technical question of building a monad whose category of algebras is naturally isomorphic to Module(X). This construction will not be used in detail later, but will instead serve as a method of producing several formal consequences. In particular, if $\mathcal{C}$ has (co)limits, the same is true of Module(X), and there is a free X-module functor left adjoint to the evident forgetful functor Module$(X) \longrightarrow \mathcal{C}$.

DEFINITION 10.21. Suppose $T(?,?)$ and (X,γ_X) as in Definition 10.18, and $M \in \mathcal{C}$. Then we use the following *split* coequalizer to define a functor $\mathcal{F}_X : \mathcal{C} \longrightarrow \mathcal{C}$
(10.8)

$$\overline{T}_X(T_X M) \underset{T(\gamma_X,\lambda_{X,M})}{\overset{\mu_{X,T_X M}}{\rightrightarrows}} T_X^2 M \xrightarrow{\lambda_{X,M}} T_X M \xrightarrow{\eta_{X,M}} \mathcal{F}_X M \ ,$$

where the adjective split refers to the natural common section of the parallel composites

$$\begin{array}{ccc} T_X M & \xrightarrow{\nu_{X,T_X M}} & T_X^2 M \\ \vdots & & \downarrow \nu_{TX,T_X^2 M} \\ \overline{T}_X(T_X M) & \xleftarrow{=} & T(TX, T_X^2 M) \ . \end{array}$$

Our goal for the remainder of the section is to establish the following replacement result.

THEOREM 10.22. *Given $T(?,?)$ a pointed extension of a monad T on $\mathcal{C}$ and (X,γ_X) an algebra over T, the functor $\mathcal{F}_X$ defines another monad on $\mathcal{C}$ equipped with a natural isomorphism of categories*

$$\mathbf{Alg}(\mathcal{F}_X) \cong \mathrm{Module}_T(X) \ .$$

We will break the proof into several steps, due to their technical nature.

LEMMA 10.23. *Given $T(?,?)$ and (X,γ_X) as in Theorem 10.22, $\mathcal{F}_X$ defines a functor $\mathcal{C} \longrightarrow \mathrm{Module}_T(X)$.*

PROOF. We first require a module structure map

$$\alpha_{\mathcal{F}_X M} : T_X(\mathcal{F}_X M) \longrightarrow \mathcal{F}_X M$$

which coequalizes $\mu_{X,\mathcal{F}_X M}$ and $T(\gamma_X, \alpha_{\mathcal{F}_X M})$. Since T_X preserves split coequalizers as a monad, the structure map is the induced right vertical in the following commutative diagram:

$$\begin{array}{ccccccc} T_X\overline{T}_X(T_X M) & \underset{T_X(T(\gamma_X,\lambda_{X,M}))}{\overset{T_X(\mu_{T_X M})}{\rightrightarrows}} & T_X^3 M & \xrightarrow{T_X(\lambda_{X,M})} & T_X^2 M & \xrightarrow{T_X(\eta_M)} & T_X(\mathcal{F}_X M) \\ \downarrow \beta_X & & \downarrow \lambda_{T(X,M)} & & \downarrow \lambda_{X,M} & & \vdots\ \alpha_{\mathcal{F}_X M} \\ \overline{T}_X(T_X M) & \underset{T(\gamma_X,\lambda_{X,M})}{\overset{\mu_{T_X M}}{\rightrightarrows}} & T_X^2 M & \xrightarrow{\lambda_{X,M}} & T_X M & \xrightarrow{\eta_M} & \mathcal{F}_X M \ , \end{array}$$

where $\beta_X = \mu_{TX,T_X^2 M} \circ T(\nu_{TX} \circ \nu_X, Id)$. Notice the middle square commutes as the associativity diagram for the monad $(T_X, \lambda_{X,?}, \nu_{X,?})$. We break the left square into upper and lower portions, since both verticals are defined as compositions. The unit condition $T\gamma_X \nu_{TX} = Id_{TX}$ implies both upper diagrams of the form

$$\begin{array}{ccc}
T_X\overline{T}_X(T_XM) & \xrightarrow{T_X(f)} & T_X^3 M \\
\Big\downarrow{\scriptstyle T(\nu_X, Id)} & & \\
T_{TX}\overline{T}_X(T_XM) & & \Big\downarrow{\scriptstyle T(\nu_X, Id)} \\
\Big\downarrow{\scriptstyle T(\nu_{TX}, Id)} & \searrow{\scriptstyle T(Id_{TX}, f)} & \\
T_{T^2X}\overline{T}_X(T_XM) & \xrightarrow{T(T\gamma_X, f)} & T_{TX}(T_X^2 M)
\end{array}$$

commute. For the lower portion, we have

$$\begin{array}{ccc}
T_X\overline{T}_X T_X M & \underset{T(T\gamma_X, T(\gamma_X, \lambda_{X,M}))}{\overset{T(T\gamma_X, \mu_{X,T_XM})}{\rightrightarrows}} & T_{TX}(T_X^2 M) \\
\Big\downarrow{\scriptstyle \mu_{TX, T_X^2 M}} & & \Big\downarrow{\scriptstyle \mu_{X, T_X M}} \\
\overline{T}_X T_X M & \underset{T(\gamma_X, \lambda_{X,M})}{\overset{\mu_{X,T_XM}}{\rightrightarrows}} & T_X^2 M
\end{array}$$

which commutes with the upper horizontals by the associativity diagram for a pointed extension, and with the lower horizontals by the naturality condition for $\mu_{X,M}$.

To see that $\alpha_{\mathcal{F}_X M}$ coequalizes the relevant pair to be a module structure map, consider the diagram

$$\begin{array}{ccccc}
\overline{T}_X(T_XM) & \underset{T(\gamma_X, \lambda_{X,M})}{\overset{\mu_{T_XM}}{\rightrightarrows}} & T_X^2 M & \xrightarrow{\lambda_{X,M}} & T_X M \\
\Big\downarrow{\scriptstyle \overline{T}_X(\eta_M)} & & \Big\downarrow{\scriptstyle T_X(\eta_M)} & & \Big\downarrow{\scriptstyle \eta_M} \\
\overline{T}_X(\mathcal{F}_X M) & \underset{T(\gamma_X, \alpha_{\mathcal{F}_X M})}{\overset{\mu_{\mathcal{F}_X M}}{\rightrightarrows}} & T_X(\mathcal{F}_X M) & \xrightarrow{\alpha_{\mathcal{F}_X M}} & \mathcal{F}_X M\ .
\end{array}$$

Here the right square commutes by construction, as the right square in the previous diagram. Then the left square using the lower maps commutes as $T(\gamma_X, ?)$ applied to the composites in the right square, while the left square with upper maps commutes by naturality of μ. Now the left vertical is an epimorphism, as η_M is a map into a split coequalizer preserved by $\overline{T}_X$, while the upper longest paths agree by η_M coequalizing them specifically. This suffices to imply the two lower paths agree, as desired.

Similarly, to check the unit condition for the structure map $\alpha_{\mathcal{F}_X M}$, consider the diagram

$$\begin{array}{ccccc}
T_X M & \xrightarrow{\nu_{X,T_XM}} & T_X^2 M & \xrightarrow{\lambda_{X,M}} & T_X M \\
\Big\downarrow{\scriptstyle \eta_M} & & \Big\downarrow{\scriptstyle T_X(\eta_M)} & & \Big\downarrow{\scriptstyle \eta_M} \\
\mathcal{F}_X M & \xrightarrow[\nu_{X,\mathcal{F}_X M}]{} & T_X(\mathcal{F}_X M) & \xrightarrow[\alpha_X]{} & \mathcal{F}_X M\ .
\end{array}$$

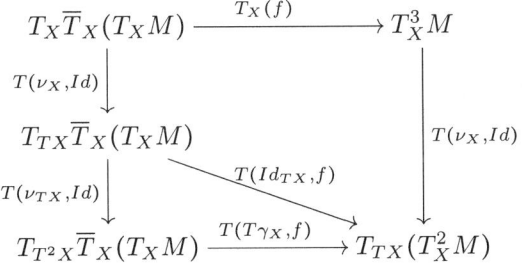

Notice the left square here commutes by naturality of ν and the right square is again the right square in the diagram defining $\alpha_{\mathcal{F}_X M}$. Again, the upper longest composite produces η_X by the unit condition on $T(?, ?)$, so by uniqueness in the definition of an induced map out of a coequalizer, the lower horizontal composite must be the identity.

Finally, we must check that $\mathcal{F}_X$ as defined is functorial, where the compatibility with structure maps, rather than the straightforward compatibility with composition, remains to be verified. Once again we will exploit an epimorphism property to verify $f : M \longrightarrow N$ in $\mathcal{C}$ induces a morphism of modules. In this case, we must verify commutativity of the lower square in the left diagram, which follows from commutativity of the right diagram with the same boundary and the epimorphism property of $T_X(\eta_M)$, the upper left vertical in the first diagram.

$$
\begin{array}{ccc}
T_X^2 M & \xrightarrow{T_X^2(f)} & T_X^2 N \\
{\scriptstyle T_X(\eta_M)}\downarrow & & \downarrow{\scriptstyle T_X(\eta_N)} \\
T_X(\mathcal{F}_X M) & \xrightarrow{T_X\mathcal{F}_X(f)} & T_X\mathcal{F}_X N \\
{\scriptstyle \alpha_{\mathcal{F}_X M}}\downarrow & & \downarrow{\scriptstyle \alpha_{\mathcal{F}_X N}} \\
\mathcal{F}_X M & \xrightarrow{\mathcal{F}_X(f)} & \mathcal{F}_X N
\end{array}
\qquad
\begin{array}{ccc}
T_X^2 M & \xrightarrow{T_X^2(f)} & T_X^2 N \\
{\scriptstyle \lambda_{X,M}}\downarrow & & \downarrow{\scriptstyle \lambda_{X,N}} \\
T_X(\mathcal{F}_X M) & \xrightarrow{T_X(f)} & T_X\mathcal{F}_X N \\
{\scriptstyle \eta_M}\downarrow & & \downarrow{\scriptstyle \eta_N} \\
\mathcal{F}_X M & \xrightarrow{\mathcal{F}_X(f)} & \mathcal{F}_X N\,.
\end{array}
$$

□

LEMMA 10.24. *The functor $\mathcal{F}_X : \mathcal{C} \longrightarrow \mathrm{Module}_T(X)$ is left adjoint to the evident forgetful functor.*

PROOF. Suppose (N, α_N) is an X-module and $f : M \longrightarrow N$ in $\mathcal{C}$. We must first show that f extends uniquely to $\overline{f} : \mathcal{F}_X M \longrightarrow N$ a morphism of X-modules. The extension as a morphism in $\mathcal{C}$ is the induced dotted arrow in the diagram

$$
\begin{array}{ccccccc}
 & & & & M & & \\
 & & & & {\scriptstyle \nu_{X,M}}\downarrow & \searrow{\scriptstyle \overline{\nu}_{X,M}} & \\
\overline{T}_X(T_X M) & \overset{\mu_{X,T_X M}}{\underset{T(\gamma_X,\lambda_{X,M})}{\rightrightarrows}} & T_X^2 M & \xrightarrow{\lambda_{X,M}} & T_X M & \xrightarrow{\eta_M} & \mathcal{F}_X M \\
{\scriptstyle \overline{T}_X(T_X(f))}\downarrow & & \downarrow{\scriptstyle T_X^2(f)} & & \downarrow{\scriptstyle T_X(f)} & & \vdots\,{\scriptstyle \overline{f}} \\
\overline{T}_X(T_X N) & \overset{\mu_{X,T_X N}}{\underset{T(\gamma_X,\lambda_{X,N})}{\rightrightarrows}} & T_X^2 N & \xrightarrow{\lambda_{X,N}} & T_X N & & \vdots \\
{\scriptstyle \overline{T}_X(\alpha_N)}\downarrow & & \downarrow{\scriptstyle T_X(\alpha_N)} & & \downarrow{\scriptstyle \alpha_N} & \swarrow & \\
\overline{T}_X N & \overset{\mu_{X,N}}{\underset{T(\gamma_X,\alpha_N)}{\rightrightarrows}} & T_X N & \xrightarrow{\alpha_N} & N\,. & &
\end{array}
$$

Here the lower right square commutes by considering its expansion to

$$
\begin{array}{ccccc}
T_X^2 N & & \xrightarrow{\lambda_{X,N}} & & T_X N \\
 & \searrow{\scriptstyle T(\nu_X,Id)} & & \nearrow{\scriptstyle \mu_{X,N}} & \\
{\scriptstyle T_X(\alpha_N)}\downarrow & & \overline{T}_X N & & \downarrow{\scriptstyle \alpha_N} \\
 & \swarrow{\scriptstyle T(\gamma_X,\alpha_N)} & & & \\
T_X N & & \xrightarrow{\alpha_N} & & N
\end{array}
$$

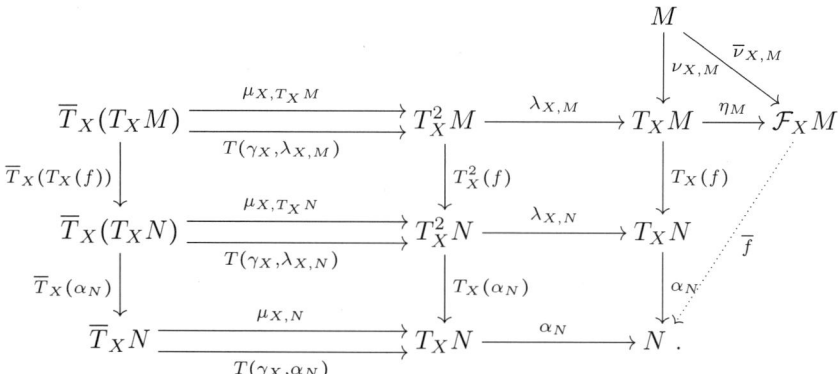

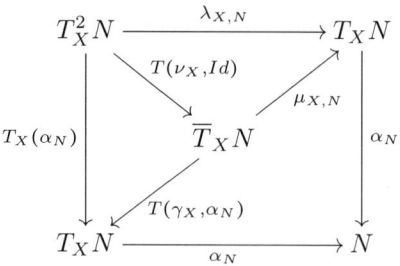

where the bottom right distorted square is the associativity square for the X-module N, the top right triangle commutes by the definition of $\lambda_{X,N}$, and the left triangle commutes by the unit condition for γ_X. This implies the lower left square using the lower maps commutes as well, by applying $T(\gamma_X, ?)$ to the matching composites. The remaining case of the lower left square commutes by naturality of μ, while both upper squares commute by naturality as well.

Now commutativity of the main diagram implies the existence of $\overline{f}$ and uniqueness follows from the universal property of the coequalizer. Commutativity of

$$\begin{array}{ccccc} M & \xrightarrow{\nu_{X,M}} & T_X M & & \\ {\scriptstyle f}\downarrow & & \downarrow{\scriptstyle T_X(f)} & & \\ N & \xrightarrow{\nu_{X,N}} & T_X N & \xrightarrow{\alpha_N} & N\ , \end{array}$$

along with the bottom composite here being the identity by N an X-module, and the large right triangle in the main diagram, imply $\overline{f}\overline{\nu}_M = f$. Also notice that for any morphism of algebras $\mathcal{F}_X M \longrightarrow N$, precomposition with $\overline{\nu}_M$ yields a morphism in $\mathcal{C}$, $M \longrightarrow \mathcal{F}_X M \longrightarrow N$. In fact, as we have just shown by uniqueness, those two processes are mutually inverse, so it remains only to show $\overline{f}$ is an algebra morphism.

We once again exploit the fact that $T_X(\eta_M)$ is an epimorphism to verify commutativity of the diagram on the left by the commutativity of the diagram on the right, which is actually part of the main diagram, with the same boundary.

$$\begin{array}{ccc} T_X^2 M & \xrightarrow{\lambda_{X,M}} & T_X M \\ {\scriptstyle T_X(\eta_M)}\downarrow & & \downarrow{\scriptstyle \eta_M} \\ T_X(\mathcal{F}_X M) & \xrightarrow{\alpha_{\mathcal{F}_X M}} & \mathcal{F}_X M \\ {\scriptstyle T_X(\overline{f})}\downarrow & & \downarrow{\scriptstyle \overline{f}} \\ T_X N & \xrightarrow{\alpha_N} & N \end{array} \qquad \begin{array}{ccc} T_X^2 M & \xrightarrow{\lambda_{X,M}} & T_X M \\ {\scriptstyle T_X^2(f)}\downarrow & & \downarrow{\scriptstyle T_X(f)} \\ T_X^2 N & \xrightarrow{\lambda_{X,N}} & T_X N \\ {\scriptstyle T_X(\alpha_N)}\downarrow & & \downarrow{\scriptstyle \alpha_N} \\ T_X N & \xrightarrow{\alpha_N} & N \end{array}$$

□

Proof of Theorem 10.22. By Lemma 10.24, $\mathcal{F}_X$ is left adjoint to the forgetful functor from X-modules to $\mathcal{C}$. Any adjoint pair produces a monad, by taking the right adjoint applied to the left adjoint, in this case just taking the underlying $\mathcal{C}$-valued functor $\mathcal{F}_X M$ and forgetting about $\alpha_{\mathcal{F}_X M}$. In this case, the unit is $\overline{\nu}_M$, and the monadic multiplication γ will be $\overline{\alpha_{\mathcal{F}_X M}}$ as in the proof of Lemma 10.24. Notice every X-module (N, α_N) will have an $\mathcal{F}_X$-algebra structure map $\overline{\alpha_N}$, and similarly $\alpha_N = \overline{\alpha_N}\eta_N$ will give each $\mathcal{F}_X$-algebra an X-module structure map. In fact, this comparison extends to the notions of morphisms by the fact that the left square in the diagram

$$\begin{array}{ccccc} T_X M & \xrightarrow{\eta_M} & \mathcal{F}_X M & \xrightarrow{\overline{\alpha_N}} & M \\ \downarrow{\scriptstyle T_X(f)} & & \downarrow{\scriptstyle \mathcal{F}_X(f)} & & \downarrow{\scriptstyle f} \\ T_X N & \xrightarrow[\eta_N]{} & \mathcal{F}_X N & \xrightarrow[\overline{\alpha_N}]{} & N \end{array}$$

commutes and the upper left map η_X is an epimorphism. Thus, it follows that the two categories of algebras will coincide up to this comparison of structure maps. □

10.3. Colored Objects, Bimodules, and Decorated Graphs

Fix a choice of S, a replete and full subcategory of $\mathcal{P}(\mathfrak{C})^{op} \times \mathcal{P}(\mathfrak{C})$, which will sometimes be called a **factor of** $\mathcal{P}(\mathfrak{C})^{op} \times \mathcal{P}(\mathfrak{C})$. In this section, we will discuss the fundamental categories of S-colored objects in $\mathcal{E}$ and $\mathbf{\Sigma}_S^{\mathcal{E}}$-bimodules, or factor bimodules, in addition to the key concept of decorated graphs.

10.3.1. Diagram Categories.

It is often useful to consider diagram categories, or functors from a small indexing category into a larger category such as Sets or Topological Spaces.

DEFINITION 10.25. Given a small category $\mathcal{D}$ and a symmetric monoidal category $\mathcal{E}$, **$\mathcal{D}$-indexed diagrams in** $\mathcal{E}$, or $\mathcal{E}^{\mathcal{D}}$, will denote the category of functors $\mathcal{D} \longrightarrow \mathcal{E}$, with natural transformations as morphisms.

A large variety of constructions in $\mathcal{E}^{\mathcal{D}}$ can be performed 'entrywise' in $\mathcal{E}$.

EXAMPLE 10.26.

(1) Given $F \in \mathcal{E}^{\mathcal{D}}$ and $e \in \mathcal{E}$, define $F \otimes e \in \mathcal{E}^{\mathcal{D}}$ with entries

$$(F \otimes e)(d) = F(d) \otimes e$$

in $\mathcal{E}$. Alternatively, this can be described as the functor $F : \mathcal{D} \longrightarrow \mathcal{E}$ followed by the functor $? \otimes e : \mathcal{E} \longrightarrow \mathcal{E}$. If $\mathcal{E}$ is symmetric monoidal closed, one can similarly apply the cotensor, or internal hom, construction entrywise to produce $\mathrm{Hom}_{\mathcal{E}}(e, F)$.

(2) Given F, $G \in \mathcal{E}^{\mathcal{D}}$, define $F \otimes G \in \mathcal{E}^{\mathcal{D}}$ with values

$$(F \otimes G)(d) = F(d) \otimes G(d)$$

in $\mathcal{E}$.

(3) Given a diagram $F : \mathcal{M} \longrightarrow \mathcal{E}^{\mathcal{D}}$, define $\lim_{\mathcal{M}} F \in \mathcal{E}^{\mathcal{D}}$ with values

$$\left(\lim_{\mathcal{M}} F\right)(d) = \lim_{\mathcal{M}}(F(d))$$

in $\mathcal{E}$, where each $F(d)$ is an induced functor $\mathcal{M} \longrightarrow \mathcal{E}$. In this fashion, all small limits and colimits exist in $\mathcal{E}^{\mathcal{D}}$ provided they exist in $\mathcal{E}$ itself.

In fact, the entrywise tensor and cotensor operations of the first example above make diagram categories tensored and cotensored as enriched categories over $\mathcal{E}$. We will not dwell on this point, as we do not use it extensively, although it can be helpful for overview discussions.

DEFINITION 10.27. Given any diagram category $\mathcal{E}^{\mathcal{D}}$ and some choice of functor $G : \mathcal{D}' \longrightarrow \mathcal{D}$, there is an induced **precomposition functor** $G^* : \mathcal{E}^{\mathcal{D}} \longrightarrow \mathcal{E}^{\mathcal{D}'}$ sending $F : \mathcal{D} \longrightarrow \mathcal{E}$ to $FG : \mathcal{D}' \longrightarrow \mathcal{D} \longrightarrow \mathcal{E}$.

In fact, such precomposition functors G^* have both left and right adjoint functors by the Kan extensions, at least when the target $\mathcal{E}$ has enough (co)limits. However, we will describe these constructions more explicitly whenever they are relevant for our purposes.

10.3.2. Colored Objects and Bimodules. We will use the notation $dis(S)$ to indicate the discrete category inside S, so the objects of S together with just their identity maps.

DEFINITION 10.28. The category of **S-colored objects in $\mathcal{E}$** will refer to

$$\mathcal{E}^{dis(S)} \cong \prod_S \mathcal{E}.$$

On the other hand, the category of **$\Sigma_S^{\mathcal{E}}$-bimodules**, or **S factor bimodules**, will refer to $\mathcal{E}^S$. For the maximal case, $S = \mathcal{P}(\mathfrak{C})^{op} \times \mathcal{P}(\mathfrak{C})$, **$\Sigma^{\mathcal{E}}$-bimodules** refers to $\mathcal{E}^{\mathcal{P}(\mathfrak{C})^{op}\times\mathcal{P}(\mathfrak{C})}$.

The ability to decompose any groupoid as a disjoint union of its components yields a canonical isomorphism of categories,

$$\Sigma^{\mathcal{E}} \cong \prod_{[\underline{d}],[\underline{c}]} \mathcal{E}^{\Sigma_{[\underline{c};\underline{d}]}}, \quad \mathsf{P} \mapsto \left\{\mathsf{P}\binom{[\underline{d}]}{[\underline{c}]}\right\}, \tag{10.9}$$

in which the product runs over all the pairs of orbit types of $\mathfrak{C}$-profiles.

The inclusion functor $dis(S) \longrightarrow S$ will induce a forgetful functor from $\Sigma_S^{\mathcal{E}}$-bimodules to S-colored objects in $\mathcal{E}$ as in Definition 10.27. The introduction of the factor S is, as in chapter 8, a mechanism for conveniently restricting the entries under consideration, such as restriction of colors or dealing with the fact that trees have only a single output.

REMARK 10.29. When $\mathcal{E}$ has an initial object $\varnothing$, notice that $\Sigma_S^{\mathcal{E}}$-bimodules is isomorphic to the full subcategory of $\Sigma^{\mathcal{E}}$-bimodules with objects satisfying

$$\mathsf{P}\binom{\underline{d}}{\underline{c}} = \varnothing \quad \text{whenever} \quad (\underline{c};\underline{d}) \notin S.$$

A similar statement also holds for $\mathcal{E}^{dis(S)}$.

Unwrapping the above definition, a $\Sigma^{\mathcal{E}}$-bimodule P consists of the following data:

(1) For any pair of profiles $(\underline{c};\underline{d}) \in \mathcal{P}(\mathfrak{C})^{op} \times \mathcal{P}(\mathfrak{C})$, it has an object

$$\mathsf{P}\binom{\underline{d}}{\underline{c}} = \mathsf{P}\binom{d_1,\ldots,d_n}{c_1,\ldots,c_m} \in \mathcal{E}.$$

This object should be thought of as the space of operations with $|\underline{c}| = m$ inputs and $|\underline{d}| = n$ outputs. The m inputs have colors $c_1,\ldots,c_m$, and the n outputs have colors $d_1,\ldots,d_n$.

(2) Given any pair of permutations $(\tau;\sigma) \in \Sigma_{|\underline{d}|} \times \Sigma_{|\underline{c}|}$, there is an isomorphism

$$\mathsf{P}\binom{\underline{d}}{\underline{c}} \xrightarrow[\cong]{(\tau;\sigma)} \mathsf{P}\binom{\sigma\underline{d}}{\underline{c}\tau} \tag{10.10}$$

in $\mathcal{E}$ such that:
- $(1;1)$ is the identity morphism, and
- $(\tau\tau';\sigma'\sigma) = (\tau';\sigma') \circ (\tau;\sigma)$.

A morphism $f\colon \mathsf{P} \longrightarrow \mathsf{Q}$ of $\Sigma^{\mathcal{E}}$-bimodules consists of color-preserving morphisms

$$\left\{\mathsf{P}\binom{\underline{d}}{\underline{c}} \xrightarrow{f} \mathsf{Q}\binom{\underline{d}}{\underline{c}} : (\underline{c};\underline{d}) \in \mathcal{P}(\mathfrak{C})^{op} \times \mathcal{P}(\mathfrak{C})\right\} \tag{10.11}$$

that respect the $\Sigma_{|\underline{d}|}$-$\Sigma_{|\underline{c}|}$ action (10.10), in the sense that each square

$$\begin{array}{ccc} \mathsf{P}\binom{\underline{d}}{\underline{c}} & \xrightarrow{f} & \mathsf{Q}\binom{\underline{d}}{\underline{c}} \\ {\scriptstyle(\tau;\sigma)}\downarrow & & \downarrow{\scriptstyle(\tau;\sigma)} \\ \mathsf{P}\binom{\sigma\underline{d}}{\underline{c}\tau} & \xrightarrow{f} & \mathsf{Q}\binom{\sigma\underline{d}}{\underline{c}\tau} \end{array}$$

is commutative.

EXAMPLE 10.30. (1) In the one-colored case, i.e. $\mathfrak{C} = \{*\}$, the category $\mathcal{P}(\mathfrak{C})^{op} \times \mathcal{P}(\mathfrak{C})$ consists of pairs of non-negative integers $(m; n)$, so a typical component of a $\mathbf{\Sigma}^{\mathcal{E}}$-bimodule is written as $\mathsf{P}\binom{n}{m}$. There is a one-colored $\mathbf{\Sigma}^{\mathcal{E}}$-bimodule in **Set** with components

$$\mathsf{P}\binom{n}{m} = \Sigma_n \times \Sigma_m^{op},$$

the left regular Σ_n-action, and the right regular Σ_m-action.

(2) In the general colored case, there is a $\mathbf{\Sigma}^{\mathcal{E}}$-bimodule in **Set** with

$$\mathsf{P}\binom{[\underline{d}]}{[\underline{c}]} = \Sigma_{[\underline{c};\underline{d}]}$$

and with the isomorphism (10.10) given by permutations of $\mathfrak{C}$-profiles. In other words, each component is a singleton,

$$\mathsf{P}\binom{\underline{d}}{\underline{c}} = \left\{\binom{\underline{d}}{\underline{c}}\right\},$$

and (10.10) takes $\binom{\underline{d}}{\underline{c}}$ to $\binom{\sigma\underline{d}}{\underline{c}\tau}$.

(3) Let $\mathtt{G}$ be a sub-groupoid of $\mathtt{Gr}_{\mathrm{w}}^{\circlearrowright}$. There is a $\mathbf{\Sigma}^{\mathcal{E}}$-bimodule in **Set** with components

$$\mathsf{P}\binom{\underline{d}}{\underline{c}} = \mathsf{G}\binom{\underline{d}}{\underline{c}}$$

and with the isomorphism (10.10) given by input and output relabeling (4.5).

(4) Suppose $\mathcal{E}$ is closed, and $X = \{X_c\} \in \mathcal{E}^{\mathfrak{C}}$. There is a $\mathbf{\Sigma}^{\mathcal{E}}$-bimodule in $\mathcal{E}$ with components

$$\mathsf{P}\binom{\underline{d}}{\underline{c}} = \mathrm{Hom}_{\mathcal{E}}\left(\bigotimes_{i=1}^{|\underline{c}|} X_{c_i}, \bigotimes_{j=1}^{|\underline{d}|} X_{d_j}\right)$$

and with the isomorphism (10.10) induced by the symmetry in $\mathcal{E}$.

Our starting point for building generalized PROPs will be $\mathcal{E}^{dis(S)}$, but every generalized PROP will automatically have an underlying $\mathbf{\Sigma}_S^{\mathcal{E}}$-bimodule structure due to the existence of permuted corollas in every pasting scheme. Using the forgetful functor $\mathcal{E}^S \longrightarrow \mathcal{E}^{dis(S)}$ above, one can instead think of an underlying S-colored object, allowing us to produce a monad on $\mathcal{E}^{dis(S)}$ whose algebras are the relevant type of generalized PROPs.

10.3.3. Decorated Graphs.

DEFINITION 10.31. Given $\mathsf{P} \in \mathcal{E}^{dis(S)}$ and a strict isomorphism class of wheeled graphs $[G]$, define the **P-decorated graph** as the unordered tensor product

$$\begin{aligned}\mathsf{P}[G] &= \bigotimes_{v \in \mathrm{Vt}(G)} \mathsf{P}\binom{\mathrm{out}(v)}{\mathrm{in}(v)} \\ &= \left(\coprod_{\mathrm{Ord}(\mathrm{Vt}(G))} \bigotimes_{j=1}^{k} \mathsf{P}\binom{\mathrm{out}(v_j)}{\mathrm{in}(v_j)} \right)_{\Sigma_k} \in \mathcal{E},\end{aligned} \tag{10.12}$$

where $k = |\mathrm{Vt}(G)|$. If $\mathrm{Vt}(G)$ is empty, then $\mathsf{P}[G] = I$, the unit of the tensor product, by convention.

Intuitively, $\mathsf{P}[G]$ is the object of decorations of the vertices of G by components of P with the correct colors. When $\mathcal{E}$ is the category of Sets with the cartesian product as $\otimes$, this statement can be taken literally.

REMARK 10.32. Notice this again differs in presentation from many in the literature, where it is common to use

$$Bij([n], \mathrm{out}(v)) \otimes_{\Sigma_n} \mathsf{P}(m; n) \otimes_{\Sigma_m} Bij(\mathrm{in}(v), [m]) \tag{10.13}$$

instead of $\mathsf{P}(m;n)$ in the one-colored case, with $\mathrm{in}(v)$ and $\mathrm{out}(v)$ treated as sets, where Bij is the set of bijections. However, this is a consequence of our choice to impose a listing on the inputs and outputs of each vertex in order to simplify this expression.

Notice, the construction of $\mathsf{P}[G]$ above is natural in the variable P. Even more, this defines an extension functor from $\mathcal{E}^{dis(S)}$ to something that takes strict isomorphism classes of graphs as inputs, and this functor of graphs just happens to have constant values across strict isomorphism classes.

EXAMPLE 10.33. Let $\mathsf{P} \in \mathcal{E}^{dis(S)}$.

(1) For the $(\underline{c}; \underline{d})$-corolla $C_{(\underline{c};\underline{d})}$ (section 1.3.1), it follows that

$$\mathsf{P}[C_{(\underline{c};\underline{d})}] = \mathsf{P}\binom{\underline{d}}{\underline{c}}.$$

(2) For the wheeled graph in Example 1.30, we have

$$\mathsf{P}[G] = \mathsf{P}\binom{a}{c_1, c_2} \odot \mathsf{P}\binom{d_1, d_2}{a} \odot \mathsf{P}\binom{\varnothing}{\varnothing} \odot \mathsf{P}\binom{\varnothing}{\varnothing}.$$

(3) The wheeled graphs G in Example 1.36 and G' in Example 1.37 satisfy

$$\mathsf{P}[G] = \mathsf{P}[G'] = \mathsf{P}\binom{c}{a} \odot \mathsf{P}\binom{a}{c}.$$

(4) For the butterfly net graph G in Example 1.38, we have

$$\mathsf{P}[G] = \mathsf{P}\binom{c}{\varnothing} \odot \mathsf{P}\binom{c}{c, c}.$$

(5) The upward ray graph G in Example 1.39 satisfies

$$\mathsf{P}[G] = \mathsf{P}\binom{c}{\varnothing}.$$

Since

$$\mathrm{Vt}(G(H_v)) = \coprod_{v \in \mathrm{Vt}(G)} \mathrm{Vt}(H_v),$$

the compatibility of decorated graphs with graph substitution is immediate.

LEMMA 10.34. *Given* $\mathsf{P} \in \mathcal{E}^{dis(S)}$ *and graph substitution data* (G, H_v, ψ_v)*, there is a natural isomorphism*

$$\mathsf{P}[G(H_v)] \cong \bigotimes_{v \in \mathrm{Vt}(G)} \mathsf{P}[H_v].$$

Later, we will also be interested in compatibility of decorated graphs with symmetric monoidal functors.

LEMMA 10.35. *If* $M : \mathcal{E} \longrightarrow \mathcal{D}$ *is a symmetric monoidal functor, then there is a natural map in* $\mathcal{D}$*, for* $\mathsf{P} \in \mathcal{E}^{dis(S)}$ *and wheeled graph* G*,*

$$\delta : (M\mathsf{P})[G] \longrightarrow M(\mathsf{P}[G]), \tag{10.14}$$

which is the identity for G *a corolla.*

PROOF. The symmetric monoidal assumption on M implies that, for each ordering of $\mathrm{Vt}(G)$, there is a natural map

$$\bigotimes_{j=1}^{n} (M\mathsf{P})\binom{\mathrm{out}(v_j)}{\mathrm{in}(v_j)} \longrightarrow M\left(\bigotimes_{j=1}^{n} \mathsf{P}\binom{\mathrm{out}(v_j)}{\mathrm{in}(v_j)}\right).$$

Since both coproducts and Σ_n-coinvariants are colimits, the induced map is the desired natural map δ. □

EXAMPLE 10.36.

(1) Since input and output relabeling is an example of graph substitution (section 6.1.2) where the outer graph has a single vertex, there is a natural isomorphism

$$\mathsf{P}[G] \cong \mathsf{P}[\sigma G \tau].$$

(2) Similarly, union is an example of graph substitution (section 6.1.3), so there is a natural isomorphism

$$\mathsf{P}\left[\coprod_{j=1}^{r} G_j\right] \cong \bigodot_{j=1}^{r} \mathsf{P}[G_j].$$

(3) When defining contraction as a graph substitution (section 6.1.5), the outer graph again has a single vertex, so we have a natural isomorphism

$$\mathsf{P}[\xi_j^i G] \cong \mathsf{P}[G]$$

if $d_i = c_j$.

(4) Defining (partial) grafting as graph substitution into a graph with two vertices (section 6.1.4) yields a natural isomorphism

$$\mathsf{P}\left[G_1 \boxtimes_{\underline{b}'}^{\underline{c}'} G_2\right] \cong \mathsf{P}[G_1] \odot \mathsf{P}[G_2]$$

provided the (partial) grafting exists, e.g. $\underline{b}' = \underline{c}'$.

(5) In particular, for G_{in} the input extension of G (6.1), there is a natural isomorphism

$$\mathsf{P}[G_{in}] \cong \mathsf{P}[G] \odot \left(\bigodot_{i=1}^{m} \mathsf{P}\binom{c_i}{c_i} \right),$$

and for G_{out} the output extension of G (6.2), there is a natural isomorphism

$$\mathsf{P}[G_{out}] \cong \left(\bigodot_{j=1}^{n} \mathsf{P}\binom{d_j}{d_j} \right) \odot \mathsf{P}[G].$$

10.4. Generalized PROPs as Monadic Algebras

In this section, we define the monad associated to a pasting scheme and use it to define generalized PROPs. Many examples of generalized PROPs are considered in the next chapter, with just a few simple examples provided here.

10.4.1. The Monad Associated to a Pasting Scheme. Fix a pasting scheme $\mathcal{G} = (S, \mathtt{G})$ (Definition 8.2). We now define the monad associated to $\mathcal{G}$.

The functor: First define the functor

$$F = F_{\mathcal{G}} : \mathcal{E}^{dis(S)} \longrightarrow \mathcal{E}^{dis(S)}$$

by

$$FP\binom{\underline{d}}{\underline{c}} = \coprod_{[G] \in \mathtt{G}_S\binom{\underline{d}}{\underline{c}}} \mathsf{P}[G] \tag{10.15}$$

for $\mathsf{P} \in \mathcal{E}^{dis(S)}$ and $(\underline{c}; \underline{d}) \in S$, where the coproduct is taken over the strict isomorphism classes in $\mathtt{G}_S\binom{\underline{d}}{\underline{c}}$. For each $[G] \in \mathtt{G}_S\binom{\underline{d}}{\underline{c}}$, there is a morphism

$$\eta_{[G]} : \mathsf{P}[G] \longrightarrow F\mathsf{P}\binom{\underline{d}}{\underline{c}}$$

in $\mathcal{E}$ defined as the inclusion of one summand, which is clearly natural in the variable P.

Intuitively, $(F\mathsf{P})\binom{\underline{d}}{\underline{c}}$ is the space of decorations of the vertices of the graph $G \in \mathtt{G}_S\binom{\underline{d}}{\underline{c}}$ by components of P with the correct colors. Notice there is a $\mathbf{\Sigma}_S$-bimodule structure on $F\mathsf{P}$ induced by the groupoid isomorphism

$$\mathtt{G}_S\binom{\underline{d}}{\underline{c}} \xrightarrow[\cong]{(\tau;\sigma)} \mathtt{G}_S\binom{\sigma\underline{d}}{\underline{c}\tau} \tag{10.16}$$

given by restricting the groupoid isomorphism (4.5), even though P itself is not assumed to carry such a bimodule structure.

Roughly speaking, to define the monadic multiplication on F, one thinks of $F^2\mathsf{P}\binom{\underline{d}}{\underline{c}}$ as consisting of bracketed P-decorated graphs, i.e., graphs whose vertices are decorated by P-decorated graphs. The monadic multiplication is then defined as erasing the brackets, at least when $\mathcal{E}$ is Sets.

Since we will use this viewpoint repeatedly below, we record the following for ease of reference later.

LEMMA 10.37. *There are canonical isomorphisms*

$$F_{\mathcal{G}}\mathsf{P}[G] \cong \coprod_{\{[H_v]\}} \mathsf{P}[G(H_v)] \tag{10.17}$$

$$F_{\mathcal{G}}F_{\mathcal{G}}\mathsf{P}\binom{\underline{d}}{\underline{c}} \cong \coprod_{G,\{[H_v]\}} \mathsf{P}[G(H_v)] \tag{10.18}$$

where $G \in \mathsf{G}_S\binom{\underline{d}}{\underline{c}}$ *and each* $H_v \in \mathsf{G}_S\binom{\mathrm{out}(v)}{\mathrm{in}(v)}$.

PROOF. For $(\underline{c};\underline{d}) \in S$ and $G \in \mathsf{G}_S\binom{\underline{d}}{\underline{c}}$ we have

$$\begin{aligned}
F\mathsf{P}[G] &= \bigotimes_{v\in \mathrm{Vt}(G)} F\mathsf{P}\binom{\mathrm{out}(v)}{\mathrm{in}(v)}\\
&= \bigotimes_{v\in \mathrm{Vt}(G)} \coprod_{[H_v]\in \mathsf{G}_S\binom{\mathrm{out}(v)}{\mathrm{in}(v)}} \mathsf{P}[H_v]\\
&\cong \coprod_{\{[H_v]\}} \bigotimes_{v\in \mathrm{Vt}(G)} \mathsf{P}[H_v]\\
&\cong \coprod_{\{[H_v]\}} \mathsf{P}[G(H_v)]
\end{aligned}$$

with the last isomorphism coming from Lemma 10.34 and the previous coming from the fact that tensor distributes over coproducts in a symmetric monoidal *closed* category.

For the second claim, as a consequence of the first we have

$$\begin{aligned}
F^2\mathsf{P}\binom{\underline{d}}{\underline{c}} &= \coprod_{[G]\in \mathsf{G}_S\binom{\underline{d}}{\underline{c}}} (F\mathsf{P}[G])\\
&\cong \coprod_{[G]\in \mathsf{G}_S\binom{\underline{d}}{\underline{c}}} \coprod_{\{[H_v]\}} \mathsf{P}[G(H_v)]\\
&\cong \coprod_{G,\{[H_v]\}} \mathsf{P}[G(H_v)]\,.
\end{aligned}$$

□

The multiplication: The required multiplication map

$$\mu_{\mathsf{P}}\colon F^2\mathsf{P}\binom{\underline{d}}{\underline{c}} \longrightarrow F\mathsf{P}\binom{\underline{d}}{\underline{c}}$$

is now defined on any given summand of the source, $\mathsf{P}[G(H_v)]$ indexed by the pair $([G],\{[H_v]\})$ to the copy of $\mathsf{P}[G(H_v)]$ in the target indexed by the graph substitution $[G(H_v)]$. Of course, this strict isomorphism class depends only on the strict isomorphism classes $[G]$ and each $[H_v]$ by Proposition 5.33.

The reader should be careful to note that with each choice of all substituting graphs $[H_v]$ the inducing maps are isomorphisms, but this definitely does *not* imply the map from the full coproduct is an isomorphism.

The unit: The monadic unit is defined by regarding $\mathsf{P}\binom{\underline{d}}{\underline{c}}$ as consisting of decorated corollas.

More precisely, the monadic unit

$$\nu_{\mathsf{P}} \colon \mathsf{P}\binom{\underline{d}}{\underline{c}} \longrightarrow F\mathsf{P}\binom{\underline{d}}{\underline{c}} \tag{10.19}$$

is defined as the composition in

$$\begin{array}{ccc} \mathsf{P}\binom{\underline{d}}{\underline{c}} & \xrightarrow{\qquad\nu_{\mathsf{P}}\qquad} & F\mathsf{P}\binom{\underline{d}}{\underline{c}} \\ & {\scriptstyle Id}\searrow \qquad \nearrow{\scriptstyle \eta_{[C_{(\underline{c};\underline{d})}]}} & \\ & \mathsf{P}[C_{(\underline{c};\underline{d})}] & \end{array}$$

in which $C_{(\underline{c};\underline{d})}$ is the $(\underline{c};\underline{d})$-corolla (section 1.3.1).

THEOREM 10.38. *For each pasting scheme $\mathcal{G} = (S, \mathtt{G})$, the constructions above define a monad $(F_{\mathcal{G}}, \mu, \nu)$ on $\mathcal{E}^{dis(S)}$.*

PROOF. Choose $\mathsf{P} \in \mathcal{E}^{dis(S)}$. First we show that μ is associative, i.e., that the left square in (10.3) is commutative. In this diagram, we view $F^3\mathsf{P}$ in two different ways to define the two maps out of it. In considering the map

$$\mu_{\mathsf{P}} \circ \mu_{F\mathsf{P}} \colon F^3\mathsf{P}\binom{\underline{d}}{\underline{c}} \longrightarrow F\mathsf{P}\binom{\underline{d}}{\underline{c}}$$

we want to think of the composite

$$F^2(F\mathsf{P})\binom{\underline{d}}{\underline{c}} \longrightarrow F(F\mathsf{P})\binom{\underline{d}}{\underline{c}} \longrightarrow F\mathsf{P}\binom{\underline{d}}{\underline{c}}$$

keeping in mind our description of F^2 above. Thus, we think of $F^3\mathsf{P}$ as isomorphic, via two instances of Lemma 10.34 and the canonical isomorphism for distributing tensors over coproducts, to

$$\coprod_{[G(H_v)],\{[I_u^v]\}} \mathsf{P}[(G(H_v))(I_u^v)]$$

and the composite includes each summand into $\coprod_K \mathsf{P}[K]$ with such a label.

On the other hand, when considering the map

$$\mu_{\mathsf{P}} \circ F\mu_{\mathsf{P}} \colon F^3\mathsf{P}\binom{\underline{d}}{\underline{c}} \longrightarrow F\mathsf{P}\binom{\underline{d}}{\underline{c}}$$

we want to think of the composite

$$F(F^2\mathsf{P})\binom{\underline{d}}{\underline{c}} \longrightarrow F(F\mathsf{P})\binom{\underline{d}}{\underline{c}} \longrightarrow F\mathsf{P}\binom{\underline{d}}{\underline{c}}.$$

In this case, the first map is isomorphic, via two slightly different applications of Lemma 10.34 and the distributivity isomorphism, to

$$\coprod_{[G],\{[H_v(I_u^v)]\}} \mathsf{P}[G(H_v(I_u^v))]$$

and the composite includes each summand with such a label. Of course, the associativity of graph substitution (Theorem 5.32) says

$$\mathsf{P}[(G(H_v))(I_u^v)] = \mathsf{P}[G(H_v(I_u^v))],$$

so each summand is included into $F\mathsf{P}\binom{\underline{d}}{\underline{c}}$ in the same way by the two maps, making the multiplication associative.

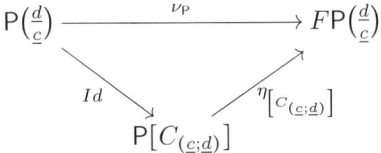

Now we need to show that F satisfies the unital axiom for a monad (i.e., the right diagram in (10.3)). In the language of Lemma 10.37, the key will be to note that $\nu_{F\mathsf{P}}$ includes the summand $\mathsf{P}[G]$ as the copy of $\mathsf{P}[C_{(\underline{c};\underline{d})}(G)]$ indexed on the pair $([C_{(\underline{c};\underline{d})}],[G])$, while $F\nu_{\mathsf{P}}$ includes the summand $\mathsf{P}[G]$ as the copy of $\mathsf{P}[G(C_v)]$ indexed on the pair $([G],\{[C_v]\})$. In both cases, the multiplication μ_{P} simply performs the graph substitution on the indices for any given summand. Thus, the composite

$$\mu_{\mathsf{P}} \circ F\nu_{\mathsf{P}}\colon F\mathsf{P}\binom{\underline{d}}{\underline{c}} \longrightarrow F\mathsf{P}\binom{\underline{d}}{\underline{c}}$$

sends the summand $\mathsf{P}[G] \longrightarrow \mathsf{P}[G(C_v)] \longrightarrow \mathsf{P}[G]$ since $[G(C_v)] = [G]$ by one side of the unit property for graph substitution (Lemma 5.31). Similarly,

$$\mu_{\mathsf{P}} \circ \nu_{F\mathsf{P}}\colon F\mathsf{P}\binom{\underline{d}}{\underline{c}} \longrightarrow F\mathsf{P}\binom{\underline{d}}{\underline{c}}$$

sends a summand $\mathsf{P}[G] \longrightarrow \mathsf{P}[C_{(\underline{c};\underline{d})}(G)] \longrightarrow \mathsf{P}[G]$ using the other side of the unit property for graph substitution. Since both composites yield the identity map on coproducts, the unital property is satisfied. □

The monad $(F_{\mathcal{G}},\mu,\nu)$ is usually abbreviated to $F_{\mathcal{G}}$ or even F if the pasting scheme $\mathcal{G}$ is clear from the context.

10.4.2. Generalized PROPs. We are now ready for the main definition of this chapter.

DEFINITION 10.39. Let $\mathcal{G} = (S,\mathtt{G})$ be a pasting scheme and $(F_{\mathcal{G}},\mu,\nu)$ be the associated monad on $\mathcal{E}^{dis(S)}$ (Theorem 10.38). We refer to $F_{\mathcal{G}}$-algebras $(\mathsf{P},\gamma : F_{\mathcal{G}}\mathsf{P} \longrightarrow \mathsf{P})$ as **$\mathcal{G}$-PROPs** in $\mathcal{E}$, or even **$\mathtt{G}$-PROPs** if S is understood, and denote the category of $F_{\mathcal{G}}$-algebras in $\mathcal{E}^{dis(S)}$ by $\mathtt{PROP}^{\mathcal{G},\mathcal{E}}$ or $\mathtt{PROP}^{\mathcal{G}}$ if $\mathcal{E}$ is understood.

We use the term *generalized PROP* to denote $\mathcal{G}$-PROP for an arbitrary pasting scheme $\mathcal{G}$. Other descriptions of generalized PROPs will be discussed in Chapter 14.

From the definitions of the monad $F_{\mathcal{G}}$ and of monadic algebras, one can distill the structure of a $\mathcal{G}$-PROP as follows.

LEMMA 10.40. *A $\mathcal{G}$-PROP consists of $\mathsf{P} \in \mathcal{E}^{dis(S)}$ equipped with structure maps*

$$\gamma_{[G]}\colon \mathsf{P}[G] \longrightarrow \mathsf{P}\binom{\underline{d}}{\underline{c}} \tag{10.20}$$

for $G \in \mathtt{G}_S\binom{\underline{d}}{\underline{c}}$ that are associative and unital in the sense that the following diagrams must commute:

Associativity:

$$\begin{array}{ccc} \bigotimes_v \mathsf{P}[H_v] & \xrightarrow{\bigotimes_v \gamma_{[H_v]}} & \bigotimes_v \mathsf{P}\binom{\mathrm{out}(v)}{\mathrm{in}(v)} = \mathsf{P}[G] \\ \cong\downarrow & & \downarrow\gamma_{[G]} \\ \mathsf{P}[G(H_v)] & \xrightarrow[\gamma_{[G(H_v)]}]{} & \mathsf{P}\binom{\underline{d}}{\underline{c}} \end{array} \tag{10.21}$$

whenever $[G] \in \mathtt{G}_S\binom{\underline{d}}{\underline{c}}$ and $[H_v] \in \mathtt{G}_S\binom{\mathrm{out}(v)}{\mathrm{in}(v)}$ for each $v \in \mathrm{Vt}(G)$, and $\bigotimes_v = \bigotimes_{v\in\mathrm{Vt}(G)}$, as well as

Unity:

$$\text{(10.22)} \qquad \begin{array}{ccc} \mathsf{P}\binom{\underline{d}}{\underline{c}} & \xrightarrow{Id} & \mathsf{P}[C_{(\underline{c};\underline{d})}] \\ & \underset{Id}{\searrow} & \downarrow{\scriptstyle \gamma_{[C]}} \\ & & \mathsf{P}\binom{\underline{d}}{\underline{c}} \end{array}$$

where $C_{(\underline{c};\underline{d})}$ is the $(\underline{c};\underline{d})$-corolla (section 1.3.1).

Furthermore, a morphism $g\colon \mathsf{P} \longrightarrow \mathsf{Q}$ of $\mathcal{G}$-PROPs is a morphism of the underlying S-colored objects that is compatible with the structure maps in the sense that the square

$$\text{(10.23)} \qquad \begin{array}{ccc} \mathsf{P}[G] & \xrightarrow{g[G]} & \mathsf{Q}[G] \\ {\scriptstyle \gamma^{\mathsf{P}}_{[G]}}\downarrow & & \downarrow{\scriptstyle \gamma^{\mathsf{Q}}_{[G]}} \\ \mathsf{P}\binom{\underline{d}}{\underline{c}} & \xrightarrow[g]{} & \mathsf{Q}\binom{\underline{d}}{\underline{c}} \end{array}$$

is commutative for each $G \in \mathtt{G}_S\binom{\underline{d}}{\underline{c}}$.

The unital condition means that for each $(\underline{c};\underline{d})$-corolla $C = C_{(\underline{c};\underline{d})} \in \mathtt{G}_S\binom{\underline{d}}{\underline{c}}$, the structure map $\gamma_{[C]}$ is the identity map on $\mathsf{P}\binom{\underline{d}}{\underline{c}}$.

Notice that, for any $\mathcal{G}$-PROP, by reversing the isomorphism in (10.21) and applying $\otimes_v \gamma_{[H_v]}$, there is an induced map, $P[G(H_v)] \longrightarrow P[G]$, sometimes also denoted $\otimes_v \gamma_{[H_v]}$ by abuse of notation, associated to any choice of substituting graphs H_v in that pasting scheme. To simplify the notation, especially within diagrams, in what follows we sometimes abbreviate $\gamma_{[G]}$ to γ_G.

10.4.3. Bi-equivariant Structure. Each $\mathcal{G}$-PROP has bi-equivariant structure induced by permuted corollas. Explicitly, for each $(\underline{c};\underline{d}) \in S$ and permutations $\sigma \in \Sigma_n$ and $\tau \in \Sigma_m$, the permuted corolla $\sigma C \tau$ with $C = C_{(\underline{c};\underline{d})}$ yields the structure map

$$(\tau;\sigma) = \gamma_{[\sigma C \tau]}\colon \mathsf{P}\binom{\underline{d}}{\underline{c}} \longrightarrow \mathsf{P}\binom{\sigma\underline{d}}{\underline{c}\tau}.$$

Equipped with these structure maps, P is a $\mathbf{\Sigma}^{\mathcal{E}}_S$-bimodule by the associativity and unity axioms above and Lemma 6.26.

REMARK 10.41. This induced bi-equivariant structure is another point of divergence between our approach and many others in the literature. In the literature, it is common to begin with a $\mathbf{\Sigma}^{\mathcal{E}}$-bimodule and to impose structure maps that satisfy associativity, unity, and compatibility axioms between the bi-equivariant structure and other structure maps. From the monadic view point, the compatibility axioms are achieved by using $Aut(G)$-coinvariants of $\mathsf{P}[G]$. In other words, in many references, the monad is defined as

$$F\mathsf{P}\binom{\underline{d}}{\underline{c}} = \coprod_{[G]} \mathsf{P}[G]_{Aut(G)}.$$

However, this definition uses the usual notion of graph, which does not have listings at the vertices, and (the colored version of) (10.13) in the definition of $\mathsf{P}[G]$.

Moreover, many existing notions of graph in the literature exclude those that have exceptional flags.

In our approach, the bi-equivariant structure is a consequence of the permuted corollas in the pasting scheme. The compatibility of the bi-equivariant structure with other structure maps is a consequence of graph substitution. In all cases of interest, when combined with our Biased Definition Theorem (Theorem 11.5), our approach produces the expected biased version of generalized PROPs. Our approach fits in with our view point that all relevant structures and properties of generalized PROPs are consequences of wheeled graphs and graph substitution.

10.4.4. Generalized PROPs for Unital Pasting Schemes. Let $\mathcal{G} = (S, \mathtt{G})$ be a unital pasting scheme (Definition 8.6) and P be a $\mathcal{G}$-PROP with structure maps $\gamma_{[G]}$ for $[G] \in \mathtt{G}_S\binom{\underline{d}}{\underline{c}}$ as in (10.20). Here we explain the structure on P given by the unital assumption on $\mathcal{G}$, which follows from the ability to factor corollas via their input and output extensions.

DEFINITION 10.42. For $c \in S$ call the map

$$\mathbf{1}_c = \gamma_{[\uparrow_c]}: \mathsf{P}[\uparrow_c] = I \longrightarrow \mathsf{P}\binom{c}{c}, \tag{10.24}$$

the **c-colored vertical unit** of P. This is well-defined because $(c;c) \in S$ and $[\uparrow_c] \in \mathtt{G}_S\binom{c}{c}$ by assumption.

Now suppose $(\underline{c};\underline{d}) \in S$ with $\underline{d} = d_{[1,n]}$ and $\underline{c} = c_{[1,m]}$, and let $C = C_{(\underline{c};\underline{d})}$ be the $(\underline{c};\underline{d})$-corolla (section 1.3.1). Using the notations in section 6.1.7, the input extension $C_{in} \in \mathtt{G}_S\binom{\underline{d}}{\underline{c}}$ has

$$\mathrm{Vt}(C_{in}) = \{w, v_1, \ldots, v_m\},$$

where w is the unique vertex in C. For $v \in \mathrm{Vt}(C_{in})$ we have

$$\binom{\mathrm{out}(v)}{\mathrm{in}(v)} = \begin{cases} \binom{\underline{d}}{\underline{c}} & \text{if } v = w,\\ \binom{c_i}{c_i} & \text{if } v = v_i.\end{cases}$$

Define

$$H_v = \begin{cases} C_{(\underline{c};\underline{d})} \in \mathtt{G}_S\binom{\underline{d}}{\underline{c}} & \text{if } v = w,\\ \uparrow_{c_i} \in \mathtt{G}_S\binom{c_i}{c_i} & \text{if } v = v_i.\end{cases}$$

Then we have the graph substitution statement

$$[C_{in}(H_v)] = [C].$$

Moreover, we have

$$\mathsf{P}[H_v] = \begin{cases} \mathsf{P}\binom{\underline{d}}{\underline{c}} & \text{if } v = w,\\ I & \text{if } v = v_i\end{cases}$$

and $\gamma_{[C]} = Id$ by (10.22). Therefore, the associativity diagram (10.21) with $G = C_{in}$ is equivalent to the commutative diagram:

$$
\begin{array}{ccc}
\mathsf{P}\binom{\underline{d}}{\underline{c}} \otimes I^{\otimes m} & \xrightarrow{Id \otimes \mathbf{1}_{c_1} \otimes \cdots \otimes \mathbf{1}_{c_m}} & \mathsf{P}\binom{\underline{d}}{\underline{c}} \otimes \bigotimes_{i=1}^{m} \mathsf{P}\binom{c_i}{c_i} \\
 & & \big\downarrow \text{natural} \\
\cong \big\downarrow & & \mathsf{P}[C_{in}] \\
 & & \big\downarrow \gamma_{[C_{in}]} \\
\mathsf{P}\binom{\underline{d}}{\underline{c}} & \xrightarrow{\quad Id \quad} & \mathsf{P}\binom{\underline{d}}{\underline{c}}
\end{array}
\tag{10.25}
$$

The top right vertical map is the natural map from the ordered tensor product to the unordered tensor product $\mathsf{P}[C_{in}]$. The commutativity of the diagram (10.25) is the first unital condition which holds for the $\mathcal{G}$-PROP P.

The other (dual) unital condition concerns the output extension $C_{out} \in \mathtt{G}_S\binom{\underline{d}}{\underline{c}}$ with

$$\mathrm{Vt}(C_{out}) = \{w, u_1, \ldots, u_n\}.$$

In this case, we have

$$\binom{\mathrm{out}(v)}{\mathrm{in}(v)} = \begin{cases} \binom{\underline{d}}{\underline{c}} & \text{if } v = w, \\ \binom{d_i}{d_i} & \text{if } v = u_i. \end{cases}$$

Define

$$H_v = \begin{cases} C_{(\underline{c};\underline{d})} \in \mathtt{G}_S\binom{\underline{d}}{\underline{c}} & \text{if } v = w, \\ \uparrow_{d_i} \in \mathtt{G}_S\binom{d_i}{d_i} & \text{if } v = u_i. \end{cases}$$

Then we have

$$\mathsf{P}[H_v] = \begin{cases} \mathsf{P}\binom{\underline{d}}{\underline{c}} & \text{if } v = w, \\ I & \text{if } v = u_i \end{cases}$$

and

$$[C_{out}(H_v)] = [C].$$

The associativity diagram (10.21) with $G = C_{out}$ is equivalent to the commutative diagram:

$$
\begin{array}{ccc}
I^{\otimes n} \otimes \mathsf{P}\binom{\underline{d}}{\underline{c}} & \xrightarrow{\mathbf{1}_{d_1} \otimes \cdots \otimes \mathbf{1}_{d_n} \otimes Id} & \left(\bigotimes_{i=1}^{n} \mathsf{P}\binom{d_i}{d_i}\right) \otimes \mathsf{P}\binom{\underline{d}}{\underline{c}} \\
 & & \big\downarrow \text{natural} \\
\cong \big\downarrow & & \mathsf{P}[C_{out}] \\
 & & \big\downarrow \gamma_{[C_{out}]} \\
\mathsf{P}\binom{\underline{d}}{\underline{c}} & \xrightarrow{\quad Id \quad} & \mathsf{P}\binom{\underline{d}}{\underline{c}}
\end{array}
\tag{10.26}
$$

The commutativity of this diagram is the second unital condition that holds for the $\mathcal{G}$-PROP P.

10.5. First Examples of Generalized PROPs

In this section, we consider a few simple examples of pasting schemes and their associated generalized PROPs.

10.5.1. Unital Linear PROPs. In the case where $\mathcal{G} = \mathtt{ULin}$ (see 8.7), the unital pasting scheme consisting of linear graphs and exceptional linear graphs, the answer is much more familiar than our current description. In fact, here a $\mathcal{G}$-PROP is simply a small $\mathcal{E}$-enriched category with object set the set of colors $\mathfrak{C}$.

Since the translation can be a bit distracting, a bit more detail is in order. The basic idea is that a linear graph with one vertex represents a primordial single non-identity morphism from the color of the input leg to the color of the output leg, while identity maps are represented by the exceptional linear graph of the relevant color. That is, $\mathsf{P}\binom{d}{c}$ of a linear corolla when $c \neq d$ (and when $c = d$ instead $\mathsf{P}\binom{c}{c} \amalg \mathsf{P}(\uparrow_c))$ is the morphism object of an enriched category. For longer linear graphs the tensor product $\mathsf{P}[G]$ represents all composable strings with the input color of G as source and the output color of G as target.

The generalized composition law then comes from $\gamma_{[G]}$, while the unital conditions come from the substitution of exceptional linear graphs and linear corollas into a longer linear graph by the associativity condition. Since providing a single associative composition map is equivalent by induction to providing an associative way of composing arbitrary strings of morphisms, this enriched category encodes the entire structure of $\mathcal{G}$-PROP on $\mathsf{P} \in \mathcal{E}^{dis(S)}$.

10.5.2. Contraction PROPs. Recall the Contractions Pasting Scheme of 8.4, consisting of (repeatedly) contracted corollas. In this case, the $\mathcal{G}$-PROPs are simply the category of diagrams in $\mathcal{E}$ indexed on a certain small indexing category $\mathcal{B}$ which contains $\mathcal{P}(\mathfrak{C})^{op} \times \mathcal{P}(\mathfrak{C})$ as its internal groupoid (or subcategory of isomorphisms).

To see this, we build the small category $\mathcal{B}$ directly. Begin with $\mathcal{P}(\mathfrak{C})^{op} \times \mathcal{P}(\mathfrak{C})$, and insert a new morphism for each contraction operation ξ^i_j as well as formal compositions of these satisfying the relations of Lemma 6.40. The only remaining condition to impose is compatibility with input and output relabeling, as implied by the statement of Lemma 6.38. The maps $\gamma_{[\xi^i_j C_{(\underline{d};\underline{c})}]}$ then provide the necessary structure maps to show any $\mathcal{G}$-PROP provides such a diagram in $\mathcal{E}$, since the listed identities are those for the $\xi^i_j C_{(\underline{d};\underline{c})}$.

On the other hand, the category $\mathcal{B}$ is really built of just those contractions needed to describe the structure of a $\mathcal{G}$-PROP, together with the relabeling that corresponds to the permutation action of the internal groupoid $\mathcal{P}(\mathfrak{C})^{op} \times \mathcal{P}(\mathfrak{C})$.

10.5.3. Horizontal and Vertical PROPs. In **[JY09]**, hPROPs and vPROPs were introduced and described as just the horizontal and vertical portions of the structure of a PROP. The point was to display each as a category of monoids in the category of bimodules, while the category of PROPs was a category of monoids in either of these. Be careful to note that such a statement includes a sense of compatibility between these two monoidal structures, which here encodes the so-called interchange law of a PROP.

First, consider the Horizontal Pasting Scheme $\mathtt{Gr}^{\uparrow}_{\mathrm{h}}$ (see 8.4), consisting of finite unions of corollas. Since the union operation can be defined in terms of graph substitution into finite unions of corollas, and any finite union can be built inductively from unions of just two corollas, we can produce a presentation of any element of this pasting scheme as a repeated substitution of unions of two corollas (and single corollas). As a consequence, the structure of a $\mathcal{G}$-PROP here reduces to choosing morphisms $\gamma_{[C_1 \amalg C_2]}$ in an associative way. It is not difficult to then describe a monoidal structure on bimodules $\boxtimes_h$ based on the idea of looking at $\mathsf{P}\binom{\underline{d}^1}{\underline{c}^1} \otimes \mathsf{P}'\binom{\underline{d}^2}{\underline{c}^2}$, with $\gamma_{[C_1 \amalg C_2]}$ an entry of a map from $\mathsf{P} \boxtimes_h \mathsf{P}$ to P. Now the associativity condition in the definition of a $\mathcal{G}$-PROP, together with the associativity of graph substitution, implies this operation is associative, and similarly the operation is unital using union with the empty graph.

Again, since the choices of such maps γ are sufficient to produce a $\mathcal{G}$-PROP structure, it follows that monoids in this $\boxtimes_h$ and $\mathcal{G}$-PROPs are equivalent.

The situation for vPROPs is similar, this time using the Vertical Pasting Scheme $\mathtt{Gr}^{\uparrow}_{\mathrm{v}}$, consisting of iterated graftings of corollas as in 8.4. Again, iterated grafting can be represented canonically in terms of grafting of pairs, which can be described completely in terms of a monoidal structure $\boxtimes_v$.

10.5.4. Generalized PROPs over Monogenic Pasting Schemes. As we saw in Definition 8.35, there is a small category associated to any monogenic pasting scheme. The next result says that generalized PROPs for monogenic pasting schemes are diagrams in the target category indexed over this small category.

PROPOSITION 10.43. *Suppose $\mathcal{G}$ is a monogenic pasting scheme. Then there is a canonical isomorphism between the categories of $\mathcal{G}$-PROPs and of $\mathcal{C}(\mathcal{G})$-diagrams in $\mathcal{E}$.*

PROOF. The isomorphism sends a $\mathcal{G}$-PROP (P, γ) to the diagram with

$$s \mapsto \mathsf{P}(s).$$

If $[G]: s \longrightarrow t$ is a morphism in $\mathcal{C}(\mathcal{G})$, then the corresponding morphism is

$$\gamma_{[G]}: \mathsf{P}(s) \longrightarrow \mathsf{P}(t)$$

in $\mathcal{E}$.

Moving in the other direction, for a *monogenic* pasting scheme the left vertical map in the associativity square (10.21) is an identity, so commutativity of these associativity squares is equivalent to the compatibility of a diagram (i.e. a functor) with the composition law of $\mathcal{C}(\mathcal{G})$ in this case. In addition, the unity condition (10.22) is simply the unit condition for a diagram. □

EXAMPLE 10.44. Both $\mathtt{Cor}$-PROPs and $\mathtt{Gr}^{\xi}$-PROPs are diagram categories in $\mathcal{E}$. In fact, $\mathtt{Cor}$-PROPs is canonically isomorphic to the category of bimodules, $\mathbf{\Sigma}_{\mathcal{S}}$.

CHAPTER 11

Biased Characterizations of Generalized PROPs

Recall that we work with a fixed non-empty set of colors $\mathfrak{C}$, so we will drop '$\mathfrak{C}$-colored' in many places below. In this chapter, we formalize what Markl calls the 'biased definition' of an operad or a PROP. The executive summary is that a biased characterization of a type of generalized PROP is a consequence of a strong generating set, although any generating set is enough to detect the appropriate notion of morphism.

The first section establishes the general result we refer to as the Biased Definition Theorem. In general, we see that the generating graphs in a generating set produce a collection of structure maps on the associated notion of generalized PROP simply by the correspondence $G \mapsto \gamma_{[G]}$. In addition, given any relaxed move, each associated generalized PROP has a commutative diagram built from the associated structure maps. For the simplest example, the bimodule structures underlying our generalized PROPs will arise from $\gamma_{[\sigma C_{(\underline{c};\underline{d})}\tau]}$, so all bimodule structures are influenced by the relaxed move of Lemma 6.26

$$(\sigma_2 C_{(\underline{c}\tau_1;\sigma_1\underline{d})}\tau_2)(\sigma_1 C_{(\underline{c};\underline{d})}\tau_1) = (\sigma_2\sigma_1)C_{(\underline{c};\underline{d})}(\tau_1\tau_2)$$

which induces a commutative triangle

$$\begin{array}{ccc} \mathsf{P}\binom{\underline{d}}{\underline{c}} & \longrightarrow & \mathsf{P}\binom{\sigma_1\underline{d}}{\underline{c}\tau_1} \\ & \searrow & \downarrow \\ & & \mathsf{P}\binom{(\sigma_2\sigma_1)\underline{d}}{\underline{c}(\tau_1\tau_2)} \end{array}$$

for P. The Biased Definition Theorem says that given a strong set $\mathcal{W}$ of relaxed moves for the generating set $\mathcal{T}$, we can describe the associated notion of generalized PROP in concrete terms using the structure morphisms $\gamma_{[G]}$ for $G \in \mathcal{T}$ and requiring commutative diagrams corresponding to each of the relaxed moves in $\mathcal{W}$.

In the second section, we point out that any generating set, even if it is not a strong generating set, suffices to determine the correct notion of morphism of generalized PROPs, which we call the Biased Morphism Theorem. This added flexibility is exploited to align our notion of morphisms of PROPs with that in common usage, since our strong generating sets sometimes carry a bit of redundant information for this question.

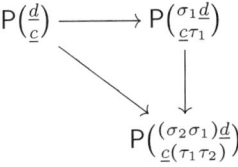

Since we have already established a large collection of pasting schemes with strong generating sets, the remainder of the chapter is devoted to a series of sections addressing the details in each of the cases of interest. Each of these later sections begins with a complete formal definition of the structure in question, which in several cases has never been made explicit in the literature. Each definition is followed by a subsection devoted to interpreting the axioms in the definition explicitly in terms of the sets $\mathcal{T}$ and $\mathcal{W}$ already chosen for that pasting scheme. Finally, each section includes the formal statements that the notions are equivalent, as consequences of the Biased Definition Theorem and the Biased Morphism Theorem.

The reader should keep in mind that there are examples not compatible with the version of the underlying graph theory presented earlier. Notably, cyclic and modular operads would require additional structure on the types of graphs allowed, so cannot be presented within the simpler framework discussed here.

11.1. Biased Definition Theorem

As before, we will sometimes talk about a set of wheeled graphs, even when we mean more precisely a set of strict isomorphism classes of wheeled graphs.

DEFINITION 11.1. Let $\mathcal{T}$ be a collection of wheeled graphs whose profiles and vertex profiles are all in a replete and full sub-groupoid S of $\mathcal{P}(\mathfrak{C})^{op} \times \mathcal{P}(\mathfrak{C})$. Let $\mathsf{P} \in \mathcal{E}^{dis(S)}$ be an S-colored object.

(1) A **$\mathcal{T}$-algebra** structure on P is a collection of maps

$$\mathsf{P}[G] \xrightarrow{\gamma[G]} \mathsf{P}\binom{\text{out}(G)}{\text{in}(G)},$$

one for each $G \in \mathcal{T}$, such that

$$\gamma[C] = Id: \mathsf{P}\binom{\underline{d}}{\underline{c}} \longrightarrow \mathsf{P}\binom{\underline{d}}{\underline{c}}$$

whenever C is the $(\underline{c};\underline{d})$-corolla.

(2) Suppose (P,γ) is a $\mathcal{T}$-algebra, and $\mathcal{H} = \mathcal{H}^{[1,n]}$ is a graph n-simplex in $\mathcal{T}$. If $\mathcal{H}^n$ is a single wheeled graph H, then we define a map

$$\gamma_{\mathcal{H}} : \mathsf{P}[\text{sub}(\mathcal{H})] \longrightarrow \mathsf{P}\binom{\text{out}(G)}{\text{in}(G)}$$

by

$$\gamma_{\mathcal{H}} = \gamma[\mathcal{H}^n] \circ \left(\bigotimes_{v \in \mathcal{H}^n} \gamma[\mathcal{H}_v^{n-1}]\right) \circ \cdots \circ \left(\bigotimes_{v \in \mathcal{H}^3} \gamma[\mathcal{H}_v^2]\right) \circ \left(\bigotimes_{v \in \mathcal{H}^2} \gamma[\mathcal{H}_v^1]\right).$$

If $\mathcal{H}^n$ is a finite set $\{\mathcal{H}_\alpha^n\}$, then define

$$\gamma_{\mathcal{H}} = \{\gamma_{\mathcal{H}_\alpha^n}\}.$$

(3) For a positive integer n, a **biased $(\mathcal{T},n)$-PROP** is a $\mathcal{T}$-algebra (P,γ) such that

$$\gamma_{\mathcal{H}} = \gamma_{\mathcal{H}'}$$

whenever $\mathcal{H}$ and $\mathcal{H}'$ are graph simplices in $\mathcal{T}$ connected by a finite sequence of relaxed n-moves within $\mathcal{T}$.

A $\mathcal{T}$-algebra is similar in essence to a magma. It has operations parametrized by the vertices of the wheeled graphs in $\mathcal{T}$. There are no axioms regarding these operations, except that the structure map corresponding to a corolla is the identity map. Going from a $\mathcal{T}$-algebra to a biased $(\mathcal{T},n)$-PROP is similar to going from a magma to a monoid, in the sense that we are now imposing associativity axioms for iterated compositions.

The following observation is immediate from finiteness and the definitions.

LEMMA 11.2. *If $\mathcal{T}$ is an n-strong generating set for some pasting scheme $(S, \mathtt{G})$, then a biased $(\mathcal{T}, n)$-PROP structure on an S-colored object P is equivalent to a $\mathcal{T}$-algebra structure such that*

$$\gamma_{\mathcal{H}} = \gamma_{\mathcal{H}'}$$

whenever $\mathcal{H}$ and $\mathcal{H}'$ are equivalent graph simplices in $\mathcal{T}$ of lengths $k \leq n$ and $l \leq n$, respectively.

Notice the equality in this lemma is more naturally interpreted as a commutative diagram, at least when both $k > 1$ and $l > 1$. The key point is that we are now focusing on commutativity of a diagram coming from the short segments in a relaxed move. There is a preliminary version of the main result of this chapter which we can now state, although our real interest lies in a strengthened version stated below.

THEOREM 11.3 (Biased Definition Theorem). *Suppose*

- *$\mathcal{G}$ is a pasting scheme, and*
- *$\mathcal{T}$ is an n-strong generating set of $\mathcal{G}$.*

Then a $\mathcal{G}$-PROP structure is equivalent to a biased $(\mathcal{T}, n)$-PROP structure.

Now we state the stronger version we will be using below, which involves a choice of set $\mathcal{W}$ which is n-strong in $\mathcal{T}$. Choosing the set of all relaxed n-moves in $\mathcal{T}$ simply recovers the case described above, so we postpone the proof for now.

DEFINITION 11.4. Let $\mathcal{T}$ be a generating set for the pasting scheme $\mathcal{G}$, and suppose the set $\mathcal{W}$ of relaxed moves in $\mathcal{T}$ is n-strong in $\mathcal{T}$. Let $\mathsf{P} \in \mathcal{E}^{dis(S)}$ be an S-colored object. A **biased $(\mathcal{T}, \mathcal{W})$-PROP** is a $\mathcal{T}$-algebra (P, γ) such that

$$\gamma_{\mathcal{H}} = \gamma_{\mathcal{H}'}$$

whenever $\mathcal{H} \sim \mathcal{H}'$ is a relaxed n-move in $\mathcal{W}$.

THEOREM 11.5 (Strong Biased Definition Theorem). *Suppose*

- *$\mathcal{G}$ is a pasting scheme,*
- *$\mathcal{T}$ is a generating set of $\mathcal{G}$, and*
- *$\mathcal{W}$ is n-strong in $\mathcal{T}$.*

Then a $\mathcal{G}$-PROP structure is equivalent to a biased $(\mathcal{T}, \mathcal{W})$-PROP structure.

PROOF. Given a $\mathcal{G}$-PROP structure, the structure maps $\gamma_{[G]}$, as discussed previously, arise as components of the structure map of the monadic algebra. Thus, it is simple to see each $\mathcal{G}$-PROP structure induces a $\mathcal{T}$-algebra structure. Commutativity of the relevant diagrams to produce a biased $(\mathcal{T}, \mathcal{W})$-PROP then follow from the fact that $\gamma_{[G]}$ is independent of any particular presentation of G by the associativity of the monadic algebra, as in (10.21).

On the other hand, a biased $(\mathcal{T}, \mathcal{W})$-PROP structure gives a tentative definition of $\gamma_{[G]}$ as $\gamma_{\mathcal{H}}$ for a chosen presentation $\mathcal{H}$ in $\mathcal{T}$ of an arbitrary wheeled graph G in the pasting scheme. Furthermore, in a biased $(\mathcal{T}, \mathcal{W})$-PROP structure the relations $\gamma_{\mathcal{H}} = \gamma_{\mathcal{H}'}$ provided, by induction, they are connected by a finite sequence of relaxed n-moves in $\mathcal{W}$, ensure that $\gamma_{[G]}$ will be the same for two such presentations in $\mathcal{T}$. However, the assumption that $\mathcal{W}$ is n-strong in $\mathcal{T}$ then implies $\gamma_{[G]}$ is defined independently of the choice of presentation in $\mathcal{T}$ and so induces the required structure map of an algebra over the monad $F_{\mathcal{G}}$. Verifying the associativity condition required of an algebra using this map is then another application of the associativity of graph substitution. □

11.2. Biased Morphism Theorem

We have seen that a biased characterization of generalized PROPs is really a consequence of a strong generating set of the pasting scheme. Here we observe that a biased characterization of a morphism of generalized PROPs is a consequence of any generating set of the pasting scheme, even if it is not a *strong* generating set.

DEFINITION 11.6. Given two $\mathcal{T}$-algebras $(\mathsf{P}, \gamma^{\mathsf{P}})$ and $(\mathsf{Q}, \gamma^{\mathsf{Q}})$, a map

$$f\colon \mathsf{P} \longrightarrow \mathsf{Q} \in \mathcal{E}^{dis(S)}$$

is a **morphism of $\mathcal{T}$-algebras** if the square

$$\begin{array}{ccc} \mathsf{P}[G] & \xrightarrow{\otimes f} & \mathsf{Q}[G] \\ \downarrow{\scriptstyle \gamma^{\mathsf{P}}} & & \downarrow{\scriptstyle \gamma^{\mathsf{Q}}} \\ \mathsf{P}\binom{\mathrm{out}(G)}{\mathrm{in}(G)} & \xrightarrow{f} & \mathsf{Q}\binom{\mathrm{out}(G)}{\mathrm{in}(G)} \end{array} \tag{11.1}$$

is commutative for each G in $\mathcal{T}$.

THEOREM 11.7 (Biased Morphism Theorem). *Let*

- $\mathcal{T}$ *be a generating set of a pasting scheme* $\mathcal{G}$ *(i.e., of* G_S*),*
- P *and* Q *be* $\mathcal{G}$*-PROPs, and*
- $f\colon \mathsf{P} \longrightarrow \mathsf{Q} \in \mathcal{E}^{dis(S)}$.

Then f is a morphism of $\mathcal{G}$-PROPs if and only if it is a morphism of $\mathcal{T}$-algebras.

PROOF. The map f is a morphism of $\mathcal{G}$-PROPs if and only if the square (11.1) is commutative for all $G \in \mathrm{G}_S$, so the only if direction is trivially true.

Now suppose f is a morphism of $\mathcal{T}$-algebras, and pick an arbitrary $G \in \mathrm{G}_S$. Since $\mathcal{T}$ is a generating set, there exists a graph simplex $\mathcal{H}^{[1,n]}$ in $\mathcal{T}$ whose substitution is strictly isomorphic to G. We have

$$\mathsf{P}[G] = \bigotimes_{\mathrm{Vt}(\mathcal{H}^n)} \cdots \bigotimes_{\mathrm{Vt}(\mathcal{H}^2)} \mathsf{P}[\mathcal{H}^1]$$

by the definition of P-decorated graph, and

$$\gamma_{[G]} = \gamma_{[\mathcal{H}^n]} \circ \Big(\bigotimes \gamma_{[\mathcal{H}^{n-1}]}\Big) \circ \cdots \circ \Big(\bigotimes \gamma_{[\mathcal{H}^2]}\Big) \circ \Big(\bigotimes \gamma_{[\mathcal{H}^1]}\Big)$$

by the associativity of the $\mathcal{G}$-PROP P. Therefore, the square (11.1) factors into a ladder diagram in which each small square is commutative because every $H \in \mathcal{H}^i$ is in $\mathcal{T}$ and f is a morphism of $\mathcal{T}$-algebras. This shows that f is actually a morphism of $\mathcal{G}$-PROPs. □

REMARK 11.8.
1. In the previous theorem, the generating set $\mathcal{T}$ is *not* required to be strong. In practice, this means that when we write down the biased axioms of a generalized PROP morphism, we are allowed to choose a generating subset of a known strong generating set. This can sometimes yield simpler axioms, for instance for morphisms of PROPs as discussed below.
2. When G runs through the permuted corollas, the commutative square (11.1) says that f is a morphism of Σ_S-bimodules.

The combination of Theorem 11.3, Theorem 11.7 and Theorem 8.29 now lead to the following.

THEOREM 11.9 (Biased Independence Theorem). *Suppose $\mathcal{T}_1$ and $\mathcal{T}_2$ are N-strong generating sets for $\mathcal{G}_1$ and $\mathcal{G}_2$ respectively that are also N-upper balanced or N-lower balanced with $N \geq 2$. Then the category of biased $(\mathcal{T}_1 \cup \mathcal{T}_2, N)$-PROPs is naturally isomorphic to the category of $\mathcal{G}_1 * \mathcal{G}_2$-PROPs.*

We now present a series of biased characterizations of various generalized PROPs and their morphisms.

11.3. Markl Non-Unital Operads as Tree-PROPs

11.3.1. Defining a Markl Non-Unital Operad. Recall that S_1 consists of pairs of profiles of the form $(\underline{c}; d)$. The following definition is actually the colored version of Markl's non-unital operad [**Mar08**].

DEFINITION 11.10. A **Markl non-unital operad** consists of

(1) a Σ_{S_1}-bimodule P and
(2) **comp-i** operations

$$\mathsf{P}\binom{d}{\underline{c}} \otimes \mathsf{P}\binom{c_i}{\underline{b}} \xrightarrow{\circ_i} \mathsf{P}\binom{d}{\underline{c} \circ_i \underline{b}},$$

satisfying the following associativity and equivariance axioms.

Associativity: The diagrams

$$\begin{array}{ccc}
\mathsf{P}\binom{d}{\underline{c}} \otimes \mathsf{P}\binom{c_i}{\underline{b}} \otimes \mathsf{P}\binom{b_j}{\underline{a}} & \xrightarrow{Id \otimes \circ_j} & \mathsf{P}\binom{d}{\underline{c}} \otimes \mathsf{P}\binom{c_i}{\underline{b} \circ_j \underline{a}} \\
\Big\downarrow {\scriptstyle \circ_i \otimes Id} & & \Big\downarrow {\scriptstyle \circ_i} \\
\mathsf{P}\binom{d}{\underline{c} \circ_i \underline{b}} \otimes \mathsf{P}\binom{b_j}{\underline{a}} & \xrightarrow{\circ_{i-1+j}} & \mathsf{P}\binom{d}{\underline{c} \circ_i (\underline{b} \circ_j \underline{a})}
\end{array}$$

and

$$\begin{array}{ccc}
\mathsf{P}\binom{d}{\underline{c}} \otimes \mathsf{P}\binom{c_i}{\underline{b}} \otimes \mathsf{P}\binom{c_j}{\underline{a}} & \xrightarrow{\text{shuffle}} & \mathsf{P}\binom{d}{\underline{c}} \otimes \mathsf{P}\binom{c_j}{\underline{a}} \otimes \mathsf{P}\binom{c_i}{\underline{b}} \\
 & & \Big\downarrow {\scriptstyle \circ_j \otimes Id} \\
\Big\downarrow {\scriptstyle \circ_i \otimes Id} & & \mathsf{P}\binom{d}{\underline{c} \circ_j \underline{a}} \otimes \mathsf{P}\binom{c_i}{\underline{b}} \\
 & & \Big\downarrow {\scriptstyle \circ_i} \\
\mathsf{P}\binom{d}{\underline{c} \circ_i \underline{b}} \otimes \mathsf{P}\binom{c_j}{\underline{a}} & \xrightarrow{\circ_{j-1+l}} & \mathsf{P}\binom{d}{(\underline{c} \circ_i \underline{b}) \circ_{j+l-1} \underline{a}}
\end{array}$$

commute, in which $i < j$ and $|\underline{b}| = l$ in the second diagram.

Equivariance: The diagram

$$\begin{array}{ccc}
\mathsf{P}\binom{d}{\underline{c}} \otimes \mathsf{P}\binom{c_i}{\underline{b}} & \xrightarrow{\tau \otimes \pi} & \mathsf{P}\binom{d}{\underline{c}\tau} \otimes \mathsf{P}\binom{c_i}{\underline{b}\pi} \\
\Big\downarrow {\scriptstyle \circ_i} & & \Big\downarrow {\scriptstyle \circ_{\tau(i)}} \\
\mathsf{P}\binom{d}{\underline{c} \circ_i \underline{b}} & \xrightarrow{\tau \circ_i \pi} & \mathsf{P}\binom{d}{\underline{c}\tau \circ_? \underline{b}\pi}
\end{array}$$

commutes.

11.3.2. Interpreting the Axioms for a Markl Non-Unital Operad. Recall the strong generating set $\mathcal{T}^{Tree}$ of Theorem 7.35, which consists of the permuted corollas with one output and the simple trees. Our assertion below will be that by Theorem 11.5 Markl's Non-Unital Operads are precisely the `Tree`-PROPs. With this in mind, we should try to characterize the structure detailed above in terms of the strong generating set and the relaxed moves in $\mathcal{W}^{Tree}$.

Part of the statement of Theorem 11.5 in this context is that the presence of the permuted corollas implies the colored object P is really a Σ_{S_1}-bimodule, while the comp-i operations arise as $\gamma_{[G]}$ for G the simple tree $C_{(\underline{c};d)} \circ_i C_{(\underline{b};c_i)}$. Then Proposition 7.36 establishes the set $\mathcal{W}^{Tree}$ consisting of the relaxed moves in Lemmas 6.89, 6.91, and 6.92 as n-strong in $\mathcal{T}^{Tree}$. Here Theorem 11.5 creates a correspondence between the two associativity diagrams above and the two relaxed moves in Lemmas 6.91 and 6.92, as well as tying the equivariance diagram above to the relaxed move of Lemma 6.89.

REMARK 11.11. In the remaining sections of this chapter, the $\mathbf{\Sigma}_S^{\mathcal{E}}$-bimodule structure will always come from the structure maps associated to the permuted corollas in a specific strong generating set. Moreover, the equivariance axioms always correspond to the consequent change of listing of a generating graph in such a strong generating set. There will also be a number of different associativity axioms, as we saw when working with partially grafted corollas, especially trees and basic dioperadic graphs.

11.3.3. Markl Non-Unital Operads Are `Tree`-PROPs. The following biased characterization of `Tree`-PROPs now follows from Theorem 11.5 using Theorem 7.35 and Proposition 7.36.

COROLLARY 11.12. `Tree`-*PROP structures and Markl non-unital operad structures are equivalent.*

The following characterization of `Tree`-PROP (i.e., Markl non-unital operad) morphisms as those Σ_{S_1}-bimodule morphisms compatible with $\circ_i$ is similarly now a consequence of Theorem 11.7 using Theorem 7.35.

COROLLARY 11.13. *Let*

- P *and* Q *be* `Tree`-*PROPs, and*
- $f\colon \mathsf{P} \longrightarrow \mathsf{Q} \in \Sigma_{S_1}$*-bimodules.*

Then f *is a morphism of* `Tree`-*PROPs if and only if the square*

$$\begin{array}{ccc} \mathsf{P}\binom{d}{\underline{c}} \otimes \mathsf{P}\binom{c_i}{\underline{b}} & \xrightarrow{f\otimes f} & \mathsf{Q}\binom{d}{\underline{c}} \otimes \mathsf{Q}\binom{c_i}{\underline{b}} \\ \circ_i \big\downarrow & & \big\downarrow \circ_i \\ \mathsf{P}\binom{d}{\underline{c}\circ_i \underline{b}} & \xrightarrow{f} & \mathsf{Q}\binom{d}{\underline{c}\circ_i \underline{b}} \end{array}$$

is commutative for every possible $\circ_i$ *operation.*

11.4. May Operads as `UTree`-PROPs

11.4.1. Defining a May Operad. We begin by recalling the definition of a May operad [**May72**] in the colored setting.

Definition 11.14. A **May operad** consists of

(1) a Σ_{S_1}-bimodule O,

(2) a c-**colored unit** map

$$I \xrightarrow{\mathbf{1}_c} \mathsf{O}\binom{c}{c}$$

for each $c \in \mathfrak{C}$, and

(3) **operad structure maps**

$$\mathsf{O}\binom{d}{\underline{c}} \otimes \bigotimes_{i=1}^{m} \mathsf{O}\binom{c_i}{\underline{b}_i} \xrightarrow{\gamma} \mathsf{O}\binom{d}{\underline{b}} \tag{11.2}$$

for all $d \in \mathfrak{C}$, $\underline{c} = (c_1, \ldots, c_m)$ with $m \geq 1$, and $\underline{b}_i \in \mathcal{P}(\mathfrak{C})$, where $\underline{b} = (\underline{b}_1, \ldots, \underline{b}_m)$.

The operad structure maps γ are assumed to be associative, unital, and equivariant as follows.

Associativity: If $\underline{b} = (b_1, \ldots, b_{P_m})$, $|\underline{b}_i| = p_i \geq 1$, $P_i = p_1 + \cdots + p_i$ with $P_0 = 0$, $\underline{a}_j \in \mathcal{P}(\mathfrak{C})$ for $j \leq P_m$, and $\underline{a} = (\underline{a}_1, \ldots, \underline{a}_{P_m})$, then the following diagram is required to be commutative.

$$\begin{array}{ccc}
\mathsf{O}\binom{d}{\underline{c}} \otimes \Big\{\bigotimes_{i=1}^{m} \mathsf{O}\binom{c_i}{\underline{b}_i}\Big\} \otimes \Big\{\bigotimes_{j=1}^{P_m} \mathsf{O}\binom{b_j}{\underline{a}_j}\Big\} & \xrightarrow{\gamma \otimes Id} & \mathsf{O}\binom{d}{\underline{b}} \otimes \bigotimes_{j=1}^{P_m} \mathsf{O}\binom{b_j}{\underline{a}_j} \\
\Big\downarrow {\scriptstyle \text{shuffle}} & & \\
\mathsf{O}\binom{d}{\underline{c}} \otimes \bigotimes_{i=1}^{m} \Big\{\mathsf{O}\binom{c_i}{\underline{b}_i} \otimes \bigotimes_{j=P_{i-1}+1}^{P_i} \mathsf{O}\binom{b_j}{\underline{a}_j}\Big\} & & \Big\downarrow {\scriptstyle \gamma} \\
\Big\downarrow {\scriptstyle Id \otimes \otimes \gamma} & & \\
\mathsf{O}\binom{d}{\underline{c}} \otimes \bigotimes_{i=1}^{m} \mathsf{O}\binom{c_i}{\underline{a}_{P_{i-1}+1}, \ldots, \underline{a}_{P_i}} & \xrightarrow{\gamma} & \mathsf{O}\binom{d}{\underline{a}}
\end{array} \tag{11.3}$$

Unity: The following two diagrams are required to be commutative.

$$\begin{array}{ccc}
\mathsf{O}\binom{d}{\underline{c}} \otimes I^{\otimes m} & \xrightarrow{\cong} & \mathsf{O}\binom{d}{\underline{c}} \\
\Big\downarrow {\scriptstyle Id \otimes \otimes \mathbf{1}_{c_i}} & \nearrow {\scriptstyle \gamma} & \\
\mathsf{O}\binom{d}{\underline{c}} \otimes \bigotimes_{i=1}^{m} \mathsf{O}\binom{c_i}{c_i} & &
\end{array}
\qquad
\begin{array}{ccc}
I \otimes \mathsf{O}\binom{d}{\underline{c}} & \xrightarrow{\cong} & \mathsf{O}\binom{d}{\underline{c}} \\
\Big\downarrow {\scriptstyle \mathbf{1}_d \otimes Id} & \nearrow {\scriptstyle \gamma} & \\
\mathsf{O}\binom{d}{d} \otimes \mathsf{O}\binom{d}{\underline{c}} & &
\end{array} \tag{11.4}$$

Equivariance: For $\sigma \in \Sigma_m$, $\tau_i \in \Sigma_{p_i}$, $\sigma\langle p_1, \ldots, p_m\rangle \in \Sigma_{P_m}$ the block permutation induced by σ that permutes the m blocks of the indicated lengths, and $\tau_1 \oplus \cdots \oplus \tau_m \in \Sigma_{P_m}$ the block sum, the following two diagrams are required to be commutative.

$$\begin{array}{ccc}
\mathsf{O}\binom{d}{\underline{c}} \otimes \bigotimes_{i=1}^{m} \mathsf{O}\binom{c_i}{\underline{b}_i} & \xrightarrow{\sigma \otimes \sigma^{-1}} & \mathsf{O}\binom{d}{\underline{c}\sigma} \otimes \bigotimes_{i=1}^{m} \mathsf{O}\binom{c_{\sigma^{-1}(i)}}{\underline{b}_{\sigma^{-1}(i)}} \\
\Big\downarrow {\scriptstyle \gamma} & & \Big\downarrow {\scriptstyle \gamma} \\
\mathsf{O}\binom{d}{\underline{b}} & \xrightarrow{\sigma\langle p_1, \ldots, p_m\rangle} & \mathsf{O}\binom{d}{\underline{b}_{\sigma^{-1}(1)}, \ldots, \underline{b}_{\sigma^{-1}(m)}}
\end{array} \tag{11.5}$$

$$\begin{array}{ccc} \mathsf{O}\binom{d}{\underline{c}} \otimes \bigotimes_{i=1}^{m} \mathsf{O}\binom{c_i}{\underline{b}_i} & \xrightarrow{Id \otimes \bigotimes \tau_i} & \mathsf{O}\binom{d}{\underline{c}} \otimes \bigotimes_{i=1}^{m} \mathsf{O}\binom{c_i}{\underline{b}_i \tau_i} \\ \gamma \downarrow & & \downarrow \gamma \\ \mathsf{O}\binom{d}{\underline{b}} & \xrightarrow{\tau_1 \oplus \cdots \oplus \tau_m} & \mathsf{O}\binom{d}{\underline{b}_1 \tau_1, \ldots, \underline{b}_m \tau_m} \end{array} \tag{11.6}$$

11.4.2. Interpreting the Axioms for a May Operad. Once again, we would like to remind the reader of the description of the structure above in terms of the strong generating set $\mathcal{T}^{UTree}$ for `UTree` established in Theorem 7.41 and the set of moves $\mathcal{W}^{UTree}$ established in Proposition 7.42. This strong generating set consists of all permuted corollas with one output, all special trees, and the exceptional edges of each color. The set of relaxed moves comes from Lemmas 6.94, 6.95, and 6.96.

As remarked above, the bimodule structure and the equivariance properties will follow from the permuted corollas in the strong generating set and consequent alterations to the listings, which will remain the case in all structures in this chapter. In a May operad, the c-colored unit $\mathbf{1}_c$ will be the map $\gamma_{[G]}$ for G the c-colored exceptional edge $\uparrow_c$, while the operad structure map γ will be $\gamma_{[G]}$ for G the special tree $T(\{\underline{b}^i\}; \underline{c}; d)$.

The associativity axiom then corresponds to the relaxed move of Lemma 6.96 and the unity axioms correspond to the two relaxed moves of Lemma 6.94. Finally, the two equivariance conditions correspond to two distinct special cases of relaxed moves from Lemma 6.95, the first with each ρ_i the identity and the second with τ the identity. Notice one could instead combine these two special cases as a single statement, which would be the general case of Lemma 6.95.

REMARK 11.15. The unity axioms involving $\mathbf{1}_c$ in a biased version of a generalized PROP, when they are present, always correspond to graph substitution involving exceptional edges.

11.4.3. May Operads are `UTree`-PROPs. The following biased characterization of `UTree`-PROPs now follows from Theorem 11.5, this time using Theorem 7.41 and Proposition 7.42.

COROLLARY 11.16. *`UTree`-PROP structures and May operad structures are equivalent.*

As before, Theorem 11.7 yields a characterization of morphisms of `UTree`-PROPs (i.e. of May operads) as Σ_{S_1}-bimodule morphisms compatible with extra structure, in this case both the unit and operad structure maps, using Theorem 7.41. In the one-colored topological case, this coincides with the original definition of an operad morphism in [**May72**].

COROLLARY 11.17. *Let*

- P *and* Q *be `UTree`-PROPs, and*
- $f \colon \mathsf{P} \longrightarrow \mathsf{Q} \in \Sigma_{S_1}$*-bimodules.*

Then f is a morphism of UTree*-PROPs if and only if all squares*

$$\begin{array}{ccc} I & = & I \\ {\scriptstyle \mathbf{1}_c}\downarrow & & \downarrow{\scriptstyle \mathbf{1}_c} \\ \mathsf{P}\binom{c}{c} & \xrightarrow{f} & \mathsf{Q}\binom{c}{c} \end{array} \tag{11.7}$$

and

$$\begin{array}{ccc} \mathsf{P}\binom{d}{\underline{c}} \otimes \bigotimes_{i=1}^{m} \mathsf{P}\binom{c_i}{\underline{b}_i} & \xrightarrow{\otimes f} & \mathsf{Q}\binom{d}{\underline{c}} \otimes \bigotimes_{i=1}^{m} \mathsf{Q}\binom{c_i}{\underline{b}_i} \\ {\scriptstyle \gamma}\downarrow & & \downarrow{\scriptstyle \gamma} \\ \mathsf{P}\binom{d}{\underline{b}} & \xrightarrow{f} & \mathsf{Q}\binom{d}{\underline{b}} \end{array} \tag{11.8}$$

are commutative.

11.5. Dioperads as $\mathtt{Gr}^{\uparrow}_{\text{di}}$-PROPs

11.5.1. Defining a Dioperad. Here is the colored version of Gan's notion of dioperad [**Gan03**]. In the following definition, we assume $|\underline{a}| = k$, $|\underline{b}| = l$, $|\underline{c}| = m$, $|\underline{d}| = n$, $|\underline{e}| = p$, and $|\underline{f}| = q$.

DEFINITION 11.18. A **dioperad** consists of

(1) a $\mathbf{\Sigma}_{S}^{\mathcal{E}}$-bimodule P,
(2) a c-colored unit

$$I \xrightarrow{\mathbf{1}_c} \mathsf{P}\binom{c}{c}$$

for each $c \in \mathfrak{C}$, and
(3) a **dioperadic composition**

$$\mathsf{P}\binom{\underline{d}}{\underline{c}} \otimes \mathsf{P}\binom{\underline{b}}{\underline{a}} \xrightarrow{{}_j\circ_i} \mathsf{P}\binom{\underline{b} \circ_j \underline{d}}{\underline{c} \circ_i \underline{a}}$$

whenever $b_j = c_i$,

such that the following unity, equivariance, and associativity axioms are satisfied.

Unity: The triangles

$$\begin{array}{ccc} \mathsf{P}\binom{\underline{d}}{\underline{c}} \otimes I & \xrightarrow{\cong} & \mathsf{P}\binom{\underline{d}}{\underline{c}} \\ {\scriptstyle Id \otimes \mathbf{1}_{c_i}}\downarrow & \nearrow_{\circ_i} & \\ \mathsf{P}\binom{\underline{d}}{\underline{c}} \otimes \mathsf{P}\binom{c_i}{c_i} & & \end{array}$$

and

$$\begin{array}{ccc} I \otimes \mathsf{P}\binom{\underline{b}}{\underline{a}} & \xrightarrow{\cong} & \mathsf{P}\binom{\underline{b}}{\underline{a}} \\ {\scriptstyle \mathbf{1}_{b_j} \otimes Id}\downarrow & \nearrow_{{}_j\circ} & \\ \mathsf{P}\binom{b_j}{b_j} \otimes \mathsf{P}\binom{\underline{b}}{\underline{a}} & & \end{array}$$

are commutative, where

$$\circ_i = {}_1\circ_i \quad \text{and} \quad {}_j\circ = {}_j\circ_1. \tag{11.9}$$

Equivariance: The diagram

(11.10)
$$
\begin{array}{ccc}
\mathsf{P}\binom{\underline{d}}{\underline{c}} \otimes \mathsf{P}\binom{\underline{b}}{\underline{a}} & \xrightarrow{\ {}_j\circ_i\ } & \mathsf{P}\binom{\underline{b}\circ_j \underline{d}}{\underline{c}\circ_i \underline{a}} \\
{\scriptstyle (\tau_1;\sigma_1)\otimes(\tau_2;\sigma_2)}\Big\downarrow & & \Big\downarrow{\scriptstyle (\tau_1\circ_i\tau_2;\sigma_2\circ_j\sigma_1)} \\
\mathsf{P}\binom{\sigma_1\underline{d}}{\underline{c}\tau_1} \otimes \mathsf{P}\binom{\sigma_2\underline{b}}{\underline{a}\tau_2} & \xrightarrow{\ {}_{\sigma_2^{-1}(j)}\circ_{\tau_1(i)}\ } & \mathsf{P}\binom{\sigma_2\underline{b}\circ_j\sigma_1\underline{d}}{\underline{c}\tau_1\circ_i\underline{a}\tau_2}
\end{array}
$$

is commutative.

Associativity: There are three associativity axioms.

(1) With $b_j = c_i$ and $d_s = e_r$, the diagram

(11.11)
$$
\begin{array}{ccc}
\mathsf{P}\binom{\underline{f}}{\underline{e}} \otimes \mathsf{P}\binom{\underline{d}}{\underline{c}} \otimes \mathsf{P}\binom{\underline{b}}{\underline{a}} & \xrightarrow{\ {}_s\circ_r\otimes Id\ } & \mathsf{P}\binom{\underline{d}\circ_s\underline{f}}{\underline{e}\circ_r\underline{c}} \otimes \mathsf{P}\binom{\underline{b}}{\underline{a}} \\
{\scriptstyle Id\otimes{}_j\circ_i}\Big\downarrow & & \Big\downarrow{\scriptstyle {}_j\circ_{r+i-1}} \\
\mathsf{P}\binom{\underline{f}}{\underline{e}} \otimes \mathsf{P}\binom{\underline{b}\circ_j\underline{d}}{\underline{c}\circ_i\underline{a}} & \xrightarrow{\ {}_{j+s-1}\circ_r\ } & \mathsf{P}\binom{\underline{b}\circ_j(\underline{d}\circ_s\underline{f})}{\underline{e}\circ_r(\underline{c}\circ_i\underline{a})}
\end{array}
$$

is commutative.

(2) With $d_s = e_r$, $b_j = e_i$, and $r < i$, the diagram

(11.12)
$$
\begin{array}{ccc}
\mathsf{P}\binom{\underline{f}}{\underline{e}} \otimes \mathsf{P}\binom{\underline{d}}{\underline{c}} \otimes \mathsf{P}\binom{\underline{b}}{\underline{a}} & \xrightarrow{\ \text{shuffle}\ } & \mathsf{P}\binom{\underline{f}}{\underline{e}} \otimes \mathsf{P}\binom{\underline{b}}{\underline{a}} \otimes \mathsf{P}\binom{\underline{d}}{\underline{c}} \\
{\scriptstyle {}_s\circ_r\otimes Id}\Big\downarrow & & \Big\downarrow{\scriptstyle {}_j\circ_i\otimes Id} \\
\mathsf{P}\binom{\underline{d}\circ_s\underline{f}}{\underline{e}\circ_r\underline{c}} \otimes \mathsf{P}\binom{\underline{b}}{\underline{a}} & & \mathsf{P}\binom{\underline{b}\circ_j\underline{f}}{\underline{e}\circ_i\underline{a}} \otimes \mathsf{P}\binom{\underline{d}}{\underline{c}} \\
{\scriptstyle {}_j\circ_{i+m-1}}\Big\downarrow & & \Big\downarrow{\scriptstyle {}_s\circ_r} \\
\mathsf{P}\binom{\underline{b}\circ_j(\underline{d}\circ_s\underline{f})}{(\underline{e}\circ_r\underline{c})\circ_{i+m-1}\underline{a}} & \xleftarrow{\ (Id;\sigma)\ } & \mathsf{P}\binom{\underline{d}\circ_s(\underline{b}\circ_j\underline{f})}{(\underline{e}\circ_r\underline{c})\circ_{i+m-1}\underline{a}}
\end{array}
$$

is commutative, where σ is the block permutation

$$((1\ 2)(4\ 5))\langle s-1, j-1, q, l-j, n-s\rangle \in \Sigma_{l+n+q-2}$$

induced by $(1\ 2)(4\ 5) \in \Sigma_5$.

(3) With $b_j = c_i$, $b_s = e_r$, and $j < s$, the diagram

$$
\begin{array}{ccc}
\mathsf{P}\binom{\underline{f}}{\underline{e}} \otimes \mathsf{P}\binom{\underline{d}}{\underline{c}} \otimes \mathsf{P}\binom{\underline{b}}{\underline{a}} & \xrightarrow{\text{shuffle}} & \mathsf{P}\binom{\underline{d}}{\underline{c}} \otimes \mathsf{P}\binom{\underline{f}}{\underline{e}} \otimes \mathsf{P}\binom{\underline{b}}{\underline{a}} \\
\Big\downarrow {\scriptstyle Id\otimes {}_j\circ_i} & & \Big\downarrow {\scriptstyle Id\otimes {}_s\circ_r} \\
\mathsf{P}\binom{\underline{f}}{\underline{e}} \otimes \mathsf{P}\binom{\underline{b}\circ_j \underline{d}}{\underline{c}\circ_i \underline{a}} & & \mathsf{P}\binom{\underline{d}}{\underline{c}} \otimes \mathsf{P}\binom{\underline{b}\circ_s \underline{f}}{\underline{e}\circ_r \underline{a}} \\
\Big\downarrow {\scriptstyle {}_{s+n-1}\circ_r} & & \Big\downarrow {\scriptstyle {}_j\circ_i} \\
\mathsf{P}\binom{(\underline{b}\circ_j \underline{d})\circ_{s+n-1} \underline{f}}{\underline{e}\circ_r (\underline{c}\circ_i \underline{a})} & \xleftarrow{(\tau;Id)} & \mathsf{P}\binom{(\underline{b}\circ_j \underline{d})\circ_{s+n-1} \underline{f}}{\underline{c}\circ_i (\underline{e}\circ_r \underline{a})}
\end{array}
\tag{11.13}
$$

is commutative, where τ is the block permutation

$$((1\ 2)(4\ 5))\langle r-1, i-1, k, m-i, p-r\rangle \in \Sigma_{k+m+p-2}$$

induced by $(1\ 2)(4\ 5) \in \Sigma_5$.

REMARK 11.19. (1) In [**Gan03**] Gan worked with the one-colored case and a linear base category. The notation for the dioperadic composition is not consistent in the literature. The dioperadic composition we write as ${}_j\circ_i$ is actually written as ${}_i\circ_j$ in [**Gan03**] and as $\circ_j^i$ in [**MMS09**]. Our notation follows [**MV09**] because the special case ${}_1\circ_i$ (11.9) represents composition in the ith place, and the most common notation for this operation is $\circ_i$. For example, this $\circ_i$ notation is used in [**Ger63, MSS02, May97**].

(2) In [**Gan03**], six associativity axioms were listed for a dioperad. Among those six axioms, three of them are redundant by a change of indices.

11.5.2. Interpreting the Axioms for a Dioperad. Recall $\mathtt{Gr}^{\uparrow}_{\mathrm{di}}$ consists of all simply-connected wheeled graphs. The strong generating set $\mathcal{T}^{di}$ established in Theorem 7.57 consists of all permuted corollas, the exceptional edge of each color, and all basic dioperadic graphs. As usual, the exceptional edges induce unit maps, the permuted corollas yield the bimodule structure on P, and here for G a basic dioperadic graph $C_{(\underline{c};\underline{d})} \boxtimes_j^i C_{(\underline{a};\underline{b})}$, the map $\gamma_{[G]}$ yields the ${}_j\circ_i$ operation.

Now the required commutative diagrams correspond to the relaxed moves of $\mathcal{W}^{di}$, established in Proposition 7.58. The unity condition corresponds to the relaxed move in Lemma 6.102, the equivariance condition corresponds to the relaxed move in Lemma 6.103, and the three associativity conditions correspond to the relaxed moves in Lemmas 6.105, 6.106, and 6.107.

11.5.3. Dioperads are $\mathtt{Gr}^{\uparrow}_{\mathrm{di}}$-PROPs. This next biased characterization is a consequence of Theorem 11.5, exploiting Theorem 7.57 and Proposition 7.58.

COROLLARY 11.20. *$\mathtt{Gr}^{\uparrow}_{di}$-PROPs and dioperads are equivalent.*

The next result is a consequence of Theorem 11.7, this time using Corollary 11.20 and Theorem 7.57. In the one-colored linear case, this biased characterization is the original definition of a dioperad morphism in [**Gan03**].

COROLLARY 11.21. *Let*

- P *and* Q *be* $\mathtt{Gr}^{\uparrow}_{di}$*-PROPs (i.e., dioperads), and*
- $f: \mathsf{P} \longrightarrow \mathsf{Q} \in \mathbf{\Sigma}_S$.

Then f *is a morphism of* $\mathtt{Gr}^{\uparrow}_{di}$*-PROPs if and only if the squares* (11.7) *and*

$$\begin{array}{ccc} \mathsf{P}\binom{d}{\underline{c}} \otimes \mathsf{P}\binom{b}{\underline{a}} & \xrightarrow{f \otimes f} & \mathsf{Q}\binom{d}{\underline{c}} \otimes \mathsf{Q}\binom{b}{\underline{a}} \\ {\scriptstyle {}_j\circ_i} \big\downarrow & & \big\downarrow {\scriptstyle {}_j\circ_i} \\ \mathsf{P}\binom{b \circ_j \underline{d}}{\underline{c} \circ_i \underline{a}} & \xrightarrow{f} & \mathsf{Q}\binom{b \circ_j \underline{d}}{\underline{c} \circ_i \underline{a}} \end{array}$$

are commutative whenever they are defined.

11.6. Half-PROPs as $\mathtt{Gr}_{\frac{1}{2}}$-PROPs

Half-PROPs were invented by Kontsevich in unpublished work communicated to Markl [**Mar06**]. These non-unital objects were studied extensively in [**Mar06, MV09**].

11.6.1. Defining a Half-PROP. First we write down the colored version of a half-PROP. Denote by $S_{1/2}$ the full sub-groupoid of $\mathcal{P}(\mathfrak{C})^{op} \times \mathcal{P}(\mathfrak{C})$ consisting of pairs $(\underline{c}; \underline{d})$ of profiles with $|\underline{c}||\underline{d}| \geq 2$. Note that $S_{1/2}$ does not contain any pair of the form $(c; d)$.

DEFINITION 11.22. A **half-PROP** consists of

(1) a $\mathbf{\Sigma}^{\mathcal{E}}_{S_{1/2}}$-bimodule P,
(2) a **comp-**i operation

$$\mathsf{P}\binom{d}{\underline{c}} \otimes \mathsf{P}\binom{c_i}{\underline{a}} \xrightarrow{\circ_i} \mathsf{P}\binom{d}{\underline{c} \circ_i \underline{a}}$$

for each $1 \leq i \leq m$, and

(3) a j**-comp** operation

$$\mathsf{P}\binom{\underline{d}}{b_j} \otimes \mathsf{P}\binom{\underline{b}}{\underline{a}} \xrightarrow{{}_j\circ} \mathsf{P}\binom{\underline{b} \circ_j \underline{d}}{\underline{a}}$$

for each $1 \leq j \leq l$,

such that the following axioms are satisfied.

Equivariance: The dioperadic equivariance diagram (11.10) commutes in both of the following cases:
- $l = 1$, which implies $j = 1$.
- $m = 1$, which implies $i = 1$.

Overlap: The operations $\circ_1$ and ${}_1\circ$, both as maps

$$\mathsf{P}\binom{\underline{d}}{b} \otimes \mathsf{P}\binom{b}{\underline{a}} \longrightarrow \mathsf{P}\binom{\underline{d}}{\underline{a}},$$

are equal.

Associativity: There are five associativity axioms, which are all restrictions of those for dioperads.

(1) The diagram (11.11) commutes in each of the following three cases:
- $l = n = 1$, which implies $j = s = 1$.
- $m = p = 1$, which implies $i = r = 1$.
- $l = p = 1$, which implies $j = r = 1$.

(2) The diagram (11.12) commutes when $l = n = 1$, which implies $j = s = 1$.
(3) The diagram (11.13) commutes when $m = p = 1$, which implies $i = r = 1$.

11.6.2. Interpreting the Axioms for a Half-PROP. The $\circ_i$ and the ${}_j\circ$ operations in a half-PROP correspond to the basic dioperadic graphs $C_{(\underline{c};\underline{d})} \boxtimes_1^i C_{(\underline{a};c_i)}$ and $C_{(b_j;\underline{d})} \boxtimes_j^1 C_{(\underline{a};\underline{b})}$, respectively, as discussed in Example 2.24. The overlap axiom recognizes that both maps correspond to the same dioperadic graph $C_{b_1;\underline{d}} \boxtimes_1^1 C_{(\underline{a};b_1)}$. In $\mathcal{T}^{1/2}$ established in Theorem 7.64, we also have the permuted $S_{1/2}$-corollas, or permuted corollas satisfying the half-graph condition, which correspond to the bimodule structure maps as usual.

Looking back at $\mathcal{W}^{1/2}$ established in Proposition 7.65, the two special cases of relaxed moves from Lemma 6.103 correspond to the equivariance conditions above. In addition, the three special cases of relaxed moves from Lemma 6.105 included in $\mathcal{W}^{1/2}$ correspond to the first three associativity conditions above, while a special case of each of the relaxed moves in Lemmas 6.106 and 6.107 included in $\mathcal{W}^{1/2}$ correspond to the remaining two associativity conditions above.

11.6.3. Half-PROPs are $\mathtt{Gr}_{\frac{1}{2}}$-PROPs. This consequence of Theorem 11.5 follows using Theorem 7.64 and Proposition 7.65.

COROLLARY 11.23. *$\mathtt{Gr}_{\frac{1}{2}}$-PROPs and half-PROPs are equivalent.*

Theorem 11.7, applying Theorem 7.64 and Corollary 11.23, yields the following.

COROLLARY 11.24. *Let*

- P *and* Q *be* $\mathtt{Gr}_{\frac{1}{2}}$*-PROPs (i.e., half-PROPs), and*
- $f : \mathsf{P} \longrightarrow \mathsf{Q} \in \Sigma^{\mathcal{E}}_{S_{1/2}}$.

Then f is a morphism of $\mathtt{Gr}_{\frac{1}{2}}$-PROPs if and only if the squares

$$\begin{array}{ccc} \mathsf{P}\binom{\underline{d}}{\underline{c}} \otimes \mathsf{P}\binom{c_i}{\underline{a}} & \xrightarrow{f \otimes f} & \mathsf{Q}\binom{\underline{d}}{\underline{c}} \otimes \mathsf{Q}\binom{c_i}{\underline{a}} \\ \downarrow{\scriptstyle \circ_i} & & \downarrow{\scriptstyle \circ_i} \\ \mathsf{P}\binom{\underline{d}}{\underline{c} \circ_i \underline{a}} & \xrightarrow{f} & \mathsf{Q}\binom{\underline{d}}{\underline{c} \circ_i \underline{a}} \end{array}$$

and

$$\begin{array}{ccc} \mathsf{P}\binom{\underline{d}}{b_j} \otimes \mathsf{P}\binom{\underline{b}}{\underline{a}} & \xrightarrow{f \otimes f} & \mathsf{Q}\binom{\underline{d}}{b_j} \otimes \mathsf{Q}\binom{\underline{b}}{\underline{a}} \\ \downarrow{\scriptstyle {}_j\circ} & & \downarrow{\scriptstyle {}_j\circ} \\ \mathsf{P}\binom{\underline{b} \circ_j \underline{d}}{\underline{a}} & \xrightarrow{f} & \mathsf{Q}\binom{\underline{b} \circ_j \underline{d}}{\underline{a}} \end{array}$$

are commutative.

11.7. Properads as $\mathtt{Gr}^{\uparrow}_{\mathrm{c}}$-PROPs

11.7.1. Defining a Properad. We begin by writing down the biased axioms of the colored version of Vallette's notion of a properad [**Val07**]. We will use the same notations and conventions as in section 6.4.1.

DEFINITION 11.25. A **properad** consists of

(1) a $\mathbf{\Sigma}_{S}^{\mathcal{E}}$-bimodule P,
(2) a c-colored unit

$$I \xrightarrow{\mathbf{1}_c} \mathsf{P}\binom{c}{c}$$

for each $c \in \mathfrak{C}$, and
(3) a **properadic composition**

$$\mathsf{P}\binom{\underline{d}}{\underline{c}} \otimes \mathsf{P}\binom{\underline{b}}{\underline{a}} \xrightarrow{\boxtimes_{\underline{b}'}^{\underline{c}'}} \mathsf{P}\binom{\underline{b} \circ_{\underline{b}'} \underline{d}}{\underline{c} \circ_{\underline{c}'} \underline{a}}$$

whenever $\underline{b}' = \underline{c}'$ are k-segments,

such that the following unity, bi-equivariance, and associativity axioms are satisfied.

Unity: The two unity axioms of a dioperad are satisfied, where we interpret $\boxtimes_{b_j}^{c_i}$ as the dioperadic composition ${}_j\circ_i$.

Outer Bi-equivariance: In the context of Lemma 6.60, the diagram

$$\begin{array}{ccc}
\mathsf{P}\binom{\underline{d}}{\underline{c}} \otimes \mathsf{P}\binom{\underline{b}}{\underline{a}} & \xrightarrow{\boxtimes_{\underline{b}'}^{\underline{c}'}} & \mathsf{P}\binom{\underline{b} \circ_{\underline{b}'} \underline{d}}{\underline{c} \circ_{\underline{c}'} \underline{a}} \\
{\scriptstyle (\tau;\sigma)\otimes(\pi;\lambda)} \downarrow & & \downarrow {\scriptstyle (\tau \circ_{\underline{c}'} \pi; \lambda \circ_{\underline{b}'} \sigma)} \\
\mathsf{P}\binom{\sigma\underline{d}}{\underline{c}\tau} \otimes \mathsf{P}\binom{\lambda\underline{b}}{\underline{a}\pi} & \xrightarrow{\boxtimes_{\underline{b}'}^{\underline{c}'}} & \mathsf{P}\binom{\lambda\underline{b} \circ_{\underline{b}'} \sigma\underline{d}}{\underline{c}\tau \circ_{\underline{c}'} \underline{a}\pi}
\end{array}$$

is commutative.

Inner Bi-equivariance: In the context of Lemma 6.63, the diagram

$$\begin{array}{ccc}
\mathsf{P}\binom{\underline{d}}{\underline{c}} \otimes \mathsf{P}\binom{\underline{b}}{\underline{a}} & \xrightarrow{(\tau;Id)\otimes(Id;\lambda^{-1})} & \mathsf{P}\binom{\underline{d}}{\underline{c}\tau} \otimes \mathsf{P}\binom{\lambda^{-1}\underline{b}}{\underline{a}} \\
{\scriptstyle \boxtimes_{\underline{b}'}^{\underline{c}'}} \downarrow & & \downarrow {\scriptstyle \boxtimes_{\theta^{-1}\underline{b}'}^{\underline{c}'\theta}} \\
\mathsf{P}\binom{\underline{b} \circ_{\underline{b}'} \underline{d}}{\underline{c} \circ_{\underline{c}'} \underline{a}} & =\!=\!= & \mathsf{P}\binom{\lambda^{-1}\underline{b} \circ_{\theta^{-1}\underline{b}'} \underline{d}}{\underline{c}\tau \circ_{\underline{c}'\theta} \underline{a}}
\end{array}$$

is commutative.

Global Bi-equivariance: In the context of Lemma 6.74, if $\sigma C \tau$ is substituted into the top vertex of $C_0 \boxtimes C_0'$ and the various graphs have the following profiles:

- $(\underline{e}; \underline{f})$ for C_1 and $(\underline{c}; \underline{d})$ for C_1' with $\underline{d}' = \underline{e}'$,
- $(\underline{g}; \underline{h})$ for C_0 and $(\underline{a}; \underline{b})$ for C_0' with $\underline{b}' = \underline{g}'$,
- $(\underline{q}; \underline{p})$ for $C_0 \boxtimes C_0'$,
- $\sigma_4(\underline{e}; \underline{f})\tau_4 = (\underline{e}_*; \underline{f}_*)$, $\sigma_3(\underline{c}; \underline{d})\tau_3 = (\underline{c}_*; \underline{d}_*)$ with $\underline{d}_*' = \underline{e}_*'$,
- $\sigma_2(\underline{a}; \underline{b})\tau_2 = (\underline{a}_*; \underline{b}_*)$ and $(\underline{g}_*; \underline{h}_*)$ for $C_3 \boxtimes C_3'$ with $\underline{b}_*' = \underline{g}_*'$,
- $(\underline{q}_*; \underline{p}_*)$ for $C_2 \boxtimes C_2'$,

then the following diagram is commutative

$$
\begin{array}{ccccc}
\mathsf{P}\binom{\underline{f}}{\underline{e}} \otimes \mathsf{P}\binom{\underline{d}}{\underline{c}} \otimes \mathsf{P}\binom{\underline{b}}{\underline{a}} & & \xrightarrow{\{(\tau_j;\sigma_j)\}} & & \mathsf{P}\binom{\underline{f}_*}{\underline{e}_*} \otimes \mathsf{P}\binom{\underline{d}_*}{\underline{c}_*} \otimes \mathsf{P}\binom{\underline{b}_*}{\underline{a}_*} \\
\Big\downarrow {\scriptstyle \boxtimes^{\underline{e}'}_{\underline{d}'} \otimes Id} & & & & \Big\downarrow {\scriptstyle \boxtimes^{\underline{e}'_*}_{\underline{d}'_*} \otimes Id} \\
\mathsf{P}\binom{\underline{d}\circ_{\underline{d}'}\underline{f}}{\underline{e}\circ_{\underline{e}'}\underline{c}} \otimes \mathsf{P}\binom{\underline{b}}{\underline{a}} & & & & \mathsf{P}\binom{\underline{h}_*}{\underline{g}_*} \otimes \mathsf{P}\binom{\underline{b}_*}{\underline{a}_*} \\
\Big\downarrow {\scriptstyle (\tau;\sigma)\otimes Id} & & & & \Big\downarrow {\scriptstyle \boxtimes^{\underline{g}'_*}_{\underline{b}'_*}} \\
\mathsf{P}\binom{\underline{h}}{\underline{g}} \otimes \mathsf{P}\binom{\underline{b}}{\underline{a}} & \xrightarrow{\boxtimes^{\underline{g}'}_{\underline{b}'}} & \mathsf{P}\binom{\underline{q}}{\underline{p}} & \xleftarrow{(\tau_1;\sigma_1)} & \mathsf{P}\binom{\underline{q}_*}{\underline{p}_*}
\end{array}
$$

There is also an analogous diagram we must require to commute if $\sigma C \tau$ is substituted into the bottom vertex of $C_0 \boxtimes C_0'$.

Associativity: There are four associativity axioms.

(1) In the context of Lemma 6.66, the diagram

$$
\begin{array}{ccc}
\mathsf{P}\binom{\underline{f}}{\underline{e}} \otimes \mathsf{P}\binom{\underline{d}}{\underline{c}} \otimes \mathsf{P}\binom{\underline{b}}{\underline{a}} & \xrightarrow{\boxtimes^{\underline{e}'}_{\underline{d}'} \otimes Id} & \mathsf{P}\binom{\underline{d}\circ_{\underline{d}'}\underline{f}}{\underline{e}\circ_{\underline{e}'}\underline{c}} \otimes \mathsf{P}\binom{\underline{b}}{\underline{a}} \\
\Big\downarrow {\scriptstyle Id \otimes \boxtimes^{\underline{c}'}_{\underline{b}'}} & & \Big\downarrow {\scriptstyle \boxtimes^{\underline{c}'}_{\underline{b}'}} \\
\mathsf{P}\binom{\underline{f}}{\underline{e}} \otimes \mathsf{P}\binom{\underline{b}\circ_{\underline{b}'}\underline{d}}{\underline{c}\circ_{\underline{c}'}\underline{a}} & \xrightarrow{\boxtimes^{\underline{e}'}_{\underline{d}'}} & \mathsf{P}\binom{(\underline{b}\circ_{\underline{b}'}\underline{d})\circ_{\underline{d}'}\underline{f}}{\underline{e}\circ_{\underline{e}'}(\underline{c}\circ_{\underline{c}'}\underline{a})}
\end{array}
$$

is commutative.

(2) In the context of Lemma 6.68, the diagram

$$
\begin{array}{ccc}
\mathsf{P}\binom{\underline{f}}{\underline{e}} \otimes \mathsf{P}\binom{\underline{d}}{\underline{c}} \otimes \mathsf{P}\binom{\underline{b}}{\underline{a}} & \xrightarrow{\text{shuffle}} & \mathsf{P}\binom{\underline{f}}{\underline{e}} \otimes \mathsf{P}\binom{\underline{b}}{\underline{a}} \otimes \mathsf{P}\binom{\underline{d}}{\underline{c}} \\
\Big\downarrow {\scriptstyle \boxtimes^{\underline{e}'}_{\underline{d}'} \otimes Id} & & \Big\downarrow {\scriptstyle \boxtimes^{\underline{e}''}_{\underline{b}'} \otimes Id} \\
\mathsf{P}\binom{\underline{d}\circ_{\underline{d}'}\underline{f}}{\underline{e}\circ_{\underline{e}'}\underline{c}} \otimes \mathsf{P}\binom{\underline{b}}{\underline{a}} & & \mathsf{P}\binom{\underline{b}\circ_{\underline{b}'}\underline{f}}{\underline{e}\circ_{\underline{e}''}\underline{a}} \otimes \mathsf{P}\binom{\underline{d}}{\underline{c}} \\
\Big\downarrow {\scriptstyle \boxtimes^{\underline{e}''}_{\underline{b}'}} & & \Big\downarrow {\scriptstyle \boxtimes^{\underline{d}'}_{\underline{e}'}} \\
\mathsf{P}\binom{\underline{b}\circ_{\underline{b}'}(\underline{d}\circ_{\underline{d}'}\underline{f})}{(\underline{e}\circ_{\underline{e}'}\underline{c})\circ_{\underline{e}''}\underline{a}} & \xleftarrow{(Id;\sigma)} & \mathsf{P}\binom{\underline{d}\circ_{\underline{d}'}(\underline{b}\circ_{\underline{b}'}\underline{f})}{(\underline{e}\circ_{\underline{e}''}\underline{a})\circ_{\underline{e}'}\underline{c}}
\end{array}
$$

is commutative.

(3) In the context of Lemma 6.70, the diagram

$$\begin{array}{ccc}
\mathsf{P}\binom{\underline{f}}{\underline{e}} \otimes \mathsf{P}\binom{\underline{d}}{\underline{c}} \otimes \mathsf{P}\binom{\underline{b}}{\underline{a}} & \xrightarrow{\text{shuffle}} & \mathsf{P}\binom{\underline{d}}{\underline{c}} \otimes \mathsf{P}\binom{\underline{f}}{\underline{e}} \otimes \mathsf{P}\binom{\underline{b}}{\underline{a}} \\
\Big\downarrow{\scriptstyle Id \otimes \boxtimes^{\underline{c}'}_{\underline{b}'}} & & \Big\downarrow{\scriptstyle Id \otimes \boxtimes^{\underline{e}'}_{\underline{b}''}} \\
\mathsf{P}\binom{\underline{f}}{\underline{e}} \otimes \mathsf{P}\binom{\underline{b} \circ_{\underline{b}'} \underline{d}}{\underline{c} \circ_{\underline{c}'} \underline{a}} & & \mathsf{P}\binom{\underline{d}}{\underline{c}} \otimes \mathsf{P}\binom{\underline{b} \circ_{\underline{b}''} \underline{f}}{\underline{e} \circ_{\underline{e}'} \underline{a}} \\
\Big\downarrow{\scriptstyle \boxtimes^{\underline{e}'}_{\underline{b}''}} & & \Big\downarrow{\scriptstyle \boxtimes^{\underline{c}'}_{\underline{b}'}} \\
\mathsf{P}\binom{(\underline{b} \circ_{\underline{b}'} \underline{d}) \circ_{\underline{b}''} \underline{f}}{\underline{e} \circ_{\underline{e}'} (\underline{c} \circ_{\underline{c}'} \underline{a})} & \xleftarrow{(\tau; Id)} & \mathsf{P}\binom{(\underline{b} \circ_{\underline{b}''} \underline{f}) \circ_{\underline{b}'} \underline{d}}{\underline{c} \circ_{\underline{c}'} (\underline{e} \circ_{\underline{e}'} \underline{a})}
\end{array}$$

is commutative.

(4) In the context of Lemma 6.72, the diagram

$$\begin{array}{ccc}
\mathsf{P}\binom{\underline{f}}{\underline{e}} \otimes \mathsf{P}\binom{\underline{d}}{\underline{c}} \otimes \mathsf{P}\binom{\underline{b}}{\underline{a}} & \xrightarrow{Id \otimes \boxtimes^{\underline{c}'}_{\underline{b}''}} & \mathsf{P}\binom{\underline{f}}{\underline{e}} \otimes \mathsf{P}\binom{\underline{b} \circ_{\underline{b}''} \underline{d}}{\underline{c} \circ_{\underline{c}'} \underline{a}} \\
\Big\downarrow{\scriptstyle \boxtimes^{\underline{e}''}_{\underline{d}'} \otimes Id} & & \Big\downarrow{\scriptstyle \boxtimes^{(\underline{e}', \underline{e}'')}_{(\underline{b}', \underline{d}')}} \\
\mathsf{P}\binom{\underline{d} \circ_{\underline{d}'} \underline{f}}{\underline{e} \circ_{\underline{e}''} \underline{c}} \otimes \mathsf{P}\binom{\underline{b}}{\underline{a}} & \xrightarrow{\boxtimes^{(\underline{e}', \underline{c}')}_{(\underline{b}', \underline{b}'')}} & \mathsf{P}\binom{\underline{b} \circ_{(\underline{b}', \underline{b}'')} (\underline{d} \circ_{\underline{d}'} \underline{f})}{(\underline{e} \circ_{\underline{e}''} \underline{c}) \circ_{(\underline{e}', \underline{c}')} \underline{a}}
\end{array}$$

is commutative.

REMARK 11.26. (1) As far as the authors are aware, the above biased axioms for a properad, even in the one-colored case, have not appeared in detail in the literature before. In [**Val07**] a (one-colored linear) properad was defined as a monoid with respect to the connected vertical composition. In [**Val07**, Sec. 4], a 'partial composition product' was introduced, along with the remark that all properadic operations can be generated in this way, but these generating axioms were not made explicit.

One advantage of the definition above over the original one or the unbiased one is that the only generating operations, besides the units and the bi-equivariance structure, are the properadic compositions $\boxtimes^{\underline{c}'}_{\underline{b}'}$, which are very easy to write down. In particular, each properadic composition arises from a partially grafted corollas, which has only two vertices. Moreover, the generating associativity axioms have very simple geometric interpretations.

(2) A dioperad is exactly a properad in which the only properadic compositions allowed are $\boxtimes^{c_i}_{b_j} = {}_j\circ_i$. In this case, the inner bi-equivariance axiom is trivial because $k = 1$, and the last associativity axiom does not occur.

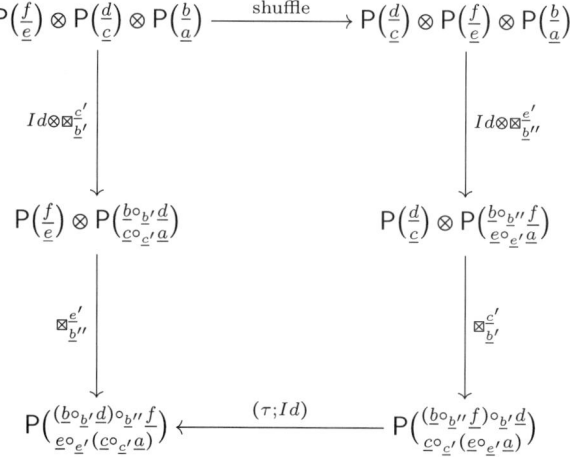

11.7.2. An Alternate Definition of a Properad. Here we provide an alternative way of defining a properad, to be consistent with our second strong generating set for the groupoid $\mathtt{Gr}^{\uparrow}_{\mathrm{c}}$.

DEFINITION 11.27. An **alternate properad** consists of

(1) a $\mathbf{\Sigma}^{\mathcal{E}}_{S}$-bimodule P,
(2) a c-colored unit

$$I \xrightarrow{\mathbf{1}_c} \mathsf{P}\binom{c}{c}$$

for each $c \in \mathfrak{C}$, and
(3) an **extended properadic composition**

$$\mathsf{P}\binom{\underline{d}}{\underline{c}} \otimes \mathsf{P}\binom{\underline{b}}{\underline{a}} \xrightarrow{\boxtimes^{\underline{c}'}_{\underline{b}'}(\tau;\sigma)} \mathsf{P}\binom{\sigma\underline{b}\circ_{\underline{b}'}\underline{d}}{\underline{c}\tau\circ_{\underline{c}'}\underline{a}}$$

whenever $\underline{b}' \subset \sigma\underline{b}$ and $\underline{c}' \subset \underline{c}\tau$ are matching non-empty segments.

such that the following unity, bi-equivariance, and associativity axioms are satisfied.

Unity: The extended properadic composition remains unital in the sense that the diagrams

$$\begin{array}{ccc}
\mathsf{P}\binom{\underline{d}}{\underline{c}} \otimes I & \xrightarrow{\quad\cong\quad} & \mathsf{P}\binom{\underline{d}}{\underline{c}} \\
{\scriptstyle Id\otimes \mathbf{1}_{c_i}}\Big\downarrow & & \Big\downarrow{\scriptstyle (\tau;1)} \\
\mathsf{P}\binom{\underline{d}}{\underline{c}} \otimes \mathsf{P}\binom{c_i}{c_i} & \xrightarrow[\boxtimes^{c_i}_{c_i}(\tau;1)]{} & \mathsf{P}\binom{\underline{d}}{\underline{c}\tau}
\end{array}$$

and

$$\begin{array}{ccc}
I \otimes \mathsf{P}\binom{\underline{b}}{\underline{a}} & \xrightarrow{\quad\cong\quad} & \mathsf{P}\binom{\underline{b}}{\underline{a}} \\
{\scriptstyle \mathbf{1}_{b_j}\otimes Id}\Big\downarrow & & \Big\downarrow{\scriptstyle (1;\sigma)} \\
\mathsf{P}\binom{b_j}{b_j} \otimes \mathsf{P}\binom{\underline{b}}{\underline{a}} & \xrightarrow[\boxtimes^{b_j}_{b_j}(1;\sigma)]{} & \mathsf{P}\binom{\sigma\underline{b}}{\underline{a}}
\end{array}$$

commute.

Bi-equivariance: In the context of Lemma 6.76, the diagram

$$\begin{array}{ccc}
\mathsf{P}\binom{\underline{d}}{\underline{c}} \otimes \mathsf{P}\binom{\underline{b}}{\underline{a}} & \xrightarrow{\boxtimes^{\underline{c}'}_{\underline{b}'}(\tau\delta;\nu\lambda)} & \mathsf{P}\binom{\nu\lambda\underline{b}\circ_{\underline{b}'}\underline{d}}{\underline{c}\tau\delta\circ_{\underline{c}'}\underline{a}} \\
{\scriptstyle (\tau;\sigma)\otimes(\rho;\lambda)}\Big\downarrow & & \Big\downarrow{\scriptstyle (1\circ_{\underline{c}'}\rho;1\circ_{\underline{b}'}\sigma)} \\
\mathsf{P}\binom{\sigma\underline{d}}{\underline{c}\tau} \otimes \mathsf{P}\binom{\lambda\underline{b}}{\underline{a}\pi} & \xrightarrow{\boxtimes^{\underline{c}'}_{\underline{b}'}(\delta;\nu)} & \mathsf{P}\binom{\nu\lambda\underline{b}\circ_{\underline{b}'}\sigma\underline{d}}{\underline{c}\tau\delta\circ_{\underline{c}'}\underline{a}\pi}
\end{array}$$

is commutative.

Associativity: There are again four associativity axioms, including the second and third associativity conditions in the definition of a properad. In addition, in the context of Lemma 6.77, the diagram

$$\begin{array}{ccc}
\mathsf{P}\binom{\underline{f}}{\underline{e}} \otimes \mathsf{P}\binom{\underline{d}}{\underline{c}} \otimes \mathsf{P}\binom{\underline{b}}{\underline{a}} & \xrightarrow{Id\otimes\boxtimes^{\underline{c}'}_{\underline{b}''}(\tau;\lambda)} & \mathsf{P}\binom{\underline{f}}{\underline{e}} \otimes \mathsf{P}\binom{\lambda\underline{b}\circ_{\underline{b}''}\underline{d}}{\underline{c}\tau\circ_{\underline{c}'}\underline{a}} \\
\Big\downarrow {\scriptstyle \boxtimes^{\underline{e}''}_{\underline{d}'}(\rho;\sigma)\otimes Id} & & \Big\downarrow {\scriptstyle \boxtimes^{\underline{e}'}_{\underline{h}'}(\rho;\nu)} \\
 & & \mathsf{P}\binom{\lambda\underline{b}\circ_{\underline{b}'}(\sigma\underline{d}\circ_{\underline{d}'}\underline{f})}{\underline{e}\rho\circ_{\underline{e}''}(\underline{c}\tau\circ_{\underline{c}'}\underline{a})} \\
 & & \Big\downarrow {\scriptstyle (\pi;1)} \\
\mathsf{P}\binom{\sigma\underline{d}\circ_{\underline{d}'}\underline{f}}{\underline{e}\rho\circ_{\underline{e}''}\underline{c}} \otimes \mathsf{P}\binom{\underline{b}}{\underline{a}} & \xrightarrow{\boxtimes^{\underline{g}'}_{\underline{b}'}(\delta;\lambda)} & \mathsf{P}\binom{\lambda\underline{b}\circ_{\underline{b}'}(\sigma\underline{d}\circ_{\underline{d}'}\underline{f})}{(\underline{e}\rho\circ_{\underline{e}''}\underline{c})\delta\circ_{\underline{g}'}\underline{a}}
\end{array}$$

commutes. There is also a variant for the context of Remark 6.78.

11.7.3. Interpreting the Axioms for a Properad. Recall $\mathtt{Gr}^{\uparrow}_{\mathrm{c}}$ consists of all connected wheel-free graphs. The strong generating set $\mathcal{T}^{\uparrow}_{\mathrm{c}}$ established in Theorem 7.67 consists of all permuted corollas, the exceptional edge of each color, and all partially grafted corollas. As usual, the exceptional edges induce unit maps and the permuted corollas yield the bimodule structure on P. Here for G a partially grafted corollas $C_{(\underline{c};\underline{d})} \boxtimes^{\underline{c}'}_{\underline{b}'} C_{(\underline{a};\underline{b})}$, the map $\gamma_{[G]}$ yields the properadic composition $\boxtimes^{\underline{c}'}_{\underline{b}'}$.

The required commutative diagrams come from the set of relaxed moves $\mathcal{W}^{\uparrow}_{\mathrm{c}}$ established in Proposition 7.68. The relaxed move in Lemma 6.26 is connected to the bimodule structure, the unity conditions correspond to the relaxed moves in Lemma 6.102, and the bi-equivariance conditions correspond to the relaxed moves in Lemmas 6.60, 6.63, and 6.74 as well as Remark 6.75. The four associativity conditions correspond to the relaxed moves in Lemmas 6.66, 6.68, 6.70, and 6.72, respectively.

11.7.4. Interpreting the Axioms for an Alternate Properad. The strong generating set $\mathcal{T}^{\uparrow}_{\mathrm{c},2}$ established in Theorem 7.81 consists of all permuted corollas, the exceptional edge of each color, and all basic properadic graphs. As usual, the exceptional edges induce unit maps and the permuted corollas yield the bimodule structure on P. Here for G a basic properadic graph $C_{(\underline{c};\underline{d})}\tau \boxtimes^{\underline{c}'}_{\underline{b}'} \sigma C_{(\underline{a};\underline{b})}$, the map $\gamma_{[G]}$ yields the extended properadic composition $\boxtimes^{\underline{c}'}_{\underline{b}'}(\tau;\sigma)$.

The required commutative diagrams come from the set of relaxed moves $\mathcal{W}^{\uparrow}_{\mathrm{c},2}$ established in Proposition 7.82. The relaxed move in Lemma 6.26 is connected to the bimodule structure, the unity conditions correspond to the relaxed moves in Cor. 6.58 with $G = C_{(\underline{c};\underline{d})}\tau$ or $G = \sigma C_{(\underline{c};\underline{d})}$, and the bi-equivariance condition corresponds to the relaxed moves in Lemma 6.76. The four associativity conditions correspond to the relaxed moves in Lemmas 6.68, 6.70, 6.77 and Remark 6.78.

11.7.5. Properads and Alternate Properads are $\mathtt{Gr}^{\uparrow}_{\mathrm{c}}$-PROPs. Theorem 11.5, using Theorem 7.67 and Proposition 7.68, along with Theorem 7.81 and Proposition 7.82, now yields the following.

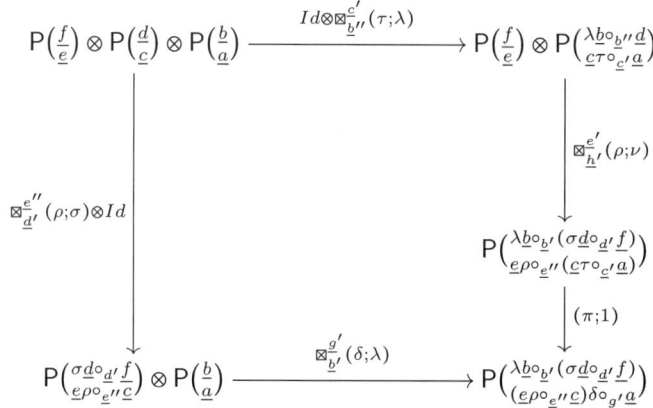

COROLLARY 11.28. *$\mathsf{Gr}^{\uparrow}_{c}$-PROPs, properads, and alternate properads are equivalent.*

In addition, Theorem 11.7, exploiting Theorem 7.67 and Corollary 11.28, gives us the following.

COROLLARY 11.29. *Let*

- *P and Q be $\mathsf{Gr}^{\uparrow}_{c}$-PROPs (i.e., properads), and*
- *$f\colon \mathsf{P} \longrightarrow \mathsf{Q} \in \mathbf{\Sigma}_S$.*

Then f is a morphism of $\mathsf{Gr}^{\uparrow}_{c}$-PROPs if and only if the squares (11.7) *and*

$$\begin{array}{ccc} \mathsf{P}\binom{\underline{d}}{\underline{c}} \otimes \mathsf{P}\binom{\underline{b}}{\underline{a}} & \xrightarrow{f\otimes f} & \mathsf{Q}\binom{\underline{d}}{\underline{c}} \otimes \mathsf{Q}\binom{\underline{b}}{\underline{a}} \\ {\scriptstyle \boxtimes^{\underline{c}'}_{\underline{b}'}}\downarrow & & \downarrow{\scriptstyle \boxtimes^{\underline{c}'}_{\underline{b}'}} \\ \mathsf{P}\binom{\underline{b}\circ_{\underline{b}'}\underline{d}}{\underline{c}\circ_{\underline{c}'}\underline{a}} & \xrightarrow{\quad f \quad} & \mathsf{Q}\binom{\underline{b}\circ_{\underline{b}'}\underline{d}}{\underline{c}\circ_{\underline{c}'}\underline{a}} \end{array}$$

are commutative whenever they are defined.

11.8. PROPs as $\mathsf{Gr}^{\uparrow}$-PROPs

11.8.1. Defining a PROP. Let us first write down the colored version of Mac Lane's PROP [**Mac63, Mac65**].

DEFINITION 11.30. A **PROP** consists of:

(1) a $\mathbf{\Sigma}_S^{\mathcal{E}}$-bimodule P;
(2) for any $\underline{a}, \underline{b}, \underline{c}, \underline{d} \in \mathcal{P}(\mathfrak{C})$, a morphism

$$\mathsf{P}\binom{\underline{d}}{\underline{c}} \otimes \mathsf{P}\binom{\underline{b}}{\underline{a}} \xrightarrow{\otimes_h} \mathsf{P}\binom{\underline{d},\underline{b}}{\underline{c},\underline{a}}, \tag{11.14}$$

called the **horizontal composition**;
(3) for any $\underline{b}, \underline{c}, \underline{d} \in \mathcal{P}(\mathfrak{C})$ with $\underline{c} \neq \varnothing$, a morphism

$$\mathsf{P}\binom{\underline{d}}{\underline{c}} \otimes \mathsf{P}\binom{\underline{c}}{\underline{b}} \xrightarrow{\otimes_v} \mathsf{P}\binom{\underline{d}}{\underline{b}}, \tag{11.15}$$

called the **vertical composition**;
(4) a morphism

$$I \xrightarrow{\mathbf{1}_\varnothing} \mathsf{P}\binom{\varnothing}{\varnothing}, \tag{11.16}$$

called the **empty unit**;
(5) for each $c \in \mathfrak{C}$, a morphism

$$I \xrightarrow{\mathbf{1}_c} \mathsf{P}\binom{c}{c}, \tag{11.17}$$

called the c**-colored unit**.

These operations are required to satisfy the following associativity, bi-equivariant, unital, and compatibility conditions.

Associativity of $\otimes_h$: The horizontal composition is associative in the sense that the diagram

$$\begin{array}{ccc} \mathsf{P}\binom{\underline{f}}{\underline{e}} \otimes \mathsf{P}\binom{\underline{d}}{\underline{c}} \otimes \mathsf{P}\binom{\underline{b}}{\underline{a}} & \xrightarrow{\otimes_h \otimes Id} & \mathsf{P}\binom{\underline{f},\underline{d}}{\underline{e},\underline{c}} \otimes \mathsf{P}\binom{\underline{b}}{\underline{a}} \\ {\scriptstyle Id\otimes\otimes_h}\downarrow & & \downarrow{\scriptstyle \otimes_h} \\ \mathsf{P}\binom{\underline{f}}{\underline{e}} \otimes \mathsf{P}\binom{\underline{d},\underline{b}}{\underline{c},\underline{a}} & \xrightarrow{\otimes_h} & \mathsf{P}\binom{\underline{f},\underline{d},\underline{b}}{\underline{e},\underline{c},\underline{a}} \end{array} \tag{11.18}$$

is commutative.

Bi-equivariance of $\otimes_h$: The horizontal composition is bi-equivariant in the sense that the following diagrams are commutative for all permutations $\sigma_1 \in \Sigma_{|\underline{d}|}$, $\sigma_2 \in \Sigma_{|\underline{b}|}$, $\tau_1 \in \Sigma_{|\underline{c}|}$, and $\tau_2 \in \Sigma_{|\underline{a}|}$, and all block permutations $\sigma = (1\ 2)\langle|\underline{d}|, |\underline{b}|\rangle \in \Sigma_{|\underline{d}|+|\underline{b}|}$ and $\tau = (1\ 2)\langle|\underline{a}|, |\underline{c}|\rangle \in \Sigma_{|\underline{c}|+|\underline{a}|}$.

$$\begin{array}{ccc} \mathsf{P}\binom{\underline{d}}{\underline{c}} \otimes \mathsf{P}\binom{\underline{b}}{\underline{a}} & \xrightarrow{\otimes_h} & \mathsf{P}\binom{\underline{d},\underline{b}}{\underline{c},\underline{a}} \\ {\scriptstyle (\tau_1;\sigma_1)\otimes(\tau_2;\sigma_2)}\downarrow & & \downarrow{\scriptstyle (\tau_1\times\tau_2;\sigma_1\times\sigma_2)} \\ \mathsf{P}\binom{\sigma_1\underline{d}}{\underline{c}\tau_1} \otimes \mathsf{P}\binom{\sigma_2\underline{b}}{\underline{a}\tau_2} & \xrightarrow{\otimes_h} & \mathsf{P}\binom{\sigma_1\underline{d},\sigma_2\underline{b}}{\underline{c}\tau_1,\underline{a}\tau_2} \end{array} \tag{11.19}$$

$$\begin{array}{ccc} \mathsf{P}\binom{\underline{d}}{\underline{c}} \otimes \mathsf{P}\binom{\underline{b}}{\underline{a}} & \xrightarrow{\otimes_h} & \mathsf{P}\binom{\underline{d},\underline{b}}{\underline{c},\underline{a}} \\ {\scriptstyle \text{switch}}\downarrow & & \downarrow{\scriptstyle (\tau;\sigma)} \\ \mathsf{P}\binom{\underline{b}}{\underline{a}} \otimes \mathsf{P}\binom{\underline{d}}{\underline{c}} & \xrightarrow{\otimes_h} & \mathsf{P}\binom{\underline{b},\underline{d}}{\underline{a},\underline{c}} \end{array} \tag{11.20}$$

Unity of $\otimes_h$: The empty unit is a two-sided unit for the horizontal composition in the sense that the diagram

$$\begin{array}{ccccc} & & \mathsf{P}\binom{\underline{d}}{\underline{c}} & & \\ & \overset{\cong}{\swarrow} & & \overset{\cong}{\searrow} & \\ I \otimes \mathsf{P}\binom{\underline{d}}{\underline{c}} & & & & \mathsf{P}\binom{\underline{d}}{\underline{c}} \otimes I \\ {\scriptstyle \mathbf{1}_\varnothing\otimes Id}\downarrow & & \downarrow{\scriptstyle Id} & & \downarrow{\scriptstyle Id\otimes\mathbf{1}_\varnothing} \\ \mathsf{P}\binom{\varnothing}{\varnothing} \otimes \mathsf{P}\binom{\underline{d}}{\underline{c}} & & & & \mathsf{P}\binom{\underline{d}}{\underline{c}} \otimes \mathsf{P}\binom{\varnothing}{\varnothing} \\ & \underset{\otimes_h}{\searrow} & & \underset{\otimes_h}{\swarrow} & \\ & & \mathsf{P}\binom{\underline{d}}{\underline{c}} & & \end{array} \tag{11.21}$$

is commutative.

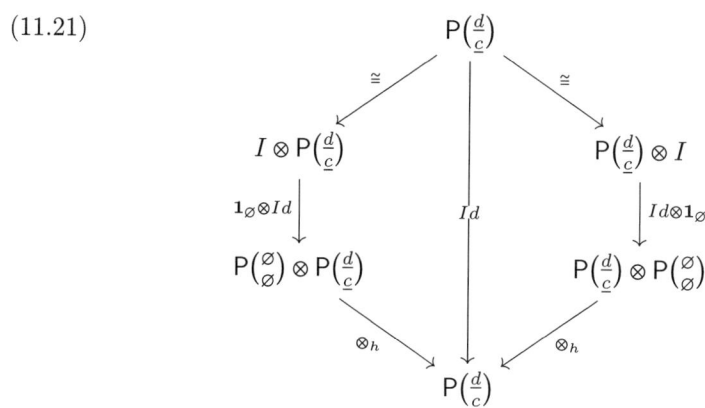

Associativity of $\otimes_v$: The vertical composition is associative in the sense that the diagram

$$
\begin{array}{ccc}
\mathsf{P}\binom{\underline{d}}{\underline{c}} \otimes \mathsf{P}\binom{\underline{c}}{\underline{b}} \otimes \mathsf{P}\binom{\underline{b}}{\underline{a}} & \xrightarrow{\otimes_v \otimes Id} & \mathsf{P}\binom{\underline{d}}{\underline{b}} \otimes \mathsf{P}\binom{\underline{b}}{\underline{a}} \\
{\scriptstyle Id \otimes \otimes_v}\downarrow & & \downarrow{\scriptstyle \otimes_v} \\
\mathsf{P}\binom{\underline{d}}{\underline{c}} \otimes \mathsf{P}\binom{\underline{c}}{\underline{a}} & \xrightarrow{\otimes_v} & \mathsf{P}\binom{\underline{d}}{\underline{a}}
\end{array}
\tag{11.22}
$$

is commutative.

Bi-equivariance of $\otimes_v$: The vertical composition is bi-equivariant in the sense that the diagram

$$
\begin{array}{ccc}
\mathsf{P}\binom{\underline{d}}{\underline{c}} \otimes \mathsf{P}\binom{\underline{c}}{\underline{b}} & \xrightarrow{\otimes_v} & \mathsf{P}\binom{\underline{d}}{\underline{b}} \\
{\scriptstyle (\tau^{-1};\sigma)\otimes(\pi;\tau)}\downarrow & & \downarrow{\scriptstyle (\pi;\sigma)} \\
\mathsf{P}\binom{\sigma\underline{d}}{\underline{c}\tau^{-1}} \otimes \mathsf{P}\binom{\tau\underline{c}}{\underline{b}\pi} & \xrightarrow{\otimes_v} & \mathsf{P}\binom{\sigma\underline{d}}{\underline{b}\pi}
\end{array}
\tag{11.23}
$$

is commutative, in which $\sigma \in \Sigma_{|\underline{d}|}$, $\tau \in \Sigma_{|\underline{c}|}$, and $\pi \in \Sigma_{|\underline{b}|}$.

Unity of $\otimes_v$: The colored units are units for the vertical composition in the sense that the following diagrams are commutative.

$$
\begin{array}{ccc}
I^{\otimes|\underline{d}|} \otimes \mathsf{P}\binom{\underline{d}}{\underline{c}} & \xrightarrow{\cong} & \mathsf{P}\binom{\underline{d}}{\underline{c}} \\
{\scriptstyle (\otimes_j \mathbf{1}_{d_j})\otimes Id}\downarrow & & \uparrow{\scriptstyle \otimes_v} \\
\left(\bigotimes_{j=1}^{|\underline{d}|} \mathsf{P}\binom{d_j}{d_j}\right) \otimes \mathsf{P}\binom{\underline{d}}{\underline{c}} & \xrightarrow{\otimes_h \otimes Id} & \mathsf{P}\binom{\underline{d}}{\underline{d}} \otimes \mathsf{P}\binom{\underline{d}}{\underline{c}}
\end{array}
\tag{11.24}
$$

$$
\begin{array}{ccc}
\mathsf{P}\binom{\underline{d}}{\underline{c}} \otimes I^{\otimes|\underline{c}|} & \xrightarrow{\cong} & \mathsf{P}\binom{\underline{d}}{\underline{c}} \\
{\scriptstyle Id\otimes(\otimes_i \mathbf{1}_{c_i})}\downarrow & & \uparrow{\scriptstyle \otimes_v} \\
\mathsf{P}\binom{\underline{d}}{\underline{c}} \otimes \bigotimes_{i=1}^{|\underline{c}|} \mathsf{P}\binom{c_i}{c_i} & \xrightarrow{Id \otimes \otimes_h} & \mathsf{P}\binom{\underline{d}}{\underline{c}} \otimes \mathsf{P}\binom{\underline{c}}{\underline{c}}
\end{array}
\tag{11.25}
$$

In (11.24) and (11.25) it is assumed that $|\underline{d}| \geq 1$ and $|\underline{c}| \geq 1$, respectively.

The Interchange Rule: The horizontal and the vertical compositions are compatible in the sense that the diagram

$$
\begin{array}{ccc}
\mathsf{P}\binom{\underline{d}_1}{\underline{c}_1} \otimes \mathsf{P}\binom{\underline{c}_1}{\underline{b}_1} \otimes \mathsf{P}\binom{\underline{d}_2}{\underline{c}_2} \otimes \mathsf{P}\binom{\underline{c}_2}{\underline{b}_2} & \xrightarrow{(\otimes_v,\otimes_v)} & \mathsf{P}\binom{\underline{d}_1}{\underline{b}_1} \otimes \mathsf{P}\binom{\underline{d}_2}{\underline{b}_2} \\
{\scriptstyle \text{switch}}\downarrow & & \\
\mathsf{P}\binom{\underline{d}_1}{\underline{c}_1} \otimes \mathsf{P}\binom{\underline{d}_2}{\underline{c}_2} \otimes \mathsf{P}\binom{\underline{c}_1}{\underline{b}_1} \otimes \mathsf{P}\binom{\underline{c}_2}{\underline{b}_2} & & \downarrow{\scriptstyle \otimes_h} \\
{\scriptstyle (\otimes_h,\otimes_h)}\downarrow & & \\
\mathsf{P}\binom{\underline{d}}{\underline{c}} \otimes \mathsf{P}\binom{\underline{c}}{\underline{b}} & \xrightarrow{\otimes_v} & \mathsf{P}\binom{\underline{d}}{\underline{b}}
\end{array}
\tag{11.26}
$$

is commutative, in which $\underline{a} = (\underline{a}_1, \underline{a}_2)$ for $\underline{a} = \underline{b}, \underline{c}, \underline{d}$.

11.8.2. Interpreting the Axioms for a PROP. The horizontal and vertical compositions correspond to the union and grafting of two corollas, respectively. Their associativity axioms correspond to the union and iterated graftings involving three corollas, Lemmas 6.32 and 6.50. The biequivariance conditions correspond to Lemmas 6.28 and 6.34 for the horizontal case, and Example 6.48 for the vertical case.

The empty unit corresponds to $\gamma_\varnothing$, and the colored units correspond to the exceptional linear graphs $\uparrow_c$. As a consequence, Lemma 6.30 correspond to the horizontal unit condition, while the two cases of Lemma 6.45 correspond to the vertical unit conditions. Finally, the interchange rule corresponds to the two ways to build the union of two grafted corollas, as in Example 6.55.

11.8.3. PROPs are $\mathtt{Gr}^\uparrow$-PROPs. The following corollary of Theorem 11.5 is a consequence of Theorem 7.27 and Proposition 7.28.

COROLLARY 11.31. *$\mathtt{Gr}^\uparrow$-PROPs and PROPs are equivalent.*

In applying Theorem 11.7 below, in conjunction with Theorem 7.27 and Corollary 11.31, we are using the generating subset of $\mathcal{T}^\uparrow$ in which $\underline{c}$-exceptional edges are replaced by c-exceptional edges for $c \in \mathfrak{C}$ when referring to the c-colored unit conditions from (11.7).

COROLLARY 11.32. *Let*

- *P and Q be $\mathtt{Gr}^\uparrow$-PROPs (i.e., PROPs), and*
- *$f: \mathsf{P} \longrightarrow \mathsf{Q} \in \Sigma_S$.*

Then f is a morphism of $\mathtt{Gr}^\uparrow$-PROPs if and only if the squares (11.7),

(11.27)
$$\begin{array}{ccc} I & = & I \\ {\scriptstyle 1_\varnothing}\downarrow & & \downarrow{\scriptstyle 1_\varnothing} \\ \mathsf{P}\binom{\varnothing}{\varnothing} & \xrightarrow{f} & \mathsf{Q}\binom{\varnothing}{\varnothing}, \end{array}$$

(11.28)
$$\begin{array}{ccc} \mathsf{P}\binom{\underline{d}}{\underline{c}} \otimes \mathsf{P}\binom{\underline{b}}{\underline{a}} & \xrightarrow{f\otimes f} & \mathsf{Q}\binom{\underline{d}}{\underline{c}} \otimes \mathsf{Q}\binom{\underline{b}}{\underline{a}} \\ {\scriptstyle \otimes_h}\downarrow & & \downarrow{\scriptstyle \otimes_h} \\ \mathsf{P}\binom{\underline{d},\underline{b}}{\underline{c},\underline{a}} & \xrightarrow{f} & \mathsf{Q}\binom{\underline{d},\underline{b}}{\underline{c},\underline{a}}, \end{array}$$

and

$$\begin{array}{ccc} \mathsf{P}\binom{\underline{d}}{\underline{c}} \otimes \mathsf{P}\binom{\underline{c}}{\underline{b}} & \xrightarrow{f\otimes f} & \mathsf{Q}\binom{\underline{d}}{\underline{c}} \otimes \mathsf{Q}\binom{\underline{c}}{\underline{b}} \\ {\scriptstyle \otimes_v}\downarrow & & \downarrow{\scriptstyle \otimes_v} \\ \mathsf{P}\binom{\underline{d}}{\underline{b}} & \xrightarrow{f} & \mathsf{Q}\binom{\underline{d}}{\underline{b}} \end{array}$$

are commutative.

11.9. Wheeled PROPs as $\mathtt{Gr}^{\circlearrowright}_{\mathtt{w}}$-PROPs

11.9.1. Defining a Wheeled PROP. Let us first write down the biased definition of a colored wheeled PROP.

DEFINITION 11.33. A **wheeled PROP** consists of:

(1) a $\Sigma_S^{\mathcal{E}}$-bimodule P;
(2) for any $\underline{a}, \underline{b}, \underline{c}, \underline{d} \in \mathcal{P}(\mathfrak{C})$, a morphism

$$\mathsf{P}\binom{\underline{d}}{\underline{c}} \otimes \mathsf{P}\binom{\underline{b}}{\underline{a}} \xrightarrow{\otimes_h} \mathsf{P}\binom{\underline{d},\underline{b}}{\underline{c},\underline{a}}, \tag{11.29}$$

called the **horizontal composition**;
(3) a morphism

$$I \xrightarrow{\mathbf{1}_\varnothing} \mathsf{P}\binom{\varnothing}{\varnothing}, \tag{11.30}$$

called the **empty unit**;
(4) for each $c \in \mathfrak{C}$, a morphism

$$I \xrightarrow{\mathbf{1}_c} \mathsf{P}\binom{c}{c}, \tag{11.31}$$

called the c-**colored unit**;
(5) for any $\underline{c} = c_{1,m}$ and $\underline{d} = d_{1,n}$ in $\mathcal{P}(\mathfrak{C})$ with $m, n \geq 1$ and $d_i = c_j$ for some $i \leq n$ and $j \leq m$, a morphism

$$\mathsf{P}\binom{\underline{d}}{\underline{c}} \xrightarrow{\xi_j^i} \mathsf{P}\binom{\underline{d} \setminus d_i}{\underline{c} \setminus c_j}, \tag{11.32}$$

called the **contraction**.

These operations are required to satisfy the following associativity, bi-equivariant, unital, commutativity, and compatibility conditions.

Associativity, bi-equivariance, and unity of $\otimes_h$: The diagrams (11.18)-(11.21) are commutative.

Unity: For any $\underline{b}, \underline{c}, \underline{d} \in \mathcal{P}(\mathfrak{C})$ with $\underline{c} \neq \varnothing$, define the **vertical composition**

$$\mathsf{P}\binom{\underline{d}}{\underline{c}} \otimes \mathsf{P}\binom{\underline{c}}{\underline{b}} \xrightarrow{\otimes_v} \mathsf{P}\binom{\underline{d}}{\underline{b}},$$

as the composition

$$\begin{array}{ccc} \mathsf{P}\binom{\underline{d}}{\underline{c}} \otimes \mathsf{P}\binom{\underline{c}}{\underline{b}} & \xrightarrow{\quad\otimes_v\quad} & \mathsf{P}\binom{\underline{d}}{\underline{b}}. \\ & {\scriptstyle \otimes_h} \searrow \qquad \nearrow {\scriptstyle \left(\xi_1^{|\underline{d}|+1}\right)^{|\underline{c}|}} & \\ & \mathsf{P}\binom{\underline{d},\underline{c}}{\underline{c},\underline{b}} & \end{array} \tag{11.33}$$

Then the colored units $\mathbf{1}_c$ are units for the vertical composition in the sense that the diagrams (11.24) and (11.25) are commutative.

Bi-equivariance of ξ: The contraction ξ_j^i is bi-equivariant in the sense that the square

$$
\begin{array}{ccc}
\mathsf{P}\binom{\underline{d}}{\underline{c}} & \xrightarrow{\ \xi_j^i\ } & \mathsf{P}\binom{\underline{d}\smallsetminus d_i}{\underline{c}\smallsetminus c_j} \\
\Big\downarrow{\scriptstyle(\sigma;\tau)} & & \Big\downarrow{\scriptstyle(\sigma^{(i)};\tau^{(j)})} \\
\mathsf{P}\binom{\sigma\underline{d}}{\underline{c}\tau} & \xrightarrow{\ \xi_{\tau(j)}^{\sigma^{-1}(i)}\ } & \mathsf{P}\binom{\sigma\underline{d}\smallsetminus d_i}{\underline{c}\tau\smallsetminus c_j}
\end{array}
\tag{11.34}
$$

is commutative for all $\sigma \in \Sigma_n$ and $\tau \in \Sigma_m$, where $|\underline{d}| = n$, $|\underline{c}| = m$,

$$
\binom{\sigma\underline{d} \smallsetminus d_i}{\underline{c}\tau \smallsetminus c_j} = \binom{d_{\sigma(1)}, \ldots, \hat{d_i}, \ldots, d_{\sigma(n)}}{c_{\tau^{-1}(1)}, \ldots, \hat{c_j}, \ldots, c_{\tau^{-1}(m)}},
$$

and $\sigma^{(i)} \in \Sigma_{n-1}$ and $\tau^{(j)} \in \Sigma_{m-1}$ are the obvious permutations induced by σ and τ.

Commutativity of ξ: Suppose $|\underline{c}| = m \geq 2$, $|\underline{d}| = n \geq 2$, $d_i = c_j$, and $d_k = c_l$ for some $i \neq k \leq n$ and $j \neq l \leq m$. Then ξ_j^i and ξ_l^k almost commute in the following sense.

(1) The square

$$
\begin{array}{ccc}
\mathsf{P}\binom{\underline{d}}{\underline{c}} & \xrightarrow{\ \xi_j^i\ } & \mathsf{P}\binom{\underline{d}\smallsetminus d_i}{\underline{c}\smallsetminus c_j} \\
\Big\downarrow{\scriptstyle\xi_l^k} & & \Big\downarrow{\scriptstyle\xi_{l-1}^{k-1}} \\
\mathsf{P}\binom{\underline{d}\smallsetminus d_k}{\underline{c}\smallsetminus c_l} & \xrightarrow{\ \xi_j^i\ } & \mathsf{P}\binom{\underline{d}\smallsetminus\{d_i,d_k\}}{\underline{c}\smallsetminus\{c_j,c_l\}}
\end{array}
\tag{11.35}
$$

is commutative if $i < k$ and $j < l$.

(2) The square

$$
\begin{array}{ccc}
\mathsf{P}\binom{\underline{d}}{\underline{c}} & \xrightarrow{\ \xi_j^i\ } & \mathsf{P}\binom{\underline{d}\smallsetminus d_i}{\underline{c}\smallsetminus c_j} \\
\Big\downarrow{\scriptstyle\xi_l^k} & & \Big\downarrow{\scriptstyle\xi_l^{k-1}} \\
\mathsf{P}\binom{\underline{d}\smallsetminus d_k}{\underline{c}\smallsetminus c_l} & \xrightarrow{\ \xi_{j-1}^i\ } & \mathsf{P}\binom{\underline{d}\smallsetminus\{d_i,d_k\}}{\underline{c}\smallsetminus\{c_l,c_j\}}
\end{array}
\tag{11.36}
$$

is commutative if $i < k$ and $j > l$.

Compatibility of $\otimes_h$ and ξ: The horizontal composition and the contraction are compatible in the following sense.

(1) The square

$$
\begin{array}{ccc}
\mathsf{P}\binom{\underline{d}}{\underline{c}} \otimes \mathsf{P}\binom{\underline{b}}{\underline{a}} & \xrightarrow{\ \otimes_h\ } & \mathsf{P}\binom{\underline{d},\underline{b}}{\underline{c},\underline{a}} \\
\Big\downarrow{\scriptstyle\xi_j^i\otimes Id} & & \Big\downarrow{\scriptstyle\xi_j^i} \\
\mathsf{P}\binom{\underline{d}\smallsetminus d_i}{\underline{c}\smallsetminus c_j} \otimes \mathsf{P}\binom{\underline{b}}{\underline{a}} & \xrightarrow{\ \otimes_h\ } & \mathsf{P}\binom{\underline{d}\smallsetminus d_i,\underline{b}}{\underline{c}\smallsetminus c_j,\underline{a}}
\end{array}
\tag{11.37}
$$

is commutative whenever $d_i = c_j$ for some $i \leq |\underline{d}|$ and $j \leq |\underline{c}|$.

(2) The square

$$\begin{array}{ccc} \mathsf{P}\binom{\underline{d}}{\underline{c}} \otimes \mathsf{P}\binom{\underline{b}}{\underline{a}} & \xrightarrow{\otimes_h} & \mathsf{P}\binom{\underline{d},\underline{b}}{\underline{c},\underline{a}} \\ {\scriptstyle Id\otimes\xi_l^k}\downarrow & & \downarrow{\scriptstyle \xi_{|\underline{c}|+l}^{|\underline{d}|+k}} \\ \mathsf{P}\binom{\underline{d}}{\underline{c}} \otimes \mathsf{P}\binom{\underline{b}\setminus b_k}{\underline{a}\setminus a_l} & \xrightarrow{\otimes_h} & \mathsf{P}\binom{\underline{d},\underline{b}\setminus b_k}{\underline{c},\underline{a}\setminus a_l} \end{array} \tag{11.38}$$

is commutative whenever $b_k = a_l$ for some $k \le |\underline{b}|$ and $l \le |\underline{a}|$.

REMARK 11.34. The one-colored linear version of a wheeled PROP was introduced in [**MMS09**]. However, the axioms regarding the generating operations of a wheeled PROP were not written down explicitly there.

11.9.2. Interpreting the Axioms for a Wheeled PROP. In terms of generating graphs, the horizontal composition, the vertical composition (by Lemma 6.52), the empty unit, and the colored units in a wheeled PROP are to be interpreted as in a PROP (Definition 11.30). The contraction ξ_j^i (11.32) is $\gamma_{[G]}$ for G the contracted corolla $C^{j,i}_{(\underline{c};\underline{d})} = \xi_j^i C_{(\underline{c};\underline{d})}$ (3.5).

As a consequence, the associativity, bi-equivariance, and unital conditions for $\otimes_h$, as well as the vertical unital conditions follow as for PROPs above. Then Lemma 6.38 corresponds to the bi-equivariance condition for ξ, the two cases of Lemma 6.40 are tied to the commutativity conditions for ξ, and the two cases discussed in Example 6.43 correspond to the compatibility of $\otimes_h$ and ξ.

11.9.3. Wheeled PROPs are $\mathtt{Gr}_{\mathtt{w}}^{\circlearrowright}$-PROPs. Here the characterization follows from Theorem 11.5 using Theorem 7.25 and Proposition 7.22.

COROLLARY 11.35. $\mathtt{Gr}_{w}^{\circlearrowright}$*-PROPs and wheeled PROPs are equivalent.*

Now apply Theorem 11.7 together with Corollary 7.21 and Corollary 11.35 to understand the natural notion of morphism here.

COROLLARY 11.36. *Let*

- P *and* Q *be* $\mathtt{Gr}_{w}^{\circlearrowright}$*-PROPs (i.e., wheeled PROPs), and*
- $f : \mathsf{P} \longrightarrow \mathsf{Q} \in \Sigma_{\mathfrak{S}}$.

Then f *is a morphism of* $\mathtt{Gr}_{w}^{\circlearrowright}$*-PROPs if and only if the squares* (11.7), (11.27), (11.28), *and*

$$\begin{array}{ccc} \mathsf{P}\binom{\underline{d}}{\underline{c}} & \xrightarrow{f} & \mathsf{Q}\binom{\underline{d}}{\underline{c}} \\ {\scriptstyle \xi_j^i}\downarrow & & \downarrow{\scriptstyle \xi_j^i} \\ \mathsf{P}\binom{\underline{d}\setminus d_i}{\underline{c}\setminus c_j} & \xrightarrow{f} & \mathsf{Q}\binom{\underline{d}\setminus d_i}{\underline{c}\setminus c_j} \end{array}$$

are commutative whenever they are defined.

11.10. Wheeled Properads as $\mathtt{Gr}_{\mathtt{c}}^{\circlearrowright}$-PROPs

11.10.1. Defining a Wheeled Properad. Let us first write down the biased axioms of the colored version of a wheeled properad. We will use the notations in Definition 1.6.

DEFINITION 11.37. A **wheeled properad** consists of:

- a $\Sigma_{S}^{\mathfrak{C}}$-bimodule P,
- a c-colored unit
$$I \xrightarrow{\mathbf{1}_c} \mathsf{P}\binom{c}{c}$$
for each $c \in \mathfrak{C}$,
- a **contraction**
$$\mathsf{P}\binom{\underline{d}}{\underline{c}} \xrightarrow{\xi_j^i} \mathsf{P}\binom{\underline{d}\setminus d_i}{\underline{c}\setminus c_j}$$
whenever $c_j = d_i$, and
- a **dioperadic composition**
$$\mathsf{P}\binom{\underline{d}}{\underline{c}} \otimes \mathsf{P}\binom{\underline{b}}{\underline{a}} \xrightarrow{{}_j\circ_i} \mathsf{P}\binom{\underline{b}\circ_j \underline{d}}{\underline{c}\circ_i \underline{a}}$$
whenever $b_j = c_i$,

such that the following conditions are satisfied.

(1) $(\mathsf{P}, {}_j\circ_i, \mathbf{1})$ is a dioperad (Definition 11.18).

(2) The contraction satisfies the commutativity and bi-equivariance axioms of ξ in a wheeled PROP (Definition 11.33).

(3) The following four associativity axioms hold.

(a) In the context of Lemma 6.84, the diagram
$$\begin{array}{ccc}
\mathsf{P}\binom{\underline{d}}{\underline{c}} \otimes \mathsf{P}\binom{\underline{b}}{\underline{a}} & \xrightarrow{{}_s\circ_r} & \mathsf{P}\binom{\underline{b}\circ_s \underline{d}}{\underline{c}\circ_r \underline{a}} \\
{\scriptstyle \xi_j^i \otimes Id}\downarrow & & \downarrow{\scriptstyle \xi_{j-1+|\underline{a}|}^{s-1+i}} \\
\mathsf{P}\binom{\underline{d}\setminus d_i}{\underline{c}\setminus c_j} \otimes \mathsf{P}\binom{\underline{b}}{\underline{a}} & \xrightarrow{{}_s\circ_r} & \mathsf{P}\binom{\underline{b}\circ_s (\underline{d}\setminus d_i)}{(\underline{c}\setminus c_j)\circ_r \underline{a}}
\end{array}$$
is commutative.

(b) In the context of Lemma 6.86, the diagram
$$\begin{array}{ccc}
\mathsf{P}\binom{\underline{d}}{\underline{c}} \otimes \mathsf{P}\binom{\underline{b}}{\underline{a}} & \xrightarrow{{}_s\circ_r} & \mathsf{P}\binom{\underline{b}\circ_s \underline{d}}{\underline{c}\circ_r \underline{a}} \\
{\scriptstyle Id\otimes \xi_j^i}\downarrow & & \downarrow{\scriptstyle \xi_{r-1+j}^{i-1+|\underline{d}|}} \\
\mathsf{P}\binom{\underline{d}}{\underline{c}} \otimes \mathsf{P}\binom{\underline{b}\setminus b_i}{\underline{a}\setminus a_j} & \xrightarrow{{}_s\circ_r} & \mathsf{P}\binom{(\underline{b}\setminus b_i)\circ_s \underline{d}}{\underline{c}\circ_r (\underline{a}\setminus a_j)}
\end{array}$$
is commutative.

(c) In the context of Lemma 6.82, the diagram
$$\begin{array}{ccc}
\mathsf{P}\binom{\underline{d}}{\underline{c}} \otimes \mathsf{P}\binom{\underline{b}}{\underline{a}} & \xrightarrow{\text{switch}} & \mathsf{P}\binom{\underline{b}}{\underline{a}} \otimes \mathsf{P}\binom{\underline{d}}{\underline{c}} \\
{\scriptstyle {}_s\circ_r}\downarrow & & \downarrow{\scriptstyle {}_i\circ_j} \\
\mathsf{P}\binom{\underline{b}\circ_s \underline{d}}{\underline{c}\circ_r \underline{a}} & & \mathsf{P}\binom{\underline{d}\circ_i \underline{b}}{\underline{a}\circ_j \underline{c}} \\
{\scriptstyle \xi_{r-1+j}^{s-1+i}}\downarrow & & \downarrow{\scriptstyle \xi_{r-1+j}^{s-1+i}} \\
\mathsf{P}\binom{(\underline{b}\circ_s \underline{d})\setminus d_i}{(\underline{c}\circ_r \underline{a})\setminus a_j} & \xleftarrow{(\tau;\sigma)} & \mathsf{P}\binom{(\underline{d}\circ_i \underline{b})\setminus b_s}{(\underline{a}\circ_j \underline{c})\setminus c_r}
\end{array}$$
is commutative.

(d) In the context of Lemma 6.79, the diagram

$$\begin{array}{ccc}
\mathsf{P}\binom{\underline{d}}{\underline{c}} \otimes \mathsf{P}\binom{\underline{b}}{\underline{a}} & \xrightarrow{\ {}_s\circ_r\ } & \mathsf{P}\binom{\underline{b}\,{}_s\circ\,\underline{d}}{\underline{c}\,\circ_r\,\underline{a}} \\
\Big\downarrow {\scriptstyle {}_j\circ_i} & & \Big\downarrow {\scriptstyle \xi_i^j} \\
\mathsf{P}\binom{\underline{b}\,{}_j\circ\,\underline{d}}{\underline{c}\,\circ_i\,\underline{a}} & & \mathsf{P}\binom{(\underline{b}\,{}_s\circ\,\underline{d})\smallsetminus b_j}{(\underline{c}\,\circ_r\,\underline{a})\smallsetminus c_i} \\
\Big\downarrow {\scriptstyle \xi_{r-1+|\underline{a}|}^{s-1+|\underline{d}|}} & & \Big\downarrow {\scriptstyle (\tau;\sigma)} \\
\mathsf{P}\binom{(\underline{b}\,{}_j\circ\,\underline{d})\smallsetminus b_s}{(\underline{c}\,\circ_i\,\underline{a})\smallsetminus c_r} & = & \mathsf{P}\binom{(\underline{b}\,{}_j\circ\,\underline{d})\smallsetminus b_s}{(\underline{c}\,\circ_i\,\underline{a})\smallsetminus c_r}
\end{array}$$

is commutative.

REMARK 11.38. (1) The above biased axioms for a wheeled properad, even in the one-colored case, have not appeared in the literature before. In [**MMS09**] a (one-colored linear) wheeled properad was defined as an algebra over a suitable monad.

(2) The underlying dioperad of a wheeled properad simply forgets the contraction maps in this biased description.

(3) To compare properads with wheeled properads, first recall from Lemma 6.79 that every partially grafted corolla can be obtained from a basic dioperadic graph via iterated contractions. Using the notations in that lemma, the properadic composition $\boxtimes_{\underline{b}'}^{\underline{c}'}$ may be expressed as the composition

$$\boxtimes_{\underline{b}'}^{\underline{c}'} = \left(\xi_{l_c+|\underline{a}|}^{l_b+|\underline{d}|}\right)^{k-1} \circ \left({}_{l_b}\circ_{l_c}\right)$$

of a dioperadic composition and $k-1$ contractions. This is how a wheeled properad forgets down to an underlying properad. However, note that in a properad the various properadic compositions are independent; that is, no properadic composition can be obtained from other properadic compositions, units, and permuations. On the other hand, in a wheeled properad, the above decomposition tells us that the underlying properadic compositions are no longer independent.

11.10.2. Interpreting the Axioms for Wheeled Properads. Recall $\mathtt{Gr}_{\mathrm{c}}^{\circlearrowright}$ consists of all connected wheeled graphs. The strong generating set $\mathcal{T}_{\mathrm{c}}^{\circlearrowright}$ established in Theorem 7.88 consists of all permuted corollas, the exceptional edge of each color, all contracted corollas, and all basic dioperadic graphs. As usual, the exceptional edges induce unit maps, and the permuted corollas yield the bimodule structure on P. For G contracted corollas and basic dioperadic graphs, the maps $\gamma_{[G]}$ yield the contraction and the dioperadic composition, respectively.

The required conditions and commutative diagrams correspond to the set of relaxed moves $\mathcal{W}_{\mathrm{c}}^{\circlearrowright}$ established in Proposition 7.89. The first two conditions, about the underlying dioperad $(\mathsf{P}, {}_j\circ_i, \mathbf{1})$ and the contraction, follow from the corresponding graph substitution lemmas as in the cases of dioperads and wheeled PROPs. The four associativity conditions correspond to the relaxed moves in Lemmas 6.84, 6.86, 6.82, and 6.79, respectively.

11.10.3. Wheeled Properads are $\mathtt{Gr}_{\mathrm{c}}^{\circlearrowright}$-PROPs. This time, we apply Theorem 11.5, using Theorem 7.88 and Proposition 7.89.

COROLLARY 11.39. *$\mathtt{Gr}_c^{\circlearrowright}$-PROPs and wheeled properads are equivalent.*

In addition, Theorem 11.7 can now be combined with Theorem 7.88 and Corollary 11.39.

COROLLARY 11.40. *Let*

- P *and* Q *be* $\mathtt{Gr}_c^{\circlearrowright}$*-PROPs (i.e., wheeled properads), and*
- $f: \mathsf{P} \longrightarrow \mathsf{Q} \in \Sigma_S$.

Then f *is a morphism of* $\mathtt{Gr}_c^{\circlearrowright}$*-PROPs if and only if:*

(1) *the vertical unit squares* (11.7) *are commutative, and*
(2) f *is compatible with the dioperadic composition in the sense of Corollary* 11.21 *and with the contraction in the sense of Corollary* 11.36.

11.11. Wheeled Operads as $\mathtt{Tree}^{\circlearrowright}$-PROPs

11.11.1. Defining a Wheeled Operad. Let us first write down the definition of a wheeled operad in the colored setting. Recall that $S_{\leq 1}$ is the full sub-groupoid of $\mathcal{P}(\mathfrak{C})^{op} \times \mathcal{P}(\mathfrak{C})$ consisting of pairs of profiles $(\underline{c}; \underline{d})$ with $|\underline{d}| \leq 1$.

DEFINITION 11.41. A **wheeled operad** consists of:

(1) a $\Sigma_{S_{\leq 1}}$-bimodule

$$\mathsf{P} = \underbrace{\left\{\mathsf{P}\binom{\varnothing}{\underline{c}}\right\}}_{\mathsf{P}_w} \sqcup \underbrace{\left\{\mathsf{P}\binom{d}{\underline{c}}\right\}}_{\mathsf{P}_o}$$

together with a specified operad structure $(\mathsf{P}_o, \gamma, \mathbf{1})$. The sub-objects P_w and P_o are called the **wheeled part** and the **operadic part**, respectively.

(2) A structure map

$$\mathsf{P}\binom{\varnothing}{\underline{c}} \otimes \bigotimes_{i=1}^m \mathsf{P}\binom{c_i}{\underline{b}_i} \xrightarrow{\ \rho\ } \mathsf{P}\binom{\varnothing}{\underline{b}},$$

called the **right P_o-action on P_w**.

(3) A structure map

$$\mathsf{P}\binom{c_j}{\underline{c}} \xrightarrow{\ \xi_j^1\ } \mathsf{P}\binom{\varnothing}{\underline{c} \setminus c_j},$$

called the **contraction**.

In addition to γ defining an operad structure on P_o, the above data is required to satisfy the following conditions, in which the notations in Definition 11.14 are used.

(1) The map ρ gives the wheeled part a structure similar to that of a right module over the operadic part, in the sense that the following associativity, right unity, and equivariance diagrams commute.

Associativity:

$$\begin{array}{ccc}
\mathsf{P}\binom{\varnothing}{\underline{c}} \otimes \left\{\bigotimes_{i=1}^{m} \mathsf{P}\binom{c_i}{\underline{b}_i}\right\} \otimes \left\{\bigotimes_{j=1}^{P_m} \mathsf{P}\binom{b_j}{\underline{a}_j}\right\} & \xrightarrow{\rho\otimes Id} & \mathsf{P}\binom{\varnothing}{\underline{b}} \otimes \bigotimes_{j=1}^{P_m} \mathsf{P}\binom{b_j}{\underline{a}_j} \\
\downarrow{\cong} & & \\
\mathsf{P}\binom{\varnothing}{\underline{c}} \otimes \bigotimes_{i=1}^{m} \left\{\mathsf{P}\binom{c_i}{\underline{b}_i} \otimes \bigotimes_{j=P_{i-1}+1}^{P_i} \mathsf{P}\binom{b_j}{\underline{a}_j}\right\} & & \downarrow{\rho} \\
\downarrow{Id\otimes\bigotimes\gamma} & & \\
\mathsf{P}\binom{\varnothing}{\underline{c}} \otimes \bigotimes_{i=1}^{m} \mathsf{P}\binom{c_i}{\underline{a}_{P_{i-1}+1},\ldots,\underline{a}_{P_i}} & \xrightarrow{\rho} & \mathsf{P}\binom{\varnothing}{\underline{a}}
\end{array}$$

Right unity:

$$\begin{array}{ccc}
\mathsf{P}\binom{\varnothing}{\underline{c}} \otimes I^{\otimes m} & \xrightarrow{\cong} & \mathsf{P}\binom{\varnothing}{\underline{c}} \\
\downarrow{Id\otimes\bigotimes \mathbf{1}_{c_i}} & \nearrow{\rho} & \\
\mathsf{P}\binom{\varnothing}{\underline{c}} \otimes \bigotimes_{i=1}^{m} \mathsf{P}\binom{c_i}{c_i} & &
\end{array}$$

Equivariance:

$$\begin{array}{ccc}
\mathsf{P}\binom{\varnothing}{\underline{c}} \otimes \bigotimes_{i=1}^{m} \mathsf{P}\binom{c_i}{\underline{b}_i} & \xrightarrow{\sigma\otimes\sigma^{-1}} & \mathsf{P}\binom{\varnothing}{\underline{c}\sigma} \otimes \bigotimes_{i=1}^{m} \mathsf{P}\binom{c_{\sigma^{-1}(i)}}{\underline{b}_{\sigma^{-1}(i)}} \\
\downarrow{\rho} & & \downarrow{\rho} \\
\mathsf{P}\binom{\varnothing}{\underline{b}} & \xrightarrow{\sigma\langle p_1,\ldots,p_m\rangle} & \mathsf{P}\binom{\varnothing}{\underline{b}_{\sigma^{-1}(1)},\ldots,\underline{b}_{\sigma^{-1}(m)}}
\end{array}$$

$$\begin{array}{ccc}
\mathsf{P}\binom{\varnothing}{\underline{c}} \otimes \bigotimes_{i=1}^{m} \mathsf{P}\binom{c_i}{\underline{b}_i} & \xrightarrow{Id\otimes\bigotimes\tau_i} & \mathsf{P}\binom{\varnothing}{\underline{c}} \otimes \bigotimes_{i=1}^{m} \mathsf{P}\binom{c_i}{\underline{b}_i\tau_i} \\
\downarrow{\rho} & & \downarrow{\rho} \\
\mathsf{P}\binom{\varnothing}{\underline{b}} & \xrightarrow{\tau_1\oplus\cdots\oplus\tau_m} & \mathsf{P}\binom{\varnothing}{\underline{b}_1\tau_1,\ldots,\underline{b}_m\tau_m}
\end{array}$$

(2) The contraction is equivariant, in the sense that the square

$$\begin{array}{ccc}
\mathsf{P}\binom{c_j}{\underline{c}} & \xrightarrow{\xi_j^1} & \mathsf{P}\binom{\varnothing}{\underline{c}\smallsetminus c_j} \\
\downarrow{\tau} & & \downarrow{\tau^{(j)}} \\
\mathsf{P}\binom{c_j}{\underline{c}\tau} & \xrightarrow{\xi_{\tau(j)}^1} & \mathsf{P}\binom{\varnothing}{\underline{c}\tau\smallsetminus c_j}
\end{array}$$

is commutative. Here $\tau \in \Sigma_m$, and $\tau^{(j)} \in \Sigma_{m-1}$ is the induced permutation.

(3) The contraction is compatible with the right P_o-module action and the operad structure map, in the sense that the following diagrams are commutative.

$$
\begin{array}{ccc}
\mathsf{P}\binom{c_j}{\underline{c}} \otimes \bigotimes_{i\neq j} \mathsf{P}\binom{c_i}{\underline{b}_i} & \xrightarrow{\xi_j^1 \otimes Id} & \mathsf{P}\binom{\varnothing}{\underline{c}\smallsetminus c_j} \otimes \bigotimes_{i\neq j} \mathsf{P}\binom{c_i}{\underline{b}_i} \\
\downarrow \cong & & \downarrow \rho \\
\mathsf{P}\binom{c_j}{\underline{c}} \otimes \Big(\bigotimes_{i=1}^{j-1} \mathsf{P}\binom{c_i}{\underline{b}_i}\Big) \otimes I \otimes \Big(\bigotimes_{i=j+1}^{m} \mathsf{P}\binom{c_i}{\underline{b}_i}\Big) & & \mathsf{P}\binom{\varnothing}{\underline{b}\smallsetminus \underline{b}_j} \\
\downarrow Id\otimes \mathbf{1}_{c_j} \otimes Id & & \uparrow \xi_j^1 \\
\mathsf{P}\binom{c_j}{\underline{c}} \otimes \bigotimes_{i=1}^{m} \mathsf{P}\binom{c_i}{\underline{b}_i} & \xrightarrow{\gamma} & \mathsf{P}\binom{c_j}{\underline{b}}
\end{array}
$$

Above $\underline{b}_j = c_j$, while below $b = c_r$ and $d = a_j$, in the context of Lemma 6.82 with $|\underline{b}| = |\underline{d}| = s = i = 1$.

$$
\begin{array}{ccc}
\mathsf{P}\binom{d}{\underline{c}} \otimes \mathsf{P}\binom{b}{\underline{a}} & \xrightarrow{\text{switch}} & \mathsf{P}\binom{b}{\underline{a}} \otimes \mathsf{P}\binom{d}{\underline{c}} \\
\downarrow Id\otimes\bigotimes \mathbf{1}_{c_s} \otimes Id \otimes \bigotimes \mathbf{1}_{c_s} & & \downarrow Id\otimes\bigotimes \mathbf{1}_{a_i} \otimes Id \otimes \bigotimes \mathbf{1}_{a_i} \\
\mathsf{P}\binom{d}{\underline{c}} \otimes \Big(\bigotimes_{s=1}^{r-1} \mathsf{P}\binom{c_s}{c_s}\Big) \otimes \mathsf{P}\binom{b}{\underline{a}} \otimes \Big(\bigotimes_{s=r+1}^{m} \mathsf{P}\binom{c_s}{c_s}\Big) & & \mathsf{P}\binom{b}{\underline{a}} \otimes \Big(\bigotimes_{i=1}^{j-1} \mathsf{P}\binom{a_i}{a_i}\Big) \otimes \mathsf{P}\binom{d}{\underline{c}} \otimes \Big(\bigotimes_{i=j+1}^{n} \mathsf{P}\binom{a_i}{a_i}\Big) \\
\downarrow \gamma & & \downarrow \gamma \\
\mathsf{P}\binom{d}{\underline{c}\circ_r \underline{a}} & & \mathsf{P}\binom{b}{\underline{a}\circ_j \underline{c}} \\
\downarrow \xi_{r-1+j}^1 & & \downarrow \xi_{r-1+j}^1 \\
\mathsf{P}\binom{\varnothing}{(\underline{c}\circ_r \underline{a})\smallsetminus a_j} & \xleftarrow{(\tau;\varnothing)} & \mathsf{P}\binom{\varnothing}{(\underline{a}\circ_j \underline{c})\smallsetminus c_r}
\end{array}
$$

REMARK 11.42. (1) Although the wheeled part is akin to a right module over the operadic part, this is not exactly a right operadic module in the usual sense. Indeed, the underlying Σ-object of the wheeled part does not belong to the category of Σ-objects underlying operads.

(2) The equivariance axiom for the contraction is simply a restriction of that in a wheeled PROP (11.34).

(3) The one-colored linear version of a wheeled operad was introduced in [**MMS09**]. However, the axioms regarding the generating operations of a wheeled operad were not written down explicitly there.

11.11.2. Interpreting the Axioms for a Wheeled Operad. The operadic part of a wheeled operad is to be interpreted exactly as in a May operad, so $\gamma = \gamma_{[G]}$ for G a special tree. The wheeled part of a wheeled operad corresponds to corollas with empty output, which are to replace the level one vertex of special trees and form truncated trees. The right P_o-action ρ on the wheeled part is the map $\gamma_{[G]}$ where G is the truncated tree $T(\{\underline{b}_i\}; \underline{c}; \varnothing)$ of Example 6.97. The contraction ξ_j^1

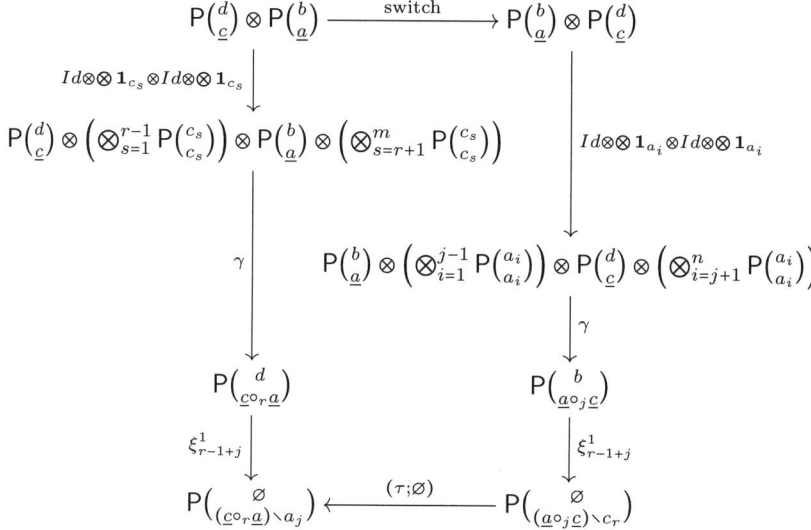

is similar, with G the contracted corolla $\xi_j^1 C_{(\underline{c};c_j)}$ (3.5). The associativity axiom involving γ and ρ corresponds to Lemma 6.100, while the right unital condition corresponds to Lemma 6.98. The two equivariance conditions are the two special cases of Lemma 6.99 corresponding to either type of permutations. The compatibility axioms between ρ, γ, and ξ_j^1 correspond to Lemma 6.101 and the special case of Lemma 6.82 with $|\underline{b}| = |\underline{d}| = s = i = 1$.

11.11.3. Wheeled Operads are $\mathtt{Tree}^{\circlearrowleft}$-PROPs. This time we pair Theorem 11.5 with Theorem 7.53 and Proposition 7.54.

COROLLARY 11.43. *$\mathtt{Tree}^{\circlearrowleft}$-PROPs and wheeled operads are equivalent.*

Theorem 11.7, when combined with Lemma 7.52 and Corollary 11.43, provides the following characterization of morphisms of wheeled operads.

COROLLARY 11.44. *Let*
- P *and* Q *be* $\mathtt{Tree}^{\circlearrowleft}$*-PROPs (i.e., wheeled operads), and*
- $f : \mathsf{P} \longrightarrow \mathsf{Q} \in \Sigma_{S_{\leq 1}}$.

Then f is a morphism of $\mathtt{Tree}^{\circlearrowleft}$-PROPs if and only if the squares (11.7), (11.8),

$$\begin{array}{ccc}
\mathsf{P}\binom{\varnothing}{\underline{c}} \otimes \bigotimes_{i=1}^{m} \mathsf{P}\binom{c_i}{\underline{b}_i} & \xrightarrow{\otimes f} & \mathsf{Q}\binom{\varnothing}{\underline{c}} \otimes \bigotimes_{i=1}^{m} \mathsf{Q}\binom{c_i}{\underline{b}_i} \\
\rho \big\downarrow & & \big\downarrow \rho \\
\mathsf{P}\binom{\varnothing}{\underline{b}} & \xrightarrow{\quad f \quad} & \mathsf{Q}\binom{\varnothing}{\underline{b}},
\end{array}$$

and

$$\begin{array}{ccc}
\mathsf{P}\binom{c_j}{\underline{c}} & \xrightarrow{f} & \mathsf{Q}\binom{c_j}{\underline{c}} \\
\xi_j^1 \big\downarrow & & \big\downarrow \xi_j^1 \\
\mathsf{P}\binom{\varnothing}{\underline{c} \setminus c_j} & \xrightarrow{f} & \mathsf{Q}\binom{\varnothing}{\underline{c} \setminus c_j}
\end{array}$$

are commutative.

REMARK 11.45. There are still other structures which can be described using this framework, but this list seems sufficiently long to illustrate the method thoroughly. For example, algebraic theories are examples of 1-colored PROPs, where the relevant monoidal product $\otimes$ must be the cartesian product in the category.

On the other hand, as referred to at the end of Section 1.4, structures like cyclic operads and modular operads are better described with a different notion of graph and underlying objects. We refer the interested reader to the material in [**MSS02**, Sec. II.5.1 and II.5.3] in those cases.

CHAPTER 12

Functors of Generalized PROPs

The purpose of this chapter is to study functors between the categories of generalized PROPs associated to various pasting schemes and base categories. The first main result in this chapter (Theorem 12.1) says that given two pasting schemes with one contained in the other, there always exists a free-forgetful adjoint pair between their associated categories of generalized PROPs. Given two specific pasting schemes, the left adjoint can often be constructed on an ad hoc basis, but the main point of Theorem 12.1 is that our left adjoint construction contains all such left adjoints in the operad/PROP literature as special cases. For example, the free operad generated by a sigma object, the free PROP generated by an operad, and the free wheeled PROP generated by a PROP are all special cases of our left adjoint. The explicit construction of the general left adjoint involves a large colimit, but when the pair of pasting schemes is well-matched, we show that the left adjoint admits a much simpler expression (Proposition 12.4) akin to the monad associated to a pasting scheme.

Rather than changing the pasting schemes, one could instead change the underlying symmetric monoidal category. In section 12.2 we provide sufficient conditions under which a symmetric monoidal functor prolongs to a functor on any category of generalized PROPs. As an application, it is shown that in the setting of Proposition 12.4 with a linear underlying category, the left adjoint commutes with homology up to natural isomorphisms (Corollary 12.18).

For this chapter, choose pasting schemes $\mathcal{G} = (S, \mathtt{G})$ and $\mathcal{G}' = (S', \mathtt{G}')$ (Definition 8.2), with $\mathcal{G} \leq \mathcal{G}'$ in the partial ordering on pasting schemes (Definition 8.3). Throughout this chapter, we work over any symmetric monoidal category $(\mathcal{E}, \otimes, I)$ with all small colimits (Convention 10.10), so $\mathtt{PROP}^{\mathcal{G}}$ is short for $\mathtt{PROP}^{\mathcal{G},\mathcal{E}}$ (Definition 10.39). The initial object (or colimit of the empty diagram) in $\mathcal{E}$ is denoted by $\varnothing$.

12.1. Adjunction Induced by an Inclusion of Pasting Schemes

THEOREM 12.1. *Let $\mathcal{G} \leq \mathcal{G}'$ be pasting schemes. Then there exists an adjoint pair*

$$L: \mathtt{PROP}^{\mathcal{G}} \rightleftarrows \mathtt{PROP}^{\mathcal{G}'} : U,$$

in which the right adjoint U is the forgetful functor.

Our goal is to give an explicit construction of the left adjoint, which is somewhat more practical to work with than, for example, the formula inherent in Freyd's adjoint functor theorem. First, we must say a few words about the existence of a forgetful functor, which will be shown to serve as a right adjoint. Then we carefully consider the special case where $\mathcal{G} \leq \mathcal{G}'$ are well-matched pasting schemes, where the left adjoint is greatly simplified. This also serves as a gradual introduction to the construction of the left adjoint in the general case.

12.1.1. The Right Adjoint. We begin by verifying the compatibility of the forgetful functor on colored objects with the monads defining generalized PROPs, in order to produce the expected forgetful functor when $\mathcal{G} \leq \mathcal{G}'$.

LEMMA 12.2. *If $\mathcal{G} \leq \mathcal{G}'$ are pasting schemes, then there is a natural forgetful functor*

$$U : \mathtt{PROP}^{\mathcal{G}'} \longrightarrow \mathtt{PROP}^{\mathcal{G}}.$$

PROOF. Let us abbreviate the functors $F_{\mathcal{G}}$ and $F_{\mathcal{G}'}$ (10.15) to F and F', respectively. First observe there is a forgetful functor

$$U : \mathcal{E}^{dis(S')} \longrightarrow \mathcal{E}^{dis(S)},$$

as discussed below Def. 10.28, so the only question is compatibility with the structure of monadic algebras. Thus, we will verify the conditions of Lemma 10.15 for this forgetful functor U. In this case, ψ is simply an inclusion of a sub-coproduct at each entry. Since the unit is also the inclusion of a summand, the commutativity of the unity diagram follows immediately. Since the restriction on graphs which can be substituted is compatible with graph substitution by the nature of a pasting scheme, the relevant muliplicativity diagram also commutes. □

12.1.2. Adjunction for a Well-Matched Pair. We now observe that when two pasting schemes are well-matched (Definition 9.13), the left adjoint L in Theorem 12.1 admits a much simpler formula than for the general case. Since the monad formulation can be opaque, let us first discuss the underlying idea.

As discussed in Lemma 10.40, a $\mathcal{G}$-PROP structure is equivalent to producing a collection of structure maps $\gamma_{[G]} : \mathsf{P}[G] \longrightarrow \mathsf{P}\binom{d}{\underline{c}}$ for each strict isomorphism class of graphs $[G] \in \mathtt{G}_S\binom{d}{\underline{c}}$, subject to associative and unital conditions. Given such maps, the question becomes how much more we must impose in order to produce similar structure maps for any $[G'] \in \mathtt{G}'_{S'}\binom{d}{\underline{c}}$ satisfying similar conditions.

In the well-matched case, the key is to observe that it suffices to define $\gamma_{[K]}$ appropriately for a representative of each weak equivalence class $K \in \mathtt{Kont}(\mathcal{G},\mathcal{G}')\binom{d}{\underline{c}}$. First, given G', the well-matched condition implies there is a unique, up to weak isomorphism, presentation as $G' = K(J_u)$ with

$$K \in \mathtt{Kont}(\mathcal{G},\mathcal{G}')\binom{d}{\underline{c}} \quad \text{and} \quad J_u \in \mathtt{G}_S\binom{\mathrm{out}(u)}{\mathrm{in}(u)}$$

for each $u \in \mathrm{Vt}(K)$. By the associativity condition we must satisfy, this tells us $\gamma_{[G']}$ must be defined as the composite of the morphism $\mathsf{P}[G'] \longrightarrow \mathsf{P}[K]$ induced by this choice of substituting graphs J_u, as discussed in (10.21) followed by $\gamma_{[K]}$. Also notice that for K' weakly equivalent to K within $\mathtt{G}'_{S'}\binom{d}{\underline{c}}$, there is a presentation $K = K'(C_u)$ for some choices of permuted corollas C_u. As above, these choices of substituting graphs induce a map $\mathsf{P}[K] \longrightarrow \mathsf{P}[K']$ and by symmetry together with the unit property of corollas in the definition of a generalized PROP, we see $\mathsf{P}[K] \cong \mathsf{P}[K']$. Thus, the same associativity condition implies it must be the case that $\gamma_{[K']}$ is determined by $\gamma_{[K]}$ and this isomorphism. The reader should be careful that we are using weak isomorphisms which fix the input and output profiles here, so we can talk about $[K]_w$ the weak isomorphism class with fixed input and output profiles to denote this relationship. In other words, we are allowed to alter the vertex profiles by substituting permuted corollas in for vertices of K, but we cannot substitute K into a permuted corolla to alter the full graph profiles. Regardless, the point is that extending to a larger well-matched pasting scheme is equivalent to defining each $\gamma_{[K]_w}$ for $[K]_w \in \mathtt{Kont}(\mathcal{G},\mathcal{G}')\binom{\underline{d}}{\underline{c}}$ in a manner compatible with the existing structure maps.

Since this can be quite technical we will separate the construction of the functor from the adjunction property.

LEMMA 12.3. *Suppose $\mathcal{G} \le \mathcal{G}'$ is a well-matched pair of pasting schemes. Then there is a functor*

$$L: \mathtt{PROP}^{\mathcal{G}} \longrightarrow \mathtt{PROP}^{\mathcal{G}'}$$

given by

$$L\mathsf{P}\binom{\underline{d}}{\underline{c}} = \coprod_{[G]_w \in Kont(\mathcal{G},\mathcal{G}')\binom{\underline{d}}{\underline{c}}} \mathsf{P}[G] \tag{12.1}$$

for $\mathsf{P} \in \mathtt{PROP}^{\mathcal{G}}$ and $(\underline{c};\underline{d}) \in S'$.

PROOF. Similar to the proof of Lemma 12.2, to verify the formula given above for L provides a functor between categories of algebras over the relevant monads, it suffices to construct a natural transformation

$$\psi : F'L \longrightarrow LF$$

that is multiplicative in the sense that the diagram

$$\begin{array}{ccccc} F'F'L\mathsf{P} & \xrightarrow{F'\psi} & F'LF\mathsf{P} & \xrightarrow{\psi F} & LFF\mathsf{P} \\ \downarrow{\mu' L} & & & & \downarrow{L\mu} \\ F'L\mathsf{P} & & \xrightarrow{\psi} & & LF\mathsf{P} \end{array} \tag{12.2}$$

commutes and satisfies a unit condition.

Notice, as in Lemma 10.37,

$$F'L\mathsf{P}\binom{\underline{d}}{\underline{c}} \cong \coprod_{G,\{H_v\}} \mathsf{P}[G(H_v)]$$

where each $[H_v]_w \in \mathtt{Kont}(\mathcal{G},\mathcal{G}')$ and $[G] \in \mathtt{G}'_{S'}\binom{\underline{d}}{\underline{c}}$ serve as strict substitution data. Similarly,

$$LF\mathsf{P}\binom{\underline{d}}{\underline{c}} \cong \coprod_{K,\{J_u\}} \mathsf{P}[K(J_u)]$$

where $[K]_w \in \mathtt{Kont}(\mathcal{G},\mathcal{G}')\binom{\underline{d}}{\underline{c}}$ and each $J_u \in \mathtt{G}_S\binom{\mathrm{out}(u)}{\mathrm{in}(u)}$ serve as strict substitution data. Given $G, \{H_v\}$ as above, the well-matched condition implies there is a unique up to weak isomorphism such $K, \{J_u\}$ with $G(H_v) = K(J_u)$. As such, our choice of ψ will be to include the summand

$$P[G(H_v)] \cong P[K(J_u)] \longrightarrow LF\mathsf{P}\binom{\underline{d}}{\underline{c}}$$

and this suffices to define the map out of the coproduct. Since ψ is really just coming from the inclusions of certain summands, the commutativity of this diagram is a mild variation on the proof of Theorem 10.38, really a consequence of the associativity of graph substitution. The unit condition is similarly a mild modification of that in the proof of Theorem 10.38, exploiting the unit property of corollas under graph substitution. □

Notice the construction of the multiplication of the monadic algebra $L\mathsf{P}$, following the proof of Lemma 10.15, is now given by

$$F'L\mathsf{P} \longrightarrow LF\mathsf{P} \longrightarrow L\mathsf{P} .$$

In components, that translates to

$$\coprod_{G',\{H_v\}} \mathsf{P}[G(H_v)] \longrightarrow \coprod_{K,\{J_u\}} \mathsf{P}[K(J_u)] \longrightarrow \coprod_K \mathsf{P}[K]$$

where the last map comes from exploiting the map

$$\otimes_u\colon \gamma_{J_u} \mathsf{P}[K(J_u)] \longrightarrow \mathsf{P}[K]$$

as in (10.21).

Now that we have constructed the functor, we can establish that it serves as the relevant left adjoint under the assumption that the pasting schemes are well-matched.

PROPOSITION 12.4. *Suppose $\mathcal{G} \le \mathcal{G}'$ is a well-matched pair of pasting schemes. Then the functor of Lemma* 12.3 *is left adjoint to the forgetful functor U of Lemma* 12.2.

PROOF. We will display the unit and counit of the adjunction, and verify the so-called triangular identities (see [**Mac98**, IV Theorem 2(v)]).

Defining the unit of adjunction $\eta\colon \mathsf{P} \longrightarrow UL\mathsf{P}$ is straightforward, since we can once again identify the target. That is, for $G \in \mathtt{G}_S$,

$$\begin{aligned} UL\mathsf{P}[G] &= \bigotimes_{v\in \mathrm{Vt}(G)} UL\mathsf{P}\binom{\mathrm{out}(v)}{\mathrm{in}(v)} \\ &\cong \bigotimes_{v\in \mathrm{Vt}(G)} \coprod_{[J_v]\in \mathtt{Kont}\binom{\mathrm{out}(v)}{\mathrm{in}(v)}} \mathsf{P}[J_v] \\ &\cong \coprod_{\{[J_v]\}} \bigotimes_{v\in \mathrm{Vt}(G)} \mathsf{P}[J_v] \\ &\cong \coprod_{\{[J_v]\}} \mathsf{P}[G(J_v)] \end{aligned}$$

so the inclusion where G and each J_v is a corolla defines our map at a fixed pair of profiles $(\underline{c};\underline{d})$. To see this provides a morphism of $\mathcal{G}$-PROPs, notice the lower path around the diagram is the map $\mathsf{P}[G] \longrightarrow \mathsf{P}\binom{\underline{d}}{\underline{c}} \longrightarrow \mathsf{P}[C_{(\underline{c};\underline{d})}]$ given by $\gamma^{\mathsf{P}}_{[G]}$ followed by the inclusion of that summand. The upper path around the diagram is instead $\mathsf{P}[G] \longrightarrow \mathsf{P}[G(C_v)] \longrightarrow \mathsf{P}[C_{(\underline{c};\underline{d})}]$, which comes from including as the summand with all of the Kontsevich graphs appropriate corollas and then applying $\gamma^{\mathsf{P}}_{[G(C_v)]}$. Since the unit property of graph substitution says $[G(C_v)] = [G]$, it follows that the diagram commutes.

On the other hand, the counit of the adjunction, traditionally written

$$\epsilon : LU\mathsf{Q} \longrightarrow \mathsf{Q},$$

for Q a $\mathcal{G}'$-PROP, requires a bit more care to define. For a given $\binom{\underline{d}}{\underline{c}}$, the required map

$$\coprod_K U\mathsf{Q}[K] \longrightarrow \mathsf{Q}\binom{\underline{d}}{\underline{c}}$$

is simply given on components by the structure map for Q as a $\mathcal{G}'$-PROP,

$$\gamma_{[K]} : \mathsf{Q}[K] \longrightarrow \mathsf{Q}\binom{d}{\underline{c}}.$$

Once again, this is clearly a morphism of colored objects, so we move to verifying the relevant diagrams for it to be a morphism of $\mathcal{G}'$-PROPs. Once again, there is a natural isomorphism as above,

$$LU\mathsf{Q}[G] \cong \coprod_{\{[H_v]\}} \mathsf{Q}[G(H_v)]$$

where each $H_v \in \mathtt{Kont}\binom{\mathrm{out}(v)}{\mathrm{in}(v)}$. However the structure map $LU\mathsf{Q}[G] \longrightarrow \mathsf{Q}\binom{d}{\underline{c}}$ restricted to a summand $\mathsf{Q}[G(H_v)]$ first identifies the weakly unique $[K]_w \in \mathtt{Kont}\binom{d}{\underline{c}}$ and $[J_u]$ such that $G(H_v) = K(J_u)$. Then $\mathsf{Q}[G(H_v)] = \mathsf{Q}[K(J_u)] \longrightarrow \mathsf{Q}[K]$, where the last map is once again $\otimes\gamma^{\mathsf{Q}}_{[J_u]}$, and this last target is a summand in $LU\mathsf{Q}\binom{d}{\underline{c}}$. At this point, the commutativity of the diagram defining a morphism of $\mathcal{G}$-PROPs follows from commutativity of

$$\begin{array}{ccc}
\mathsf{Q}[G(H_v)] & \xrightarrow{\otimes\gamma^{\mathsf{Q}}_{[H_v]}} & \mathsf{Q}[G] \\
{\scriptstyle =}\downarrow & & \\
\mathsf{Q}[K(J_u)] & & \downarrow{\scriptstyle \gamma^{\mathsf{Q}}_{[G]}} \\
{\scriptstyle \otimes\gamma^{\mathsf{Q}}_{[J_u]}}\downarrow & & \\
\mathsf{Q}[K] & \xrightarrow[\gamma^{\mathsf{Q}}_{[K]}]{} & \mathsf{Q}\binom{d}{\underline{c}}\ .
\end{array}$$

Finally, we must verify the triangular identities themselves. However, the fact that the composites

$$U\epsilon_{\mathsf{Q}} \circ \eta_{U\mathsf{Q}} : U\mathsf{Q} \longrightarrow ULU\mathsf{Q} \longrightarrow U\mathsf{Q}$$

and

$$\epsilon_{L\mathsf{P}} \circ L\eta_{\mathsf{P}} : L \longrightarrow ULU \longrightarrow L$$

are the respective identity maps then follows from the unital condition for a generalized PROP. ☐

The previous proposition applies, in particular, to all the well-matched pairs of pasting schemes in section 9.2, using the descriptions of the Kontsevich groupoids in section 9.1.4.

12.1.3. The General Left Adjoint. Next we consider the left adjoint in the general case. The most important point of what follows is that we can construct the left adjoint explicitly, rather than just appealing to an existence result. As a consequence, we are able later to verify some technical properties of the construction, which might not be possible from the formal existence of the left adjoint. Unfortunately, the construction in the general case is relatively opaque, so we again separate the construction from the question of adjointness. Recall $\mathtt{G}'_S\binom{d}{\underline{c}}$ is the replete and full subgroupoid of $\mathtt{G}'\binom{d}{\underline{c}}$ whose vertices all have profiles in the smaller S.

DEFINITION 12.5. Given a pair of profiles $(\underline{c};\underline{d}) \in S'$ for pasting schemes $\mathcal{G} \leq \mathcal{G}'$, define **the extension category** $\mathcal{D}\binom{d}{\underline{c}}$, a small category, whose objects are strict isomorphism classes of wheeled graphs in $\mathtt{G}'_S\binom{d}{\underline{c}}$, where a morphism $[K] \longrightarrow [G]$ consists of strict substitution data $\{[H_v]\}$ such that $[K] = [G(H_v)]$ and $H_v \in \mathtt{G}_S\binom{\mathrm{out}(v)}{\mathrm{in}(v)}$, while the composition law comes from the associativity of graph substitution as follows.

If $\{[I_w]\} : [K'] \longrightarrow [K]$ as well, then $[K'] = [K(I_w)]$ and the identification $w \in \mathrm{Vt}(K) = \coprod_v \mathrm{Vt}(H_v)$ allows us to partition the vertices and reindex the $I_w = I^v_u$. Then

$$[K'] = [K(I_w)] = [(G(H_v))(I^v_u)] = [G(H_v(I^v_u))]$$

so $\{[H_v(I^v_u)]\}$ provides strict substitution data into G, each lies in $\mathtt{G}_S\binom{\mathrm{out}(v)}{\mathrm{in}(v)}$ by the hereditary property of this pasting scheme, and the substitution is isomorphic to K'. The corresponding morphism $[K'] \longrightarrow [G]$ will be the composition. With this notion of composition, it follows that the strict substitution data consisting of (strict isomorphism classes of) corollas are the identity maps. Even more, the morphisms coming from (strict isomorphism classes of) permuted corollas, that is the weak isomorphisms preserving the full graph profiles, are precisely the isomorphisms in this category.

The reader should be aware that, as with the indexing category for bimodules, this category $\mathcal{D}\binom{d}{\underline{c}}$ can have many morphisms with the same names, that are distinguished only by considering the target.

As we saw in Lemma 10.40, a $\mathcal{G}$-PROP structure is equivalent to choosing $\gamma_{[G]}$ for each $G \in \mathtt{G}\binom{d}{\underline{c}}$ subject to associativity and unit conditions. However, the associativity condition can be described by first defining $\otimes_v \gamma_{[H_v]}$ for v running over the vertices of G, and then requiring

$$\gamma_{[G]}\left(\otimes_v \gamma_{[H_v]}\right) \cong \gamma_{[G(H_v)]}$$

as in (10.21). In particular, when each $H_v \in \mathtt{G}_S\binom{\mathrm{out}(v)}{\mathrm{in}(v)}$, it follows that $\otimes_v \gamma_{[H_v]}$ is well-defined for P, and compatible with graph substitution by the combination of the hereditary condition for the smaller pasting scheme and this associativity condition, which means P induces a functor $\mathcal{D}\binom{d}{\underline{c}} \longrightarrow \mathcal{E}$. We exploit this functor, in particular the colimit of such a diagram, in the construction below.

LEMMA 12.6. *If $\mathcal{G} \leq \mathcal{G}'$ are pasting schemes, then there is a functor*

$$L : \mathtt{PROP}^{\mathcal{G}} \longrightarrow \mathtt{PROP}^{\mathcal{G}'}.$$

with entries given by

$$L\mathsf{P}\binom{d}{\underline{c}} = \operatorname*{colim}_{[H]\in\mathcal{D}\binom{d}{\underline{c}}} \mathsf{P}[H].$$

PROOF. This clearly produces an S'-colored object, since $(\underline{c};\underline{d}) \in S'$ above, although in $\mathcal{D}$ we have restricted attention to graphs with these full graph profiles whose vertex profiles instead lie within the smaller S by the definition of $\mathtt{G}'_S\binom{d}{\underline{c}}$. In order to define the structure map $\gamma'_{[G]}$ for the proposed $\mathcal{G}'$-PROP, for a choice of

$[G] \in \mathtt{G}_{S'}\binom{d}{\underline{c}}$, we must define a map

$$\begin{array}{ccccc} L\mathsf{P}[G] & \xrightarrow{=} & \bigotimes_{v\in \mathrm{Vt}(G)} L\mathsf{P}\binom{\mathrm{out}(v)}{\mathrm{in}(v)} & \xrightarrow{\cong} & \mathrm{colim}_{\{[H_v]\}\in\prod_v \mathcal{D}\binom{\mathrm{out}(v)}{\mathrm{in}(v)}} \bigotimes_{v\in\mathrm{Vt}(G)} \mathsf{P}[H_v] \\ \downarrow & & & & \vdots \\ L\mathsf{P}\binom{d}{\underline{c}} & & \xrightarrow{\cong} & & \mathrm{colim}_{[K]\in\mathcal{D}\binom{d}{\underline{c}}} \mathsf{P}[K]. \end{array}$$

However, in this case, $\bigotimes_{v\in\mathrm{Vt}(G)} \mathsf{P}[H_v] \cong \mathsf{P}[G(H_v)]$, with $G(H_v)$ also in $\mathtt{G}'_S\binom{d}{\underline{c}}$. Hence, there is a natural map, from choosing $K = G(H_v)$,

$$\mathsf{P}[G(H_v)] \longrightarrow \operatorname*{colim}_{[K]\in\mathcal{D}\binom{d}{\underline{c}}} \mathsf{P}[K]$$

as well. The key observation is that the collection of all such maps forms a compatible system of maps

$$\bigotimes_{v\in\mathrm{Vt}(G)} \mathsf{P}[H_v] \longrightarrow \operatorname*{colim}_{[K]\in\mathcal{D}\binom{d}{\underline{c}}} \mathsf{P}[K]$$

under a morphism of the unordered tuple $\{[H_v]\}$, an object of $\prod_v \mathcal{D}\binom{\mathrm{out}(v)}{\mathrm{in}(v)}$, and so a map out of the colimit of the diagram in the source. That is, by the nature of a morphism in $\mathcal{D}$, if each $H_v = K_v(J_u^v)$ with each J_u^v in the smaller pasting scheme $\mathtt{G}_S$, then

$$[G(H_v)] = [G(K_v(J_{v,u}))] = [(G(K_v))(J_{v,u})]$$

by the associativity of graph substitution. As a consequence, the full collection $\{[J_u^v]\}$ defines a map in the target colimiting system and the associativity condition for P as a $\mathtt{G}$-PROP implies the composite

$$\mathsf{P}[(G(K_v))(J_{v,u})] \longrightarrow \mathsf{P}[G(K_v)] \longrightarrow \operatorname*{colim}_{[K]\in\mathcal{D}\binom{d}{\underline{c}}} \mathsf{P}[K]$$

agrees with the direct structure map

$$\mathsf{P}[G(H_v)] \longrightarrow \operatorname*{colim}_{[K]\in\mathcal{D}\binom{d}{\underline{c}}} \mathsf{P}[K].$$

All of this means the following diagram commutes

$$\begin{array}{ccc} \bigotimes_{v\in\mathrm{Vt}(G)} \mathsf{P}[H_v] & \xrightarrow{\cong} & \mathsf{P}[G(H_v)] \\ \downarrow{\scriptstyle \bigotimes_{v\in\mathrm{Vt}(G)}\bigotimes_{u\in\mathrm{Vt}(H_v)}\gamma[J_u^v]} & & \downarrow{\scriptstyle \bigotimes_{(v,u)\in\mathrm{Vt}(G(K_v))}\gamma[J_u^v]} \\ \bigotimes_{v\in\mathrm{Vt}(G)} \mathsf{P}[K_v] & \xrightarrow{\cong} & \mathsf{P}[G(K_v)] \\ & & \downarrow \\ & & \mathrm{colim}_{[K]\in\mathcal{D}\binom{d}{\underline{c}}} \mathsf{P}[K] \end{array}$$

so there is an induced map

$$\operatorname*{colim}_{\{[H_v]\}\in\prod_v \mathcal{D}\binom{\mathrm{out}(v)}{\mathrm{in}(v)}} \bigotimes_{v\in\mathrm{Vt}(G)} \mathsf{P}[H_v] \longrightarrow \operatorname*{colim}_{[K]\in\mathcal{D}\binom{d}{\underline{c}}} \mathsf{P}[K]$$

we can call $\gamma'_{[G]}$. The associativity and unital conditions for these maps then follow as in the proof of the monad structure Theorem 10.38 from the associativity and unity of graph substitution. □

REMARK 12.7. When the pasting schemes are well-matched, this colimit is equivalent to the coproduct used to define the left adjoint of Lemma 12.3. The reason is that the inclusion of the full subcategory consisting of the strict isomorphism classes $[K]$ with $K \in \mathtt{Kont}\binom{\underline{d}}{\underline{c}}$ is a final subcategory by assumption. Since this subcategory is a small groupoid, again by the assumption, with only the isomorphisms coming from substitution by permuted corollas, the colimit is isomorphic to taking the coproduct of a value for each orbit, while orbits correspond to the weak isomorphism classes $[K]_w$.

Our restatement of Theorem 12.1 now becomes the following.

LEMMA 12.8. *If $\mathcal{G} \le \mathcal{G}'$ are pasting schemes, then the forgetful functor U in Lemma 12.2 has the functor L in Lemma 12.6 as a left adjoint.*

PROOF. In order to establish the adjunction, this time we will display the natural correspondence between mapping sets directly. Thus, suppose $\mathsf{P} \in \mathcal{G}$-PROP, while $\mathsf{Q} \in \mathcal{G}'$-PROP and $g : L\mathsf{P} \longrightarrow \mathsf{Q}$ a morphism of $\mathcal{G}'$-PROPs. Then we define the adjoint map $\tilde{g} : \mathsf{P} \longrightarrow U\mathsf{Q}$ with components coming from the composite

$$\mathsf{P}\binom{\underline{d}}{\underline{c}} \cong \mathsf{P}[C_{(\underline{c};\underline{d})}] \longrightarrow L\mathsf{P}\binom{\underline{d}}{\underline{c}} \xrightarrow{g\binom{\underline{d}}{\underline{c}}} \mathsf{Q}\binom{\underline{d}}{\underline{c}}$$

since $L\mathsf{P}$ is defined as a colimit over objects including $C_{(\underline{c};\underline{d})}$. In order to verify this is a morphism of $\mathcal{G}$-PROPs, we observe it is clearly a morphism of S-colored objects, so it suffices to verify the following diagram commutes for any $[G] \in \mathtt{G}\binom{\underline{d}}{\underline{c}}$

$$\begin{array}{ccccc}
\mathsf{P}[G] & \longrightarrow & L\mathsf{P}[G] & \xrightarrow{g[G]} & \mathsf{Q}[G] \\
\downarrow{\scriptstyle \gamma_{[G]}} & & \downarrow{\scriptstyle \gamma'_{[G]}} & & \downarrow{\scriptstyle \gamma'_{[G]}} \\
\mathsf{P}\binom{\underline{d}}{\underline{c}} & \longrightarrow & L\mathsf{P}\binom{\underline{d}}{\underline{c}} & \xrightarrow{g\binom{\underline{d}}{\underline{c}}} & \mathsf{Q}\binom{\underline{d}}{\underline{c}}.
\end{array}$$

The right square commutes as an example of g a morphism of $\mathcal{G}'$-PROPs. In the left square, the construction of γ' in the previous proof implies the 'upper path' is simply given by the natural inclusion of $\mathsf{P}[G]$ into the colimit. On the other hand, since $[G]$ lies in the smaller pasting scheme by assumption, $\gamma_{[G]}$ induces a morphism in the colimiting category, hence this agrees with first applying $\gamma_{[G]} : \mathsf{P}[G] \longrightarrow \mathsf{P}\binom{\underline{d}}{\underline{c}}$ and then including the latter as $\mathsf{P}[C_{(\underline{c};\underline{d})}]$ by the natural inclusion into the colimit. Thus, the left square, and so the whole diagram, commutes, which makes $\tilde{g}$ a morphism of $\mathcal{G}$-PROPs.

Now suppose P and Q as above, with $f : \mathsf{P} \longrightarrow U\mathsf{Q}$ a morphism of $\mathcal{G}$-PROPs. Then by appropriate tensor products of maps of the form $f\binom{\mathrm{out}(v)}{\mathrm{in}(v)}$, there is an induced map $f[G] : \mathsf{P}[G] \longrightarrow \mathsf{Q}[G]$ even when $[G]$ does not lie in the smaller pasting scheme $\mathtt{G}$, or even in the restriction of the larger pasting scheme depending on the vertex profiles $\mathtt{G}'_S$. Of course, this is compatible with morphisms in $\mathcal{D}$ in the sense that it defines an induced map

$$\operatorname{colim} f : \operatorname*{colim}_{[H] \in \mathcal{D}\binom{\underline{d}}{\underline{c}}} \mathsf{P}[H] \longrightarrow \operatorname*{colim}_{[H] \in \mathcal{D}\binom{\underline{d}}{\underline{c}}} \mathsf{Q}[H],$$

that is really just $L(f)\binom{d}{\underline{c}}$. The structure maps of Q also form a natural map

$$\overline{\gamma}^{\mathsf{Q}} : \operatorname*{colim}_{[H]\in\mathcal{D}\binom{d}{\underline{c}}} \mathsf{Q}[H] \longrightarrow \mathsf{Q}\binom{d}{\underline{c}}$$

and we will take the composite $\overline{\gamma}^{\mathsf{Q}} \operatorname{colim} f$ as our definition of the components of $\tilde{f}$. Again, this is clearly a morphism of S'-colored objects, so we must next verify the commutativity of the diagram

$$\begin{array}{ccccc}
L\mathsf{P}[G] & \xrightarrow{L(f)[G]} & L\mathsf{Q}[G] & \longrightarrow & \mathsf{Q}[G] \\
\downarrow{\scriptstyle L\gamma_{[G]}} & & \downarrow{\scriptstyle LU\overline{\gamma}^{\mathsf{Q}}_{[G]}} & & \downarrow{\scriptstyle \overline{\gamma}^{\mathsf{Q}}_{[G]}} \\
L\mathsf{P}\binom{d}{\underline{c}} & \xrightarrow{L(f)\binom{d}{\underline{c}}} & L\mathsf{Q}\binom{d}{\underline{c}} & \longrightarrow & \mathsf{Q}\binom{d}{\underline{c}}.
\end{array}$$

The left square commutes as a consequence of L being a functor which lands in $\mathcal{G}'$-PROPs. However, associativity of the original structure map γ^{Q} implies for each $[G] \in \mathsf{G}'_{S'}$, there is a commutative diagram

$$\begin{array}{ccccc}
\bigotimes_{v\in \mathrm{Vt}(G)} \mathsf{Q}[H_v(I^v_u)] & \longrightarrow & \bigotimes_{v\in \mathrm{Vt}(G)} \mathsf{Q}[H_v] & \xrightarrow{\otimes\gamma^{\mathsf{Q}}\binom{\mathrm{out}(v)}{\mathrm{in}(v)}} & \bigotimes_{v\in \mathrm{Vt}(G)} \mathsf{Q}\binom{\mathrm{out}(v)}{\mathrm{in}(v)} \\
\downarrow{\scriptstyle \cong} & & \downarrow{\scriptstyle \cong} & & \downarrow{\scriptstyle \gamma^{\mathsf{Q}}_{[G]}} \\
\mathsf{Q}[(G(H_v))(I^v_u)] & \longrightarrow & \mathsf{Q}[G(H_v)] & \xrightarrow{\gamma^{\mathsf{Q}}_{[G(H_v)]}} & \mathsf{Q}\binom{d}{\underline{c}},
\end{array}$$

where the undecorated arrows come from components of the form $\gamma^{\mathsf{Q}}_{(I^v_u)}$. This implies the induced horizontal maps provide a commutative diagram

$$\begin{array}{ccc}
\bigotimes_{v\in \mathrm{Vt}(G)} \operatorname*{colim}_{[H_v]\in\mathcal{D}\binom{\mathrm{out}(v)}{\mathrm{in}(v)}} \mathsf{Q}[H_v] & \xrightarrow{\otimes_{v\in\mathrm{Vt}(G)}\overline{\gamma}^{\mathsf{Q}}\binom{\mathrm{out}(v)}{\mathrm{in}(v)}} & \bigotimes_{v\in \mathrm{Vt}(G)} \mathsf{Q}\binom{\mathrm{out}(v)}{\mathrm{in}(v)} \\
\downarrow{\scriptstyle \cong} & & \\
\operatorname{colim}_{\{[H_v]\}\in\prod_v \mathcal{D}\binom{\mathrm{out}(v)}{\mathrm{in}(v)}} \mathsf{Q}[G(H_v)] & & \downarrow{\scriptstyle \gamma^{\mathsf{Q}}_{[G]}} \\
\downarrow & & \\
\operatorname{colim}_{\{[K]\}\in\mathcal{D}\binom{d}{\underline{c}}} \mathsf{Q}[K] & \xrightarrow{\overline{\gamma}^{\mathsf{Q}}\binom{d}{\underline{c}}} & \mathsf{Q}\binom{d}{\underline{c}}
\end{array}$$

which is the right of the two squares above. Thus, we have shown $\tilde{f}$ is a morphism of $\mathcal{G}'$-PROPs.

Finally, we must observe that these two transitions are mutually inverse, which follows from the unital property of corollas. □

12.1.4. Examples of the Left Adjoint. A special case of Theorem 12.1 gives the free $\mathcal{G}$-PROP functor of a $\mathbf{\Sigma}^{\mathcal{E}}_{S}$-bimodule.

COROLLARY 12.9. *Let $\mathcal{G} = (S, \mathtt{G})$ be a pasting scheme. Then there is an adjoint pair*

$$L\colon \mathbf{\Sigma}_S^{\mathcal{E}} \rightleftarrows \mathtt{PROP}^{\mathcal{G}}\colon U,$$

where the right adjoint U is the forgetful functor.

PROOF. This is the special case of Theorem 12.1 when applied to the pasting schemes

$$Min(S) \le \mathcal{G},$$

where $Min(S)$ is the smallest pasting scheme corresponding to S. It contains only permuted corollas. ☐

EXAMPLE 12.10. Theorem 12.1 can be applied to the pasting scheme inclusions in (8.1) and section 8.1.3. In particular, we have the following free object functors.

(1) For the inclusion

$$\mathtt{Gr}_{\mathrm{c}}^{\circlearrowleft} \subseteq \mathtt{Gr}_{\mathrm{w}}^{\circlearrowleft},$$

the left adjoint $L\colon \mathtt{PROP}^{\mathtt{Gr}_{\mathrm{c}}^{\circlearrowleft}} \longrightarrow \mathtt{PROP}^{\mathtt{Gr}_{\mathrm{w}}^{\circlearrowleft}}$ sends a $\mathfrak{C}$-colored wheeled properad to the $\mathfrak{C}$-colored wheeled PROP generated by it [**MMS09**].

(2) For the inclusion

$$\mathtt{Tree}^{\circlearrowleft} \subseteq \mathtt{Gr}_{\mathrm{w}}^{\circlearrowleft},$$

the left adjoint $L\colon \mathtt{PROP}^{\mathtt{Tree}^{\circlearrowleft}} \longrightarrow \mathtt{PROP}^{\mathtt{Gr}_{\mathrm{w}}^{\circlearrowleft}}$ sends a $\mathfrak{C}$-colored wheeled operad to the $\mathfrak{C}$-colored wheeled PROP generated by it.

(3) For the inclusion

$$\mathtt{Gr}^{\uparrow} \subseteq \mathtt{Gr}_{\mathrm{w}}^{\circlearrowleft},$$

the left adjoint $L\colon \mathtt{PROP}^{\mathtt{Gr}^{\uparrow}} \longrightarrow \mathtt{PROP}^{\mathtt{Gr}_{\mathrm{w}}^{\circlearrowleft}}$ is the wheeled completion functor on $\mathfrak{C}$-colored PROPs [**MMS09**].

(4) For the inclusion

$$\mathtt{Gr}_{\mathrm{c}}^{\uparrow} \subseteq \mathtt{Gr}^{\uparrow},$$

the left adjoint $L\colon \mathtt{PROP}^{\mathtt{Gr}_{\mathrm{c}}^{\uparrow}} \longrightarrow \mathtt{PROP}^{\mathtt{Gr}^{\uparrow}}$ sends a $\mathfrak{C}$-colored properad to the $\mathfrak{C}$-colored PROP generated by it [**Val07**].

(5) For the inclusion

$$\mathtt{Gr}_{\mathrm{di}}^{\uparrow} \subseteq \mathtt{Gr}_{\mathrm{c}}^{\uparrow},$$

the left adjoint $L\colon \mathtt{PROP}^{\mathtt{Gr}_{\mathrm{di}}^{\uparrow}} \longrightarrow \mathtt{PROP}^{\mathtt{Gr}_{\mathrm{c}}^{\uparrow}}$ sends a $\mathfrak{C}$-colored dioperad to the $\mathfrak{C}$-colored properad generated by it [**Val07**].

(6) For the inclusion

$$\mathtt{Gr}_{\frac{1}{2}} \subseteq \mathtt{Gr}_{\mathrm{di}}^{\uparrow},$$

the left adjoint $L\colon \mathtt{PROP}^{\mathtt{Gr}_{\frac{1}{2}}} \longrightarrow \mathtt{PROP}^{\mathtt{Gr}_{\mathrm{di}}^{\uparrow}}$ sends a $\mathfrak{C}$-colored $\frac{1}{2}$-PROP to the $\mathfrak{C}$-colored dioperad generated by it [**MV09**].

(7) For the inclusion

$$\mathtt{UTree} \subseteq \mathtt{Gr}_{\mathrm{di}}^{\uparrow},$$

the left adjoint $L\colon \mathtt{PROP}^{\mathtt{Tree}} \longrightarrow \mathtt{PROP}^{\mathtt{Gr}_{\mathrm{di}}^{\uparrow}}$ sends a $\mathfrak{C}$-colored operad to the $\mathfrak{C}$-colored dioperad generated by it.

(8) For the inclusion of S' generated by a subset $\mathfrak{C}'$ of the colors, the left adjoint L is the change of colors adjunction at the level of each of the structures.

12.2. Generalized PROPS under a Change of Base Category

In this section, we discuss how generalized PROPs can be transferred from one symmetric monoidal category to another.

12.2.1. Transferring Generalized PROPs.

THEOREM 12.11. *Let $M: \mathcal{D} \longrightarrow \mathcal{E}$ be a symmetric monoidal functor between two symmetric monoidal categories. Then the following statements hold.*

(1) *M prolongs to a functor*

$$M: \mathtt{PROP}^{\mathcal{G},\mathcal{D}} \longrightarrow \mathtt{PROP}^{\mathcal{G},\mathcal{E}}$$

for every pasting scheme $\mathcal{G}$.

(2) *Suppose further that $M': \mathcal{D} \longrightarrow \mathcal{E}$ is another symmetric monoidal functor, and $\theta: M \longrightarrow M'$ is a monoidal natural transformation. Then θ prolongs to a natural transformation between the prolonged functors of M and M'.*

PROOF. Let P be a $\mathcal{G}$-PROP in $\mathcal{D}$. The S-colored object $M\mathsf{P}$ is defined entry-wise as

$$M\mathsf{P}\binom{\underline{d}}{\underline{c}} = M\left(\mathsf{P}\binom{\underline{d}}{\underline{c}}\right)$$

for all $(\underline{c};\underline{d}) \in S$. For $[G] \in \mathtt{G}_S(\underline{c};\underline{d})$ the structure map $\gamma'_{[G]}$ for $M\mathsf{P}$ is defined as the composition

$$\begin{array}{ccc} (M\mathsf{P})[G] & \xrightarrow{\quad \gamma'_{[G]} \quad} & M\mathsf{P}\binom{\underline{d}}{\underline{c}} \\ & {\scriptstyle\delta}\searrow \qquad \nearrow{\scriptstyle M(\gamma_{[G]})} & \\ & M(\mathsf{P}[G]), & \end{array} \tag{12.3}$$

in which the natural map δ is that of Lemma 10.35.

That γ'_C for C a corolla is the identity map follows from the fact that γ_C is the identity and the second conclusion of Lemma 10.35. The associativity diagram of γ'_G (orientation reversed from the description in Lemma 10.40) is the diagram

$$\begin{array}{ccccc} \otimes(M\mathsf{P})[H_v] & & \xrightarrow{\qquad\cong\qquad} & & (M\mathsf{P})[G(H_v)] \\ \Big\downarrow {\scriptstyle\otimes\delta} & & & & \Big\downarrow {\scriptstyle\delta} \\ \otimes M(\mathsf{P}[H_v]) & \xrightarrow{\delta} & M(\otimes \mathsf{P}[H_v]) & \xrightarrow{M(\cong)} & M(\mathsf{P}[G(H_v)]) \\ \Big\downarrow {\scriptstyle\otimes M(\gamma_{H_v})} & & \Big\downarrow {\scriptstyle M(\otimes\gamma_{H_v})} & & \Big\downarrow {\scriptstyle M(\gamma_{G(H_v)})} \\ (M\mathsf{P})[G] & \xrightarrow{\delta} & M(\mathsf{P}[G]) & \xrightarrow{M(\gamma_G)} & M\mathsf{P}\binom{\underline{d}}{\underline{c}}, \end{array}$$

$$\otimes(M\mathsf{P})[H_v] \xrightarrow{\otimes\gamma'_{H_v}} (M\mathsf{P})[G], \qquad (M\mathsf{P})[G(H_v)] \xrightarrow{\gamma'_{G(H_v)}} M\mathsf{P}\binom{\underline{d}}{\underline{c}}, \qquad (M\mathsf{P})[G] \xrightarrow{\gamma'_G} M\mathsf{P}\binom{\underline{d}}{\underline{c}}$$

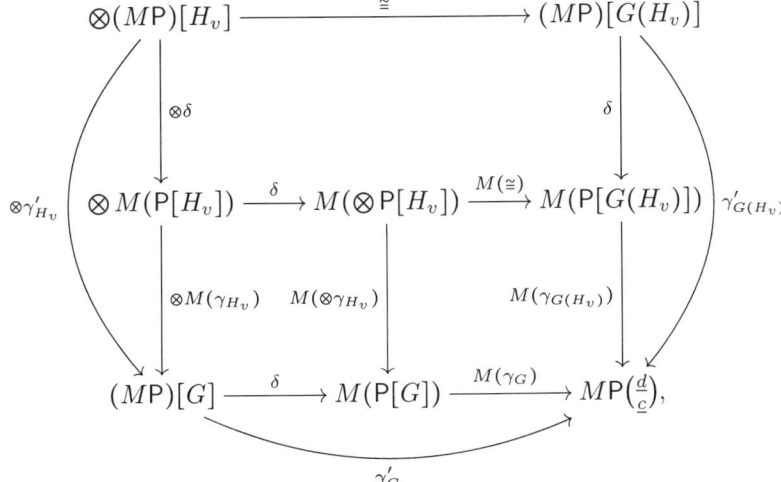

where $[G] \in \mathsf{G}_S(\underline{c};\underline{d})$, $\otimes = \otimes_{v \in \mathrm{Vt}(G)}$, and $[H_v] \in \mathsf{G}_S\binom{\mathrm{out}(v)}{\mathrm{in}(v)}$ for each $v \in \mathrm{Vt}(G)$. The three outer strips are commutative by definition (12.3). The top rectangle and the bottom left square are commutative by naturality. The bottom right square is commutative as M applied to the associativity square (10.21) of P. This shows that $M\mathsf{P}$ is a $\mathcal{G}$-PROP in $\mathcal{E}$ by Lemma 10.40.

Now consider the assertion about the natural transformation θ. Suppose P is a $\mathcal{G}$-PROP in $\mathcal{D}$. To see that the entry-wise extension

$$\theta \colon M\mathsf{P} \longrightarrow M'\mathsf{P}$$

is a morphism of $\mathcal{G}$-PROPs in $\mathcal{E}$, consider the following commutative diagram for $[G] \in \mathsf{G}_S\binom{\underline{d}}{\underline{c}}$:

$$\begin{array}{ccc}
(M\mathsf{P})[G] & \xrightarrow{\theta} & (M'\mathsf{P})[G] \\
{\scriptstyle \delta}\downarrow & & \downarrow{\scriptstyle \delta} \\
M(\mathsf{P}[G]) & \xrightarrow{\theta} & M'(\mathsf{P}[G]) \\
{\scriptstyle M(\gamma_G)}\downarrow & & \downarrow{\scriptstyle M'(\gamma_G)} \\
M\mathsf{P}\binom{\underline{d}}{\underline{c}} & \xrightarrow{\theta} & M'\mathsf{P}\binom{\underline{d}}{\underline{c}} .
\end{array}$$

The top rectangle is commutative because δ is natural, and the bottom rectangle is commutative because θ is a natural transformation with γ_G a morphism in $\mathcal{E}$. The left (resp., right) vertical composition is the structure map γ'_G for $M\mathsf{P}$ (resp., $M'\mathsf{P}$). Therefore, the entry-wise extension of θ is a morphism of $\mathcal{G}$-PROPs in $\mathcal{E}$ by Lemma 10.40. Finally, the extended θ remains a natural transformation because θ, M, and M' are all defined entry-wise. □

EXAMPLE 12.12. The homology functor from chain complexes to graded modules is symmetric monoidal, so it satisfies the hypotheses of Theorem 12.11.

Recall that a strong symmetric monoidal left adjoint has a right adjoint which is also symmetric monoidal (Lemma 10.9), providing possible applications of the following.

COROLLARY 12.13. *Suppose $L\colon \mathcal{D} \rightleftarrows \mathcal{E} : R$ is an adjoint pair of symmetric monoidal functors. Then the prolonged functors remain an adjoint pair*

$$L\colon \mathsf{PROP}^{\mathcal{G},\mathcal{D}} \rightleftarrows \mathsf{PROP}^{\mathcal{G},\mathcal{E}} \colon R.$$

PROOF. Using Theorem 12.11, these adjoints both prolong to functors of PROP categories $\mathsf{PROP}^{\mathcal{G},\mathcal{D}} \rightleftarrows \mathsf{PROP}^{\mathcal{G},\mathcal{E}}$, and the natural transformations characterizing this adjunction via the 'triangular identities' also prolong to verify the result remains an adjoint [**Mac98**, Theorem IV.2(v)]. □

EXAMPLE 12.14. Here are some adjoint pairs where both functors are symmetric monoidal, to which Corollary 12.13 can be applied.

(1) The geometric realization functor and the singular functor

$$|-|\colon \mathbf{SSet} \rightleftarrows \mathbf{Top} \colon Sing(-)$$

are symmetric monoidal adjoints (e.g.[**Hov99**, Proposition 4.2.17]).

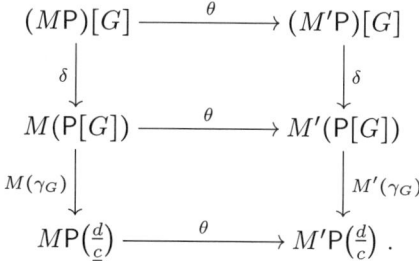

(2) The disjoint basepoint functor and the forgetting basepoints functor

$$(-)_*\colon \mathbf{SSet} \rightleftarrows \mathbf{SSet}_*\colon U$$

between simplicial sets and pointed simplicial sets under the smash product are symmetric monoidal adjoints (e.g. [**Hov99**, p.113]).

(3) Suppose $f\colon R \longrightarrow T$ is a map of associative and commutative rings. Then tensoring up and restriction of scalars along f

$$- \otimes_R T\colon \mathbf{Ch}(R) \rightleftarrows \mathbf{Ch}(T) : (-)_f$$

on the categories of chain complexes are symmetric monoidal adjoints (e.g. [**Hov99**, p.114]).

(4) For a symmetric monoidal category $\mathcal{E}$, the free object functor

$$F\colon \mathbf{Set} \rightleftarrows \mathcal{E}, \quad FX = \coprod_{x\in X} I \tag{12.4}$$

is strong symmetric monoidal, whose right adjoint is the underlying set functor

$$\mathcal{E}(I,-)\colon \mathcal{E} \rightleftarrows \mathbf{Set}. \tag{12.5}$$

For instance, when $\mathcal{E} = \mathbf{Mod}$, F is the free module functor, whose right adjoint is the underlying set functor. Once again, whenever the left adjoint is strong symmetric monoidal, the right adjoint is also a symmetric monoidal functor.

12.2.2. Changing Pasting Scheme and Base Category. We can think of the pasting scheme $\mathcal{G}$ and the symmetric monoidal category $\mathcal{E}$ as variables of the category $\mathsf{PROP}^{\mathcal{G},\mathcal{E}}$. Theorems 12.1 and 12.11 describe the change-of-variable functors for $\mathcal{G}$ and $\mathcal{E}$, respectively. The next observation describes the relationship between these two change-of-variable functors.

COROLLARY 12.15. *Let $M\colon \mathcal{D} \longrightarrow \mathcal{E}$ be a symmetric monoidal functor between two symmetric monoidal categories, and let $\mathcal{G} \leq \mathcal{G}'$ be pasting schemes. Then there is a natural transformation*

$$\begin{array}{ccc} \mathsf{PROP}^{\mathcal{G},\mathcal{D}} & \xrightarrow{L} & \mathsf{PROP}^{\mathcal{G}',\mathcal{D}} \\ {\scriptstyle M}\downarrow & \overset{\Psi}{\Longrightarrow} & \downarrow{\scriptstyle M} \\ \mathsf{PROP}^{\mathcal{G},\mathcal{E}} & \xrightarrow[L]{} & \mathsf{PROP}^{\mathcal{G}',\mathcal{E}} \end{array} \tag{12.6}$$

from LM to ML, in which L is the left adjoint in Theorem 12.1, *and M is the prolonged functor in Theorem* 12.11.

PROOF. The definition of the general L says

$$(LM\mathsf{P})\binom{\underline{d}}{\underline{c}} = \operatorname*{colim}_{[G]\in\mathcal{D}\binom{\underline{d}}{\underline{c}}} (M\mathsf{P})[G]$$

for each $\mathsf{P} \in \mathsf{PROP}^{\mathcal{G},\mathcal{D}}$ and each $(\underline{c};\underline{d}) \in S'$. On the other hand, there is a natural map

$$\operatorname*{colim}_{[G]\in\mathcal{D}\binom{\underline{d}}{\underline{c}}} M(\mathsf{P}[G]) \longrightarrow M\left(\operatorname*{colim}_{[G]\in\mathcal{D}\binom{\underline{d}}{\underline{c}}} \mathsf{P}[G]\right) = (ML\mathsf{P})\binom{\underline{d}}{\underline{c}}$$

by first applying M to the structure maps of the colimiting system for P itself and then applying the universal property of the colimit in the source. Thus, it suffices to observe that the natural map

$$\delta\colon (M\mathsf{P})[G] \longrightarrow M(\mathsf{P}[G]), \tag{12.7}$$

in Lemma 10.35 provides the intermediate natural map. □

EXAMPLE 12.16. Corollary 12.15 applies to all the symmetric monoidal functors in Examples 12.12 and 12.14.

COROLLARY 12.17. *In the context of Corollary* 12.15, *suppose further that the following two conditions hold.*

- *M is strong symmetric monoidal.*
- *Either*
  1. *M preserves all small colimits, or*
  2. *the pair $\mathcal{G} \leq \mathcal{G}'$ is well-matched, and M preserves coproducts and coinvariants for finite group actions.*

Then

$$\Psi\colon LM \longrightarrow ML$$

is a natural isomorphism.

PROOF. If M is strong symmetric monoidal and preserves all small colimits, then the natural map Ψ is an isomorphism because L is constructed using colimits and finite tensor products. In the well-matched case, use the coproduct formula (12.1) of L, which involves coproducts, finite tensor products, and coinvariants for finite group actions (for the unordered tensor product $\mathsf{P}[G]$ (10.12)). □

12.2.3. An Application to Homology. Denote by **Ch** and **Mod** the categories of chain complexes and of graded modules over a characteristic 0 field.

COROLLARY 12.18. *Suppose $\mathcal{G} \leq \mathcal{G}'$ is a well-matched pair of pasting schemes. Then the left adjoint*

$$L\colon \mathsf{PROP}^{\mathcal{G},\mathbf{Ch}} \longrightarrow \mathsf{PROP}^{\mathcal{G}',\mathbf{Ch}}$$

commutes with homology up to natural isomorphisms.

PROOF. The assertion means that there is a natural isomorphism between the composites in the diagram

$$\begin{array}{ccc} \mathsf{PROP}^{\mathcal{G},\mathbf{Ch}} & \xrightarrow{L} & \mathsf{PROP}^{\mathcal{G}',\mathbf{Ch}} \\ {\scriptstyle H}\big\downarrow & \Longrightarrow & \big\downarrow{\scriptstyle H} \\ \mathsf{PROP}^{\mathcal{G},\mathbf{Mod}} & \xrightarrow[L]{} & \mathsf{PROP}^{\mathcal{G}',\mathbf{Mod}} \end{array}$$

in which H denotes the prolonged functor of the homology functor

$$H\colon \mathbf{Ch} \longrightarrow \mathbf{Mod}.$$

Apply Corollary 12.17. The homology functor preserves coproducts and under the characteristic 0 assumption, it is strong symmetric monoidal by the Künneth formula [**Wei94**] (Theorem 3.6.3). Moreover, H preserves coinvariants for finite group actions by Maschke's Theorem [**JL01**] (Chapter 8). □

EXAMPLE 12.19. Corollary 12.18 applies, in particular, to all the well-matched pairs of pasting schemes in section 9.2.

12.3. Notes

The special cases of Theorem 12.1 and Corollary 12.18 with $\mathcal{E} = \mathbf{Ch}$ for the pasting schemes

$$\mathtt{Gr}_{\frac{1}{2}} \leq \mathtt{Gr}^{\uparrow}_{\mathrm{di}} \quad \text{and} \quad \mathtt{Gr}_{\frac{1}{2}} \leq \mathtt{Gr}^{\uparrow}$$

are in [**MV09**]. Elements in the Kontsevich groupoids for these two pairs of pasting schemes are said to be *reduced* in [**MV09**]. The special cases of Theorem 12.1 for the pasting schemes

$$\mathtt{UTree} \leq \mathtt{Gr}^{\uparrow}$$

with general $\mathcal{E}$ and

$$\mathtt{Gr}^{\uparrow}_{\mathrm{c}} \leq \mathtt{Gr}^{\uparrow}$$

with $\mathcal{E} = \mathbf{Ch}$ are in [**JY09**] and [**Val07**], respectively.

CHAPTER 13

Algebras over Generalized PROPs

PROPs and their variants are important in part because of their algebras. The purpose of this chapter is to study algebras over generalized PROPs, both in unbiased and biased forms. The traditional definition of an algebra over $\mathsf{P} \in \mathcal{G}$-PROP begins with a colored object $X_c \in \prod_c \mathcal{E}$, and extends it using tensor products to allow profiles as input $X_{\underline{c}} = X_{c_1} \otimes \cdots \otimes X_{c_m}$. Then one builds an endomorphism $\mathcal{G}$-PROP E_X associated to X with value

$$\mathcal{E}(X_{\underline{c}}, X_{\underline{d}})$$

at the pair of profiles $(\underline{c}; \underline{d})$ and permutation actions given by permuting tensor factors from X. Notice this endomorphism object contains all potential maps between tensor products of entries of X, including switch maps, so provides a convenient home for any form of algebraic structure one might want to impose on X. We then call X a P-algebra if there is a morphism of $\mathcal{G}$-PROPs, $\lambda : \mathsf{P} \longrightarrow \mathsf{E}_X$, which by adjunction produces a series of maps

$$\mathsf{P}\binom{\underline{d}}{\underline{c}} \otimes X_{\underline{c}} \longrightarrow X_{\underline{d}} \ . \tag{13.1}$$

The technical difficulty with implementing this approach for wheeled PROPs stems from the fact that there is no natural method of producing contractions for such endomorphism objects without imposing some conditions, usually some form of finiteness. As such, this approach generally fails for $\mathcal{G} = \mathtt{Gr}_{\mathrm{w}}^{\circlearrowright}$ and some other related cases. In order to be precise on this point, we introduce in Section 13.1 the idea of a pasting scheme which admits an endomorphism object for a given $\mathcal{E}$, and observe that the property restricts along the partial ordering of pasting schemes. Then we show $\mathtt{Gr}^{\uparrow}$ admits a pasting scheme over any $\mathcal{E}$, which suggests that contraction operations, or non-trivial wheels, are the only real impediment. A few issues related to enrichments, and other observations about admitting an endomorphism object are also included.

Given that the construction of E_X is both covariant and contravariant in the argument X, it is subtle to see how a morphism of colored objects $f : X \longrightarrow Y$ might be used to define a morphism of algebras over P. As a consequence, Section 13.1 also contains a detailed investigation of the so-called relative endomorphism objects which are the key to resolving this issue, in particular with a near-pullback property (Theorem 13.16) vital to producing the composition of algebra morphisms.

With these technical issues in place, Section 13.2 provides the unbiased definition of the category of algebras over a $\mathcal{G}$-PROP in $\mathcal{E}$ whenever $\mathcal{G}$ admits an endomorphism object in $\mathcal{E}$. Of particular note are the alternative description of a morphism of algebras (Proposition 13.21) in terms of the adjoint structure maps of (13.1), and the characterization of algebra morphisms in terms of the relative endomorphism object (Corollary 13.23).

Section 13.3 is devoted to studying the category of algebras over a generalized PROP under a change of pasting scheme, $\mathcal{G}$-PROP P, or underlying category, and this last involves a compatibility restriction between P and the adjoint pair of underlying categories. Finally, Section 13.4 provides the Biased Algebra Theorem and details the biased definitions in each of the pasting schemes of interest. In this way, we have concrete structural descriptions of algebras for the broad collection of pasting schemes we have featured until now, once again by exploiting strong generating sets established in Chapter 7.

CONVENTION 13.1. From this point forward, unless otherwise specified, all the symmetric monoidal closed categories under consideration are assumed to have all small colimits and pullbacks, and $\mathcal{E}$ denotes such a category.

As before, $(S, \mathtt{G}) = \mathcal{G} \leq \mathcal{G}' = (S', \mathtt{G}')$ are pasting schemes (Definition 8.2).

13.1. Endomorphism Objects

Endomorphism objects are usually considered for a single underlying category. In this section, we discuss endomorphism objects in the slightly more general enriched setting.

DEFINITION 13.2. A **symmetric monoidal $\mathcal{E}$-category** is a symmetric monoidal closed category $\mathcal{M}$ together with an adjunction

$$i\colon \mathcal{E} \rightleftarrows \mathcal{M} \colon r$$

of symmetric monoidal functors in which i is strong.

Our convention is that if $\mathcal{M}$ is not mentioned, then $\mathcal{M} = \mathcal{E}$ and $i = r = Id_{\mathcal{E}}$.

REMARK 13.3. Since the left adjoint i is strong symmetric monoidal, the right adjoint r is automatically symmetric monoidal (Lemma 10.9).

13.1.1. Hom-Tensor Adjunction. Recall the internal hom in a symmetric monoidal closed category $\mathcal{M}$ is denoted by $\mathrm{Hom}_{\mathcal{M}}$. Given a symmetric monoidal $\mathcal{E}$-category $\mathcal{M}$, we usually suppress i and r in the notations. In particular, for $A \in \mathcal{E}$ and $X, Y \in \mathcal{M}$, we abbreviate $i(A) \otimes X$ to $A \otimes X$ and $r(\mathrm{Hom}_{\mathcal{M}}(X, Y))$ to $\mathrm{Hom}_{\mathcal{M}}(X, Y)$. There is a natural isomorphism

$$\mathcal{E}(A, \mathrm{Hom}_{\mathcal{M}}(X, Y)) \cong \mathcal{M}(A \otimes X, Y), \tag{13.2}$$

called the **hom-tensor adjunction**. Given any morphism

$$f\colon A \longrightarrow \mathrm{Hom}_{\mathcal{M}}(X, Y) \in \mathcal{E},$$

denote its adjoint map under (13.2) by

$$f^{\#}\colon A \otimes X \longrightarrow Y \in \mathcal{M},$$

and vice versa.

13.1.2. Endomorphism Objects. *For the rest of this section, $\mathcal{M}$ denotes a symmetric monoidal $\mathcal{E}$-category.*

DEFINITION 13.4. Suppose $X = \{X_c\}$ and $Y = \{Y_c\}$ are two objects in $\mathcal{M}^{\mathfrak{C}}$, while S is a replete and full sub-category of $\mathcal{P}(\mathfrak{C})^{op} \times \mathcal{P}(\mathfrak{C})$.

(1) For a profile $\underline{c} = c_{[1,m]} \in \mathcal{P}(\mathfrak{C})$, define the object

$$X_{\underline{c}} = X_{c_1} \otimes \cdots \otimes X_{c_m} \in \mathcal{M}. \tag{13.3}$$

If $f: X \longrightarrow Y$ is a morphism in $\mathcal{M}^{\mathfrak{C}}$, then define the morphism

$$f_{\underline{c}} = f_{c_1} \otimes \cdots \otimes f_{c_m}: X_{\underline{c}} \longrightarrow Y_{\underline{c}} \in \mathcal{M}. \tag{13.4}$$

(2) Define the **mixed endomorphism object** $\mathsf{E}_{X,Y} \in \mathbf{\Sigma}_S$ by

$$\begin{aligned} \mathsf{E}_{X,Y}\binom{\underline{d}}{\underline{c}} &= \mathrm{Hom}_{\mathcal{M}}(X_{\underline{c}}, Y_{\underline{d}}) \\ &= \mathrm{Hom}_{\mathcal{M}}\left(X_{c_1} \otimes \cdots \otimes X_{c_m}, Y_{d_1} \otimes \cdots \otimes Y_{d_n}\right) \in \mathcal{E} \end{aligned}$$

for each $(\underline{c}; \underline{d}) \in S$, where the $\mathbf{\Sigma}_S$-bimodule structure on $\mathsf{E}_{X,Y}$ comes from permutations of the tensor factors in $X_{\underline{c}}$ and $Y_{\underline{d}}$.

(3) Define the **endomorphism object** of X in $\mathbf{\Sigma}_S$ as

$$\mathsf{E}_X = \mathsf{E}_{X,X},$$

so we have

$$\mathsf{E}_X\binom{\underline{d}}{\underline{c}} = \mathrm{Hom}_{\mathcal{M}}(X_{\underline{c}}, X_{\underline{d}}) \in \mathcal{E}.$$

We have specified a function on objects of $\mathcal{M}^{\mathfrak{C}}$, which we may view instead as a functor defined on the discrete subcategory $dis(\mathcal{M}^{\mathfrak{C}}) \subset \mathcal{M}^{\mathfrak{C}}$ whose only morphisms are identities

$$\mathsf{E}: dis(\mathcal{M}^{\mathfrak{C}}) \longrightarrow \mathbf{\Sigma}_S, \quad X \longmapsto \mathsf{E}_X.$$

Note that E_X is *not* functorial for $X \in \mathcal{M}^{\mathfrak{C}}$, since a morphism $f: X \longrightarrow Y \in \mathcal{M}^{\mathfrak{C}}$ does not in general yield a non-trivial morphism $\mathsf{E}_X \longrightarrow \mathsf{E}_Y$. The point is that such a map $f \neq Id$ induces $f_* : \mathsf{E}_{X,X} \longrightarrow \mathsf{E}_{X,Y}$ and $f^* : \mathsf{E}_{Y,Y} \longrightarrow \mathsf{E}_{X,Y}$ but no natural map from $\mathsf{E}_{X,X}$ to $\mathsf{E}_{Y,Y}$. To relate the endomorphism objects E_X and E_Y, we instead consider the following construction, which we study further in 13.1.5.

DEFINITION 13.5. Let $f: X \longrightarrow Y \in \mathcal{M}^{\mathfrak{C}}$ be a morphism. Define the **relative endomorphism object** $\mathsf{E}_f \in \mathbf{\Sigma}_S$ using the component-wise pullback

$$\begin{array}{ccc} \mathsf{E}_f\binom{\underline{d}}{\underline{c}} & \xrightarrow{\overline{f}^*} & \mathsf{E}_X\binom{\underline{d}}{\underline{c}} \\ {\scriptstyle \overline{f}_*}\downarrow & & \downarrow{\scriptstyle f_*} \\ \mathsf{E}_Y\binom{\underline{d}}{\underline{c}} & \xrightarrow[f^*]{} & \mathsf{E}_{X,Y}\binom{\underline{d}}{\underline{c}}, \end{array} \tag{13.5}$$

in which

$$f_* = \mathrm{Hom}_{\mathcal{M}}(X_{\underline{c}}, f_{\underline{d}}) \quad \text{and} \quad f^* = \mathrm{Hom}_{\mathcal{M}}(f_{\underline{c}}, Y_{\underline{d}}) \tag{13.6}$$

are the induced maps. In particular, the $\mathbf{\Sigma}_S$-bimodule structure on E_f arises by compatibility of the permutation of tensor factors in the (mixed) endomorphism objects.

EXAMPLE 13.6. If $\mathcal{E}$ is the category **Set**, **Mod**, or **Ch**, then the component $\mathsf{E}_X\binom{\underline{d}}{\underline{c}}$ of the endomorphism object consists of maps

$$X_{\underline{c}} \longrightarrow X_{\underline{d}} \in \mathcal{E}\ ,$$

while for **SSet**, or **SMod**, since we are using the simplicial mapping space, it consists of maps

$$\Delta^n \otimes X_{\underline{c}} \longrightarrow X_{\underline{d}} \in \mathcal{E}\ .$$

In the non-simplicial cases, the component $\mathsf{E}_f\binom{\underline{d}}{\underline{c}}$ of the relative endomorphism object consists of pairs

$$(g,h) \in \mathsf{E}_X\binom{\underline{d}}{\underline{c}} \times \mathsf{E}_Y\binom{\underline{d}}{\underline{c}}$$

such that the square

$$\begin{array}{ccc} X_{\underline{c}} & \xrightarrow{g} & X_{\underline{d}} \\ {\scriptstyle f_{\underline{c}}}\downarrow & & \downarrow{\scriptstyle f_{\underline{d}}} \\ Y_{\underline{c}} & \xrightarrow{h} & Y_{\underline{d}} \end{array}$$

commutes, while for the simplicial cases we would instead require the commutativity of

$$\begin{array}{ccc} \Delta^n \otimes X_{\underline{c}} & \xrightarrow{g} & X_{\underline{d}} \\ {\scriptstyle Id\otimes f_{\underline{c}}}\downarrow & & \downarrow{\scriptstyle f_{\underline{d}}} \\ \Delta^n \otimes Y_{\underline{c}} & \xrightarrow{h} & Y_{\underline{d}}\ . \end{array}$$

13.1.3. Pasting Scheme Admitting an Endomorphism Object. Endomorphism objects are generally used to define algebras, but the problem is that the contractions of a wheeled PROP generally fail to exist without some form of finiteness condition on the objects in $\mathcal{E}$. On the other hand, endomorphism PROPs naturally exist in any context. Thus, we must carefully consider when endomorphism objects exist as generalized PROPs.

The following definition is a common generalization of the familiar endomorphism operad and endomorphism PROP. We begin by specifying the values of the functor solely on objects of $\mathcal{M}^{\mathfrak{C}}$ by choosing the restriction to the discrete subcategory $dis(\mathcal{M}^{\mathfrak{C}}) \subset \mathcal{M}^{\mathfrak{C}}$.

DEFINITION 13.7. Let $\mathcal{G}$ be a pasting scheme. We say that $\mathcal{G}$ **admits an endomorphism object** in $(\mathcal{M},\mathcal{E})$ if the following two conditions hold:

(1) The functor $\mathsf{E}\colon dis(\mathcal{M}^{\mathfrak{C}}) \longrightarrow \mathbf{\Sigma}_S$ lifts to $\mathtt{PROP}^{\mathcal{G}}$ in the sense that there is a commutative triangle

$$\begin{array}{ccc} & & \mathtt{PROP}^{\mathcal{G}} \\ & {\scriptstyle \mathsf{E}}\nearrow & \downarrow{\scriptstyle U} \\ dis(\mathcal{M}^{\mathfrak{C}}) & \xrightarrow[\mathsf{E}]{} & \mathbf{\Sigma}_S \end{array} \tag{13.7}$$

where U is the forgetful functor associated to the pasting schemes $Min(S) \le \mathcal{G}$ (Lemma 12.2).

(2) For each morphism $f: X \longrightarrow Y \in \mathcal{M}^{\mathfrak{C}}$ and each $G \in \mathtt{G}_S\binom{\underline{d}}{\underline{c}}$, the following diagram in $\mathcal{E}$ commutes

$$\begin{array}{ccccc} \mathsf{E}_f[G] & \xrightarrow{\otimes_v \overline{f}^*} & \mathsf{E}_X[G] & \xrightarrow{\gamma_{X,G}} & \mathsf{E}_X\binom{\underline{d}}{\underline{c}} \\ \downarrow{\scriptstyle \otimes_v \overline{f}_*} & & & & \downarrow{\scriptstyle f_*} \\ \mathsf{E}_Y[G] & \xrightarrow[\gamma_{Y,G}]{} & \mathsf{E}_Y\binom{\underline{d}}{\underline{c}} & \xrightarrow[f^*]{} & \mathsf{E}_{X,Y}\binom{\underline{d}}{\underline{c}} \end{array} \tag{13.8}$$

where $\otimes_v = \otimes_{v \in \mathrm{Vt}(G)}$, the objects $\mathsf{E}_f[G]$, $\mathsf{E}_X[G]$, and $\mathsf{E}_Y[G]$ are as defined in Definition 10.31, and the maps $\gamma^X_{[G]}$ and $\gamma^Y_{[G]}$ are $\mathcal{G}$-PROP structure maps of E_X and E_Y respectively.

We call E_X the **endomorphism $\mathcal{G}$-PROP** of X.

REMARK 13.8. The mixed endomorphism object $\mathsf{E}_{X,Y}$ is usually *not* a $\mathcal{G}$-PROP.

The following example is the primary reason we need to be very careful with this existence question.

EXAMPLE 13.9. The pasting scheme $\mathtt{Gr}^{\circlearrowright}_{\mathrm{w}}$ of all wheeled graphs does not admit an endomorphism object for general $\mathcal{E}$.

However, for $\mathcal{E}$ the category of finite dimensional vector spaces over a field of characteristic 0, $\mathtt{Gr}^{\circlearrowright}_{\mathrm{w}}$ does admit an endomorphism object [**MMS09**] (Example 2.1.1), called the **endomorphism wheeled PROP**. Finite dimensionality ensures that there is a non-trivial contraction (11.32) on E_X given by the trace. Technically, this category $\mathcal{E}$ does not satisfy Convention 13.1. However, we can still define wheeled PROPs as in Definition 11.33, and Definition 13.7 still makes sense.

13.1.4. Examples of Endomorphism Objects.

We begin with the colored version of the endomorphism PROP (see e.g. [**JY09**] or [**FMY09**]).

PROPOSITION 13.10. *The pasting scheme* $\mathtt{Gr}^{\uparrow}$ *of wheel-free graphs admits an endomorphism object in every* $(\mathcal{M}, \mathcal{E})$.

PROOF. For $X \in \mathcal{M}^{\mathfrak{C}}$, denote by

$$ev: \mathsf{E}_X\binom{\underline{d}}{\underline{c}} \otimes X_{\underline{c}} \longrightarrow X_{\underline{d}} \in \mathcal{M}$$

the adjoint map of the identity map of $\mathsf{E}_X\binom{\underline{d}}{\underline{c}}$ under the hom-tensor adjunction (13.2). In terms of the generating operations (Definition 11.30), the vertical composition

$$\mathsf{E}_X\binom{\underline{d}}{\underline{c}} \otimes \mathsf{E}_X\binom{\underline{c}}{\underline{a}} \longrightarrow \mathsf{E}_X\binom{\underline{d}}{\underline{a}}$$

is defined as the adjoint of the composition

$$\begin{array}{ccc} \mathsf{E}_X\binom{\underline{d}}{\underline{c}} \otimes \mathsf{E}_X\binom{\underline{c}}{\underline{a}} \otimes X_{\underline{a}} & \xrightarrow{\qquad\qquad} & X_{\underline{d}} \\ & {\scriptstyle Id \otimes ev} \searrow \qquad \nearrow {\scriptstyle ev} & \\ & \mathsf{E}_X\binom{\underline{d}}{\underline{c}} \otimes X_{\underline{c}} & \end{array}$$

in $\mathcal{M}$. Likewise, the horizontal composition

$$\mathsf{E}_X\binom{\underline{d}}{\underline{c}} \otimes \mathsf{E}_X\binom{\underline{b}}{\underline{a}} \longrightarrow \mathsf{E}_X\binom{\underline{d},\underline{b}}{\underline{c},\underline{a}}$$

is defined as the adjoint of the composition

$$\begin{array}{ccc} \mathsf{E}_X\binom{\underline{d}}{\underline{c}} \otimes \mathsf{E}_X\binom{\underline{b}}{\underline{a}} \otimes X_{\underline{c},\underline{a}} & \longrightarrow & X_{\underline{d},\underline{b}} \\ \cong\downarrow & & \uparrow\cong \\ \mathsf{E}_X\binom{\underline{d}}{\underline{c}} \otimes X_{\underline{c}} \otimes \mathsf{E}_X\binom{\underline{b}}{\underline{a}} \otimes X_{\underline{a}} & \xrightarrow{ev \otimes ev} & X_{\underline{d}} \otimes X_{\underline{b}}. \end{array}$$

The empty unit

$$\mathbf{1}_\varnothing : I \longrightarrow \mathsf{E}_X\binom{\varnothing}{\varnothing}$$

is the adjoint of the isomorphism $I \otimes I \cong I$. The c-colored unit

$$\mathbf{1}_c : I \longrightarrow \mathsf{E}_X\binom{c}{c}$$

is the adjoint of the isomorphism $I \otimes X_c \cong X_c$. These structure maps are all natural, in the sense that they are defined using only the hom-tensor adjunction (13.2) and composition in $\mathcal{M}$. The commutativity of the diagram (13.8) follows since a tensor product of compositions is also a composition of maps between the tensor products. □

The PROP E_X is called the **endomorphism PROP** of X. Note that there is no restriction on the categories $\mathcal{M}$ and $\mathcal{E}$ here, so the endomorphism PROP exists in every symmetric monoidal closed category with all small colimits and pullbacks.

Next we observe that endomorphism objects respect restriction of pasting schemes.

PROPOSITION 13.11. *Suppose $\mathcal{G} \le \mathcal{G}'$ are pasting schemes, and $\mathcal{G}'$ admits an endomorphism object in $(\mathcal{M}, \mathcal{E})$. Then $\mathcal{G}$ also admits an endomorphism object in $(\mathcal{M}, \mathcal{E})$ by restriction.*

PROOF. This follows from the fact that the forgetful functors

$$\mathtt{PROP}^{\mathcal{G}'} \longrightarrow \mathtt{PROP}^{\mathcal{G}} \quad \text{and} \quad \mathbf{\Sigma}_{S'} \longrightarrow \mathbf{\Sigma}_S$$

are both given by forgetting the components not in S (Lemma 12.2). □

Combining the previous two propositions, we obtain our main existence result, that any pasting scheme contained inside $\mathtt{Gr}^\uparrow$ admits an endomorphism object in arbitrary $(\mathcal{M}, \mathcal{E})$. In all such cases, we will use this particular choice of endomorphism object exclusively.

COROLLARY 13.12. *Every pasting scheme $\mathcal{G} \le \mathtt{Gr}^\uparrow$ admits an endomorphism object in every $(\mathcal{M}, \mathcal{E})$ by restricting the endomorphism PROP.*

EXAMPLE 13.13. Corollary 13.12 applies in particular to all the pasting schemes contained in $\mathtt{Gr}^\uparrow$ in (8.1). For example, with $\mathcal{G} = \mathtt{Gr}^\uparrow_{\mathrm{c}}$, $\mathtt{Gr}^\uparrow_{\mathrm{di}}$, $\mathtt{Gr}_{\frac{1}{2}}$, $\mathtt{UTree}$, $\mathtt{Gr}^\uparrow_{\mathrm{h}}$, and $\mathtt{Gr}^\uparrow_{\mathrm{v}}$, we obtain the usual endomorphism properad, dioperad, half-PROP, operad, hPROP, and vPROP.

EXAMPLE 13.14. The coendomorphism operad is an endomorphism $\mathcal{G}$-PROP for a pasting scheme $\mathcal{G} = \mathtt{UTree}^{op}$. Indeed, in the one colored case, the **coendomorphism operad** of an object X [**MSS02**] is the operad with components

$$\mathsf{E}_X^c(n) = \mathrm{Hom}_{\mathcal{M}}(X, X^{\otimes n})$$

and structure maps that are dual to those in the endomorphism operad. There is also a colored version of the coendomorphism operad. Define $\mathtt{UTree}^{op}$ as the full sub-groupoid of $\mathtt{Gr}_{\mathrm{c}}^{\uparrow}$ in which every vertex has exactly one incoming flag. Equivalently, $\mathtt{UTree}^{op}$ consists of the exceptional wheeled graphs $\uparrow_c$ and the upside-down level-trees, i.e., level-trees with the direction δ reversed. With the replete and full sub-category

$$S^1 = \{(c;\underline{d}) : \underline{d} \in \mathcal{P}(\mathfrak{C}), c \in \mathfrak{C}\},$$

$\mathtt{UTree}^{op}$ becomes a unital pasting scheme. Since $\mathtt{UTree}^{op} \leq \mathtt{Gr}^{\uparrow}$, for each object $X \in \mathcal{M}^{\mathfrak{C}}$, the endomorphism PROP E_X restricts to the endomorphism $\mathtt{UTree}^{op}$-PROP, which is denoted by E_X^c.

Via the identification

$$\mathsf{P}^{op}\binom{\underline{c}}{\underline{d}} = \mathsf{P}\binom{\underline{d}}{\underline{c}}, \tag{13.9}$$

there is a canonical isomorphism between the categories of $\mathtt{UTree}$-PROPs and $\mathtt{UTree}^{op}$-PROPs. When considered as a $\mathtt{UTree}$-PROP, E_X^c is precisely the coendomorphism operad of X.

13.1.5. Relative Endomorphism Object. We now establish some basic properties of the endomorphism objects. First we observe that the relative endomorphism object is a $\mathcal{G}$-PROP.

THEOREM 13.15. *Suppose $f : X \longrightarrow Y \in \mathcal{M}^{\mathfrak{C}}$ is a morphism and $\mathcal{G}$ admits an endomorphism object in $(\mathcal{M}, \mathcal{E})$. Then the relative endomorphism object $\mathsf{E}_f \in \Sigma_S$ has a canonical $\mathcal{G}$-PROP structure such that both maps in*

$$\begin{array}{ccc} \mathsf{E}_f & \xrightarrow{\overline{f}^*} & \mathsf{E}_X \\ {\scriptstyle \overline{f}_*}\downarrow & & \\ \mathsf{E}_Y & & \end{array}$$

are morphisms of $\mathcal{G}$-PROPs in $\mathcal{E}$.

PROOF. First we equip E_f with a $\mathcal{G}$-PROP structure. For any $G \in \mathtt{G}_S\binom{\underline{d}}{\underline{c}}$, define the $\mathcal{G}$-PROP structure map

$$\gamma_{f,G} : \mathsf{E}_f[G] \longrightarrow \mathsf{E}_f\binom{\underline{d}}{\underline{c}}$$

as the unique dotted arrow in the commutative diagram:

$$
\begin{array}{ccccc}
\mathsf{E}_f[G] & & \xrightarrow{\quad \otimes_v \overline{f}^* \quad} & & \mathsf{E}_X[G] \\
 & \searrow^{\gamma_{f,G}} & & & \Big\downarrow{\scriptstyle \gamma_{X,G}} \\
\Big\downarrow{\scriptstyle \otimes_v \overline{f}_*} & & \mathsf{E}_f\binom{d}{\underline{c}} & \xrightarrow{\overline{f}^*} & \mathsf{E}_X\binom{d}{\underline{c}} \\
 & & \Big\downarrow{\scriptstyle \overline{f}_*} & & \Big\downarrow{\scriptstyle f_*} \\
\mathsf{E}_Y[G] & \xrightarrow[\gamma_{Y,G}]{} & \mathsf{E}_Y\binom{d}{\underline{c}} & \xrightarrow[f^*]{} & \mathsf{E}_{X,Y}\binom{d}{\underline{c}}
\end{array}
\tag{13.10}
$$

The big square in (13.10) is commutative by (13.8), and the small square is commutative by the construction of E_f (13.5). Since $\mathsf{E}_f\binom{d}{\underline{c}}$ is the pullback of the small square, there exists a unique dotted arrow $\gamma_{f,G}$ that makes both the left and the top trapezoids commutative. The associativity (10.21) and unity (10.22) of the maps $\gamma_{f,G}$ follow from those of the endomorphism $\mathcal{G}$-PROPs E_X and E_Y and the uniqueness of γ_f. Indeed, if G is a corolla C, then both $\gamma_{X,C}$ and $\gamma_{Y,C}$ are identity maps, so the big square coincides with the small square in (13.10). The uniqueness of $\gamma_{f,C}$ then implies that it is the identity map.

To prove the associativity of the maps $\gamma_{f,G}$, we must show that the square

$$
\begin{array}{ccc}
\bigotimes_v \mathsf{E}_f[H_v] & \xrightarrow{\quad \otimes_v \gamma_{f,H_v} \quad} & \bigotimes_v \mathsf{E}_f\binom{\mathrm{out}(v)}{\mathrm{in}(v)} = \mathsf{E}_f[G] \\
\Big\downarrow{\scriptstyle \cong} & & \Big\downarrow{\scriptstyle \gamma_{f,G}} \\
\mathsf{E}_f[G(H_v)] & \xrightarrow{\quad \gamma_{f,G(H_v)} \quad} & \mathsf{E}_f\binom{d}{\underline{c}}
\end{array}
\tag{13.11}
$$

is commutative for any $G \in \mathsf{G}_S\binom{d}{\underline{c}}$ and $H_v \in \mathsf{G}\binom{\mathrm{out}(v)}{\mathrm{in}(v)}$ for $v \in \mathrm{Vt}(G)$. For the rest of this proof, $\otimes$ means $\otimes_{v\in \mathrm{Vt}(G)}$. Consider the diagram:

$$
\begin{array}{ccccc}
\otimes \mathsf{E}_f[H_v] & \xrightarrow{\otimes \overline{f}^*(H_v)} & \otimes \mathsf{E}_X[H_v] & \xrightarrow{\cong} & \mathsf{E}_X[G(H_v)] \\
\Big\downarrow{\scriptstyle \otimes \overline{f}_*(H_v)} & & & & \Big\downarrow{\scriptstyle \gamma_{X,G(H_v)}} \\
\otimes \mathsf{E}_Y[H_v] & & \mathsf{E}_f\binom{d}{\underline{c}} & \xrightarrow{\overline{f}^*} & \mathsf{E}_X\binom{d}{\underline{c}} \\
\Big\downarrow{\scriptstyle \cong} & & \Big\downarrow{\scriptstyle \overline{f}_*} & & \Big\downarrow{\scriptstyle f_*} \\
\mathsf{E}_Y[G(H_v)] & \xrightarrow{\gamma_{Y,G(H_v)}} & \mathsf{E}_Y\binom{d}{\underline{c}} & \xrightarrow{f^*} & \mathsf{E}_{X,Y}\binom{d}{\underline{c}}
\end{array}
\tag{13.12}
$$

Using the universal property of the pullback $\mathsf{E}_f\binom{d}{\underline{c}}$, to prove the commutativity of (13.11), it suffices to prove the following two statements:

(1) The solid-arrow diagram in (13.12) is commutative.
(2) Both compositions in (13.11) can be used as the dotted arrow in (13.12) to make the entire diagram commutative.

To prove the statements above, first consider the following diagram:
(13.13)

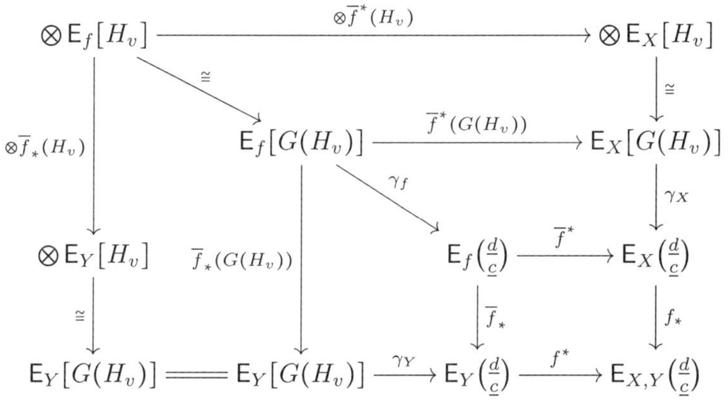

The left and the top trapezoids in (13.13) are commutative by naturality. The rest of (13.13) is the commutative diagram (13.10) with $G(H_v)$ in place of G, so the entire diagram (13.13) is commutative. It follows that the solid-arrow diagram (13.12) is commutative and that the lower composition in (13.11) can be used as the dotted arrow in (13.12) to make the whole diagram commutative.

It remains to show that the upper composition in (13.11) can also be used as the dotted arrow in (13.12) to make the whole diagram commutative. This is proved by considering the following commutative diagram:

(13.14)

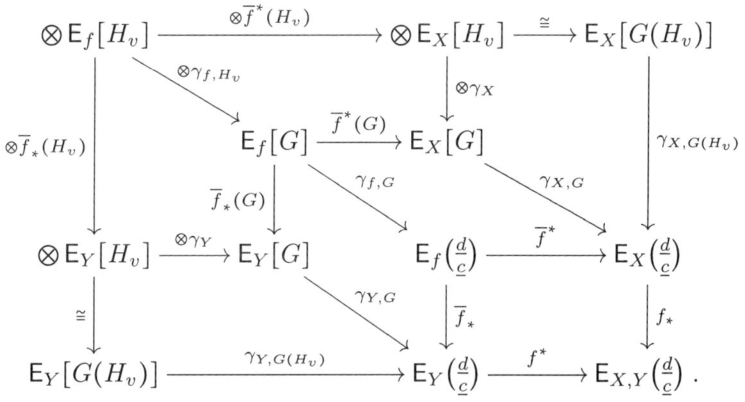

The two upper left trapezoids in (13.14) (both involving $\bigotimes \mathsf{E}_f(H_v)$) are commutative by the definition of γ_{f,H_v}. The bottom left (resp., top right) trapezoid is commutative by the associativity of the endomorphism $\mathcal{G}$-PROP E_Y (resp., E_X). The rest of (13.14) is the commutative diagram (13.10). Therefore, the upper composition in (13.11) can also be used as the dotted arrow in (13.12) to make the whole diagram commutative. We have shown that the maps $\gamma_{f,G}$ are associative, so the relative endomorphism object E_f is a $\mathcal{G}$-PROP with structure maps $\gamma_{f,G}$.

Finally, the commutativity of the two trapezoids in (13.10) implies that $\overline{f}^*$ and $\overline{f}_*$ are both morphisms of $\mathcal{G}$-PROPS. $\square$

Since the mixed endomorphism object $\mathsf{E}_{X,Y}$ is not in general a non-trivial $\mathcal{G}$-PROP, the relative endomorphism object E_f is not a pullback in $\mathtt{PROP}^{\mathcal{G}}$. However, the following observation says that the relative endomorphism object is close to being a pullback in $\mathtt{PROP}^{\mathcal{G}}$, since the diagram could be completed by inserting $\mathsf{E}_{X,Y}$.

THEOREM 13.16. *Suppose $\mathcal{G}$ admits an endomorphism object in $(\mathcal{M},\mathcal{E})$, $\mathsf{P} \in \mathtt{PROP}^{\mathcal{G}}$, $f\colon X \longrightarrow Y \in \mathcal{M}^{\mathfrak{C}}$, and*

$$\begin{array}{ccccc} \mathsf{P} & & \xrightarrow{\theta_X} & & \\ & \searrow^{\theta} & & & \\ & & \mathsf{E}_f & \xrightarrow{\overline{f}^*} & \mathsf{E}_X \\ & {\scriptstyle \theta_Y}\searrow & \downarrow{\scriptstyle \overline{f}_*} & & \\ & & \mathsf{E}_Y & & \end{array}$$

is a commutative diagram in $\mathcal{E}^{dis(S)}$. Then θ is a morphism of $\mathcal{G}$-PROPs if and only if both θ_X and θ_Y are morphisms of $\mathcal{G}$-PROPs.

PROOF. The "only if" direction follows from Proposition 13.15.

For the "if" direction, suppose θ_X and θ_Y are morphisms of $\mathcal{G}$-PROPs. In particular, the commutative square (10.23) holds for $(\mathsf{Q}, g) = (\mathsf{E}_X, \theta_X)$ and (E_Y, θ_Y). To show that θ is a morphism of $\mathcal{G}$-PROPs, we must prove the commutativity of (10.23) for (E_f, θ), i.e.,

$$\begin{array}{ccc} \mathsf{P}[G] & \xrightarrow{\theta[G]} & \mathsf{E}_f[G] \\ {\scriptstyle \gamma_G}\downarrow & & \downarrow{\scriptstyle \gamma_{f,G}} \\ \mathsf{P}\binom{d}{\underline{c}} & \xrightarrow{\theta} & \mathsf{E}_f\binom{d}{\underline{c}} \end{array} \tag{13.15}$$

for $G \in \mathsf{G}_S\binom{d}{\underline{c}}$. Consider the diagram

$$\begin{array}{ccccc} \mathsf{P}[G] & & \xrightarrow{\otimes\theta_X} & & \mathsf{E}_X[G] \\ & \ddots & & & \downarrow{\scriptstyle \gamma_{X,G}} \\ {\scriptstyle \otimes\theta_Y}\downarrow & & \mathsf{E}_f\binom{d}{\underline{c}} & \xrightarrow{\overline{f}^*} & \mathsf{E}_X\binom{d}{\underline{c}} \\ & & {\scriptstyle \overline{f}_*}\downarrow & & \downarrow{\scriptstyle f_*} \\ \mathsf{E}_Y[G] & \xrightarrow{\gamma_{Y,G}} & \mathsf{E}_Y\binom{d}{\underline{c}} & \xrightarrow{f^*} & \mathsf{E}_{X,Y}\binom{d}{\underline{c}} \end{array} \tag{13.16}$$

in which $\otimes = \otimes_{v \in \mathrm{Vt}(G)}$. Using the universal property of the pullback $\mathsf{E}_f\binom{d}{\underline{c}}$, to prove the commutativity of (13.15), it suffices to prove the following two statements:

(1) The solid-arrow diagram in (13.16) is commutative.
(2) Both compositions in (13.15) can be used as the dotted arrow in (13.16) to make the entire diagram commutative.

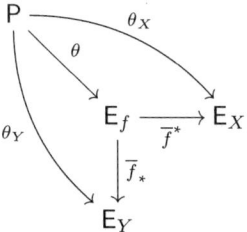

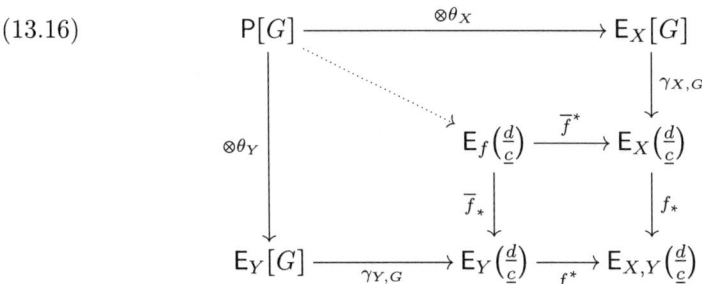

To prove these statements, first consider the diagram:

(13.17)

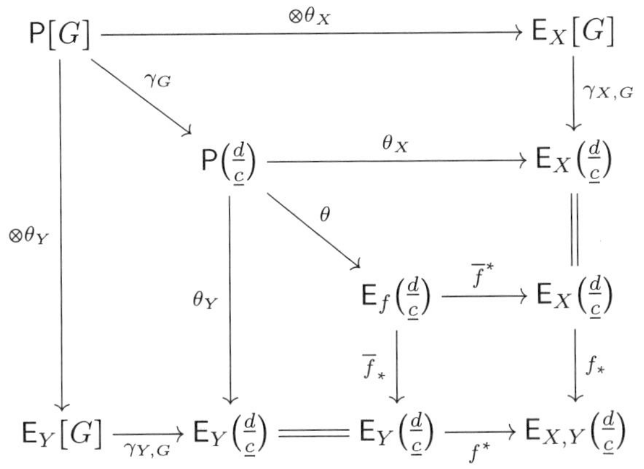

The top and the left trapezoids in (13.17) are commutative because θ_X and θ_Y are assumed to be morphisms of $\mathcal{G}$-PROPs (10.23). The rest of (13.17) is commutative by the definitions of θ_X, θ_Y, and E_f. Therefore, (13.17) is a commutative diagram. This shows that the solid-arrow diagram in (13.16) is commutative and that $\theta \circ \gamma_G$ can be used as the dotted arrow in (13.16) to make the whole diagram commutative.

It remains to show that the upper composition in (13.15) can also be used as the dotted arrow in (13.16) to make the whole diagram commutative. This is proved by considering the following commutative diagram:

(13.18)

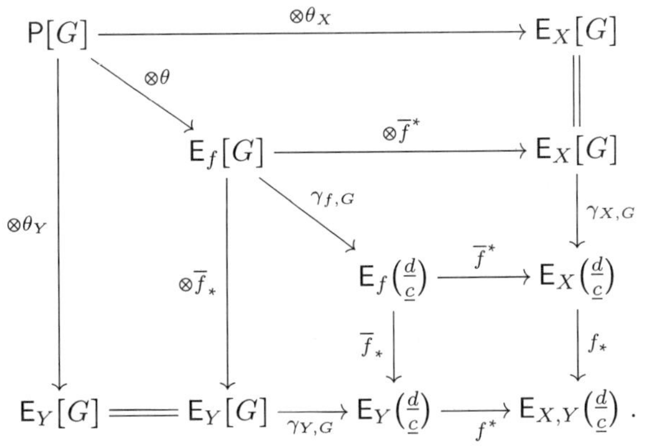

The top and the left trapezoids in (13.18) are commutative by the definitions of θ_X and θ_Y. The rest of (13.18) is the commutative diagram (13.10), so (13.18) is commutative. $\square$

13.2. Unbiased Algebras

Now that we have established the required comfort level with the technical question of the existence and properties of endomorphism objects, we can introduce the standard definition of algebras in terms of endomorphism objects, along with a few variants. Throughout this section, $\mathcal{G}$ denotes a pasting scheme that admits an endomorphism object in $(\mathcal{M}, \mathcal{E})$.

13.2.1. Unbiased Definition of Algebras.

DEFINITION 13.17. Let P be a $\mathcal{G}$-PROP in $\mathcal{E}$.

(1) A **P-algebra** in $\mathcal{M}$ is a pair (X, λ) consisting of an object $X \in \mathcal{M}^{\mathfrak{C}}$ and a morphism
$$\lambda : \mathsf{P} \longrightarrow \mathsf{E}_X \in \mathtt{PROP}^{\mathcal{G}},$$
called the **structure map**.

(2) Suppose $f : X \longrightarrow Y \in \mathcal{M}^{\mathfrak{C}}$, while X and Y are P-algebras in $\mathcal{M}$ with structure maps λ_X and λ_Y, respectively. Then f is called a **P-algebra morphism** if the diagram

$$\begin{array}{ccc} \mathsf{P} & \xrightarrow{\lambda_X} & \mathsf{E}_X \\ {\scriptstyle \lambda_Y}\big\downarrow & & \big\downarrow{\scriptstyle f_*} \\ \mathsf{E}_Y & \xrightarrow{f^*} & \mathsf{E}_{X,Y} \end{array} \tag{13.19}$$

in $\mathcal{E}^{dis(S)}$ is commutative.

With such an indirect definition of morphism, it may be a bit odd to think about composition of algebra morphisms, or identities thereof.

LEMMA 13.18. *With the notion of morphism of P-algebras above, there is a category of P-algebras inheriting units and composition from $\mathcal{M}^{\mathfrak{C}}$.*

PROOF. Suppose $f : X \longrightarrow Y$ and $g : Y \longrightarrow Z$ are both algebra morphisms with the same algebra structure on Y. Then in the diagram

$$\begin{array}{ccccc} \mathsf{P} & \longrightarrow & \longrightarrow & \longrightarrow & \mathsf{E}_X \\ \big\downarrow & \searrow & & & \big\downarrow{\scriptstyle f_*} \\ \big\downarrow & & \mathsf{E}_Y & \xrightarrow{f^*} & \mathsf{E}_{X,Y} \\ \big\downarrow & & \big\downarrow{\scriptstyle g_*} & & \big\downarrow{\scriptstyle g_*} \\ \mathsf{E}_Z & \xrightarrow{g^*} & \mathsf{E}_{Y,Z} & \xrightarrow{f^*} & \mathsf{E}_{X,Z} \end{array}$$

the top and left trapezoids commute by assumption. However, since the bottom square also commutes by inspection, while the long right vertical is $(g \circ f)_*$ and the long low horizontal is $(g \circ f)^*$, the resulting commutativity of the outer square describes $g \circ f$ as a P-algebra morphism. This verifies that composition in $\mathcal{M}^{\mathfrak{C}}$ induces the composition of P-algebra morphisms.

To verify that the underlying identity map in $\mathcal{M}^{\mathfrak{C}}$ satisfies the composition property to be an identity of algebras, consider the case above with $f = Id_Y$ and the same algebra structure map on source and target. Then the diagram above becomes

$$\begin{array}{ccccc} \mathsf{P} & \longrightarrow & \longrightarrow & \longrightarrow & \mathsf{E}_Y \\ \big\downarrow & \searrow & & & \big\downarrow{\scriptstyle Id} \\ \big\downarrow & & \mathsf{E}_Y & \xrightarrow{Id} & \mathsf{E}_Y \\ \big\downarrow & & \big\downarrow{\scriptstyle g_*} & & \big\downarrow{\scriptstyle g_*} \\ \mathsf{E}_Z & \xrightarrow{g^*} & \mathsf{E}_{Y,Z} & \xrightarrow{Id} & \mathsf{E}_{Y,Z} \end{array}$$

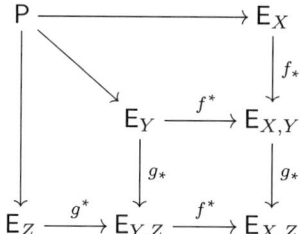

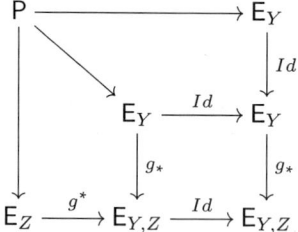

which is clearly equivalent to the left trapezoid, and similarly for when g is the underlying identity map. □

The **category of P-algebras in $\mathcal{M}$** of Lemma 13.18 is denoted by $\mathbf{Alg}_{\mathcal{M}}(\mathsf{P})$, or $\mathbf{Alg}(\mathsf{P})$ if $\mathcal{M} = \mathcal{E}$.

EXAMPLE 13.19. If P is the constant $\mathcal{G}$-PROP on the initial object in $\mathcal{E}$, then each $X \in \mathcal{M}^{\mathfrak{C}}$ has a unique P-algebra structure. In fact, every diagram of the form

$$\begin{array}{ccc} \varnothing & \longrightarrow & \mathsf{E}_X \\ \downarrow & & \downarrow f_* \\ \mathsf{E}_Y & \xrightarrow{f^*} & \mathsf{E}_{X,Y} \end{array}$$

commutes, and similarly for the composition diagrams, so $\mathbf{Alg}_{\mathcal{M}}(\mathsf{P})$ is simply $\mathcal{M}^{\mathfrak{C}}$ itself.

13.2.2. Alternative Descriptions of Algebras.

The following observation gives the adjoint form of a P-algebra.

PROPOSITION 13.20. *Suppose P is a $\mathcal{G}$-PROP in $\mathcal{E}$,*

$$\lambda: \mathsf{P} \longrightarrow \mathsf{E}_X \in \mathcal{E}^{dis(S)}$$

is a morphism for some $X \in \mathcal{M}^{\mathfrak{C}}$, and

$$\lambda^{\#}: \mathsf{P}\binom{\underline{d}}{\underline{c}} \otimes X_{\underline{c}} \longrightarrow X_{\underline{d}} \in \mathcal{M} \tag{13.20}$$

for $(\underline{c};\underline{d}) \in S$ are the adjoint maps. Then (X,λ) is a P-algebra if and only if the square

$$\begin{array}{ccc} \mathsf{P}[G] \otimes X_{\underline{c}} & \xrightarrow{\lambda[G]\otimes Id} & \mathsf{E}_X[G] \otimes X_{\underline{c}} \\ {\scriptstyle \gamma_G \otimes Id}\downarrow & & \downarrow {\scriptstyle \gamma^{\#}_{X,G}} \\ \mathsf{P}\binom{\underline{d}}{\underline{c}} \otimes X_{\underline{c}} & \xrightarrow{\lambda^{\#}} & X_{\underline{d}} \end{array} \tag{13.21}$$

is commutative for all $G \in \mathsf{G}_S(\underline{c};\underline{d})$.

PROOF. This square is the adjoint form of the square (10.23) defining a morphism of $\mathcal{G}$-PROPs. □

A similar exercise in adjunction gives the following description of a morphism of P-algebras.

PROPOSITION 13.21. *Suppose (X,λ_X) and (Y,λ_Y) are P-algebras. Then*

$$f: X \longrightarrow Y \in \mathcal{M}^{\mathfrak{C}}$$

is a P-algebra morphism if and only if the square

$$\begin{array}{ccc} \mathsf{P}\binom{\underline{d}}{\underline{c}} \otimes X_{\underline{c}} & \xrightarrow{\lambda^{\#}_X} & X_{\underline{d}} \\ {\scriptstyle Id\otimes f_{\underline{c}}}\downarrow & & \downarrow {\scriptstyle f_{\underline{d}}} \\ \mathsf{P}\binom{\underline{d}}{\underline{c}} \otimes Y_{\underline{c}} & \xrightarrow{\lambda^{\#}_Y} & Y_{\underline{d}} \end{array} \tag{13.22}$$

is commutative in $\mathcal{M}$ for all $(\underline{c};\underline{d}) \in S$.

EXAMPLE 13.22. In Definition 13.17 if $\mathcal{G} = \mathtt{Gr}^{\uparrow}$, $\mathtt{Gr}^{\uparrow}_{\mathrm{c}}$, $\mathtt{Gr}^{\uparrow}_{\mathrm{di}}$, $\mathtt{Gr}_{\frac{1}{2}}$, or $\mathtt{UTree}$, then we recover the usual definitions of the category of algebras over a PROP, a properad, a dioperad, a half-PROP, or an operad as discussed in more detail in Section 13.4.

The following corollary of Theorem 13.16 says that a P-algebra morphism is uniquely determined by a morphism of $\mathcal{G}$-PROPs from P to a relative endomorphism object.

COROLLARY 13.23. *Suppose* $\mathsf{P} \in \mathtt{PROP}^{\mathcal{G}}$ *and* $f{:}\,X \longrightarrow Y \in \mathcal{M}^{\mathfrak{C}}$. *Then the following two statements are equivalent:*

(1) *Both* X *and* Y *are* P*-algebras in* $\mathcal{M}$ *with structure maps*

$$\lambda_X{:}\,\mathsf{P} \longrightarrow \mathsf{E}_X \quad \text{and} \quad \lambda_Y{:}\,\mathsf{P} \longrightarrow \mathsf{E}_Y,$$

respectively, and f *is a* P*-algebra morphism.*

(2) *There exists a morphism*

$$\lambda{:}\,\mathsf{P} \longrightarrow \mathsf{E}_f \in \mathtt{PROP}^{\mathcal{G}}$$

such that the diagram

(13.23)

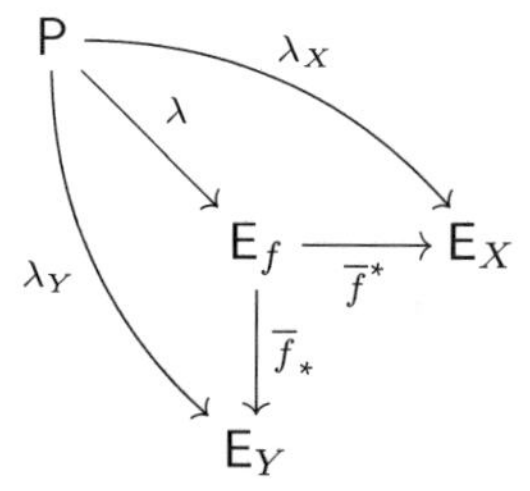

is commutative in $\mathtt{PROP}^{\mathcal{G}}$.

PROOF. By Theorem 13.16 the commutativity of the diagram (13.19) is equivalent to the existence of a morphism

$$\lambda{:}\,\mathsf{P} \longrightarrow \mathsf{E}_f \in \mathtt{PROP}^{\mathcal{G}}$$

that makes the diagram (13.23) commutative. □

Notice the map λ in the second case can be used to define the structure maps λ_X and λ_Y, although a more interesting context would be a factorization result on one side then used to define the other side. In particular, this makes clear how to address questions such as which algebra structures on X will make a previously chosen underlying map a morphism of algebras, given a fixed algebra structure on Y. In particular, in [**JY09**] a lifting property against one of the outer maps is exploited to prove a homotopy invariance result for the existence of P-algebra structures given a particularly nice choice of P.

13.3. Algebras under Change of Pasting Scheme or Base Category

In this section, we consider the category of algebras under a change of pasting scheme or the base category.

13.3.1. Change of Pasting Scheme.

The following result says that the category of algebras over a $\mathcal{G}$-PROP is isomorphic to the category of algebras over the associated free $\mathcal{G}'$-PROP whenever $\mathcal{G} \leq \mathcal{G}'$.

THEOREM 13.24. *Suppose*

- $\mathcal{M}$ *is a symmetric monoidal* $\mathcal{E}$*-category,*
- $\mathcal{G} \leq \mathcal{G}'$ *are pasting schemes such that* $\mathcal{G}'$ *admits an endomorphism object, and*
- $\mathsf{P} \in \mathtt{PROP}^{\mathcal{G}}$.

Then there is a canonical isomorphism of categories

$$\mathbf{Alg}_{\mathcal{M}}(\mathsf{P}) \cong \mathbf{Alg}_{\mathcal{M}}(L\mathsf{P}),$$

where L *is the left adjoint in Theorem* 12.1.

PROOF. For an object $X = \{X_c\} \in \mathcal{M}^{\mathfrak{C}}$, we have the endomorphism $\mathcal{G}'$-PROP E_X of X. Applying the forgetful functor

$$U: \mathtt{PROP}^{\mathcal{G}'} \longrightarrow \mathtt{PROP}^{\mathcal{G}},$$

we obtain the endomorphism $\mathcal{G}$-PROP $U\mathsf{E}_X$ of X by Proposition 13.11. The adjunction (L, U) in Theorem 12.1 gives a natural bijection

$$\mathtt{PROP}^{\mathcal{G}}(\mathsf{P}, U\mathsf{E}_X) \cong \mathtt{PROP}^{\mathcal{G}'}(L\mathsf{P}, \mathsf{E}_X), \tag{13.24}$$

that implies there is a natural bijection between the $L\mathsf{P}$-algebra structures over $\mathcal{G}'$ on X and the P-algebra structures over $\mathcal{G}$ on X.

To see that the bijection (13.24) is compatible with algebra morphisms, suppose that $f: X \longrightarrow Y$ is a morphism in $\mathcal{M}^{\mathfrak{C}}$. The relative endomorphism object E_f is a $\mathcal{G}'$-PROP by Theorem 13.15, so there is a natural bijection

$$\mathtt{PROP}^{\mathcal{G}}(\mathsf{P}, U\mathsf{E}_f) \cong \mathtt{PROP}^{\mathcal{G}'}(L\mathsf{P}, \mathsf{E}_f).$$

In fact, $U\mathsf{E}_f$ remains the relative endomorphism object over $\mathcal{G}$, since U preserves mixed endomorphism objects and the entry-wise pullback used to define E_f. Now apply Corollary 13.23 keeping in mind that the adjunction implies

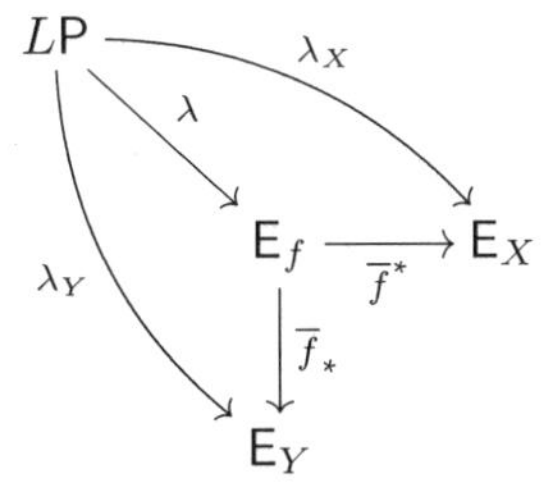

commutes in $\mathcal{G}'$-PROP precisely when

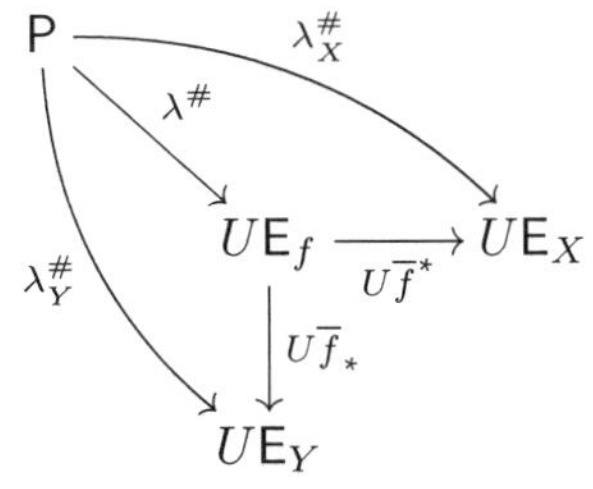

commutes in $\mathcal{G}$-PROP. Compatibility with the underlying identity map with the same algebra structure on source and target is immediate.

In order to verify compatibility with composition, one proceeds as above, this time using Corollary 13.23 to identify a composition of P-algebra maps with a commutative diagram in $\mathcal{G}$-PROP of the form

$$\begin{array}{ccccc} P & \longrightarrow & \mathsf{E}_f & \longrightarrow & \mathsf{E}_X \\ \downarrow & & \downarrow & & \\ \mathsf{E}_g & \longrightarrow & \mathsf{E}_Y & & \\ \downarrow & & & & \\ \mathsf{E}_Z. & & & & \end{array}$$

□

EXAMPLE 13.25. Theorem 13.24 can be used when $\mathcal{G}' \leq \mathtt{Gr}^{\uparrow}$ by Corollary 13.12. For example, if P is an operad, a dioperad, a half-PROP, or a properad, and if $L\mathsf{P}$ is the PROP generated by P, then the previous theorem says that P and $L\mathsf{P}$ have canonically isomorphic categories of algebras.

13.3.2. Change of $\mathcal{G}$-PROP. Here we assume $\mathsf{P} \longrightarrow \mathsf{P}'$ is a morphism of $\mathcal{G}$-PROPs and consider the resulting change of $\mathcal{G}$-PROP functor on the level of algebras.

LEMMA 13.26. *Each morphism $f : \mathsf{P} \longrightarrow \mathsf{P}'$ of $\mathcal{G}$-PROPs induces a functor*

$$f^* \colon \mathbf{Alg}_{\mathcal{M}}(\mathsf{P}') \longrightarrow \mathbf{Alg}_{\mathcal{M}}(\mathsf{P}).$$

PROOF. Precomposing structure maps

$$\mathsf{P} \longrightarrow \mathsf{P}' \longrightarrow \mathsf{E}_X$$

gives a P-algebra structure to any P'-algebra, which by Corollary 13.23 is compatible with algebra morphisms. Once again, as in the proof of 13.24, we verify compatibility with composition by using Corollary 13.23 to identify a composition of P-algebra maps with a commutative diagram in $\mathcal{G}$-PROP of the form

$$\begin{array}{ccccc} P & \longrightarrow & \mathsf{E}_f & \longrightarrow & \mathsf{E}_X \\ \downarrow & & \downarrow & & \\ \mathsf{E}_g & \longrightarrow & \mathsf{E}_Y & & \\ \downarrow & & & & \\ \mathsf{E}_Z. & & & & \end{array}$$

□

13.3.3. Change of Base Category. Next we consider the category of algebras under a change of the underlying category. Let

$$L \colon \mathcal{E} \rightleftarrows \mathcal{D} \colon R$$

be an adjoint pair between symmetric monoidal closed categories. Since we have only discussed applying symmetric monoidal functors to generalized PROPs in

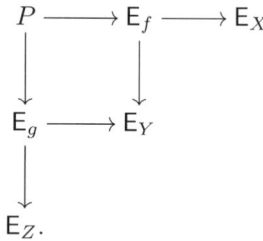

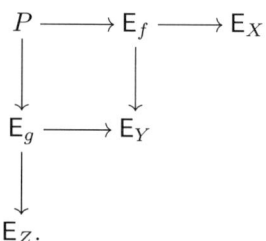

Corollary 12.13, we must assume that both L and R are symmetric monoidal. This is the case when L is strong symmetric monoidal by Lemma 10.9. However, in order to know that R is compatible with the extension of colored objects to take profiles as inputs, $X_{\underline{c}}$, it will be necessary below to assume R is also a strong monoidal functor, so throughout this discussion we assume both are strong monoidal functors. Of course, we also want $\mathcal{G}$ to be a pasting scheme that admits an endomorphism object in both $\mathcal{E}$ and $\mathcal{D}$ (Definition 13.7), and as usual $\mathsf{P} \in \mathtt{PROP}^{\mathcal{G},\mathcal{E}}$.

Once again, we will need to impose a technical condition about endomorphism objects once we move outside of $\mathtt{Gr}^{\uparrow}$, but the condition will be automatically satisfied for any pasting scheme $\mathcal{G} \leq \mathtt{Gr}^{\uparrow}$. We must begin by introducing some notation.

LEMMA 13.27. *Let $K : \mathcal{M} \longrightarrow \mathcal{M}'$ be a strong symmetric monoidal functor, with P a $\mathcal{G}$-PROP with values in $\mathcal{M}$ and $Z \in \mathbf{Alg}_{\mathcal{M}}(\mathsf{P})$ with structure map $\lambda{:}\, \mathsf{P} \longrightarrow \mathsf{E}_Z$. Then there is an induced morphism*

$$\overline{\lambda}{:}\, K\mathsf{P} \longrightarrow \mathsf{E}_{KZ} \tag{13.25}$$

in $\Sigma^{\mathcal{M}}$-bimodules, which we call the partial structure map.

PROOF. Apply K entrywise to the adjoint map $\lambda^{\#}$ (13.20), and exploit the strong symmetric monoidal isomorphism to construct maps

$$K\mathsf{P}(\underline{d};\underline{c}) \otimes (KZ)_{\underline{c}} \cong K(\mathsf{P}(\underline{d};\underline{c}) \otimes Z_{\underline{c}}) \longrightarrow K(Z_{\underline{d}}) \cong (KZ)_{\underline{d}}.$$

Via the hom-tensor adjunction (13.2) this induces a series of morphisms

$$K\mathsf{P}(\underline{d};\underline{c}) \longrightarrow \mathsf{E}_{KZ}$$

which are compatible with permutations of profiles by construction. Hence, they fit together to form a morphism in $\Sigma^{\mathcal{M}}$-bimodules as required. □

REMARK 13.28. If K is not *strong* symmetric monoidal above, then the last map

$$K(Z_{\underline{d}}) \longleftarrow (KZ)_{\underline{d}}$$

goes in the wrong direction, hence our assumption that both L and R are strong above.

DEFINITION 13.29. We say that $\mathsf{P} \in \mathtt{PROP}^{\mathcal{G},\mathcal{E}}$ is **compatible with** the adjoint pair $L{:}\,\mathcal{E} \rightleftarrows \mathcal{D}{:}\,R$ of strong symmetric monoidal functors if the following two conditions hold.

(1) For each $X \in \mathbf{Alg}(\mathsf{P})$, the partial structure map $\overline{\lambda}$ of Lemma 13.27 using $L{:}\,\mathcal{E} \longrightarrow \mathcal{D}$ is a morphism of $\mathcal{G}$-PROPs in $\mathcal{D}$.
(2) For each $Y \in \mathbf{Alg}(L\mathsf{P})$, the partial structure map $\overline{\gamma}$ of Lemma 13.27 using $R{:}\,\mathcal{D} \longrightarrow \mathcal{E}$ is a morphism of $\mathcal{G}$-PROPs in $\mathcal{E}$.

REMARK 13.30. Notice the unit of adjunction $\mathsf{P} \longrightarrow RL\mathsf{P}$ is already a morphism of $\mathcal{G}$-PROPs in $\mathcal{E}$ by Theorem 12.11. Hence, the second condition above is equivalent to instead saying the composite $\mathsf{P} \longrightarrow RL\mathsf{P} \longrightarrow \mathsf{E}_{RY}$ is a morphism of $\mathcal{G}$-PROPs in $\mathcal{E}$.

LEMMA 13.31. *If $\mathsf{P} \in \mathtt{PROP}^{\mathcal{G},\mathcal{E}}$ is compatible with the adjoint pair of strong symmetric monoidal functors $L{:}\,\mathcal{E} \rightleftarrows \mathcal{D}{:}\,R$, then there are induced functors*

$$\mathbf{Alg}(\mathsf{P}) \rightleftarrows \mathbf{Alg}(L\mathsf{P}) .$$

PROOF. For $\mathbf{Alg}(\mathsf{P}) \longrightarrow \mathbf{Alg}(L\mathsf{P})$, the construction is basically given in the proof of Lemma 13.27. The remaining issue is compatibility with composition of algebra morphisms, which we verify using Proposition 13.21 and the construction itself.

For $\mathbf{Alg}(L\mathsf{P}) \longrightarrow \mathbf{Alg}(RL\mathsf{P}) \longrightarrow \mathbf{Alg}(\mathsf{P})$, first proceed as above to define $\mathbf{Alg}(L\mathsf{P}) \longrightarrow \mathbf{Alg}(RL\mathsf{P})$. Then follow with the change of $\mathcal{G}$-PROP functor of Lemma 13.26 using the unit morphism $\eta: \mathsf{P} \longrightarrow RL\mathsf{P}$ of the adjunction from Corollary 12.13. □

LEMMA 13.32. *Suppose $\mathcal{G} \leq \mathtt{Gr}^{\uparrow}$, $\mathsf{P} \in \mathtt{PROP}^{\mathcal{G},\mathcal{E}}$ and $L: \mathcal{E} \rightleftarrows \mathcal{D}: R$ is any adjoint pair of strong symmetric monoidal functors. Then P is compatible with the adjoint pair.*

PROOF. For the case $\mathcal{G} = \mathtt{Gr}^{\uparrow}$ of MacLane's PROPs, the endomorphism PROP is defined naturally from the evaluation maps $\lambda^{\#}$ via the hom-tensor adjunction (13.2). Any strong symmetric monoidal functor will be compatible with evaluation maps, which are defined as the hom-tensor adjoint of the identity map on the relevant mapping object. As such, endomorphism PROPs are compatible with every adjoint pair of strong symmetric monoidal functors.

Since endomorphism objects restrict, for $\mathcal{G} \leq \mathtt{Gr}^{\uparrow}$ we see the maps in question will be the restrictions of morphisms of PROPs. The existence of the forgetful functor implies they will then be morphisms of $\mathcal{G}$-PROPs as well, thereby verifying the condition whenever $\mathcal{G} \leq \mathtt{Gr}^{\uparrow}$. □

THEOREM 13.33. *Suppose $\mathsf{P} \in \mathtt{PROP}^{\mathcal{G},\mathcal{E}}$ is compatible with the adjoint pair*

$$L: \mathcal{E} \rightleftarrows \mathcal{D}: R$$

of strong symmetric monoidal functors (e.g. $\mathcal{G} \leq \mathtt{Gr}^{\uparrow}$). Then there is an induced adjoint pair

$$L: \mathbf{Alg}(\mathsf{P}) \rightleftarrows \mathbf{Alg}(L\mathsf{P}): R \tag{13.26}$$

that is defined entry-wise.

PROOF. The compatibility of P with (L, R) implies that there are well-defined induced functors between the categories of algebras by Lemma 13.31. To check that they form an adjoint pair, suppose $X \in \mathbf{Alg}(\mathsf{P})$, $Y \in \mathbf{Alg}(L\mathsf{P})$, and

$$f: X \longrightarrow RY$$

is a morphism in $\mathcal{E}^{\mathfrak{C}}$. Let $f^{\#}: LX \longrightarrow Y \in \mathcal{D}^{\mathfrak{C}}$ be the entry-wise adjoint map of f. Using the strong symmetric monoidal assumptions on L and R, the commutativity of

$$\begin{array}{ccc}
\mathsf{P}\binom{\underline{d}}{\underline{c}} \otimes X_{\underline{c}} & \xrightarrow{\lambda^{\#}} & X_{\underline{d}} \\
\downarrow {\scriptstyle Id \otimes f_{\underline{c}}} & & \\
\mathsf{P}\binom{\underline{d}}{\underline{c}} \otimes (RY)_{\underline{c}} & & \downarrow {\scriptstyle f_{\underline{d}}} \\
\downarrow {\scriptstyle \eta\binom{\underline{d}}{\underline{c}} \otimes Id} & & \\
RL\mathsf{P}\binom{\underline{d}}{\underline{c}} \otimes (RY)_{\underline{c}} & \xrightarrow{\overline{\gamma}^{\#}} & (RY)_{\underline{d}}
\end{array}$$

in $\mathcal{E}$ is equivalent to the commutativity of

$$\begin{array}{ccc} L\mathsf{P}\binom{\underline{d}}{\underline{c}} \otimes (LX)_{\underline{c}} & \xrightarrow{\overline{\lambda}^{\#}} & (LX)_{\underline{d}} \\ {\scriptstyle Id\otimes f^{\#}_{\underline{c}}}\downarrow & & \downarrow{\scriptstyle f^{\#}_{\underline{d}}} \\ L\mathsf{P}\binom{\underline{d}}{\underline{c}} \otimes Y_{\underline{c}} & \xrightarrow{\gamma^{\#}} & Y_{\underline{d}} \end{array}$$

in $\mathcal{D}$. This shows that the induced functors between the algebra categories form an adjoint pair. □

13.4. Biased Algebras

In this section, we discuss biased versions of algebras over generalized PROPs.

13.4.1. Biased Algebra Theorem.

13.4.1. Biased Algebra Theorem. Our general statement here is the following application of Theorem 11.7 with $\mathsf{Q} = \mathsf{E}_X$.

THEOREM 13.34 (Biased Algebra Theorem). *Let*

- $\mathcal{T}$ *be a generating set of a pasting scheme* $\mathcal{G}$ *that admits an endomorphism object in* $(\mathcal{M}, \mathcal{E})$,
- P *be a* $\mathcal{G}$*-PROP, and*
- $f\colon \mathsf{P} \longrightarrow \mathsf{E}_X \in \mathcal{E}^{dis(S)}$ *for some* $X \in \mathcal{M}^{\mathfrak{C}}$.

Then (X, f) *is a* P*-algebra if and only if the square*

$$\begin{array}{ccc} \mathsf{P}[G] & \xrightarrow{\otimes f} & \mathsf{E}_X[G] \\ {\scriptstyle \gamma^P_{[G]}}\downarrow & & \downarrow{\scriptstyle \gamma^{\mathsf{E}_X}_{[G]}} \\ \mathsf{P}\binom{\mathrm{out}(G)}{\mathrm{in}(G)} & \xrightarrow{f} & \mathsf{E}_X\binom{\mathrm{out}(G)}{\mathrm{in}(G)}, \end{array}$$

or equivalently the square (13.21), *is commutative for each* G *in* $\mathcal{T}$.

REMARK 13.35. Observe that when G runs through the permuted corollas, the commutative square in Theorem 13.34 simply says that f is a map of $\mathbf{\Sigma}_S$-bimodules.

To simplify notations, a map $\mathsf{P} \longrightarrow \mathsf{E}_X$ and its adjoint components will often be denoted by the same symbol for the remainder of this chapter.

13.4.2. Algebras over a Markl Non-Unital Operad. The following instance of Theorem 13.34, based upon Theorem 7.35 and Corollary 11.12, provides a more concrete characterization of algebras over a Markl non-unital operad.

COROLLARY 13.36. *Let*

- P *be a* `Tree`*-PROP (i.e., a Markl non-unital operad), and*
- $f\colon \mathsf{P} \longrightarrow \mathsf{E}_X \in \Sigma_{S_1}$ *for some* $X \in \mathcal{M}^{\mathfrak{C}}$.

Then (X,f) is a P-algebra if and only if the square

$$\begin{array}{ccc}
\mathsf{P}\binom{d}{\underline{c}} \otimes \mathsf{P}\binom{c_i}{\underline{b}} \otimes X_{\underline{e}} & \xrightarrow{\cong} & \mathsf{P}\binom{d}{\underline{c}} \otimes X_{\underline{c}'} \otimes \Big(\mathsf{P}\binom{c_i}{\underline{b}} \otimes X_{\underline{b}}\Big) \otimes X_{\underline{c}''} \\
 & & \big\downarrow {\scriptstyle (Id,f,Id)} \\
{\scriptstyle (\circ_i, Id)} \big\downarrow & & \mathsf{P}\binom{d}{\underline{c}} \otimes X_{\underline{c}} \\
 & & \big\downarrow {\scriptstyle f} \\
\mathsf{P}\binom{d}{\underline{c}\circ_i \underline{b}} \otimes X_{\underline{e}} & \xrightarrow{\quad f \quad} & X_d
\end{array}$$

is commutative for every possible $\circ_i$ operation, where $\underline{c}' = c_{[1,i-1]}$, $\underline{c}'' = c_{[i+1,m]}$, and $\underline{e} = (\underline{c}', \underline{b}, \underline{c}'') = \underline{c} \circ_i \underline{b}$.

13.4.3. Algebras over a May Operad. Another application of Theorem 13.34, here based on Theorem 7.41 and Corollary 11.16, characterizes the algebras over a May operad.

COROLLARY 13.37. *Let*

- P *be a* $\mathtt{UTree}$*-PROP (i.e., a May operad), and*
- $f\colon \mathsf{P} \longrightarrow \mathsf{E}_X \in \Sigma_{S_1}$ *for some* $X \in \mathcal{M}^{\mathfrak{C}}$.

Then (X,f) is a P-algebra if and only if the squares

$$\begin{array}{ccc}
I \otimes X_c & \xrightarrow{\cong} & X_c \\
{\scriptstyle (\mathbf{1}_c, Id)} \big\downarrow & & \big\Vert \\
\mathsf{P}\binom{c}{c} \otimes X_c & \xrightarrow{f} & X_c
\end{array} \tag{13.27}$$

and

$$\begin{array}{ccc}
\mathsf{P}\binom{d}{\underline{c}} \otimes \Big(\bigotimes_{i=1}^m \mathsf{P}\binom{c_i}{\underline{b}_i}\Big) \otimes X_{\underline{b}} & \xrightarrow{\cong} & \mathsf{P}\binom{d}{\underline{c}} \otimes \bigotimes_{i=1}^m \Big(\mathsf{P}\binom{c_i}{\underline{b}_i} \otimes X_{\underline{b}_i}\Big) \\
 & & \big\downarrow {\scriptstyle (Id, \otimes f)} \\
{\scriptstyle (\gamma, Id)} \big\downarrow & & \mathsf{P}\binom{d}{\underline{c}} \otimes X_{\underline{c}} \\
 & & \big\downarrow {\scriptstyle f} \\
\mathsf{P}\binom{d}{\underline{b}} \otimes X_{\underline{b}} & \xrightarrow{\quad f \quad} & X_d
\end{array} \tag{13.28}$$

are commutative.

13.4.4. Algebras over a Dioperad. This time we apply Theorem 13.34 exploiting Theorem 7.57 and Corollary 11.20 to characterize the algebras over a dioperad.

COROLLARY 13.38. *Let*

- P *be a* $\mathtt{Gr}^{\uparrow}_{di}$*-PROP (i.e., a dioperad), and*
- $f\colon \mathsf{P} \longrightarrow \mathsf{E}_X \in \mathbf{\Sigma}_S$ *for some* $X \in \mathcal{M}^{\mathfrak{C}}$.

Then (X, f) *is a* P*-algebra if and only if the unit squares* (13.27) *and*

$$\begin{array}{ccc}
\mathsf{P}\binom{\underline{d}}{\underline{c}} \otimes \mathsf{P}\binom{\underline{b}}{\underline{a}} \otimes X_{\underline{r}} & \xrightarrow{\cong} & \mathsf{P}\binom{\underline{d}}{\underline{c}} \otimes X_{\underline{c}'} \otimes \left(\mathsf{P}\binom{\underline{b}}{\underline{a}} \otimes X_{\underline{a}}\right) \otimes X_{\underline{c}''} \\
 & & \downarrow{\scriptstyle (Id,f,Id)} \\
 & & \mathsf{P}\binom{\underline{d}}{\underline{c}} \otimes X_{\underline{e}} \\
\downarrow{\scriptstyle ({}_j\circ_i, Id)} & & \downarrow{\scriptstyle \cong} \\
 & & X_{\underline{b}'} \otimes \left(\mathsf{P}\binom{\underline{d}}{\underline{c}} \otimes X_{\underline{c}}\right) \otimes X_{\underline{b}''} \\
 & & \downarrow{\scriptstyle (Id,f,Id)} \\
\mathsf{P}\binom{\underline{b} \circ_j \underline{d}}{\underline{c} \circ_i \underline{a}} \otimes X_{\underline{r}} & \xrightarrow{f} & X_{\underline{s}}
\end{array}$$

are commutative whenever they are defined, where $\underline{c}' = c_{[1,i-1]}$, $\underline{c}'' = c_{[i+1,m]}$, $\underline{e} = (\underline{c}', \underline{b}, \underline{c}'') = \underline{c} \circ_i \underline{b}$, $\underline{r} = (\underline{c}', \underline{a}, \underline{c}'') = \underline{c} \circ_i \underline{a}$, $\underline{b}' = b_{[1,j-1]}$, $\underline{b}'' = b_{[j+1,l]}$, *and* $\underline{s} = (\underline{b}', \underline{d}, \underline{b}'') = \underline{b} \circ_j \underline{d}$.

13.4.5. Algebras over a Half-PROP. To characterize the algebras over a half-PROP, we apply Theorem 13.34 based upon Theorem 7.64 and Corollary 11.23.

COROLLARY 13.39. *Let*

- P *be a* $\mathtt{Gr}_{\frac{1}{2}}$*-PROP (i.e., a half-PROP), and*
- $f : \mathsf{P} \longrightarrow \mathsf{E}_X \in \Sigma^{\mathcal{E}}_{S_{1/2}}$ *for some* $X \in \mathcal{M}^{\mathfrak{C}}$.

Then (X, f) *is a* P*-algebra if and only if the squares*

$$\begin{array}{ccc}
\mathsf{P}\binom{\underline{d}}{\underline{c}} \otimes \mathsf{P}\binom{c_i}{\underline{a}} \otimes X_{\underline{r}} & \xrightarrow{\cong} & \mathsf{P}\binom{\underline{d}}{\underline{c}} \otimes X_{\underline{c}'} \otimes \left(\mathsf{P}\binom{c_i}{\underline{a}} \otimes X_{\underline{a}}\right) \otimes X_{\underline{c}''} \\
 & & \downarrow{\scriptstyle (Id,f,Id)} \\
\downarrow{\scriptstyle (\circ_i, Id)} & & \mathsf{P}\binom{\underline{d}}{\underline{c}} \otimes X_{\underline{c}} \\
 & & \downarrow{\scriptstyle f} \\
\mathsf{P}\binom{\underline{d}}{\underline{c} \circ_i \underline{a}} \otimes X_{\underline{r}} & \xrightarrow{f} & X_{\underline{d}}
\end{array}$$

and

$$\begin{array}{ccc}
\mathsf{P}\binom{\underline{d}}{b_j} \otimes \mathsf{P}\binom{\underline{b}}{\underline{a}} \otimes X_{\underline{a}} & \xrightarrow{(Id,f)} & \mathsf{P}\binom{\underline{d}}{b_j} \otimes X_{\underline{b}} \\
 & & \downarrow{\scriptstyle \cong} \\
\downarrow{\scriptstyle ({}_j\circ, Id)} & & X_{\underline{b}'} \otimes \left(\mathsf{P}\binom{\underline{d}}{b_j} \otimes X_{b_j}\right) \otimes X_{\underline{b}''} \\
 & & \downarrow{\scriptstyle (Id,f,Id)} \\
\mathsf{P}\binom{\underline{b} \circ_j \underline{d}}{\underline{a}} \otimes X_{\underline{a}} & \xrightarrow{f} & X_{\underline{s}}
\end{array}$$

are commutative, where the notations are as in Corollary 13.38.

13.4.6. Algebras over a Properad. This time we apply Theorem 13.34 exploiting Theorem 7.67 and Corollary 11.28 to characterize the algebras over a properad.

COROLLARY 13.40. *Let*

- P *be a* $\mathtt{Gr}^{\uparrow}_{c}$*-PROP (i.e., a properad), and*
- $f \colon \mathsf{P} \longrightarrow \mathsf{E}_X \in \mathbf{\Sigma}_S$ *for some* $X \in \mathcal{M}^{\mathfrak{C}}$.

Then (X, f) *is a* P*-algebra if and only if the unit squares* (13.27) *and*

$$\begin{tikzcd}
\mathsf{P}\binom{\underline{d}}{\underline{c}} \otimes \mathsf{P}\binom{\underline{b}}{\underline{a}} \otimes X_{\underline{c}^0 \circ_{\underline{c}'} \underline{a}} \arrow[r, "\cong"] \arrow[ddd, "\boxtimes^{\underline{c}'}_{\underline{b}'} \otimes Id"'] & \mathsf{P}\binom{\underline{d}}{\underline{c}} \otimes X_{\underline{c}^0} \otimes \Big(\mathsf{P}\binom{\underline{b}}{\underline{a}} \otimes X_{\underline{a}}\Big) \otimes X_{\underline{c}^1} \arrow[d, "(Id,f,Id)"] \\
& \mathsf{P}\binom{\underline{d}}{\underline{c}} \otimes X_{\underline{c}^0 \circ_{\underline{c}'} \underline{b}} \arrow[d, "\cong"] \\
& X_{\underline{b}^0} \otimes \Big(\mathsf{P}\binom{\underline{d}}{\underline{c}} \otimes X_{\underline{c}}\Big) \otimes X_{\underline{b}^1} \arrow[d, "(Id,f,Id)"] \\
\mathsf{P}\binom{\underline{b}^0 \circ_{\underline{b}'} \underline{d}}{\underline{c}^0 \circ_{\underline{c}'} \underline{a}} \otimes X_{\underline{c}^0 \circ_{\underline{c}'} \underline{a}} \arrow[r, "f"] & X_{\underline{b}^0 \circ_{\underline{b}'} \underline{d}}
\end{tikzcd}$$

are commutative whenever they are defined, where $\underline{b} = (\underline{b}^0, \underline{b}', \underline{b}^1)$ *and* $\underline{c} = (\underline{c}^0, \underline{c}', \underline{c}^1)$.

13.4.7. Algebras over a PROP. To characterize the algebras over a PROP, in applying Theorem 13.34 we use a slight variation of the generating set $\mathcal{T}^{\uparrow}$ of Theorem 7.27 and Corollary 11.31. Rather than using all of $\mathcal{T}^{\uparrow}$, we use only c-exceptional edges for $c \in \mathfrak{C}$ rather than all $\underline{c}$-exceptional edges.

COROLLARY 13.41. *Let*

- P *be a* $\mathtt{Gr}^{\uparrow}$*-PROP (i.e., a PROP), and*
- $f \colon \mathsf{P} \longrightarrow \mathsf{E}_X \in \mathbf{\Sigma}_S$ *for some* $X \in \mathcal{M}^{\mathfrak{C}}$.

Then (X, f) *is a* P*-algebra if and only if the unit squares* (13.27),

(13.29)
$$\begin{tikzcd}
I \otimes I \arrow[r, equal] \arrow[d, "(\mathbf{1}_{\varnothing}, Id)"'] & I \otimes I \arrow[d, "\cong"] \\
\mathsf{P}\binom{\varnothing}{\varnothing} \otimes I \arrow[r, "f"] & I,
\end{tikzcd}$$

(13.30)
$$\begin{tikzcd}
\mathsf{P}\binom{\underline{d}}{\underline{c}} \otimes \mathsf{P}\binom{\underline{b}}{\underline{a}} \otimes X_{(\underline{c}, \underline{a})} \arrow[r, "\cong"] \arrow[d, "(\otimes_h, Id)"'] & \Big(\mathsf{P}\binom{\underline{d}}{\underline{c}} \otimes X_{\underline{c}}\Big) \otimes \Big(\mathsf{P}\binom{\underline{b}}{\underline{a}} \otimes X_{\underline{a}}\Big) \arrow[d, "(f,f)"] \\
\mathsf{P}\binom{\underline{d}, \underline{b}}{\underline{c}, \underline{a}} \otimes X_{(\underline{c}, \underline{a})} \arrow[r, "f"] & X_{(\underline{d}, \underline{b})},
\end{tikzcd}$$

and

$$\begin{array}{ccc} \mathsf{P}\binom{\underline{d}}{\underline{c}} \otimes \mathsf{P}\binom{\underline{c}}{\underline{b}} \otimes X_{\underline{b}} & \xrightarrow{(Id,f)} & \mathsf{P}\binom{\underline{d}}{\underline{c}} \otimes X_{\underline{c}} \\ \big\downarrow{\scriptstyle (\otimes_v, Id)} & & \big\downarrow{\scriptstyle f} \\ \mathsf{P}\binom{\underline{d}}{\underline{b}} \otimes X_{\underline{b}} & \xrightarrow{f} & X_{\underline{d}} \end{array}$$

are commutative.

13.4.8. Algebras over a Wheeled PROP. In this instance, we apply Theorem 13.34 to characterize the algebras over a wheeled PROP using Corollary 7.21 and Corollary 11.35.

COROLLARY 13.42. *Suppose* $\mathtt{Gr}_w^{\circlearrowleft}$ *admits an endomorphism object (e.g., when* $\mathcal{E}$ *is the category of finite dimensional vector spaces). Let*

- P *be a* $\mathtt{Gr}_w^{\circlearrowleft}$*-PROP (i.e., a wheeled PROP), and*
- $f\colon \mathsf{P} \longrightarrow \mathsf{E}_X \in \mathbf{\Sigma}_S$ *for some* $X \in \mathcal{M}^{\mathfrak{C}}$.

Then (X, f) *is a* P*-algebra if and only if the unit squares* (13.27), *and* (13.29), *as well as the horizontal operation squares* (13.30), *and the contraction squares*

$$\begin{array}{ccc} \mathsf{P}\binom{\underline{d}}{\underline{c}} & \xrightarrow{f} & \mathsf{E}_X\binom{\underline{d}}{\underline{c}} \\ \big\downarrow{\scriptstyle \xi_j^i} & & \big\downarrow{\scriptstyle \xi_j^i} \\ \mathsf{P}\binom{\underline{d}\smallsetminus d_i}{\underline{c}\smallsetminus c_j} & \xrightarrow{f} & \mathsf{E}_X\binom{\underline{d}\smallsetminus d_i}{\underline{c}\smallsetminus c_j} \end{array}$$

are commutative whenever they are defined.

13.4.9. Algebras over a Wheeled Properad. This time we apply Theorem 13.34 exploiting Theorem 7.88 and Corollary 11.39 to characterize the algebras over a wheeled properad.

COROLLARY 13.43. *Suppose* $\mathtt{Gr}_c^{\circlearrowleft}$ *admits an endomorphism object (e.g., when* $\mathcal{E}$ *is the category of finite dimensional vector spaces). Let*

- P *be a* $\mathtt{Gr}_c^{\circlearrowleft}$*-PROP (i.e., a wheeled properad), and*
- $f\colon \mathsf{P} \longrightarrow \mathsf{E}_X \in \mathbf{\Sigma}_S$ *for some* $X \in \mathcal{M}^{\mathfrak{C}}$.

Then (X, f) *is a* P*-algebra if and only if*

(1) (X, f) *is an algebra over the dioperad* $(\mathsf{P}, {}_j\circ_i, \mathbf{1})$, *and*
(2) *the contraction squares in Corollary* 13.42 *are commutative whenever they are defined.*

13.4.10. Algebras over a Wheeled Operad. Theorem 13.34 together with Lemma 7.52 and Corollary 11.43 can also be used to characterize the algebras over a wheeled operad.

COROLLARY 13.44. *Suppose* $\mathtt{Tree}^{\circlearrowleft}$ *admits an endomorphism object (e.g., when* $\mathcal{E}$ *is the category of finite dimensional vector spaces). Let*

- P *be a* $\mathtt{Tree}^{\circlearrowleft}$*-PROP (i.e., a wheeled operad), and*
- $f\colon \mathsf{P} \longrightarrow \mathsf{E}_X \in \Sigma_{S_{\le 1}}$ *for some* $X \in \mathcal{M}^{\mathfrak{C}}$.

Then (X,f) *is a* P*-algebra if and only if the operad squares* (13.27), (13.28), *as well as*

$$\begin{CD}
\mathsf{P}\binom{\varnothing}{\underline{c}} \otimes \Bigl(\bigotimes_{i=1}^m \mathsf{P}\binom{c_i}{\underline{b}_i}\Bigr) \otimes X_{\underline{b}} @>{\cong}>> \mathsf{P}\binom{\varnothing}{\underline{c}} \otimes \Bigl(\bigotimes_{i=1}^m \mathsf{P}\binom{c_i}{\underline{b}_i} \otimes X_{\underline{b}_i}\Bigr)\\
@V{(\rho,Id)}VV @VV{(Id,\otimes f)}V\\
@. \mathsf{P}\binom{\varnothing}{\underline{c}} \otimes X_{\underline{c}}\\
@. @VV{f}V\\
\mathsf{P}\binom{\varnothing}{\underline{b}} \otimes X_{\underline{b}} @>{f}>> I
\end{CD}$$

and

$$\begin{CD}
\mathsf{P}\binom{c_j}{\underline{c}} @>{f}>> \mathsf{E}_X\binom{c_j}{\underline{c}}\\
@V{\xi_j^1}VV @VV{\xi_j^1}V\\
\mathsf{P}\binom{\varnothing}{\underline{c}\smallsetminus c_j} @>{f}>> \mathsf{E}_X\binom{\varnothing}{\underline{c}\smallsetminus c_j}
\end{CD}$$

are commutative.

13.5. Notes

The endomorphism object E_X is an abstraction of the endomorphism operad, originally defined in the topological setting in [**May72**]. The definition of a P-algebra as a morphism $\mathsf{P} \longrightarrow \mathsf{E}_X$ is also abstracted from [**May72**].

In the operad setting, the relative endomorphism object E_f for a morphism $f\colon X \longrightarrow Y$ goes back at least to [**Rez96**]. The special cases of Theorems 13.15 and 13.16 and Corollary 13.23 for $\mathcal{G} = \mathtt{Gr}^{\uparrow}$ are in [**JY09**], along with the special case of Theorem 13.24 for $\mathcal{G} = \mathtt{UTree}$ and $\mathcal{G}' = \mathtt{Gr}^{\uparrow}$.

The direction-reversing process discussed in Example 13.14 can be performed for every pasting scheme $\mathcal{G}$. Then one has

$$(\mathcal{G}^{op})^{op} = \mathcal{G},$$

and the identification (13.9) gives a canonical isomorphism

$$\mathtt{PROP}^{\mathcal{G}} \cong \mathtt{PROP}^{\mathcal{G}^{op}}$$

of categories.

CHAPTER 14

Alternative Descriptions of Generalized PROPs

Given a pasting scheme $\mathcal{G}$ (Definition 8.2), a $\mathcal{G}$-PROP is defined as an algebra over the monad $F_{\mathcal{G}}$ (Definition 10.39). The purpose of this chapter is to give two alternative characterizations of $\mathcal{G}$-PROPs. In Theorem 14.1 it is shown that there is an S-colored *operad* $\overline{\mathsf{U}}_{\mathcal{G}}$ whose algebras are $\mathcal{G}$-PROPs. Notice that regardless of what the pasting scheme $\mathcal{G}$ is, the object $\overline{\mathsf{U}}_{\mathcal{G}}$ is still a colored operad, instead of a bigger object such as a PROP or a wheeled PROP. It is also important to notice that $\mathcal{G}$-PROPs are $\mathfrak{C}$-colored, while $\overline{\mathsf{U}}_{\mathcal{G}}$ is instead S-colored, which is a significant increase in the size of the set of colors.

In Theorem 14.12 it is shown that $\mathcal{G}$-PROPs are precisely the enriched multicategorical functors from a small enriched multicategory into the base category. This is the appropriate generalization of the fact that $\mathcal{G}$-PROPs form a diagram category when $\mathcal{G}$ is monogenic (Proposition 10.43). If we think of $\mathcal{E}$ as a variable in $\mathtt{PROP}^{\mathcal{G},\mathcal{E}}$, then this result can be interpreted as saying that the functor $\mathtt{PROP}^{\mathcal{G},?}$ is corepresentable (Corollary 14.13).

For this chapter, choose any pasting scheme $\mathcal{G} = (S, \mathtt{G})$, while $\mathcal{E}$ denotes a symmetric monoidal closed category with all small colimits and pullbacks (Convention 13.1) still holds, with $\mathtt{PROP}^{\mathcal{G}}$ again short for $\mathtt{PROP}^{\mathcal{G},\mathcal{E}}$.

14.1. Generalized PROPs as Operadic Algebras

The purpose of this section is to show that $\mathtt{PROP}^{\mathcal{G}}$ is isomorphic to the category of algebras (Definition 13.17) over an S-colored operad (Definition 11.14). In other words, there exists an S-colored operad of $\mathcal{G}$-PROPs. Recall from Corollary 11.16 that the categories of $\mathfrak{C}$-colored operads and of $\mathtt{UTree}$-PROPs are canonically isomorphic, while $\mathbf{Alg}(\overline{\mathsf{U}}_{\mathcal{G}})$ is the category of $\overline{\mathsf{U}}_{\mathcal{G}}$-algebras in $\mathcal{E}$.

THEOREM 14.1. *Suppose $\mathcal{G}$ is a pasting scheme. Then there exist*

- *an S-colored operad $\overline{\mathsf{U}}_{\mathcal{G}}$ in $\mathcal{E}$, and*
- *a canonical isomorphism of categories*

$$\mathtt{PROP}^{\mathcal{G}} \cong \mathbf{Alg}(\overline{\mathsf{U}}_{\mathcal{G}}) \ . \tag{14.1}$$

PROOF. Let **Set** be the symmetric monoidal closed category of sets under cartesian product, where the singleton is the unit. In Lemma 14.2 below, we will construct an S-colored operad $\mathsf{U}_{\mathcal{G}}$ in **Set** that is made up of ordered wheeled graphs in $\mathcal{G}$. Applying the strong symmetric monoidal functor

$$\mathbf{Set} \longrightarrow \mathcal{E}, \quad X \longmapsto \coprod_{x \in X} I, \tag{14.2}$$

in (12.4) to $\mathsf{U}_{\mathcal{G}}$ yields an S-colored operad $\overline{\mathsf{U}}_{\mathcal{G}}$ in $\mathcal{E}$ by Theorem 12.11 for the pasting scheme $\mathtt{UTree}$ and $\mathcal{D} = \mathbf{Set}$. Then Lemma 14.4 will show that the S-colored operad $\overline{\mathsf{U}}_{\mathcal{G}}$ in $\mathcal{E}$ has the desired property (14.1). □

14.1.1. The Operad Associated to a Pasting Scheme. Recall from section 1.4.2 the notion of an ordered wheeled graph, which is a wheeled graph together with an ordering of its vertices. A strict isomorphism between two ordered wheeled graphs is a strict isomorphism of the wheeled graphs that preserves the ordering.

As described in the proof of Theorem 14.1, we begin by introducing an S-colored operad $\mathsf{U}_{\mathcal{G}}$ in **Set**.

Underlying sets: For $n \geq 0$, $t = \binom{d}{\underline{c}}$, $s_i = \binom{d_i}{\underline{c}_i}$ $(1 \leq i \leq n)$ in S, and $\underline{s} = s_{[1,n]} \in \mathcal{P}(S)$, the set

$$\mathsf{U}_{\mathcal{G}}\binom{t}{\underline{s}} \tag{14.3}$$

will consist of strict isomorphism classes of ordered wheeled graphs $[G, \sigma]$ with $G \in \mathsf{G}_S\binom{d}{\underline{c}}$ and

$$\binom{\text{out}(v)}{\text{in}(v)} = s_{\sigma(v)}$$

for each $v \in \text{Vt}(G)$. In other words, $\mathsf{U}_{\mathcal{G}}\binom{t}{\underline{s}}$ is the set of strict isomorphism classes of ordered wheeled graphs in G_S with full graph profiles t, in which $\underline{s}$ represents the profiles of the vertices, so the profiles of v are $s_{\sigma(v)}$.

Equivariance: Given $\rho \in \Sigma_n$, to define the effect of acting from the right by ρ on $\mathsf{U}_{\mathcal{G}}\binom{t}{\underline{s}}$, simply send $[G, \sigma]$ to $[G, \rho^{-1} \circ \sigma]$. The result is then an element of $\mathsf{U}_{\mathcal{G}}\binom{t}{\underline{s}\rho}$, since $\underline{s}\rho_{\rho^{-1}\circ\sigma(v)} = \underline{s}_{\sigma(v)}$ still yields the profiles of v. In other words, the right Σ-action simply relabels the vertices in the ordered wheeled graph.

Units: For $t = \binom{d}{\underline{c}} \in S$, the t-colored unit

$$\mathbf{1}_t \colon \{*\} \longrightarrow \mathsf{U}_{\mathcal{G}}\binom{t}{t}$$

is given by

$$\mathbf{1}_t(*) = \left[C_{(\underline{c};\underline{d})}\right], \tag{14.4}$$

where $C_{(\underline{c};\underline{d})}$ is the $(\underline{c};\underline{d})$-corolla and there is a unique ordering of the single vertex.

Operad Structure Map: Suppose t, s_i, and $\underline{s}$ are as above, and $\underline{r}_i \in \mathcal{P}(S)$ with concatenation $\underline{r} = (\underline{r}_1, \ldots, \underline{r}_n)$. The operad structure map

$$\gamma \colon \mathsf{U}_{\mathcal{G}}\binom{t}{\underline{s}} \times \prod_{i=1}^{n} \mathsf{U}_{\mathcal{G}}\binom{s_i}{\underline{r}_i} \longrightarrow \mathsf{U}_{\mathcal{G}}\binom{t}{\underline{r}}$$

is defined as the ordered graph substitution

$$\gamma\big([G,\sigma]; \{[H_i,\sigma_i]\}_{i=1}^{n}\big) = [G(H_i), \nu], \tag{14.5}$$

where ν is induced by σ and the σ_i using the lexicographical ordering.

LEMMA 14.2. *The data (14.3)–(14.5) defines an S-colored operad $\mathsf{U}_{\mathcal{G}}$ in* **Set**.

PROOF. That $\mathsf{U}_{\mathcal{G}}$ is a Σ_{S_1}-(bi)module in **Set** follows immediately from the definition of the right Σ-action (14.1.1). The associativity and unity of the operad structure map γ (14.5) follow from those of ordered graph substitution. Both equivariance diagrams (11.5) and (11.6) for $\mathsf{U}_{\mathcal{G}}$ commute because ν is naturally induced by σ and the σ_i using the lexicographical ordering. □

DEFINITION 14.3. Let $\overline{\mathsf{U}}_{\mathcal{G}}$ be the S-colored operad in $\mathcal{E}$ obtained by applying the strong symmetric monoidal discrete enrichment functor (14.2) to the S-colored operad $\mathsf{U}_{\mathcal{G}}$. We call $\overline{\mathsf{U}}_{\mathcal{G}}$ the **operad associated to** the pasting scheme $\mathcal{G}$.

14.1.2. The Colored Operad of $\mathcal{G}$-PROPs. We now show that the operad associated to the pasting scheme $\mathcal{G}$ has the $\mathcal{G}$-PROPs as its algebras. It will be important to note that the S-colored operad $\overline{\mathsf{U}}_{\mathcal{G}}$ in $\mathcal{E}$ has components

$$\overline{\mathsf{U}}_{\mathcal{G}}\binom{t}{\underline{s}} = \coprod_{[G,\sigma]\in \mathsf{U}_{\mathcal{G}}\binom{t}{\underline{s}}} I \tag{14.6}$$

for $t \in S$ and $\underline{s} \in \mathcal{P}(S)$, where I is the monoidal unit in $\mathcal{E}$.

LEMMA 14.4. *There is a canonical isomorphism of categories*

$$\mathsf{PROP}^{\mathcal{G}} \cong \mathbf{Alg}(\overline{\mathsf{U}}_{\mathcal{G}}) .$$

PROOF. First, we observe that both $\mathcal{G}$-PROPs with colors $\mathfrak{C}$ and algebras over an S-colored operad begin with S-colored objects in $\mathcal{E}$. Now we would like to compare their basic structure maps.

Suppose $X = \{X_s\} \in \mathcal{E}^S$ is a $\overline{\mathsf{U}}_{\mathcal{G}}$-algebra in $\mathcal{E}$ with adjoint structure maps

$$\lambda \colon \overline{\mathsf{U}}_{\mathcal{G}}\binom{t}{\underline{s}} \otimes X_{\underline{s}} \longrightarrow X_t.$$

The domain of this structure map is

$$\begin{aligned}\overline{\mathsf{U}}_{\mathcal{G}}\binom{t}{\underline{s}} \otimes X_{\underline{s}} &= \left[\coprod_{[G,\sigma]\in \mathsf{U}_{\mathcal{G}}\binom{t}{\underline{s}}} I\right] \otimes X_{\underline{s}}\\ &\cong \coprod_{[G,\sigma]\in \mathsf{U}_{\mathcal{G}}\binom{t}{\underline{s}}} X_{\underline{s}},\end{aligned}$$

so λ is equivalent to a collection of maps

$$\lambda_{[G,\sigma]} : X_{\underline{s}} \longrightarrow X_t.$$

Furthermore, the adjoint of λ is required to be a morphism of S_1-(bi)modules, where the action of a permutation ρ from the right on the endomorphism object at $(\underline{s};t)$ leaves t invariant and acts on $\underline{s}$ from the right. Since the permutation ρ acts on $\overline{\mathsf{U}}_{\mathcal{G}}$ from the right simply by acting on the ordering of each ordered graph, the shuffle of tensor factors induced by ρ must then fit into a commutative diagram

$$\begin{array}{ccc} X_{\underline{s}} & & \\ & \searrow^{\lambda_{[G,\sigma]}} & \\ \Big\downarrow{\scriptstyle \mathrm{shuffle}(\rho)} & & X_t \\ & \nearrow_{\lambda_{[G,\rho^{-1}\sigma]}} & \\ X_{\underline{s}\rho}. & & \end{array}$$

Applying the universal property of the unordered tensor product, this is equivalent to saying that for the various orderings on a fixed G, the maps $\lambda_{[G,\sigma]}$ all come from a single map

$$\lambda_{[G]} : \odot X_{s_i} \longrightarrow X_t.$$

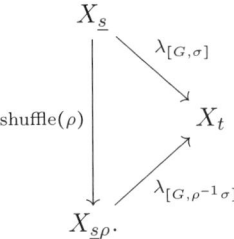

We can now see that a $\mathcal{G}$-PROP, consisting of an S-colored P and structure maps $\gamma_{[G]}$ from an unordered tensor product of the values at the pairs of profiles of the vertices, is very similar to a $\overline{\mathsf{U}}_{\mathcal{G}}$-algebra.

To see the two structures can be identified, we must show that the commutativity of all diagrams (13.28) (for X) is equivalent to the commutativity of all diagrams (10.21) (for P), provided they have the same underlying colored objects and their structure maps are compatible as described in the previous paragraph. First, notice the image of the left map in (13.28) is the sub-coproduct consisting of factors of the form $[G(H_v), \nu]$ with $G \in \overline{\mathsf{U}}_{\mathcal{G}}\binom{t}{\underline{s}}$ and $H_v \in \overline{\mathsf{U}}_{\mathcal{G}}\binom{s_{\sigma(v)}}{\underline{r}_{\sigma(v)}}$. Similarly, the composite of the top and upper right maps in the same diagram is really a tensor product of various λ maps. In particular, this composite is isomorphic to a coproduct indexed again over tuples of $H_v \in \overline{\mathsf{U}}_{\mathcal{G}}\binom{s_{\sigma(v)}}{\underline{r}_{\sigma(v)}}$ of maps $\lambda_{[H_v]}$. Finally, the lower right vertical in this diagram is a coproduct of maps $\lambda_{[G]}$. Taken together, and keeping in mind the last paragraph, we see the commutativity of this one diagram is equivalent to the commutativity of each diagram of the form (10.21) where $G \in \overline{\mathsf{U}}_{\mathcal{G}}\binom{t}{\underline{s}}$.

Similarly, we must show that the unit diagrams for an algebra and for a $\mathcal{G}$-PROP correspond, but this follows from the fact that the image of the unit map in $\mathsf{U}_{\mathcal{G}}$ is the corolla on the given pair of profiles. Now to see the categories are isomorphic, we must verify that the notions of morphism correspond. In other words, we notice diagram (13.22) consists, via coproducts, of a collection of diagrams of the form (10.23) (other than switching the orientations of the squares as depicted). □

14.2. Generalized PROPs as Multicategorical Functors

The purpose of this section is to observe that $\mathcal{G}$-PROPs in $\mathcal{E}$ can be equivalently described as functors from a certain enriched multicategory into $\mathcal{E}$.

14.2.1. Defining an Enriched Multicategory. Let us first recall some definitions regarding enriched multicategories. We use the abbreviations

$$x_{[i,j]} = (x_i, x_{i+1}, \ldots, x_j) \quad \text{and} \quad f(x_{[i,j]}) = (f(x_i), f(x_{i+1}), \ldots, f(x_j)).$$

DEFINITION 14.5. An **$\mathcal{E}$-multicategory** $\mathcal{C}$ consists of:

(1) a class $\mathrm{Ob}(\mathcal{C})$ of **objects**;
(2) an $\mathcal{E}$-object

$$\mathcal{C}((x_1, \ldots, x_m), y) = \mathcal{C}(x_{[1,m]}, y) \in \mathcal{E}$$

of **multi-morphisms** for each $m \geq 0$ and $x_1, \ldots, x_m, y \in \mathrm{Ob}(\mathcal{C})$;
(3) an **identity**

$$\mathbf{1}_x : I \longrightarrow \mathcal{C}(x, x) \in \mathcal{E}$$

for each $x \in \mathrm{Ob}(\mathcal{C})$;
(4) a right Σ-action

$$\mathcal{C}((x_1, \ldots, x_m), y) \xrightarrow{\tau} \mathcal{C}\big((x_{\tau^{-1}(1)}, \ldots, x_{\tau^{-1}(m)}), y\big) \in \mathcal{E}$$

for $m \geq 0$, $x_1, \ldots, x_m, y \in \mathrm{Ob}(\mathcal{C})$, and $\tau \in \Sigma_m$;

(5) a multi-morphism **composition**

$$\begin{array}{c} \mathcal{C}(x_{[1,m]}, y) \otimes \bigotimes_{i=1}^m \mathcal{C}\left(w_{[N_{i-1}+1,N_i]}, x_i\right) \\ \big\downarrow \gamma \\ \mathcal{C}\left(w_{[1,N_m]}, y\right) \end{array}$$

in $\mathcal{E}$, where $N_0 = 0$, $N_i = n_1 + \cdots + n_i$, and $w_j, x_i, y \in \mathrm{Ob}(\mathcal{C})$;

such that the multi-morphism composition γ satisfies exactly the same associative, equivariant, and unity axioms as in a colored operad (Definition 11.14 with $\mathrm{Ob}(\mathcal{C})$ as $\mathfrak{C}$). An $\mathcal{E}$-multicategory $\mathcal{C}$ is **small** if $\mathrm{Ob}(\mathcal{C})$ is a set.

EXAMPLE 14.6. A $\mathfrak{C}$-colored operad O in $\mathcal{E}$ (Definition 11.14) yields a small $\mathcal{E}$-multicategory $\mathcal{D}$ with

$$\mathrm{Ob}(\mathcal{D}) = \mathfrak{C}$$

and multi-morphisms

$$\mathcal{D}\left((c_1, \ldots, c_m), d\right) = \mathsf{O}\binom{d}{c_1, \ldots, c_m} \in \mathcal{E}$$

for $c_1, \ldots, c_m, d \in \mathfrak{C}$. The identity $\mathbf{1}_x$ is the x-colored unit of $\mathcal{D}$ and the right Σ-action on $\mathcal{D}((c_1, \ldots, c_m), d)$ is the one on $\mathsf{O}\binom{d}{\underline{c}}$, while the multi-morphism composition is the operad structure map on O. *Our convention is that a $\mathfrak{C}$-colored operad in $\mathcal{E}$ is also considered as a small $\mathcal{E}$-multicategory in this way.*

In what follows, to save space we will sometimes write $[-,-]$ for the internal hom $\mathrm{Hom}_{\mathcal{E}}(-,-)$.

EXAMPLE 14.7. The symmetric monoidal closed category $\mathcal{E}$ can be regarded as an $\mathcal{E}$-multicategory with the same objects as $\mathcal{E}$ and multi-morphisms

$$\mathcal{E}((x_1, \ldots, x_m), y) = \mathrm{Hom}_{\mathcal{E}}(x_1 \otimes \cdots \otimes x_m, y) \in \mathcal{E}.$$

The identity

$$\mathbf{1}_x : I \longrightarrow \mathrm{Hom}_{\mathcal{E}}(x, x) \in \mathcal{E}$$

is the adjoint of the identity map $Id_x \in \mathcal{E}(x, x)$. The right Σ-action is given by permutation of the tensor factors in the source. The multi-morphism composition γ is defined as the composition

$$\begin{array}{ccc} \left[\bigotimes_{i=1}^m x_i, y\right] \otimes \bigotimes_{i=1}^m \left[\bigotimes_{j=N_{i-1}+1}^{N_i} w_j, x_i\right] & \xrightarrow{\gamma} & \left[\bigotimes_{j=1}^{N_m} w_j, y\right] \\ \big\downarrow & \nearrow & \\ \left[\bigotimes_{i=1}^m x_i, y\right] \otimes \left[\bigotimes_{j=1}^{N_m} w_j, \bigotimes_{i=1}^m x_i\right] & & \end{array} \tag{14.7}$$

of natural maps. *Our convention is that $\mathcal{E}$ is considered as an $\mathcal{E}$-multicategory as above.*

14.2.2. Functors of Enriched Multicategories.

DEFINITION 14.8. Let $\mathcal{C}$ and $\mathcal{D}$ be $\mathcal{E}$-multicategories. An $\mathcal{E}$-multicategory **functor** $F{:}\,\mathcal{C} \longrightarrow \mathcal{D}$ consists of:

(1) a function

$$F{:}\,\mathrm{Ob}(\mathcal{C}) \longrightarrow \mathrm{Ob}(\mathcal{D})$$

on objects and

(2) an $\mathcal{E}$-morphism

$$F_{x_{[1,m]},y}{:}\,\mathcal{C}(x_{[1,m]}, y) \longrightarrow \mathcal{D}(F(x_{[1,m]}), F(y)) \in \mathcal{E}$$

on multi-morphisms for each $m \geq 0$ and $x_1, \ldots, x_m, y \in \mathrm{Ob}(\mathcal{C})$,

such that the $\mathcal{E}$-morphisms $F_{x_{[1,m]},y}$ are compatible with the identity, the right Σ-action, and the multi-morphism composition as follows, where the notations are as in Definition 14.5.

Compatibility with the identity: The diagram

$$(14.8)\qquad \begin{array}{ccc} I & \xrightarrow{\mathbf{1}_x} & \mathcal{C}(x,x) \\ & {\scriptstyle \mathbf{1}_{F(x)}}\searrow & \downarrow{\scriptstyle F_{x,x}} \\ & & \mathcal{D}(F(x), F(x)) \end{array}$$

is commutative.

Compatibility with the Σ-action: The square

$$(14.9)\qquad \begin{array}{ccc} \mathcal{C}(x_{[1,m]}, y) & \xrightarrow{\tau} & \mathcal{C}\big((x_{\tau^{-1}(1)}, \ldots, x_{\tau^{-1}(m)}), y\big) \\ {\scriptstyle F}\downarrow & & \downarrow{\scriptstyle F} \\ \mathcal{D}(F(x_{[1,m]}), F(y)) & \xrightarrow{\tau} & \mathcal{D}\big((F(x_{\tau^{-1}(1)}), \ldots, F(x_{\tau^{-1}(m)})), F(y)\big) \end{array}$$

is commutative.

Compatibility with the composition: The square

$$(14.10)\qquad \begin{array}{ccc} \mathcal{C}(x_{[1,m]}, y) \otimes \bigotimes_{i=1}^{m} \mathcal{C}\big(w_{[N_{i-1}+1,N_i]}, x_i\big) & \xrightarrow{\gamma} & \mathcal{C}\big(w_{[1,N_m]}, y\big) \\ {\scriptstyle F\otimes\bigotimes F}\downarrow & & \downarrow{\scriptstyle F} \\ \mathcal{D}(F(x_{[1,m]}), F(y)) \otimes \bigotimes_{i=1}^{m} \mathcal{D}\big(F(w_{[N_{i-1}+1,N_i]}), F(x_i)\big) & \xrightarrow{\gamma} & \mathcal{D}\big(F(w_{[1,N_m]}), F(y)\big) \end{array}$$

is commutative.

DEFINITION 14.9. Let $\mathcal{C}$ be an $\mathcal{E}$-multicategory and let $F, G{:}\,\mathcal{C} \longrightarrow \mathcal{E}$ be $\mathcal{E}$-multicategory functors. A **natural transformation** $\theta{:}\,F \longrightarrow G$ consists of an $\mathcal{E}$-morphism

$$(14.11)\qquad \theta_x{:}\,I \longrightarrow \mathrm{Hom}_{\mathcal{E}}(F(x), G(x)) \in \mathcal{E}$$

for each $x \in \mathrm{Ob}(\mathcal{C})$ such that the following diagram is commutative, where the notations are as in Definitions 14.5 and 14.8.

(14.12)

$$\begin{array}{ccc}
\mathcal{C}(x_{[1,m]}, y) & \xrightarrow{\cong} & I \otimes \mathcal{C}(x_{[1,m]}, y) \\
{\scriptstyle\cong}\downarrow & & \downarrow{\scriptstyle \theta_y \otimes F} \\
\mathcal{C}(x_{[1,m]}, y) \otimes I^{\otimes m} & & [F(y), G(y)] \otimes [\otimes_i F(x_i), F(y)] \\
{\scriptstyle G \otimes \theta_{\underline{x}}}\downarrow & & \downarrow{\scriptstyle \text{natural}} \\
[\otimes_i G(x_i), G(y)] \otimes [\otimes_i F(x_i), \otimes_i G(x_i)] & \xrightarrow{\text{natural}} & [\otimes_i F(x_i), G(y)]
\end{array}$$

In the diagram above, we abbreviated $\otimes_{i=1}^m$ to $\otimes_i$ and the map $\theta_{\underline{x}}$ is the composition

$$\begin{array}{ccc}
I^{\otimes m} & \xrightarrow{\theta_{\underline{x}}} & [\otimes_i F(x_i), \otimes_i G(x_i)] \\
{\scriptstyle \otimes_i \theta_{x_i}}\downarrow & \nearrow{\scriptstyle \text{natural}} & \\
\otimes_{i=1}^m [F(x_i), G(x_i)] . & &
\end{array}$$

EXAMPLE 14.10. When $\mathcal{E}$ is a symmetric monoidal *closed* category, e.g. the category **Ch**, **SSet**, **Top**, **SMod**, **Mod**, or **Set**, by adjunction the commutativity of the diagram (14.12) is equivalent to that of the square

$$\begin{array}{ccc}
\bigotimes_i F(x_i) & \xrightarrow{\otimes_i \theta_{x_i}} & \bigotimes_i G(x_i) \\
{\scriptstyle F(f)}\downarrow & & \downarrow{\scriptstyle G(f)} \\
F(y) & \xrightarrow{\theta_y} & G(y)
\end{array}$$

for all multi-morphisms $f \in \mathcal{C}(x_{[1,m]}, y)$. The $m = 1$ case of this commutative square is exactly the condition that defines a natural transformation over ordinary categories.

LEMMA 14.11. *Let $\mathcal{C}$ be a small $\mathcal{E}$-multicategory. Then there exists a category* $\mathtt{Fun}_{\mathcal{E}}(\mathcal{C}, \mathcal{E})$ *with*

- *objects the $\mathcal{E}$-multicategory functors $\mathcal{C} \longrightarrow \mathcal{E}$ and*
- *morphisms the natural transformations between such functors.*

PROOF. The category axioms are easy to check because θ_x in (14.11) is adjoint to a morphism

$$\theta_x^{ad}: F(x) \longrightarrow G(x) \in \mathcal{E}$$

by the hom-tensor adjunction. The smallness of $\mathcal{C}$ is used to make sure that, given functors $F, G: \mathcal{C} \longrightarrow \mathcal{E}$, there is a set of natural transformations $\theta: F \longrightarrow G$. $\square$

14.2.3. Corepresenting $\mathcal{G}$-PROPs. Here is the main result of this section, which implies that the functor

$$\mathcal{E} \longmapsto \mathtt{PROP}^{\mathcal{G}, \mathcal{E}}$$

is corepresentable, although strictly speaking that implication is a corollary.

THEOREM 14.12. *Suppose $\mathcal{G}$ is a pasting scheme. Then there exists a canonical isomorphism of categories,*

$$\mathtt{PROP}^{\mathcal{G},\mathcal{E}} \cong \mathtt{Fun}_{\mathcal{E}}(\overline{\mathsf{U}}_{\mathcal{G}}, \mathcal{E})$$

where

- $\overline{\mathsf{U}}_{\mathcal{G}}$ *is the operad associated to* $\mathcal{G}$*, regarded as a small* $\mathcal{E}$*-multicategory, and*
- $\mathtt{Fun}(\overline{\mathsf{U}}_{\mathcal{G}}, \mathcal{E})$ *is the category in Lemma* 14.11.

PROOF. We begin by observing that the object function underlying such an $\mathcal{E}$-multicategorical functor F is equivalent to an S-colored object P in $\mathcal{E}$.

Next we would like to understand how the $\mathcal{E}$-morphisms of F on multi-morphism spaces and their compatibility with the right action should correspond to the existence of $\mathcal{G}$-PROP structure maps for P. First, we notice the pair $(x_{[1,m]}, y)$ in Definition 14.8 corresponds to the pair $(\underline{s}; t)$ for P, so

$$F_{x_{[1,m]},y} : \mathcal{C}(x_{[1,m]}, y) \longrightarrow \mathcal{D}(F(x_{[1,m]}), F(y)) \in \mathcal{E}$$

corresponds to an $\mathcal{E}$-morphism

$$\begin{array}{c} \overline{\mathsf{U}}_{\mathcal{G}}\binom{t}{\underline{s}} = \coprod_{[G,\sigma]\in \mathsf{U}_{\mathcal{G}}\binom{t}{\underline{s}}} I \\ \Big\downarrow{\scriptstyle P_{\underline{s},t}} \\ \mathrm{Hom}_{\mathcal{E}}\left(\bigotimes_{i=1}^{n} \mathsf{P}\binom{\underline{d}_i}{\underline{c}_i}, \mathsf{P}\binom{\underline{d}}{\underline{c}}\right) . \end{array}$$

The restriction to the summand indexed by $[G,\sigma]$ is then adjoint to a map

$$\overline{\gamma}_{[G,\sigma]} : \bigotimes_{i=1}^{n} \mathsf{P}\binom{\underline{d}_i}{\underline{c}_i} \longrightarrow \mathsf{P}\binom{\underline{d}}{\underline{c}}$$

and as in the proof of Lemma 14.4, compatibility of F with the right action will translate as the statement that all such maps for a fixed G are really coming from a single structure map out of the unordered tensor product and ignoring the ordering σ of the vertices, so $\gamma_{[G]} : \mathsf{P}[G] \longrightarrow \mathsf{P}\binom{\underline{d}}{\underline{c}}$.

Now we notice that in the diagram for compatibility of F with the identity, the image of the composite is really just the corolla on some pair of profiles $s \in S$, while the diagonal map in that diagram picks out the identity morphism of $F(s)$. As a consequence, this is equivalent to the unit condition for P, namely that $\gamma_{[C_s]}$ is the identity map on $\mathsf{P}(s)$. It remains to show that the commutativity of all diagrams of the form (14.10) (for F) correspond to the associativity condition for P given by commutativity of all diagrams of the form (10.21).

From the definition of $\overline{\mathsf{U}}_{\mathcal{G}}$, the upper left corner of (14.10) in this case is isomorphic to a coproduct of copies of the unit I indexed on pairs of the form $\big([G,\sigma], \{[H_v,\sigma_i]\}_{v\in \mathrm{Vt}(G)}\big)$ (where $i = \sigma(v)$) which constitute strict substitution data. Then the left vertical map restricted to this summand will be adjoint to the pair consisting of $\big(\overline{\gamma}_{[G,\sigma]}, \{\overline{\gamma}_{[H_v,\sigma_i]}\}\big)$ and the bottom horizontal will send this to (the adjoint of) $\overline{\gamma}_{[G,\sigma]} \circ \otimes_v \overline{\gamma}_{[H_v,\sigma_i]}$. Once again, compatibility of F with the right action then implies the lower path around (14.10) corresponds to all of the possible composites, with fixed vertex profiles $\underline{s}$ and full graph profiles t of G, of the form

$$\gamma_{[G]} \circ \otimes_v \gamma_{[H_v]} ; \otimes_v \mathsf{P}[H_v] \longrightarrow \mathsf{P}[G] \longrightarrow \mathsf{P}(t).$$

On the other hand, the upper horizontal will send this pair to the single ordered graph $[G(H_v), \nu]$ where ν is built from σ and the σ_i by the lexicographical ordering. Then the right vertical map restricted to this summand will be adjoint to $\overline{\gamma}_{[G(H_v),\nu]}$ and once again the compatibility of F with the right action will imply this corresponds to a single map

$$\gamma_{[G(H_v)]} : \mathsf{P}[G(H_v)] \longrightarrow \mathsf{P}(t).$$

As a consequence, the commutativity of (14.10) with a choice of $\underline{s}$ and t is equivalent to the commutativity of (10.21) for all choices of substitution data with the vertex profiles of $[G, \sigma]$ just $\underline{s}$ and full graph profiles t.

Now given two such multicategorical functors F and G, corresponding to the $\mathcal{G}$-PROPs P and Q as just described, we would like to understand how natural transformations $\theta : F \longrightarrow G$ correspond to morphisms of $\mathcal{G}$-PROPs $f : \mathsf{P} \longrightarrow \mathsf{Q}$. It is straightforward to see that for a given $t \in S$, the $\mathcal{E}$-morphism

$$\theta_t : I \longrightarrow \mathrm{Hom}_{\mathcal{E}}\left(\mathsf{P}\binom{\underline{d}}{\underline{c}}, \mathsf{Q}\binom{\underline{d}}{\underline{c}}\right)$$

should be the adjoint of the component map

$$f(t) : \mathsf{P}(t) \longrightarrow \mathsf{Q}(t).$$

Now we would like to see how commutativity of all diagrams of the form (14.12) and of all diagrams of the form (10.23) are equivalent. In this case, (14.12) is the diagram

$$\begin{array}{ccc} \overline{\mathsf{U}}_{\mathcal{G}}\binom{t}{\underline{s}} = \coprod_{[G,\sigma]\in \mathsf{U}_{\mathcal{G}}\binom{t}{\underline{s}}} I & \longrightarrow & [F(t), G(t)] \otimes [\bigotimes F(s_i), F(t)] \\ \downarrow & & \downarrow \\ [\bigotimes G(s_i), G(t)] \otimes [\bigotimes F(s_i), \bigotimes G(s_i)] & \longrightarrow & [\bigotimes F(s_i), G(t)], \end{array}$$

which is adjoint to the square

$$\begin{array}{ccc} \bigotimes_{i=1}^n F(s_i) & \xrightarrow{\otimes_i f(s_i)} & \bigotimes_{i=1}^n G(s_i) \\ {\scriptstyle \overline{\gamma}_{[G]}}\downarrow & & \downarrow{\scriptstyle \overline{\gamma}_{[G]}} \\ F(t) & \xrightarrow{f(t)} & G(t)\,. \end{array} \tag{14.13}$$

Once again, compatibility of $\theta : F \longrightarrow G$ with the right actions implies the vertical maps naturally factor through the unordered tensor products. Thus, we have a diagram

$$\begin{array}{ccc} \bigotimes_{i=1}^n F(s_i) & \xrightarrow{\otimes_i f(s_i)} & \bigotimes_{i=1}^n G(s_i) \\ \downarrow & & \downarrow \\ \mathsf{P}[G] & \xrightarrow{f(G)} & \mathsf{Q}[G] \\ \downarrow{\scriptstyle \gamma_{[G]}} & & {\scriptstyle \gamma_{[G]}}\downarrow \\ \mathsf{P}(t) & \xrightarrow{f(t)} & \mathsf{Q}(t) \end{array} \qquad \text{(outer curved arrows } \overline{\gamma}_{[G]}\text{)} \tag{14.14}$$

where the top square commutes by these natural factorizations and the top vertical maps are epimorphisms. Hence, the distorted outer square, equivalent to the previous diagram, commutes precisely when the lower square commutes. As the distorted outer square corresponds to (14.12) and the bottom square is (10.23), while the definitions of composition in the two categories are identical, this completes the proof. □

Following standard usage, when $\mathcal{E} = Set$, we simply remove the $\mathcal{E}$-subscripts, etc. The usefulness of the following extension of the theorem is that there is a single object $\mathsf{U}_{\mathcal{G}}$, rather than a family of objects, all called $\overline{\mathsf{U}}_{\mathcal{G}}$ that depend on the choice of $\mathcal{E}$ at this point.

COROLLARY 14.13. *For any pasting scheme $\mathcal{G}$, there is a corepresenting object $\mathsf{U}_{\mathcal{G}}$ in Set-multicategories for the functor $\mathcal{E} \longmapsto \mathtt{PROP}^{\mathcal{G},\mathcal{E}}$, i.e. a canonical isomorphism of categories*

$$\mathtt{PROP}^{\mathcal{G},\mathcal{E}} \cong \mathtt{Fun}(\mathsf{U}_{\mathcal{G}}, \mathcal{E})$$

viewing $\mathcal{E}$ as an ordinary multicategory with morphism sets of the form

$$\mathrm{Hom}_{\mathcal{E}}(I, \mathcal{E}(x_{[1,m]}, y)) \ .$$

PROOF. In addition to Theorem 14.12, the point is that we have a natural isomorphism of categories

$$\mathtt{Fun}_{\mathcal{E}}(\overline{\mathsf{U}}_{\mathcal{G}}, \mathcal{E}) \longrightarrow \mathtt{Fun}(\mathsf{U}_{\mathcal{G}}, \mathcal{E})$$

by the adjunction $\mathsf{U}_{\mathcal{G}} \longmapsto \overline{\mathsf{U}}_{\mathcal{G}}$, which is left adjoint to the underlying multicategory functor as described. □

14.3. Notes

The special case of Theorem 14.1 for the case of (unital) operads, $\mathcal{G} = \mathtt{UTree}$, is in [**BM07**] (1.5.6 and 1.5.7).

In [**BM08**], Borisov and Manin defined their generalized operads as monoidal functors from a graph category to the base category satisfying some additional conditions. This is similar to Theorem 14.12, where generalized PROPs are characterized as certain enriched multicategorical functors, or even more to Corollary 14.13 using a graph multicategory.

CHAPTER 15

Modules over Generalized PROPs

Here we define modules over any generalized PROP. One main reason for introducing modules over generalized PROPs is that they include, as special cases, Markl's modules over operads and PROPs that are used in deformation theory. The essential mechanism is a pointed variant of the monad $F_{\mathcal{G}}$ associated to a pasting scheme, where we index over *pointed* wheeled graphs rather than wheeled graphs. Recall entries of $F_{\mathcal{G}}$ are built from decorated graphs $\mathsf{P}[G]$, whereas here we will use $(\mathsf{P}, M)[G, v]$, where the distinguished vertex is decorated with M while all other vertices are decorated by P. Thus, a map from a coproduct of such constructions to M incorporates a series of maps from a tensor product of many copies of P and a single factor of M to a value of M. This is a generalization of the bimodule version of Example 10.20, where any product involving a single factor from M appears in the source, thereby tracking all iterated left and/or right multiplications by P at once. One reason we need to work with analogs of bimodules, rather than just left modules, is that without restricting our graphs dramatically, there is no reasonable notion of a rightmost (or last) vertex in a graph.

One might like to instead consider a notion of module following Quillen, as an Abelian group object in a comma category. It is not, at present, clear how such a definition compares with the current definition, although it seems reasonable that the two would coincide when imposing sufficient restrictions on the target category. In the next chapter, the pointed monad definition of modules is shown to coincide with an approach following May, as modules over an algebra over an operad. Thus, comparing either with the Quillen approach would clarify the situation.

Section 15.1 sets the stage by discussing pointed decorated graphs, and all of the constructions related to the pointed variant $F_{\mathcal{G}}(\mathsf{P}, M)$ of $F_{\mathcal{G}}$. Then Section 15.2 provides the unbiased definition of a module over a $\mathcal{G}$-PROP in terms of a structure map

$$\lambda : F_{\mathcal{G}}(\mathsf{P}, M) \longrightarrow M$$

making an appropriate associativity diagram commute, and a morphism of modules needs to be compatible with the components of these structure maps. In addition, a graphical interpretation of these constructions is provided. At this point, Section 15.3 produces the analogs of results in Sections 12.2 or 13.3 for PROPs or Algebras over PROPs, about changes of pasting scheme, base category, or $\mathcal{G}$-PROP P. Finally, Section 15.4 includes the Biased Module Theorem (Theorem 15.10), Biased Module Morphism Theorem (Theorem 15.12), and the details of the biased presentations of modules for all of our standard pasting schemes.

15.1. Pointed Decorated Graphs and a Monad Variation

The reader may want to review 1.4.1 and material surrounding Lemma 5.37 concerning pointed graphs and the slight variant of substitution for them.

15.1.1. Pointed Decorated Graphs. To simplify typography, for a vertex u in a wheeled graph, we will often abbreviate the pair of profiles $\binom{\mathrm{out}(u)}{\mathrm{in}(u)}$ to just $\widehat{u}$.

Fix a pasting scheme $\mathcal{G} = (S, \mathtt{G})$. Suppose $(G, v) \in \mathtt{G}_{S*}$ is a pointed wheeled graph, while P and M are S-colored objects in $\mathcal{E}$. Define the object

$$(\mathsf{P}, M)[G, v] = \left\{\bigotimes_{u \neq v} \mathsf{P}(\widehat{u})\right\} \otimes M(\widehat{v}), \tag{15.1}$$

in which the tensor products are unordered, and u runs through $\mathrm{Vt}(G) \smallsetminus \{v\}$. In particular, if G has a single vertex w, then $(\mathsf{P}, M)[G, w] = M(\widehat{w})$.

Note that

$$(\mathsf{P}, \mathsf{P})[G, v] = \bigotimes_{u \in \mathrm{Vt}(G)} \mathsf{P}(\widehat{u}) = \mathsf{P}[G],$$

the usual P-decorated graph. We think of $(\mathsf{P}, M)[G, v]$ as the space of decorations of the pointed wheeled graphs (G, v) by P in all the vertices except the distinguished one, which is decorated by M.

15.1.2. A Pointed Extension of the Monad $F_{\mathcal{G}}$. To describe the axioms of a module over a generalized PROP, we will build a structure which is not quite a monad on $\mathcal{E}^{dis(S)}$. This was the point of describing pointed extensions of monads carefully in Section 10.2, particularly the associated monad construction (Theorem 10.22) which provides us with a free module functor and the existence of (co)limits.

The analogs of the unit and multiplication map for a monad are the following constructions.

Define the bifunctor

$$F_{\mathcal{G}}(-,-) \colon \mathcal{E}^{dis(S)} \times \mathcal{E}^{dis(S)} \longrightarrow \mathcal{E}^{dis(S)} \tag{15.2}$$

by setting

$$F_{\mathcal{G}}(\mathsf{P}, M)\binom{d}{\underline{c}} = \coprod_{[G,v] \in \mathtt{G}_{S*}\binom{d}{\underline{c}}} (\mathsf{P}, M)[G, v].$$

Now define the analog of a unit map

$$\nu_{\mathsf{P},M} \colon M \longrightarrow F_{\mathcal{G}}(\mathsf{P}, M) \in \mathcal{E}^{dis(S)}$$

with components

$$\begin{array}{ccc} M\binom{d}{\underline{c}} & \xrightarrow{\quad \nu \quad} & F_{\mathcal{G}}(\mathsf{P}, M)\binom{d}{\underline{c}} \\ \Big\| & \nearrow_{\text{inclusion}} & \\ (\mathsf{P}, M)[C_{(\underline{c};\underline{d})}, w] & & \end{array}$$

where w denotes the unique vertex in the corolla $C_{(\underline{c};\underline{d})}$.

Finally, define the analog of the monoidal multiplication map

$$\mu_{\mathsf{P},M} \colon F_{\mathcal{G}}\left(F_{\mathcal{G}}\mathsf{P}, F_{\mathcal{G}}(\mathsf{P}, M)\right) \longrightarrow F_{\mathcal{G}}(\mathsf{P}, M) \in \mathcal{E}^{dis(S)}$$

as the one induced by pointed graph substitution. More precisely, the $(\underline{c};\underline{d})$ component of the domain of $\mu_{\mathsf{P},M}$ is:

$$
\begin{aligned}
&F_{\mathcal{G}}\left(F_{\mathcal{G}}\mathsf{P}, F_{\mathcal{G}}(\mathsf{P},M)\right)\binom{\underline{d}}{\underline{c}} \\
&= \coprod_{[G,v]\in \mathsf{G}_{S*}\binom{\underline{d}}{\underline{c}}} \left(F_{\mathcal{G}}\mathsf{P}, F_{\mathcal{G}}(\mathsf{P},M)\right)[G,v] \\
&= \coprod_{[G,v]\in \mathsf{G}_{S*}\binom{\underline{d}}{\underline{c}}} \left[\left\{\bigotimes_{\substack{u\neq v\\ \text{in } \mathrm{Vt}(G)}} (F_{\mathcal{G}}\mathsf{P})(\widehat{u})\right\} \otimes F_{\mathcal{G}}(\mathsf{P},M)(\widehat{v})\right] \\
&= \coprod_{[G,v]\in \mathsf{G}_{S*}\binom{\underline{d}}{\underline{c}}} \left[\left\{\bigotimes_{\substack{u\neq v\\ \text{in } \mathrm{Vt}(G)}} \coprod_{[H_u]\in \mathsf{G}_S(\widehat{u})} \mathsf{P}[H_u]\right\} \otimes \left\{\coprod_{[H_v,w]\in \mathsf{G}_{S*}(\widehat{v})} (\mathsf{P},M)[H_v,w]\right\}\right] \\
&\cong \coprod_{[G,v]\in \mathsf{G}_{S*}\binom{\underline{d}}{\underline{c}}} \coprod_{\substack{u\neq v\\ [H_u]\in \mathsf{G}_S(\widehat{u})}} \coprod_{[H_v,w]\in \mathsf{G}_{S*}(\widehat{v})} \left[\left(\bigotimes_{u\neq v} \mathsf{P}[H_u]\right) \otimes (\mathsf{P},M)[H_v,w]\right] \\
&\cong \coprod_{\substack{([G,v],\{[H_u]\}_{u\neq v},[H_v,w])\\ \text{in } \mathsf{G}_{S*}\binom{\underline{d}}{\underline{c}}\times\prod \mathsf{G}_S(\widehat{u})\times \mathsf{G}_{S*}(\widehat{v})}} (\mathsf{P},M)[G(H_x),w]
\end{aligned}
$$

where the last transition comes from a natural isomorphism

$$
\left[\left(\bigotimes_{u\neq v} \mathsf{P}[H_u]\right) \otimes (\mathsf{P},M)[H_v,w]\right] \cong (\mathsf{P},M)[G(H_x),w], \tag{15.3}
$$

with $(G(H_x),w)$ the pointed graph substitution.

On the other hand, the $(\underline{c};\underline{d})$ component of the target of $\mu_{\mathsf{P},M}$ is

$$
F_{\mathcal{G}}(\mathsf{P},M)\binom{\underline{d}}{\underline{c}} = \coprod_{[K,w]\in \mathsf{G}_{S*}\binom{\underline{d}}{\underline{c}}} (\mathsf{P},M)[K,w],
$$

so choosing $(K,w) = (G(H_x),w)$ gives a natural map from each summand in the domain to the target. Thus, applying the universal property of the coproduct in the domain, we have an induced map $\mu_{\mathsf{P},M}$ as desired.

PROPOSITION 15.1. *Given a pasting scheme $\mathcal{G}$ and a $\mathcal{G}$-PROP (P,γ), the constructions $(F_{\mathcal{G}}(\mathsf{P},M), \mu_{\mathsf{P},M}, \nu_{\mathsf{P},M})$ define a pointed extension of the monad $F_{\mathcal{G}}$ associated to the pasting scheme.*

PROOF. The fact that $F_{\mathcal{G}}(\mathsf{P},?)$ forms a monad of its own can eventually be reduced to consequences of the fact that graph substitution is both associative and unital (from both sides) when working with isomorphism classes of (pointed) graphs. This is a tedious exercise, but not particularly difficult. The broader unit condition (10.6) can be verified directly from the unit condition for corollas after similar manipulations. Thus, it remains only to verify the broader associativity condition (10.7).

Thus, we consider the diagram (writing F for $F_{\mathcal{G}}$ and F_{P} for $F_{\mathcal{G}}(\mathsf{P},?)$)

$$\begin{array}{ccc} F(F^2\mathsf{P}, F(F\mathsf{P}, F_{\mathsf{P}}^2 M)) & \xrightarrow{\mu_{F\mathsf{P}, F_{\mathsf{P}}^2 M}} & F(F\mathsf{P}, F_{\mathsf{P}}^2 M) \\ {\scriptstyle F(F(\gamma), \mu_{\mathsf{P}, F_{\mathsf{P}} M})}\downarrow & & \downarrow{\scriptstyle \mu_{\mathsf{P}, F_{\mathsf{P}} M}} \\ F(F\mathsf{P}, F_{\mathsf{P}}^2 M) & \xrightarrow{\mu_{\mathsf{P}, F_{\mathsf{P}} M}} & F_{\mathsf{P}}^2 M\,. \end{array}$$

Here we must exploit a variety of presentations in the spirit of Lemma 10.37,

$$F_{\mathsf{P}}^2 M(\widehat{v}_3) \cong \coprod_{[H^3, v_4], [H^4_{v_4}, v_5]} (\mathsf{P}, M)\left[H^3\left(\{C_u\}, H^4_{v_4}\right)\right]$$

$$F\left(F\mathsf{P}, F_{\mathsf{P}}^2 M\right)(\widehat{v}_2) \cong \coprod_{[H^2, v_3], \{[H^3_w]\}, [H^3, v_4], [H^4_{v_4}, v_5]} (\mathsf{P}, M)\left[H^2\left(\{[H^3_w]\}, H^3\left(\{C_u\}, H^4_{v_4}\right)\right)\right]$$

$$F\left(F^2\mathsf{P}, F\left(F\mathsf{P}, F_{\mathsf{P}}^2 M\right)\right)(\widehat{v}_1) \cong \coprod (\mathsf{P}, M)\left[H^1\left(\{H^2_u\left(H^{3,u}_w\right)\}, H^2\left(\{[H^3_w]\}, H^3\left(\{C_u\}, H^4_{v_4}\right)\right)\right)\right]$$

where the last coproduct is indexed over tuples of the form

$$[H^1, v_2], \left\{[H^2_u], [H^{3,u}_w]\right\}, [H^2, v_3], \left\{[H^3_w]\right\}, [H^3, v_4], [H^4_{v_4}, v_5]\,.$$

As we saw before, such identifications exploit the definition of decorated (pointed) graphs, using the distributivity of tensors over coproducts and regrouping tensor products using the decomposition of vertices in a graph substitution. In this notation, the maps denoted μ simply include each summand, as does the map $F(F(\gamma), \mu_{\mathsf{P}, F_{\mathsf{P}} M})$. To see this last part, recall that $F(\gamma)$ should be applied on the unpointed part, here denoted $\left\{H^2_u\left(H^{3,u}_w\right)\right\}$, which has already been rewritten in terms of the graph substitution. As a consequence, in the current language, both composites simply include each summand as indicated, while performing the indicated (pointed) graph substitutions on the indexing sets. Thus, looking at one summand at a time, it is straightforward to verify the diagram commutes in this notation. □

15.2. Unbiased Modules over a $\mathcal{G}$-PROP

15.2.1. Unbiased Definition of Modules over a $\mathcal{G}$-PROP. Here is the unbiased definition of a module over a generalized PROP.

DEFINITION 15.2. Let $(\mathsf{P}, \gamma\colon F_{\mathcal{G}}\mathsf{P} \longrightarrow \mathsf{P})$ be a $\mathcal{G}$-PROP in $\mathcal{E}$.

(1) A **P-module** is a pair (M, λ) with $M \in \mathcal{E}^{dis(S)}$ and

$$\lambda\colon F_{\mathcal{G}}(\mathsf{P}, M) \longrightarrow M \in \mathcal{E}^{dis(S)}$$

such that the associativity square

$$\begin{array}{ccc} F_{\mathcal{G}}(F_{\mathcal{G}}\mathsf{P}, F_{\mathcal{G}}(\mathsf{P}, M)) & \xrightarrow{\mu_{\mathsf{P},M}} & F_{\mathcal{G}}(\mathsf{P}, M) \\ {\scriptstyle F_{\mathcal{G}}(\gamma,\lambda)}\downarrow & & \downarrow{\scriptstyle \lambda} \\ F_{\mathcal{G}}(\mathsf{P}, M) & \xrightarrow{\lambda} & M \end{array} \tag{15.4}$$

and the unity diagram

$$\begin{array}{ccc} M & \xrightarrow{\nu_{\mathsf{P},M}} & F_{\mathcal{G}}(\mathsf{P}, M) \\ & {\scriptstyle Id}\searrow & \downarrow{\scriptstyle \lambda} \\ & & M \end{array} \tag{15.5}$$

are commutative.

(2) A **morphism** $f: (M, \lambda^M) \longrightarrow (N, \lambda^N)$ of P-modules is a morphism $f: M \longrightarrow N \in \mathcal{E}^{dis(S)}$ such that the square

$$\begin{array}{ccc} F_{\mathcal{G}}(\mathsf{P}, M) & \xrightarrow{\lambda^M} & M \\ {\scriptstyle F_{\mathcal{G}}(Id_{\mathsf{P}}, f)}\downarrow & & \downarrow{\scriptstyle f} \\ F_{\mathcal{G}}(\mathsf{P}, N) & \xrightarrow{\lambda^N} & N \end{array} \tag{15.6}$$

is commutative.

The category of P-modules is denoted by $\mathbf{Mod}(\mathsf{P})$.

EXAMPLE 15.3. Suppose $f: (\mathsf{P}, \gamma^{\mathsf{P}}) \longrightarrow (\mathsf{Q}, \gamma^{\mathsf{Q}})$ is a morphism of $\mathcal{G}$-PROPs. Then Q becomes a P-module whose structure map λ is defined by the composition

$$\begin{array}{ccc} (\mathsf{P}, \mathsf{Q})[G, v] & \xrightarrow{\lambda} & \mathsf{Q}\binom{\underline{d}}{\underline{c}} \\ {\scriptstyle (\otimes_{u \neq v} f) \otimes Id}\downarrow & & \uparrow{\scriptstyle \gamma^{\mathsf{Q}}} \\ (\mathsf{Q}, \mathsf{Q})[G, v] & = & \mathsf{Q}[G]. \end{array}$$

In particular, assuming $\mathcal{G}$ admits an endomorphism object, if A is a P-algebra, then the endomorphism $\mathcal{G}$-PROP E_A becomes a P-*module* via the P-algebra structure map $\mathsf{P} \longrightarrow \mathsf{E}_A$.

By construction, the category of modules over P is precisely $\mathrm{Module}_{F_{\mathcal{G}}}(\mathsf{P})$ of Definition 10.16, using the pointed extension of the monad associated to a pasting scheme from Proposition 15.1. Thus we can apply Theorem 10.22 to construct a free module functor, and to verify that $\mathbf{Mod}(\mathsf{P})$ has all small (co)limits as the algebras over a monad in $\mathcal{E}^{dis(S)}$.

COROLLARY 15.4. *For each $\mathcal{G}$-PROP P, there is a natural isomorphism of categories*

$$\mathbf{Mod}(\mathsf{P}) \cong \mathrm{Module}_{F_{\mathcal{G}}}(\mathsf{P})$$

and a left adjoint $\mathcal{F} : \mathcal{E}^{dis(S)} \longrightarrow \mathbf{Mod}(\mathsf{P})$ to the evident forgetful functor, with $\mathbf{Mod}(\mathsf{P})$ monadic over this adjunction.

15.2.2. Bi-equivariant Structure of a Module. Essentially as in section 10.4.3, a P-module (M,λ) acquires the structure of a $\boldsymbol{\Sigma}_S$-bimodule via the structure map restricted to permuted corollas. More explicitly, if C is the $(\underline{c};\underline{d})$-corolla and σ and τ are suitable permutations, then one coproduct summand of $F_{\mathcal{G}}(\mathsf{P},M)\binom{\sigma\underline{d}}{\underline{c}\tau}$ is

$$(\mathsf{P},M)[\sigma C\tau,w] = M(\widehat{w}) = M\binom{\underline{d}}{\underline{c}}.$$

As a consequence, the restriction of $\lambda : F_{\mathcal{G}}(\mathsf{P},M)\binom{\sigma\underline{d}}{\underline{c}\tau} \longrightarrow M\binom{\sigma\underline{d}}{\underline{c}\tau}$ to these summands

$$(\tau;\sigma) = \lambda_{[\sigma C\tau,w]} \colon M\binom{\underline{d}}{\underline{c}} \longrightarrow M\binom{\sigma\underline{d}}{\underline{c}\tau}$$

give M the structure of a $\boldsymbol{\Sigma}_S$-bimodule.

15.2.3. Graphical Interpretation. Let us provide a graphical presentation of the constructions above. Suppose (G,v) is a pointed wheeled graph such that $G \in \mathtt{G}_S\binom{\underline{d}}{\underline{c}}$, which we draw generically as:

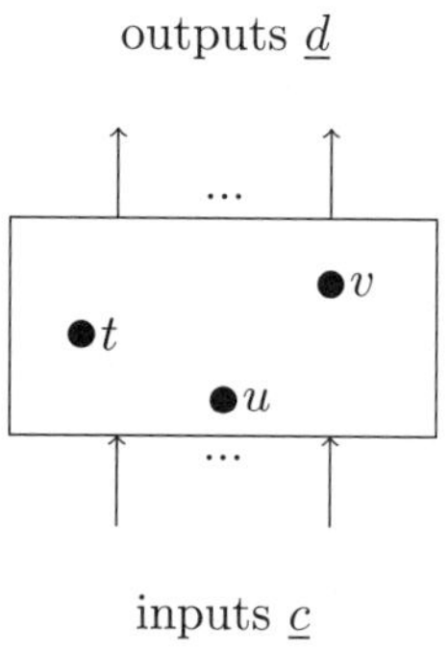

(15.7)

We emphasize that G does not need to be connected, and it may have directed cycles and exceptional connected components. Besides the distinguished vertex v, we draw two other generic vertices t and u.

Then the $\binom{\underline{d}}{\underline{c}}$-component of $F_{\mathcal{G}}(\mathsf{P},M)$ is the space of decorated pointed graphs

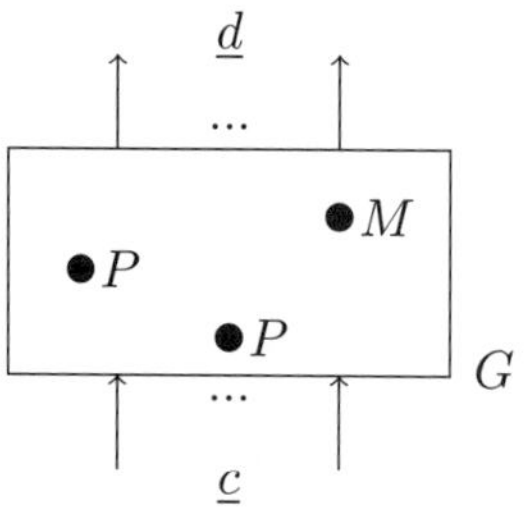

(15.8)

with (G,v) running through the strict isomorphism classes of pointed wheeled graphs in $\mathtt{G}_S\binom{\underline{d}}{\underline{c}}$. Here the label P next to the vertex t means that that vertex is labeled by the component $\mathsf{P}(t)$ of P, and similarly for the other labels.

Similarly, the $\binom{\underline{d}}{\underline{c}}$-component of M is drawn as the decorated corolla

(15.9)

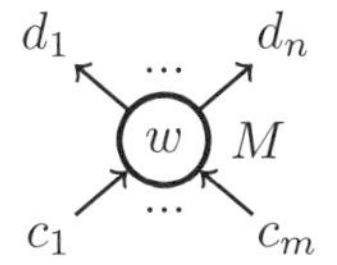

in which the unique vertex is decorated by $M\binom{\underline{d}}{\underline{c}}$. The $\binom{\underline{d}}{\underline{c}}$-component of a map $\lambda: F_{\mathcal{G}}(\mathsf{P}, M) \longrightarrow M$ will then be graphically regarded as shrinking all of the decorated graphs (15.8) to the decorated corolla (15.9).

With (G, v) as in (15.7), the $\binom{\underline{d}}{\underline{c}}$-component of $F_{\mathcal{G}}(F_{\mathcal{G}}\mathsf{P}, F_{\mathcal{G}}(\mathsf{P}, M))$ is the space of bracketed decorated graphs

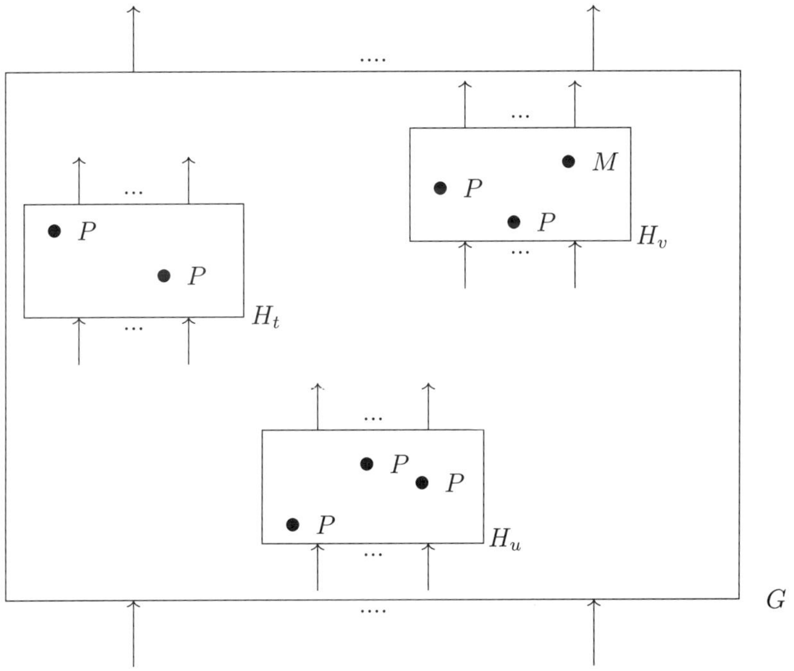

(15.10)

with (G, v) running through the strict isomorphism classes of pointed wheeled graphs in $\mathtt{G}_S\binom{\underline{d}}{\underline{c}}$. For each vertex $u \neq v$ in G, H_u runs through the strict isomorphism classes of wheeled graphs in $\mathtt{G}_S(u)$. For the distinguished vertex v, (H_v, w) runs through the strict isomorphism classes of pointed wheeled graphs in $\mathtt{G}_S(v)$.

The left-hand side of (15.3) is represented by the bracketed decorated graph in (15.10). The natural isomorphism (15.3), then corresponds to simply erasing the brackets, i.e., performing pointed graph substitution. The associativity square (15.4) says that, starting from the bracketed decorated graph in (15.10), two ways of shrinking it to a decorated corolla agree. The upper path around the associativity square (15.4) erases the inside brackets and then shrinks the resulting graph to a decorated corolla using the P-module structure map λ. On the other hand, the lower path around that associativity square shrinks each of the inner decorated graphs to a decorated corolla, using the $\mathcal{G}$-PROP structure on P and the P-module structure on M, and then shrinks the resulting graph to a decorated corolla again using the P-module structure on M.

15.3. Modules under Change of Pasting Scheme, $\mathcal{G}$-PROP, or Base Category

In this section, we consider the category of modules under a change of pasting scheme, $\mathcal{G}$-PROP, or the base category.

15.3.1. Change of Pasting Scheme. First we have a restriction or forgetful functor for modules.

LEMMA 15.5. *Suppose*

- $\mathcal{E}$ *is a symmetric monoidal category,*
- $\mathcal{G} \leq \mathcal{G}'$ *are pasting schemes, and*
- P *is a* $\mathcal{G}'$*-PROP.*

Then there is a restriction functor

$$\mathbf{Mod}_{\mathcal{G}'}(\mathsf{P}) \longrightarrow \mathbf{Mod}_{\mathcal{G}}(U\mathsf{P}),$$

where U is the forgetful functor from $\mathcal{G}'$-PROPs to $\mathcal{G}$-PROPs.

PROOF. Since $i : F_{\mathcal{G}}(U\mathsf{P}, M) \longrightarrow F_{\mathcal{G}'}(\mathsf{P}, M)$ is the inclusion of a natural sub-coproduct at each entry, the following diagram commutes.

$$\begin{array}{ccccc} F_{\mathcal{G}}(U\mathsf{P},M) & \xleftarrow{F_{\mathcal{G}}(\gamma_{U\mathsf{P}},\lambda i)} & F_{\mathcal{G}}(F_{\mathcal{G}}U\mathsf{P},F_{\mathcal{G}}(U\mathsf{P},M)) & \xrightarrow{\mu_{U\mathsf{P},M}} & F_{\mathcal{G}}(U\mathsf{P},M) \\ \downarrow{\scriptstyle i} & & \downarrow{\scriptstyle i_*} & & \downarrow{\scriptstyle i} \\ F_{\mathcal{G}'}(\mathsf{P},M) & \xleftarrow{F_{\mathcal{G}'}(\gamma_{\mathsf{P}},\lambda)} & F_{\mathcal{G}'}(F_{\mathcal{G}'}\mathsf{P},F_{\mathcal{G}'}(\mathsf{P},M)) & \xrightarrow{\mu_{\mathsf{P},M}} & F_{\mathcal{G}'}(\mathsf{P},M) \end{array}$$

Together with the associativity condition for $(M,\lambda) \in \mathbf{Mod}(\mathsf{P})$, this verifies the associativity condition for λi. Since the unit map factors through the sub-coproduct, the unit condition for λi also follows. Finally, naturality of i implies the restriction is compatible with morphisms of modules, and their composition, as well. □

15.3.2. Change of $\mathcal{G}$-PROP. Here we assume $\mathsf{P} \longrightarrow \mathsf{P}'$ is a morphism of $\mathcal{G}$-PROPs and describe the resulting change of $\mathcal{G}$-PROP functor on the level of modules.

LEMMA 15.6. *Each morphism $f : \mathsf{P} \longrightarrow \mathsf{P}'$ of $\mathcal{G}$-PROPs induces a functor*

$$f^*\colon \mathbf{Mod}(\mathsf{P}') \longrightarrow \mathbf{Mod}(\mathsf{P}).$$

PROOF. Suppose (M,λ) is a module over P', and define

$$\overline{\lambda} : F_{\mathcal{G}}(\mathsf{P},M) \xrightarrow{F_{\mathcal{G}}(f,Id)} F_{\mathcal{G}}(\mathsf{P}',M) \xrightarrow{\lambda} M$$

as the module structure map over P. The unit condition for $\overline{\lambda}$ follows from commutativity of the diagram

$$\begin{array}{ccc} & & F_{\mathcal{G}}(\mathsf{P},M) \\ & \nearrow^{\nu_{\mathsf{P},M}} & \downarrow{\scriptstyle F_{\mathcal{G}}(f,Id)} \\ M & \xrightarrow{\nu_{\mathsf{P}',M}} & F_{\mathcal{G}}(\mathsf{P}',M) \\ & \searrow_{Id} & \downarrow{\scriptstyle\lambda} \\ & & M \end{array}$$

where the upper triangle commutes by P actually being irrelevant in the definition of $\nu_{\mathsf{P},M}$ and the lower triangle is the assumed unit condition for λ. Similarly, the associativity diagram for $\overline{\lambda}$ follows from that for λ and the commutative diagram

$$\begin{array}{ccccc} F_{\mathcal{G}}(\mathsf{P},M) & \xleftarrow{F_{\mathcal{G}}(\gamma_{\mathsf{P}},\overline{\lambda})} & F_{\mathcal{G}}(F_{\mathcal{G}}\mathsf{P},F_{\mathcal{G}}(\mathsf{P},M)) & \xrightarrow{\mu_{\mathsf{P},M}} & F_{\mathcal{G}}(\mathsf{P},M) \\ \downarrow{\scriptstyle F_{\mathcal{G}}(f,Id)} & & \downarrow{\scriptstyle F_{\mathcal{G}}(F_{\mathcal{G}}f,F_{\mathcal{G}}(f,Id))} & & \downarrow{\scriptstyle F_{\mathcal{G}}(f,Id)} \\ F_{\mathcal{G}}(\mathsf{P}',M) & \xleftarrow{F_{\mathcal{G}}(\gamma_{\mathsf{P}'},\lambda)} & F_{\mathcal{G}}(F_{\mathcal{G}}\mathsf{P}',F_{\mathcal{G}}(\mathsf{P}',M)) & \xrightarrow{\mu_{\mathsf{P}',M}} & F_{\mathcal{G}}(\mathsf{P}',M) \end{array}$$

Here the left square commutes since f is compatible with γ_{P} and $\gamma_{\mathsf{P}'}$ as a morphism of $\mathcal{G}$-PROPs, along with the definition of $\overline{\lambda}$ in the second argument. Commutativity of the right square follows from naturality of all constructions in the variable P.

To see this construction is compatible with the notions of morphism, consider the commutative diagram

$$\begin{array}{ccccc} F_{\mathcal{G}}(\mathsf{P},M) & \xrightarrow{F_{\mathcal{G}}(f,Id_M)} & F_{\mathcal{G}}(\mathsf{P}',N) & \xrightarrow{\lambda^M} & M \\ \downarrow{\scriptstyle F_{\mathcal{G}}(Id_{\mathsf{P}},g)} & & \downarrow{\scriptstyle F_{\mathcal{G}}(Id_{\mathsf{P}'},g)} & & \downarrow{\scriptstyle g} \\ F_{\mathcal{G}}(\mathsf{P},N) & \xrightarrow{F_{\mathcal{G}}(f,Id_N)} & F_{\mathcal{G}}(\mathsf{P}',N) & \xrightarrow{\lambda^N} & N \end{array}$$

where the right square says g is a morphism of P'-modules and the left square commutes by naturality. Compatibility with composition is automatic, so the proof is complete. □

15.3.3. Change of Base Category. Next we consider the category of modules under a change of the underlying category.

LEMMA 15.7. *If $K : \mathcal{M} \longrightarrow \mathcal{M}'$ is strong symmetric monoidal, while P is a $\mathcal{G}$-PROP with values in $\mathcal{M}$, then K induces a functor*

$$\mathbf{Mod}_{\mathcal{M}}(\mathsf{P}) \longrightarrow \mathbf{Mod}_{\mathcal{M}'}(K\mathsf{P})$$

entrywise.

PROOF. The key observation is that the strong monoidal condition implies $K((\mathsf{P},M)[G,v]) \cong (K\mathsf{P},KM)[G,v]$, so together with the natural map associated to a coproduct, there is a natural map

$$F_{\mathcal{G}}(K\mathsf{P},KM) \longrightarrow K(F_{\mathcal{G}}(\mathsf{P},M)) \ .$$

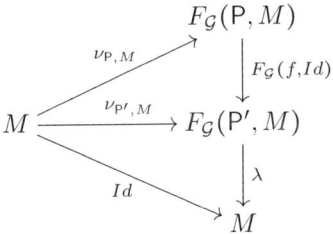

As a consequence, simply apply K entrywise and use naturality of the map associated to a coproduct to verify the associativity condition for the composite structure map consisting of $K\lambda$ and this natural map. For the unit condition, consider the commutative diagram

$$\begin{array}{ccc} & & F_{\mathcal{G}}(K\mathsf{P}, KM) \\ & \nearrow^{\nu_{K\mathsf{P},KM}} & \big\downarrow \text{natural} \\ KM & \xrightarrow{K(\nu_{\mathsf{P},M})} & K\left(F_{\mathcal{G}}(\mathsf{P}, M)\right) \\ & \searrow_{Id} & \big\downarrow K(\lambda) \\ & & KM\ . \end{array}$$

Finally, naturality of the displayed map also implies compatibility with morphisms of induced modules, as well as composition. □

THEOREM 15.8. *Suppose* $\mathsf{P} \in \mathsf{PROP}^{\mathcal{G},\mathcal{E}}$ *along with an adjoint pair*

$$L{:}\,\mathcal{E} \rightleftarrows \mathcal{D}{:}\,R$$

of strong symmetric monoidal functors. Then there is an induced adjoint pair

$$L{:}\,\mathbf{Mod}_{\mathcal{E}}(\mathsf{P}) \rightleftarrows \mathbf{Mod}_{\mathcal{D}}(L\mathsf{P}){:}\,R \tag{15.11}$$

that is defined entry-wise.

PROOF. The fact that L produces such a functor follows immediately from Lemma 15.7. To see R also produces such a functor follows from the combination of Lemma 15.7, to land in $\mathbf{Mod}(RL\mathsf{P})$, followed by Lemma 15.6 using the unit of adjunction $\mathsf{P} \longrightarrow RL\mathsf{P}$, which is a $\mathcal{G}$-PROP map by Corollary 12.13. The adjunction property then follows by prolonging the "triangular identities" and the fact that the second map is still R applied entrywise at the level of $\mathcal{E}^{dis(S)}$. □

15.4. Biased Characterizations of Modules

The above definition of a P-module is unbiased, in the sense that $F_{\mathcal{G}}$ involves *all* the strict isomorphism classes of pointed wheeled graphs in G_{S*}. As for $\mathcal{G}$-PROPs and algebras over a $\mathcal{G}$-PROP, here we describe P-modules in biased terms. In specific examples, this means that we describe P-modules using a few generating operations that satisfy a few basic relations. Not surprisingly, our main tool is once again strong generating sets of pasting schemes. The reader may want to review 7.1.3 concerning pointed graph simplices at this point.

15.4.1. Biased Module Theorem.

DEFINITION 15.9. Suppose $\mathcal{T}$ is a collection of wheeled graphs whose full graph profiles and vertex profiles are all in a replete and full sub-groupoid S inside of $\mathcal{P}(\mathfrak{C})^{op} \times \mathcal{P}(\mathfrak{C})$, while P and M are S-colored objects in $\mathcal{E}$.

(1) A **$(\mathsf{P},\mathcal{T})$-module** structure on M is a collection of maps

$$(\mathsf{P}, M)[G, v] \xrightarrow{\lambda_{[G,v]}} M\binom{\text{out}(G)}{\text{in}(G)},$$

one for each $(G, v) \in \mathcal{T}_*$, such that

$$\lambda_{[C,w]} = Id : M\binom{\underline{d}}{\underline{c}} \longrightarrow M\binom{\underline{d}}{\underline{c}}$$

whenever $C = C_{(\underline{c};\underline{d})}$ is a corolla.

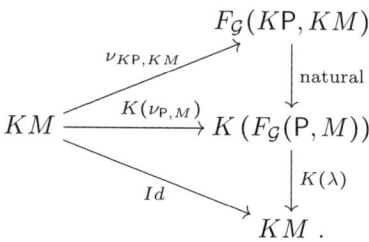

(2) Suppose:
- $\mathcal{H}$ is a pointed graph simplex in $\mathcal{T}$,
- (P,γ) is a $\mathcal{T}$-algebra, and
- (M,λ) is a $(\mathsf{P},\mathcal{T})$-module.

Define

$$\gamma_{\mathcal{H}} : (\mathsf{P},M)[\mathrm{sub}(\mathcal{H})] \longrightarrow M\binom{\mathrm{out}(\mathrm{sub}(\mathcal{H}))}{\mathrm{in}(\mathrm{sub}(\mathcal{H}))}$$

as in 11.1 with the exception that λ must be used in place of γ for those wheeled graphs which are pointed.

(3) Suppose (P,γ) is a $\mathcal{T}$-algebra, and $\mathcal{W}$ is a strong set of relaxed moves in $\mathcal{T}$. A **biased $(\mathsf{P},\mathcal{T},\mathcal{W})$-module** structure on M is a $(\mathsf{P},\mathcal{T})$-module (M,λ) such that

$$\gamma_{\mathcal{H}} = \gamma_{\mathcal{H}'}$$

whenever $\mathcal{H}$ and $\mathcal{H}'$ are pointed graph simplices in $\mathcal{T}$ that are connected by a finite number of relaxed moves in $\mathcal{W}$.

THEOREM 15.10 (Biased Module Theorem). *Let:*
- *$\mathcal{G}$ be a pasting scheme,*
- *$(\mathcal{T},\mathcal{W})$ be a strong generating set of $\mathtt{G}_S$,*
- *P be a $\mathcal{G}$-PROP, and*
- *$M \in \mathcal{E}^{dis(S)}$.*

Then a P-module structure on M is equivalent to a biased $(\mathsf{P},\mathcal{T},\mathcal{W})$-module structure on M.

PROOF. This is essentially identical to the proof of Theorem 11.5, except that pointed wheeled graphs are used in the pointed parts of pointed graph simplices. □

DEFINITION 15.11. Given (M,λ^M) and (N,λ^N) both $(\mathsf{P},\mathcal{T})$-modules, a **morphism of $(\mathsf{P},\mathcal{T})$-modules** $f : M \longrightarrow N$ is a morphism in $\mathcal{E}^{dis(S)}$ making the following diagram commute

$$\begin{array}{ccc} (\mathsf{P},M)[G,v] & \xrightarrow{\lambda^M_{[G,v]}} & M\binom{\mathrm{out}(G)}{\mathrm{in}(G)} \\ \downarrow{\scriptstyle (1,f)} & & \downarrow{\scriptstyle f} \\ (\mathsf{P},N)[G,v] & \xrightarrow{\lambda^N_{[G,v]}} & N\binom{\mathrm{out}(G)}{\mathrm{in}(G)} \end{array}$$

where $(1,f) : (\mathsf{P},M)[G,v] \longrightarrow (\mathsf{P},N)[G,v]$ indicates the map which is the identity on all P factors and f on the single M factor.

THEOREM 15.12 (Biased Module Morphism Theorem). *Let:*
- *$\mathcal{G}$ be a pasting scheme,*
- *$\mathcal{T}$ be any generating set of $\mathtt{G}_S$,*
- *P be a $\mathcal{G}$-PROP, and*
- *(M,λ^M) and (N,λ^N) both P-modules, with*
- *$f : M \longrightarrow N$ a morphism in $\mathcal{E}^{dis(S)}$.*

Then f is a morphism of P-modules precisely when it is a morphism of $(\mathsf{P},\mathcal{T})$-modules.

PROOF. In this case, we imitate the proof of Theorem 11.7. □

15.4.2. Modules over a Markl Non-Unital Operad. Fix a Markl non-unital operad $(\mathsf{P}, \circ_i)$ as in section 11.3.

DEFINITION 15.13. A **biased P-module** is a triple $(M, \circ_i^l, \circ_i^r)$ in which:

- $M \in \Sigma_{S_1}$,
- $\circ_i^l$ is a map

$$\mathsf{P}\binom{d}{\underline{c}} \otimes M\binom{c_i}{\underline{b}} \xrightarrow{\circ_i^l} M\binom{d}{\underline{c} \circ_i \underline{b}},$$

  and
- $\circ_i^r$ is a map

$$M\binom{d}{\underline{c}} \otimes \mathsf{P}\binom{c_i}{\underline{b}} \xrightarrow{\circ_i^r} M\binom{d}{\underline{c} \circ_i \underline{b}}.$$

The required commutative diagrams for this structure are variants of those for a Markl non-unital operad (Definition 11.10), in which each commutative diagram now begins with exactly one tensor factor of M and all other tensor factors are still P.

A **morphism** of biased P-modules is a morphism in Σ_{S_1} that respects both $\circ_i^l$ and $\circ_i^r$. The category of biased P-modules is denoted by Module(P).

EXAMPLE 15.14. Suppose $f: \mathsf{P} \longrightarrow \mathsf{Q}$ is a morphism of Markl non-unital operads. Then Q becomes a biased P-module with structure maps

$$\begin{array}{ccc}
\mathsf{P}\binom{d}{\underline{c}} \otimes \mathsf{Q}\binom{c_i}{\underline{b}} & \xrightarrow{\circ_i^l} & \mathsf{Q}\binom{d}{\underline{c} \circ_i \underline{b}} \\
{\scriptstyle (f, Id)} \downarrow & \nearrow {\scriptstyle \circ_i} & \\
\mathsf{Q}\binom{d}{\underline{c}} \otimes \mathsf{Q}\binom{c_i}{\underline{b}} & &
\end{array}$$

and

$$\begin{array}{ccc}
\mathsf{Q}\binom{d}{\underline{c}} \otimes \mathsf{P}\binom{c_i}{\underline{b}} & \xrightarrow{\circ_i^r} & \mathsf{Q}\binom{d}{\underline{c} \circ_i \underline{b}} \\
{\scriptstyle (Id, f)} \downarrow & \nearrow {\scriptstyle \circ_i} & \\
\mathsf{Q}\binom{d}{\underline{c}} \otimes \mathsf{Q}\binom{c_i}{\underline{b}} & &
\end{array}$$

respectively. In particular, if A is a P-algebra, then the endomorphism non-unital operad E_A becomes a biased P-module via the structure map $\mathsf{P} \longrightarrow \mathsf{E}_A$. This is the way in which one-colored linear P-modules are used in [**Mar96a**] to study deformation theory and minimal models.

The next Corollary of Theorems 15.10 and 15.12 now follows with the use of Theorem 7.35 and Corollary 11.12.

COROLLARY 15.15. *Given an $M \in \Sigma_{S_1}$ and a Markl non-unital operad P, regarded as a* `Tree`*-PROP, a biased P-module structure on M is equivalent to a P-module structure on M. Furthermore, this extends to a natural isomorphism of these biased and unbiased module categories.*

15.4.3. Modules over a May Operad. Fix a May operad $(\mathsf{P},\gamma,\mathbf{1})$ as in Definition 11.14. Let us first recall the biased definition of a P-module [**MSS02**] (Definition 3.28).

DEFINITION 15.16. A **biased P-module** is a triple (M,γ^l,γ^r) in which:

- $M \in \Sigma_{S_1}$;
- γ^l is a map

$$\mathsf{P}\binom{d}{\underline{c}} \otimes \underbrace{\mathsf{P}\binom{c_1}{\underline{b}_1} \otimes \cdots \otimes M\binom{c_i}{\underline{b}_i} \otimes \cdots \otimes \mathsf{P}\binom{c_m}{\underline{b}_m}}_{\text{exactly one tensor factor of } M} \xrightarrow{\gamma^l} M\binom{d}{\underline{b}} \tag{15.12}$$

  for each $d \in \mathfrak{C}$, $\underline{c} \in \mathcal{P}(\mathfrak{C})$ with $|\underline{c}| = m \geq 1$, $\underline{b}_j \in \mathcal{P}(\mathfrak{C})$, and $1 \leq i \leq m$;
- γ^r is a map

$$M\binom{d}{\underline{c}} \otimes \bigotimes_{i=1}^m \mathsf{P}\binom{c_i}{\underline{b}_i} \xrightarrow{\gamma^r} M\binom{d}{\underline{b}}. \tag{15.13}$$

Variants of the commutative diagrams required in the definition of a May operad (Definition 11.14) are also required, in which each commutative diagram begins with one tensor factor of M and all other tensor factors P.

A **morphism** of biased P-modules is a morphism in Σ_{S_1} that respects both γ^l and γ^r.

Now Theorems 15.10 and 15.12, in this case using Theorem 7.41 and Corollary 11.16, yield the following.

COROLLARY 15.17. *Given $M \in \Sigma_{S_1}$ and a May operad* P*, regarded as a* `UTree`*-PROP, a biased* P*-module structure on M is equivalent to a* P*-module structure on M. Furthermore, this extends to a natural isomorphism of these biased and unbiased module categories.*

15.4.4. Modules over a Dioperad. Fix a dioperad $(\mathsf{P},{}_j\circ_i,\mathbf{1})$ as in Definition 11.18.

DEFINITION 15.18. A **biased P-module** is a triple $(M,{}_j\circ_i{}^l,{}_j\circ_i{}^r)$ in which:

- $M \in \mathbf{\Sigma}_S$,
- ${}_j\circ_i{}^l$ is a map

$$\mathsf{P}\binom{\underline{d}}{\underline{c}} \otimes M\binom{\underline{b}}{\underline{a}} \xrightarrow{{}_j\circ_i{}^l} M\binom{\underline{b}\circ_j\underline{d}}{\underline{c}\circ_i\underline{a}}$$

  whenever $b_j = c_i$, and
- ${}_j\circ_i{}^r$ is a map

$$M\binom{\underline{d}}{\underline{c}} \otimes \mathsf{P}\binom{\underline{b}}{\underline{a}} \xrightarrow{{}_j\circ_i{}^r} M\binom{\underline{b}\circ_j\underline{d}}{\underline{c}\circ_i\underline{a}}$$

  whenever $b_j = c_i$.

In addition, one requires the commutativity of all variants of the commutative diagrams in the definition of a dioperad (Definition 11.18) beginning with one tensor factor of M and all other tensor factors P.

A **morphism** of biased P-modules is a morphism in $\mathbf{\Sigma}_S$ that respects both ${}_j\circ_i{}^l$ and ${}_j\circ_i{}^r$.

Note that there are 9 associativity axioms for a module over a dioperad, since each of the three associativity diagrams for a dioperad begins with three factors of P.

In this case, Theorems 15.10 and 15.12 are paired with Theorem 7.57 and Corollary 11.20 to deduce the following.

COROLLARY 15.19. *Given $M \in \Sigma_S$ and a dioperad P, regarded as a $\mathtt{Gr}^{\uparrow}_{di}$-PROP, a biased P-module structure on M is equivalent to a P-module structure on M. Furthermore, this extends to a natural isomorphism of these biased and unbiased module categories.*

15.4.5. Modules over a Half-PROP. Fix a half-PROP $(\mathsf{P}, \circ_i, {}_j\circ)$ as in Definition 11.22.

DEFINITION 15.20. A **biased P-module** is a tuple $(M, \circ_i^l, \circ_i^r, {}_j\circ^l, {}_j\circ^r)$ in which:

- $M \in \Sigma^{\mathcal{E}}_{S_{1/2}}$,
- $\circ_i^l$ is a map
$$\mathsf{P}\binom{d}{\underline{c}} \otimes M\binom{c_i}{\underline{a}} \xrightarrow{\circ_i^l} M\binom{d}{\underline{c} \circ_i \underline{a}}$$
for each $1 \le i \le m$,
- $\circ_i^r$ is a map
$$M\binom{d}{\underline{c}} \otimes \mathsf{P}\binom{c_i}{\underline{a}} \xrightarrow{\circ_i^r} M\binom{d}{\underline{c} \circ_i \underline{a}}$$
for each $1 \le i \le m$,
- ${}_j\circ^l$ is a map
$$\mathsf{P}\binom{\underline{d}}{b_j} \otimes M\binom{\underline{b}}{\underline{a}} \xrightarrow{{}_j\circ^l} M\binom{\underline{b} \circ_j \underline{d}}{\underline{a}}$$
for each $1 \le j \le l$, and
- ${}_j\circ^r$ is a map
$$M\binom{\underline{d}}{b_j} \otimes \mathsf{P}\binom{\underline{b}}{\underline{a}} \xrightarrow{{}_j\circ^r} M\binom{\underline{b} \circ_j \underline{d}}{\underline{a}}$$
for each $1 \le j \le l$.

Additionally, one must require the commutativity of each variant of the diagrams in the definition of a half-PROP (Definition 11.22) in which each commutative diagram begins with one tensor factor of M.

A **morphism** of biased P-modules is a morphism in $\Sigma^{\mathcal{E}}_{S_{1/2}}$ that respects $\circ_i^l$, $\circ_i^r$, ${}_j\circ^l$, and ${}_j\circ^r$.

Note that there are 15 associativity axioms for a module over a half-PROP, since there are five associativity diagrams in the definition of a half-PROP, each beginning with three factors of P.

This time we apply Theorems 15.10 and 15.12, relying upon Theorem 7.64 and Corollary 11.23, to produce the following.

COROLLARY 15.21. *Given $M \in \Sigma^{\mathcal{E}}_{S_{1/2}}$ and a half-PROP P, regarded as a $\mathtt{Gr}_{\frac{1}{2}}$-PROP, a biased P-module structure on M is equivalent to a P-module structure on M. Furthermore, this extends to a natural isomorphism of these biased and unbiased module categories.*

15.4.6. Modules over a Properad. Fix a properad $\left(\mathsf{P}, \boxtimes_{\underline{b}'}^{\underline{c}'}, \mathbf{1}\right)$ as in Definition 11.25.

DEFINITION 15.22. A **biased** P**-module** is a triple $\left(M, {}^{l}\boxtimes_{\underline{b}'}^{\underline{c}'}, {}^{r}\boxtimes_{\underline{b}'}^{\underline{c}'}\right)$ consisting of:

- $M \in \mathbf{\Sigma}_S$,
- a map
$$\mathsf{P}\binom{\underline{d}}{\underline{c}} \otimes M\binom{\underline{b}}{\underline{a}} \xrightarrow{{}^{l}\boxtimes_{\underline{b}'}^{\underline{c}'}} M\binom{\underline{b} \circ_{\underline{b}'} \underline{d}}{\underline{c} \circ_{\underline{c}'} \underline{a}}$$
whenever $\underline{b}' = \underline{c}'$, and
- a map
$$M\binom{\underline{d}}{\underline{c}} \otimes \mathsf{P}\binom{\underline{b}}{\underline{a}} \xrightarrow{{}^{r}\boxtimes_{\underline{b}'}^{\underline{c}'}} M\binom{\underline{b} \circ_{\underline{b}'} \underline{d}}{\underline{c} \circ_{\underline{c}'} \underline{a}}$$
whenever $\underline{b}' = \underline{c}'$.

As above, one must also require commutativity of the variants of each diagram in the definition of a properad (Definition 11.25) in which each commutative diagram begins with one tensor factor of M.

A **morphism** of biased P-modules is a morphism in $\mathbf{\Sigma}_S$ that respects both ${}^{l}\boxtimes_{\underline{b}'}^{\underline{c}'}$ and ${}^{r}\boxtimes_{\underline{b}'}^{\underline{c}'}$.

Note that for a module over a properad, there are 7 bi-equivariance axioms and 12 associativity axioms.

In this case the input for Theorems 15.10 and 15.12 is Theorem 7.67 and Corollary 11.28, to produce the following.

COROLLARY 15.23. *Given $M \in \mathbf{\Sigma}_S$ and a properad P, regarded as a $\mathtt{Gr}_c^{\uparrow}$-PROP, a biased P-module structure on M is equivalent to a P-module structure on M. Furthermore, this extends to a natural isomorphism of these biased and unbiased module categories.*

15.4.7. Modules over a PROP. Fix a PROP $(\mathsf{P}, \otimes_h, \otimes_v, \mathbf{1})$ as in Definition 11.30. In the one-colored (resp., multi-colored) linear case, biased versions of P-modules were used in [**Mar96b**] (resp., [**FMY09**]) to study deformations of algebras over linear PROPs.

DEFINITION 15.24. A **biased** P**-module** $(M, \otimes_h^l, \otimes_h^r, \otimes_v^l, \otimes_v^r)$ consists of:

- an object $M = \{M\binom{\underline{d}}{\underline{c}}\} \in \mathbf{\Sigma}_S$,
- a left horizontal P-action
$$\mathsf{P}\binom{\underline{d}}{\underline{c}} \otimes M\binom{\underline{b}}{\underline{a}} \xrightarrow{\otimes_h^l} M\binom{\underline{d},\underline{b}}{\underline{c},\underline{a}},$$
- a right horizontal P-action
$$M\binom{\underline{d}}{\underline{c}} \otimes \mathsf{P}\binom{\underline{b}}{\underline{a}} \xrightarrow{\otimes_h^r} M\binom{\underline{d},\underline{b}}{\underline{c},\underline{a}},$$
- a left vertical P-action
$$\mathsf{P}\binom{\underline{d}}{\underline{c}} \otimes M\binom{\underline{c}}{\underline{b}} \xrightarrow{\otimes_v^l} M\binom{\underline{d}}{\underline{b}},$$

and

- a right vertical P-action

$$M\binom{\underline{d}}{\underline{c}} \otimes \mathsf{P}\binom{\underline{c}}{\underline{b}} \xrightarrow{\otimes_v^r} M\binom{\underline{d}}{\underline{b}}.$$

These operations satisfy the analogs of the diagrams for a $\mathfrak{C}$-colored PROP in which each diagram begins with one tensor factor of M.

A **morphism** of biased P-modules is a morphism of the underlying $\mathbf{\Sigma}_S$-bimodules that is compatible with the four P-module actions.

For example, there are four interchange rules for a P-module M corresponding to (11.26), one of which is the commutative diagram:

$$\begin{array}{ccc}
\mathsf{P}\binom{\underline{d}_1}{\underline{c}_1} \otimes M\binom{\underline{c}_1}{\underline{b}_1} \otimes \mathsf{P}\binom{\underline{d}_2}{\underline{c}_2} \otimes \mathsf{P}\binom{\underline{c}_2}{\underline{b}_2} & \xrightarrow{(\otimes_v^l,\otimes_v)} & M\binom{\underline{d}_1}{\underline{b}_1} \otimes \mathsf{P}\binom{\underline{d}_2}{\underline{b}_2} \\
\Big\downarrow{\scriptstyle \text{shuffle}} & & \\
\mathsf{P}\binom{\underline{d}_1}{\underline{c}_1} \otimes \mathsf{P}\binom{\underline{d}_2}{\underline{c}_2} \otimes M\binom{\underline{c}_1}{\underline{b}_1} \otimes \mathsf{P}\binom{\underline{c}_2}{\underline{b}_2} & & \Big\downarrow{\scriptstyle \otimes_h^r} \\
\Big\downarrow{\scriptstyle (\otimes_h,\otimes_h^r)} & & \\
\mathsf{P}\binom{\underline{d}_1,\underline{d}_2}{\underline{c}_1,\underline{c}_2} \otimes M\binom{\underline{c}_1,\underline{c}_2}{\underline{b}_1,\underline{b}_2} & \xrightarrow{\otimes_v^l} & M\binom{\underline{d}_1,\underline{d}_2}{\underline{b}_1,\underline{b}_2}.
\end{array} \tag{15.14}$$

EXAMPLE 15.25. Suppose $f\colon \mathsf{P} \longrightarrow \mathsf{Q}$ is a morphism of $\mathfrak{C}$-colored PROPs. Then Q becomes a biased P-module whose left horizontal P-action is the composition

$$\begin{array}{ccc}
\mathsf{P}\binom{\underline{d}}{\underline{c}} \otimes \mathsf{Q}\binom{\underline{b}}{\underline{a}} & \xrightarrow{\otimes_h^l} & \mathsf{Q}\binom{\underline{d},\underline{b}}{\underline{c},\underline{a}} \\
\Big\downarrow{\scriptstyle f\otimes Id} & \nearrow{\scriptstyle \otimes_h} & \\
\mathsf{Q}\binom{\underline{d}}{\underline{c}} \otimes \mathsf{Q}\binom{\underline{b}}{\underline{a}}. & &
\end{array}$$

The other three P-action maps are defined similarly. In particular, if A is a P-algebra, then the endomorphism PROP E_A becomes a biased P-module via the structure map $\mathsf{P} \longrightarrow \mathsf{E}_A$. This is the way in which P-modules are used in [**FMY09**] to study deformation theory.

Here we apply Theorems 15.10 and 15.12 using Theorem 7.27 and Corollary 11.31 to produce the following.

COROLLARY 15.26. *Given $M \in \mathbf{\Sigma}_S$ and a $\mathfrak{C}$-colored PROP P, regarded as a $\mathtt{Gr}^{\uparrow}$-PROP, a biased P-module structure on M is equivalent to a P-module structure on M. Furthermore, this extends to a natural isomorphism of these biased and unbiased module categories.*

15.4.8. Modules over a Wheeled PROP. Fix a wheeled PROP $(\mathsf{P},\otimes_h,\xi_j^i,\mathbf{1})$ as in Definition 11.33.

DEFINITION 15.27. A **biased P-module** $(M,\otimes_h^l,\otimes_h^r,\xi_j^i)$ consists of:

- an object $M \in \mathbf{\Sigma}_S$,

- a left horizontal P-action

$$\mathsf{P}\binom{\underline{d}}{\underline{c}} \otimes M\binom{\underline{b}}{\underline{a}} \xrightarrow{\otimes_h^l} M\binom{\underline{d},\underline{b}}{\underline{c},\underline{a}},$$

- a right horizontal P-action

$$M\binom{\underline{d}}{\underline{c}} \otimes \mathsf{P}\binom{\underline{b}}{\underline{a}} \xrightarrow{\otimes_h^r} M\binom{\underline{d},\underline{b}}{\underline{c},\underline{a}},$$

and

- a contraction

$$M\binom{\underline{d}}{\underline{c}} \xrightarrow{\xi_j^i} M\binom{\underline{d} \setminus d_i}{\underline{c} \setminus c_j}$$

whenever $c_j = d_i$.

As expected, one must also require the commutativity of each analog of the diagrams in Definition 11.33 in which each diagram begins with one tensor factor of M.

A **morphism** of biased P-modules is a morphism of the underlying $\mathbf{\Sigma}_S$-bimodules that is compatible with $\otimes_h^l$, $\otimes_h^r$, and ξ_j^i.

In this context, Theorems 15.10 and 15.12 pair with Theorem 7.25 and Corollary 11.35 to produce the following.

COROLLARY 15.28. *Given $M \in \mathbf{\Sigma}_S$ and a wheeled PROP P, regarded as a $\mathtt{Gr}_w^{\circlearrowleft}$-PROP, a P-module structure on M is equivalent to a P-module structure on M. Furthermore, this extends to a natural isomorphism of these biased and unbiased module categories.*

15.4.9. Modules over a Wheeled Properad. Fix a choice of wheeled properad $\left(\mathsf{P}, \xi_j^i, {}_j\circ_i, \mathbf{1}\right)$ as in Definition 11.37.

DEFINITION 15.29. A **biased P-module** is a quadruple $\left(M, \xi_j^i, {}_j\circ_i{}^l, {}_j\circ_i{}^r\right)$ consisting of:

- $M \in \mathbf{\Sigma}_S$,
- a contraction

$$M\binom{\underline{d}}{\underline{c}} \xrightarrow{\xi_j^i} M\binom{\underline{d} \setminus d_i}{\underline{c} \setminus c_j}$$

whenever $c_j = d_i$,

- a map

$$\mathsf{P}\binom{\underline{d}}{\underline{c}} \otimes M\binom{\underline{b}}{\underline{a}} \xrightarrow{{}_j\circ_i{}^l} M\binom{\underline{b} \circ_j \underline{d}}{\underline{c} \circ_i \underline{a}}$$

whenever $b_j = c_i$,

- a map

$$M\binom{\underline{d}}{\underline{c}} \otimes \mathsf{P}\binom{\underline{b}}{\underline{a}} \xrightarrow{{}_j\circ_i{}^r} M\binom{\underline{b} \circ_j \underline{d}}{\underline{c} \circ_i \underline{a}}$$

whenever $b_j = c_i$.

In addition, one must require that all diagrams altered from those in Definition 11.37 by replacing one tensor factor of P with M in the common source should commute.

A **morphism** of biased P-modules is a morphism in $\mathbf{\Sigma}_S$ that respects the operations ${}_j\circ_i{}^l$, ${}_j\circ_i{}^r$, and ξ_j^i.

This time it is Theorem 7.88 and Corollary 11.39 which serve as the input for Theorems 15.10 and 15.12 to produce the following.

COROLLARY 15.30. *Given $M \in \Sigma_S$ and a wheeled properad* P, *regarded as a* $\mathtt{Gr}_c^{\circlearrowright}$*-PROP, a* P*-module structure on* M *is equivalent to a* P*-module structure on* M*. Furthermore, this extends to a natural isomorphism of these biased and unbiased module categories.*

15.4.10. Modules over a Wheeled Operad. Fix a choice of wheeled operad $(\mathsf{P}, \gamma, \mathbf{1}, \rho, \xi_j^1)$ as in Definition 11.41.

DEFINITION 15.31. A **biased** P**-module** $(M, \gamma^l, \gamma^r, \rho^l, \rho^r, \xi_j^1)$ consists of:

- an object $M \in \Sigma_{S_{\le 1}}$,
- maps γ^l and γ^r as in Definition 15.16,
- a map ρ^l

$$\mathsf{P}\binom{\varnothing}{\underline{c}} \otimes \underbrace{\mathsf{P}\binom{c_1}{\underline{b}_1} \otimes \cdots \otimes M\binom{c_i}{\underline{b}_i} \otimes \cdots \otimes \mathsf{P}\binom{c_m}{\underline{b}_m}}_{\text{exactly one tensor factor of } M} \xrightarrow{\rho^l} M\binom{\varnothing}{\underline{b}},$$

- a map ρ^r

$$M\binom{\varnothing}{\underline{c}} \otimes \bigotimes_{i=1}^m \mathsf{P}\binom{c_i}{\underline{b}_i} \xrightarrow{\rho^r} M\binom{\varnothing}{\underline{b}},$$

  and
- a contraction

$$M\binom{c_j}{\underline{c}} \xrightarrow{\xi_j^1} M\binom{\varnothing}{\underline{c} \setminus c_j}.$$

These operations satisfy the variations on the diagrams required of a wheeled operad (Definition 11.41) in which each diagram begins with one tensor factor of M.

A **morphism** of biased P-modules is a morphism of the underlying $\Sigma_{S_{\le 1}}$-bimodules that is compatible with γ^l, γ^r, ρ^l, ρ^r, and ξ_j^1.

The following result is a consequence of Theorems 15.10 and 15.12, using Theorem 7.53 and Corollary 11.43.

COROLLARY 15.32. *Given $M \in \Sigma_{S_{\le 1}}$ and a wheeled operad* P, *regarded as a* $\mathtt{Tree}^{\circlearrowright}$*-PROP, a biased* P*-module structure on* M *is equivalent to a* P*-module structure on* M*. Furthermore, this extends to a natural isomorphism of these biased and unbiased module categories.*

CHAPTER 16

May Modules over Algebras over Operads

The main purpose of this final chapter is to establish that the notion of May module agrees with the usual notions of modules over an operad, or over a PROP (Definition 15.2), used in deformation theory (e.g. [**FMY09**]). Along the way, it is shown that May modules are monadic algebras (Theorem 16.18), so come equipped with a free module construction, as well as standard (co)limit constructions. May modules are defined over an algebra A over an operad O. When choosing $\mathsf{O} = \overline{\mathsf{U}}_{\mathcal{G}}$ and $A = \mathsf{P}$ a $\mathcal{G}$-PROP viewed as a $\overline{\mathsf{U}}_{\mathcal{G}}$-algebra (Theorem 14.1), the result is another description of the category of P-modules as the category of May P-modules (Theorem 16.20).

Recall the pointed extension $F_{\mathcal{G}}(\mathsf{P}, M)$ of the monad associated to a pasting scheme was built from a coproduct of copies of $(\mathsf{P}, M)[G, v]$, decorating a pointed graph at the distinguished vertex by M and everywhere else by P. In this section, the analog is even simpler, where we instead look at a tree and decorate the *final* input by M. Since this still recovers things like classical bimodules, it should not be surprising that here we will need to work more with equivariant machinery than we have done so far. Thus, the first section starts with some technical background on groupoid-indexed colimits, before presenting the classical monad associated to an operad (Definition 16.3). Then its pointed extension used to define May modules (Proposition 16.9) and the associated monad $\mathcal{F}$ (Definition 16.10) are presented, along with a nice connection to the universal enveloping algebra in the one color case (Proposition 16.11 and Remark 16.12).

The second section begins with a concrete definition of May modules, before establishing them as the modules with respect to the pointed extension just constructed (Lemma 16.16). At this point, the results of Section 10.2 imply the category of May modules is the category of algebras over $\mathcal{F}$, implying the existence of a free May module functor and the existence of (co)limits of May modules. The chapter concludes with the verification that there is a canonical isomorphism of categories between the May P-modules over the S-colored operad $\overline{\mathsf{U}}_{\mathcal{G}}$ and the P-modules with respect to $F_{\mathcal{G}}$ (Theorem 16.20).

16.1. Preliminaries on Modules over an Algebra over an Operad

The goal of this section is to produce a pointed extension of the monad associated to the pasting scheme for May operads, `UTree`, similar to that used to define modules over a $\mathcal{G}$-PROP. In Chapter 15, the basic idea was to take the monad associated to a pasting scheme, built from a coproduct of tensor products of terms of P, and replace exactly one tensor factor instead with a term of M. In the context of colored operads, we can be even more specific, and replace exactly the *last* tensor factor with a term from M, at the cost of dealing with more equivariant machinery than we have dealt with until now. Thus, we begin with a brief discussion of tensoring over a groupoid and the examples of greatest interest to us.

16.1.1. Groupoid-Indexed Colimit. Let $\mathtt{G}$ be a small, non-empty groupoid. Let $X: \mathtt{G}^{op} \longrightarrow \mathcal{E}$ and $Y: \mathtt{G} \longrightarrow \mathcal{E}$ be two diagrams in $\mathcal{E}$. Define the diagram

$$X \otimes Y: \mathtt{G} \longrightarrow \mathcal{E}$$

as follows. For an object $g \in \mathtt{G}$, set

$$(X \otimes Y)(g) = X(g) \otimes Y(g).$$

For an isomorphism $\sigma: g \longrightarrow h \in \mathtt{G}$, set

$$(X \otimes Y)(\sigma) = X(\sigma^{-1}) \otimes Y(\sigma): (X \otimes Y)(g) \longrightarrow (X \otimes Y)(h).$$

Note contravariance in the X variable is the reason for the restriction to considering this construction only for groupoids. Define the colimit

$$X \otimes_{\mathtt{G}} Y = \operatorname*{colim}_{\mathtt{G}}(X \otimes Y) \in \mathcal{E}. \tag{16.1}$$

Note that the object $X \otimes_{\mathtt{G}} Y$ is represented by $(X \otimes Y)(g)$ for any object $g \in \mathtt{G}$, in the sense that a morphism

$$f: X \otimes_{\mathtt{G}} Y \longrightarrow Z \in \mathcal{E}$$

is uniquely determined by the restriction $\overline{f}$ in

$$\begin{array}{ccc} (X \otimes Y)(g) & \xrightarrow{\;\cong\;} & X \otimes_{\mathtt{G}} Y \\ & \overline{f} \searrow \quad \swarrow f & \\ & Z. & \end{array}$$

Our primary interest for this construction will be when $\mathtt{G} = \Sigma_{\underline{c}}$ for some $\underline{c} \in \mathcal{P}(\mathfrak{C})$.

DEFINITION 16.1. (1) If $A = \{A_c\}_{c \in \mathfrak{C}} \in \mathcal{E}^{\mathfrak{C}}$, define the diagram

$$A_{[\underline{c}]}: \Sigma_{\underline{c}} \longrightarrow \mathcal{E} \tag{16.2}$$

on objects by setting

$$A_{[\underline{c}]}(\sigma\underline{c}) = A_{c_{\sigma(1)}} \otimes \cdots \otimes A_{c_{\sigma(m)}},$$

with arrows coming from permutations of the tensor factors using the symmetry of the monoidal structure in $\mathcal{E}$.

(2) If O is a Σ_{S_1}-bimodule and $[\underline{c}]$ is an orbit type in $\mathcal{P}(\mathfrak{C})$, define the diagram

$$\mathsf{O}\binom{d}{[\underline{c}]}: \Sigma_{\underline{c}}^{op} \longrightarrow \mathcal{E}, \tag{16.3}$$

for any $d \in \mathfrak{C}$, on objects by setting

$$\mathsf{O}\binom{d}{[\underline{c}]}(\underline{c}\sigma^{-1}) = \mathsf{O}\binom{d}{\underline{c}\sigma^{-1}}$$

with arrows coming from the Σ_{S_1}-bimodule structure maps of O.

Combining this construction with the previous one, we obtain the object

$$\mathsf{O}\binom{d}{[\underline{c}]} \otimes_{\Sigma_{\underline{c}}} A_{[\underline{c}]} \in \mathcal{E}.$$

(3) If O is a Σ_{S_1}-bimodule and $[\underline{c}]$ is an orbit type in $\mathcal{P}(\mathfrak{C})$, define the diagram

$$\mathsf{O}\binom{d}{[\underline{c}],b} : \Sigma_{\underline{c}}^{op} \longrightarrow \mathcal{E},$$

for any $b, d \in \mathfrak{C}$, on objects by setting

$$\mathsf{O}\binom{d}{[\underline{c}],b}(\underline{c}\sigma^{-1}) = \mathsf{O}\binom{d}{\underline{c}\sigma^{-1},b}.$$

with arrows again coming from the Σ_{S_1}-bimodule structure maps of O.

Again combining this construction with the first one, we obtain the object

$$\mathsf{O}\binom{d}{[\underline{c}],b} \otimes_{\Sigma_{\underline{c}}} A_{[\underline{c}]} \quad \in \mathcal{E}.$$

In essence, the constructions just described allow us to work with $\mathfrak{C}$-colored and S-colored objects a full orbit at a time.

EXAMPLE 16.2. In the one-colored case,

$$\mathsf{O}\binom{d}{[\underline{c}]} \otimes_{\Sigma_{\underline{c}}} A_{[\underline{c}]} \text{ specifies to } \mathsf{O}(n) \otimes_{\Sigma_n} A^{\otimes n},$$

and

$$\mathsf{O}\binom{d}{[\underline{c}],b} \otimes_{\Sigma_{\underline{c}}} A_{[\underline{c}]} \text{ specifies to } \mathsf{O}(n+1) \otimes_{\Sigma_n} A^{\otimes n},$$

where Σ_n acts on $\mathsf{O}(n+1)$ via the inclusion $\Sigma_n \hookrightarrow \Sigma_{n+1}$ as permutations fixing the last entry.

16.1.2. A Pointed Monad Extension for Modules over an Algebra over a Colored Operad. We begin here by introducing the original formula of the monad associated to an operad, followed by the variant thereof with the last tensor factor always M. We will then establish enough technical machinery to verify that the second of these produces a pointed extension of the first monad, whose modules are precisely the modules over an algebra over an operad. Then, applying Theorem 10.22 will produce a monad whose algebras are the modules in question, so a free module functor in this context.

DEFINITION 16.3. Let O be a Σ_{S_1}-bimodule, while A and M are objects in $\mathcal{E}^{\mathfrak{C}}$.

(1) Define the algebra functor

$$F_{\mathsf{O}}(-) : \mathcal{E}^{\mathfrak{C}} \longrightarrow \mathcal{E}^{\mathfrak{C}}$$

by setting

$$F_{\mathsf{O}}(A)_d = \coprod_{[\underline{c}] \in \mathcal{P}(\mathfrak{C})} \mathsf{O}\binom{d}{[\underline{c}]} \otimes_{\Sigma_{\underline{c}}} A_{[\underline{c}]} \in \mathcal{E}. \tag{16.4}$$

(2) Define the module functor

$$F_{\mathsf{O}}(-,-) : \mathcal{E}^{\mathfrak{C}} \times \mathcal{E}^{\mathfrak{C}} \longrightarrow \mathcal{E}^{\mathfrak{C}}$$

by setting

$$F_{\mathsf{O}}(A, M)_d = \coprod_{\substack{[\underline{c}] \in \mathcal{P}(\mathfrak{C}) \\ b \in \mathfrak{C}}} \left[\mathsf{O}\binom{d}{[\underline{c}],b} \otimes_{\Sigma_{\underline{c}}} A_{[\underline{c}]} \right] \otimes M_b. \tag{16.5}$$

The coproduct is taken over all orbit types of $\underline{c} \in \mathcal{P}(\mathfrak{C})$ and all colors $b \in \mathfrak{C}$.

EXAMPLE 16.4. For example, in the one-colored case, we have

$$F_{\mathsf{O}}(A) = \coprod_{n\geq 0} \mathsf{O}(n) \otimes_{\Sigma_n} A^{\otimes n},$$

which is the original formula of the monad associated to an operad, if O is such, in [**May72**]. In addition, still in the one-colored case, we have

$$F_{\mathsf{O}}(A,M) = \coprod_{n\geq 0} \left[\mathsf{O}(n+1) \otimes_{\Sigma_n} A^{\otimes n}\right] \otimes M.$$

As expected, there is a unit map $M \longrightarrow F_{\mathsf{O}}(A,M)$ provided O has the unit maps of an operad.

DEFINITION 16.5. Suppose $(\mathsf{O}, \gamma, \mathbf{1})$ is a $\mathfrak{C}$-colored operad, while A and M are objects in $\mathcal{E}^{\mathfrak{C}}$. Define the natural map

$$M \xrightarrow{\nu_{A,M}} F_{\mathsf{O}}(A,M) \tag{16.6}$$

in $\mathcal{E}^{\mathfrak{C}}$ whose d-colored component is the composition

$$\begin{array}{ccc} M_d & \xrightarrow{\nu_{A,M}} & F_{\mathsf{O}}(A,M)_d \\ \cong\big\downarrow & & \big\uparrow \text{natural} \\ I \otimes M_d & \xrightarrow{(\mathbf{1}_d, Id)} & \mathsf{O}\binom{d}{d} \otimes M_d. \end{array}$$

Now we observe that both A itself and $F_{\mathsf{O}}(A)$ could potentially act on $F_{\mathsf{O}}(A,M)$.

DEFINITION 16.6. Let O be a Σ_{S_1}-bimodule, while A and M are objects in $\mathcal{E}^{\mathfrak{C}}$.

(1) Define the functor $F^2_{\mathsf{O}} : \mathcal{E}^{\mathfrak{C}} \times \mathcal{E}^{\mathfrak{C}} \longrightarrow \mathcal{E}^{\mathfrak{C}}$ as

$$F^2_{\mathsf{O}}(A,M) = F_{\mathsf{O}}(A, F_{\mathsf{O}}(A,M)) \in \mathcal{E}^{\mathfrak{C}}. \tag{16.7}$$

(2) Define the functor $\overline{F}_{\mathsf{O}} : \mathcal{E}^{\mathfrak{C}} \times \mathcal{E}^{\mathfrak{C}} \longrightarrow \mathcal{E}^{\mathfrak{C}}$ as

$$\overline{F}_{\mathsf{O}}(A,M) = F_{\mathsf{O}}\left(F_{\mathsf{O}}A, F_{\mathsf{O}}(A,M)\right) \in \mathcal{E}^{\mathfrak{C}}. \tag{16.8}$$

We can visualize $F^2_{\mathsf{O}}(A,M)_d$ as the space of decorated simple trees

(16.9)

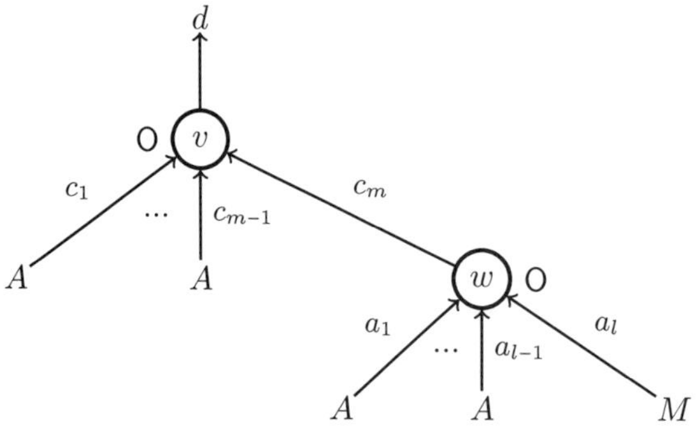

with one d-colored output and $m-1+l$ inputs with colors $(c_{[1,m-1]}, a_{[1,l]})$. All but the last inputs are decorated by components of A with the corresponding colors, while the last input is decorated by M_{a_l}. The top and the bottom vertices are decorated by $\mathsf{O}\binom{d}{\underline{c}}$ and $\mathsf{O}\binom{c_m}{\underline{a}}$, respectively.

Note that for $d \in \mathfrak{C}$, the d-colored component of $F^2_{\mathsf{O}}(A, M)$ is

$$\begin{aligned}&\coprod_{[c_{[1,m-1]}],c_m} \left[\mathsf{O}\binom{d}{[c_{[1,m-1]}],c_m} \otimes_{\Sigma_{c_{[1,m-1]}}} A_{[c_{[1,m-1]}]}\right] \otimes F_{\mathsf{O}}(A,M)_{c_m}\\ &\cong \coprod \left[\mathsf{O}\binom{d}{\underline{c}} \otimes A_{c_{[1,m-1]}}\right] \otimes \left[\mathsf{O}\binom{c_m}{a_{[1,l]}} \otimes A_{a_{[1,l-1]}} \otimes M_{a_l}\right].\end{aligned}$$

Similarly, the d-colored component of $\overline{F}_{\mathsf{O}}(A, M)$ is

$$\begin{aligned}&\coprod_{[c_{[1,m-1]}],c_m} \left[\mathsf{O}\binom{d}{[c_{[1,m-1]}],c_m} \otimes_{\Sigma_{c_{[1,m-1]}}} (F_{\mathsf{O}}A)_{[c_{[1,m-1]}]}\right] \otimes F_{\mathsf{O}}(A,M)_{c_m}\\ &\cong \coprod \left[\mathsf{O}\binom{d}{\underline{c}} \otimes \bigotimes_{i=1}^{m-1} \mathsf{O}\binom{c_i}{\underline{b}_i} \otimes A_{\underline{b}_i}\right] \otimes \left[\mathsf{O}\binom{c_m}{\underline{b}_m} \otimes A_{b_{[P_{m-1}+1,P-1]}} \otimes M_{b_P}\right].\end{aligned}$$

DEFINITION 16.7. Suppose $(\mathsf{O}, \gamma, \mathbf{1})$ is a $\mathfrak{C}$-colored operad, while A and M are objects in $\mathcal{E}^{\mathfrak{C}}$. Define the natural map

$$\overline{F}_{\mathsf{O}}(A, M) \xrightarrow{\mu_{A,M}} F_{\mathsf{O}}(A, M) \tag{16.10}$$

in $\mathcal{E}^{\mathfrak{C}}$ whose d-colored component is induced by the composition

$$\begin{array}{c} \mathsf{O}\binom{d}{\underline{c}} \otimes \left[\bigotimes_{i=1}^{m-1} \mathsf{O}\binom{c_i}{\underline{b}_i} \otimes A_{\underline{b}_i}\right] \otimes \left[\mathsf{O}\binom{c_m}{\underline{b}_m} \otimes A_{b_{[P_{m-1}+1,P-1]}} \otimes M_{b_P}\right] \\ \downarrow \cong \\ \left[\mathsf{O}\binom{d}{\underline{c}} \otimes \bigotimes_{i=1}^{m} \mathsf{O}\binom{c_i}{\underline{b}_i}\right] \otimes A_{b_{[1,P-1]}} \otimes M_{b_P} \\ \downarrow (\gamma, Id) \\ \mathsf{O}\binom{d}{\underline{b}} \otimes A_{b_{[1,P-1]}} \otimes M_{b_P} \\ \downarrow \text{natural} \\ F_{\mathsf{O}}(A, M)_d. \end{array}$$

We can now verify that these constructions provide a pointed extension of F_{O}. In fact, we can define the composite λ in this case explicitly as follows.

DEFINITION 16.8. Suppose $(\mathsf{O}, \gamma, \mathbf{1})$ is a $\mathfrak{C}$-colored operad, while A and M are objects in $\mathcal{E}^{\mathfrak{C}}$. Define the natural map

$$F^2(A, M) \xrightarrow{\lambda_{A,M}} F(A, M) \tag{16.11}$$

in $\mathcal{E}^{\mathfrak{C}}$ whose d-colored component is induced by the composition

$$\begin{array}{c}
\left[\mathsf{O}\binom{d}{\underline{c}} \otimes A_{c_{[1,m-1]}}\right] \otimes \left[\mathsf{O}\binom{c_m}{\underline{a}} \otimes A_{a_{[1,l-1]}} \otimes M_{a_l}\right] \\
\Big\downarrow \cong \\
\left[\mathsf{O}\binom{d}{\underline{c}} \otimes I^{\otimes m-1} \otimes \mathsf{O}\binom{c_m}{\underline{a}}\right] \otimes A_{c_{[1,m-1]},a_{[1,l-1]}} \otimes M_{a_l} \\
\Big\downarrow (Id \otimes \mathbf{1}_{c_i}, Id) \\
\left[\mathsf{O}\binom{d}{\underline{c}} \otimes \left(\otimes_{i=1}^{m-1} \mathsf{O}\binom{c_i}{c_i}\right) \otimes \mathsf{O}\binom{c_m}{\underline{a}}\right] \otimes A_{c_{[1,m-1]},a_{[1,l-1]}} \otimes M_{a_l} \\
\Big\downarrow (\gamma, Id) \\
\mathsf{O}\binom{d}{c_{[1,m-1]},\underline{a}} \otimes A_{c_{[1,m-1]},a_{[1,l-1]}} \otimes M_{a_l} \\
\Big\downarrow \text{natural} \\
F(A,M)_d.
\end{array}$$

PROPOSITION 16.9. *Given a $\mathfrak{C}$-colored operad $(\mathsf{O}, \gamma, \mathbf{1})$ and (A, θ) an algebra over O, there is an associated pointed extension of the monad F_{O} given by*

$$(F_{\mathsf{O}}(A, M), \nu_{A,M}, \mu_{A,M}) \ .$$

PROOF. It is relatively straightforward to deduce that $F_{\mathsf{O}}(A, ?)$ is a monad from the respective diagrams for (O, γ) itself as an operad, considered as an algebra over the monad associated to the pasting scheme $\mathcal{G}$ = `UTree`. That is, the associativity condition for $F_{\mathsf{O}}(A, ?)$ follows as a consequence of some manipulation and the associativity of γ, while the two unit triangles follow from the related unit conditions for γ. The general unit condition (10.6) then follows by inspection from the unit condition for γ once again, so it remains to verify the associativity diagram (10.7) commutes.

Thus, we consider the diagram (writing F for F_{O} and F_A for $F_{\mathsf{O}}(A, ?)$)

$$\begin{array}{ccc}
F(F^2 A, F(FA, F_A^2 M)) & \xrightarrow{\mu_{FA, F_A^2 M}} & F(FA, F_A^2 M) \\
{\scriptstyle F(F(\theta), \mu_{A, F_A M})}\Big\downarrow & & \Big\downarrow{\scriptstyle \mu_{A, F_A M}} \\
F(FA, F_A^2 M) & \xrightarrow{\mu_{A, F_A M}} & F_A^2 M \ .
\end{array}$$

Here the commutativity follows by combining the unity and associativity of the operad structure map γ with the unity of the O-algebra structure map θ for A. □

The associated monad of Theorem 10.22 is now given by the following.

DEFINITION 16.10. Define the object

$$\mathcal{F}(A, M) = \text{coequalizer}\left\{F_{\mathsf{O}}(A, F_{\mathsf{O}}(A, M)) \rightrightarrows F_{\mathsf{O}}(A, M)\right\} \tag{16.12}$$

as a coequalizer in $\mathcal{E}^{\mathfrak{C}}$ of the two compositions in (10.8). The natural map from $F_{\mathsf{O}}(A, M)$ to $\mathcal{F}(A, M)$ is denoted by $\eta_{A,M}$, η_M, or simply η.

In the one-colored case, there is an easier description of the coequalizer, which unfortunately does *not* generalize. As a consequence, the following result is not used elsewhere, and provides an indication of why the universal enveloping algebra does not generalize.

PROPOSITION 16.11. *Suppose* $\mathfrak{C} = \{*\}$ *and* $A, M \in \mathcal{E}^{\mathfrak{C}}$. *Then there is a natural isomorphism*

$$F_{\mathsf{O}}(A, M) \cong F_{\mathsf{O}}(A, I) \otimes M.$$

Moreover, if (A, θ) *is an* O*-algebra, then there is a natural isomorphism*

$$\mathcal{F}(A, M) \cong \mathcal{F}(A, I) \otimes M.$$

PROOF. To prove the first isomorphism, one uses the fact that the monoidal product $\otimes$ preserves coproducts and computes as follows:

$$\begin{aligned} F(A, M) &= \coprod_{m \geq 0} \left[\mathsf{O}(m+1) \otimes_{\Sigma_m} A^{\otimes m}\right] \otimes M \\ &\cong \left[\coprod_{m \geq 0} \mathsf{O}(m+1) \otimes_{\Sigma_m} A^{\otimes m}\right] \otimes M \\ &= F(A, I) \otimes M. \end{aligned}$$

The second isomorphism is proved similarly, using the fact that $\otimes$ preserves split coequalizers. □

REMARK 16.12. In the one-colored linear case, the object $\mathcal{F}(A, I)$ is called the universal enveloping algebra of A and is sometimes denoted by $U(A)$ [**KM95, GK94**]. In the general colored case, the natural isomorphisms in Proposition 16.11 do *not* hold, as can be seen directly from the definition (16.5) of $F(A, M)$.

16.2. May Modules over an Operadic Algebra

Fix a $\mathfrak{C}$-colored operad $(\mathsf{O}, \gamma, \mathbf{1})$ (Definition 11.14) and an O-algebra (A, θ) (Corollary 13.37) in $\mathcal{E}$ for this section. We first define a May A-module in terms of structure maps and commutative diagrams, but then verify this is equivalent to the notion of module associated to the pointed extension $F_{\mathsf{O}}(A, M)$ just constructed. Then we can appeal to the results of Section 10.2 to produce a monad with the category of May A-modules as its category of algebras. Specializing to $\mathsf{O} = \overline{\mathsf{U}}_{\mathcal{G}}$ and $A = \mathsf{P}$ we then identify the May A-modules with the modules over the $\mathcal{G}$-PROP P in the sense of the last chapter (Definition 15.2). In particular, this identifies the usual modules over operads or PROPs in deformation theory as May modules.

16.2.1. Definition of a May Module.

The one-colored case of the following definition can be found in, e.g., [**May97**].

DEFINITION 16.13. A **May** A**-module** is a pair (M, λ) in which

- $M = \{M_c\} \in \mathcal{E}^{\mathfrak{C}}$, and
- λ is a map

$$(\mathsf{O}, A, M)(\underline{c}; d) \overset{\text{def}}{=} \mathsf{O}\binom{d}{\underline{c}} \otimes A_{c_{[1,m-1]}} \otimes M_{c_m} \xrightarrow{\lambda} M_d \tag{16.13}$$

in $\mathcal{E}$ for each color d and each $\mathfrak{C}$-profile $\underline{c}$ with $|\underline{c}| = m \geq 1$, such that the following associativity, unity, and equivariant conditions are required to hold.

Associativity: Suppose $\underline{b}_i \in \mathcal{P}(\mathfrak{C})$ for $1 \leq i \leq m$ with

$$|\underline{b}_i| = p_i \geq 1, \quad P_i = p_1 + \cdots + p_i, \quad P_0 = 0, \quad P = P_m,$$

and

$$\underline{b} = (\underline{b}_1, \ldots, \underline{b}_m) = (b_1, \ldots, b_P).$$

Then the following diagram is required to be commutative.

(16.14)

$$\begin{array}{ccc}
\mathsf{O}\binom{d}{\underline{c}} \otimes \Big[\bigotimes_{i=1}^{m} \mathsf{O}\binom{c_i}{\underline{b}_i}\Big] \otimes A_{b_{[1,P-1]}} \otimes M_{b_P} & \xrightarrow{\gamma\otimes Id} & (\mathsf{O}, A, M)(\underline{b}; d) \\
\downarrow{\scriptstyle \text{shuffle}\ \cong} & & \\
\mathsf{O}\binom{d}{\underline{c}} \otimes \Big[\bigotimes_{i=1}^{m-1} (\mathsf{O}, A, A)(\underline{b}_i; c_i)\Big] \otimes (\mathsf{O}, A, M)(\underline{b}_m; c_m) & & \downarrow{\scriptstyle \lambda} \\
\downarrow{\scriptstyle (Id, \theta^{m-1}, \lambda)} & & \\
(\mathsf{O}, A, M)(\underline{c}; d) & \xrightarrow{\lambda} & M_d
\end{array}$$

Unity: The square

(16.15)

$$\begin{array}{ccc}
M_d & \xrightarrow{Id} & M_d \\
\downarrow{\scriptstyle \cong} & & \uparrow{\scriptstyle \lambda} \\
I \otimes M_d & \xrightarrow{\mathbf{1}_d \otimes Id} & \mathsf{O}\binom{d}{d} \otimes M_d
\end{array}$$

is required to be commutative for each $d \in \mathfrak{C}$.

Equivariance: For each permutation $\tau \in \Sigma_{m-1} \subseteq \Sigma_m$, the triangle

(16.16)

$$\begin{array}{ccc}
\mathsf{O}\binom{d}{\underline{c}} \otimes A_{c_{[1,m-1]}} \otimes M_{c_m} & \xrightarrow{\tau\otimes\tau^{-1}\otimes Id} & \mathsf{O}\binom{d}{\underline{c}\tau} \otimes A_{c_{[1,m-1]}\tau} \otimes M_{c_m} \\
& {\scriptstyle \lambda} \searrow \quad \swarrow {\scriptstyle \lambda} & \\
& M_d &
\end{array}$$

is required to be commutative, in which τ^{-1} acts on $A_{c_{[1,m-1]}}$ from the left by permuting the tensor factors.

A **morphism** $f: M \longrightarrow N$ of May A-modules is a morphism

$$f: M \longrightarrow N \in \mathcal{E}^{\mathfrak{C}}$$

such that the square

(16.17)

$$\begin{array}{ccc}
\mathsf{O}\binom{d}{\underline{c}} \otimes A_{c_{[1,m-1]}} \otimes M_{c_m} & \xrightarrow{\lambda} & M_d \\
\downarrow{\scriptstyle Id\otimes f} & & \downarrow{\scriptstyle f} \\
\mathsf{O}\binom{d}{\underline{c}} \otimes A_{c_{[1,m-1]}} \otimes N_{c_m} & \xrightarrow{\lambda} & N_d
\end{array}$$

is commutative. The category of May A-modules is denoted by $\mathsf{Mod}_{\mathsf{O}}(A)$.

EXAMPLE 16.14. Suppose O is the 1-colored operad for monoids in $\mathcal{E}$, and A is an O-algebra. Then a May A-module is a bimodule over the monoid A in the usual sense [**Mac98**] (VII.4).

EXAMPLE 16.15. Suppose $f:(A,\theta^A) \longrightarrow (B,\theta^B)$ is a morphism of O-algebras. Then B becomes a May A-module whose structure map is the composition

$$\begin{array}{ccc} \mathsf{O}\binom{d}{\underline{c}} \otimes A_{c_{[1,m-1]}} \otimes B_{c_m} & \xrightarrow{\lambda} & B_d \\ \Big\downarrow {\scriptstyle (Id_{\mathsf{O}},f^{m-1},Id_{B_{c_m}})} & \nearrow_{\theta^B} & \\ \mathsf{O}\binom{d}{\underline{c}} \otimes B_{\underline{c}}. & & \end{array}$$

As we will see below, the largest class of examples of May A-modules will arise from choosing $\mathsf{O} = \overline{\mathsf{U}}_{\mathcal{G}}$ and $A = \mathsf{P}$ for a $\mathcal{G}$-PROP, where $\mathbf{Mod}(\mathsf{P})$ will agree with the category of May P-modules.

16.2.2. Alternative Description of a May Module. The following observation gives a more compact description of an A-module using the constructions in the previous section.

LEMMA 16.16. *Suppose O is a $\mathfrak{C}$-colored operad, and (A,θ) is an O-algebra. Then a May A-module structure is equivalent to the structure of an A-module with respect to F_{O} (Definition* 10.18*). Furthermore, this extends to a canonical isomorphism of categories*

$$\mathrm{Mod}_{\mathsf{O}}(A) \cong \mathbf{Mod}_{F_{\mathsf{O}}}(A) \ .$$

Expanded a bit, the claim says a May A-module consists of a pair (M,λ) with $M \in \mathcal{E}^{\mathfrak{C}}$ and

$$F_{\mathsf{O}}(A,M) \xrightarrow{\lambda} M \in \mathcal{E}^{\mathfrak{C}}$$

such that the associativity square

$$\begin{array}{ccc} \overline{F}_{\mathsf{O}}(A,M) & \xrightarrow{\mu_{A,M}} & F_{\mathsf{O}}(A,M) \\ {\scriptstyle F_{\mathsf{O}}(\theta,\lambda)}\Big\downarrow & & \Big\downarrow{\scriptstyle \lambda} \\ F_{\mathsf{O}}(A,M) & \xrightarrow{\lambda} & M \end{array} \tag{16.18}$$

and the unity diagram

$$\begin{array}{ccc} M & \xrightarrow{\nu_{A,M}} & F_{\mathsf{O}}(A,M) \\ & {\scriptstyle Id}\searrow & \Big\downarrow{\scriptstyle \lambda} \\ & & M \end{array} \tag{16.19}$$

are commutative in $\mathcal{E}^{\mathfrak{C}}$.

Moreover, suppose (M,λ^M) and (N,λ^N) are A-modules. Then a morphism $f: M \longrightarrow N \in \mathcal{E}^{\mathfrak{C}}$ is a morphism of A-modules if and only if the square

$$\begin{array}{ccc} F_{\mathsf{O}}(A,M) & \xrightarrow{\lambda^M} & M \\ {\scriptstyle F_{\mathsf{O}}(Id_A,f)}\Big\downarrow & & \Big\downarrow{\scriptstyle f} \\ F_{\mathsf{O}}(A,N) & \xrightarrow{\lambda^N} & N \end{array} \tag{16.20}$$

is commutative.

PROOF. From the construction of $\otimes_{\Sigma_{\underline{c}}}$, the map $\lambda: F_{\mathsf{O}}(A, M) \longrightarrow M$ is uniquely determined by maps

$$\mathsf{O}\binom{d}{\underline{c}} \otimes A_{c_{[1,m-1]}} \otimes M_{c_m} \xrightarrow{\lambda} M_d \tag{16.21}$$

in $\mathcal{E}$ for $\underline{c} \in \mathcal{P}(\mathfrak{C})$ with $|\underline{c}| = m \geq 1$. With respect to the maps in (16.21), the diagrams (16.14) and (16.15) in Definition 16.13 are equivalent to the diagrams (16.18) and (16.19), respectively. Moreover, the square (16.17) is equivalent to the square (16.20). □

REMARK 16.17. In the one-colored case, the description of a May module as described after the statement of Lemma 16.16 appeared in [**GH00**] as the definition of a module over an algebra over an operad in a symmetric monoidal closed *abelian* category.

16.2.3. May Modules are Monadic Algebras. Combining Theorem 10.22, Lemma 16.16, and the fact that the definition of $\mathcal{F}(A,-)$ is that of Definition 10.21, we have now established the following.

THEOREM 16.18. *Suppose* O *is a* $\mathfrak{C}$*-colored operad, and* (A,θ) *is an* O*-algebra. Then there is a monad* $(\mathcal{F}(A,-),\mu,\nu)$ *in* $\mathcal{E}^{\mathfrak{C}}$ *whose category of algebras is canonically isomorphic to the category of May* A*-modules. In particular, the forgetful functor*

$$U: \mathrm{Mod}_{\mathsf{O}}(A) \longrightarrow \mathcal{E}^{\mathfrak{C}}, \quad (M,\lambda) \longmapsto M$$

admits a left adjoint $\mathcal{F}(A,-)$ (16.12).

REMARK 16.19. The one-colored case of Theorem 16.18, with the further assumption that $\mathcal{E}$ be an abelian category, is [**GH00**, Proposition 1.10].

16.2.4. Modules Over Generalized PROPs are May Modules. Fix a pasting scheme $\mathcal{G} = (S, \mathtt{G})$ and a $\mathcal{G}$-PROP (P,θ). For O we will use $\overline{\mathsf{U}}_{\mathcal{G}}$, the S-colored operad associated to $\mathcal{G}$ (Definition 14.3). Its algebras are equivalent to $\mathcal{G}$-PROPs by Theorem 14.1. Recall from section 14.1.1 that

$$\overline{\mathsf{U}}_{\mathcal{G}}\binom{t}{\underline{s}} = \coprod_{[G,\sigma]} I,$$

where the coproduct runs over the strict isomorphism classes of ordered wheeled graphs (G,σ) with $G \in \mathtt{G}_S$, $\binom{\mathrm{out}(G)}{\mathrm{in}(G)} = t$, and $s_{\sigma(v)}$ the pair of profiles of the vertex v.

In the following identification of the two definitions of P-modules, both ordered wheeled graphs and pointed wheeled graphs will be used. Ordered wheeled graphs (G,σ) appear in the S-colored operad O, which is part of the definition of a May P-module. Pointed wheeled graphs (G,v) are used in the definition of P-modules based upon the pointed extension $F_{\mathcal{G}}(\mathsf{P},M)$ of the monad $F_{\mathcal{G}}$ associated to a pasting scheme.

Recall the notation that $\mathbf{Mod}(\mathsf{P})$ is the category of modules over the generalized PROP P as originally defined in Definition 15.2, while $\mathrm{Mod}_{\overline{\mathsf{U}}_{\mathcal{G}}}(\mathsf{P})$ is the category of May P-modules when P is regarded as a $\overline{\mathsf{U}}_{\mathcal{G}}$-algebra (Definition 16.13).

THEOREM 16.20. *Suppose* (P,θ) *is a* $\mathcal{G}$*-PROP. Then there is a canonical isomorphism of categories*

$$\mathbf{Mod}(\mathsf{P}) \cong \mathrm{Mod}_{\overline{\mathsf{U}}_{\mathcal{G}}}(\mathsf{P}).$$

PROOF. Suppose (M, λ) is a P-module as in Definition 15.2. The P-module structure map

$$\lambda\colon F_{\mathcal{G}}(\mathsf{P}, M) \longrightarrow M \in \mathcal{E}^{dis(S)} \tag{16.22}$$

is uniquely determined by the component maps

$$\begin{array}{c} [\bigotimes_{u \neq v} \mathsf{P}(\widehat{u})] \otimes M(\widehat{v}) \cong (\mathsf{P}, M)[G, v] \\ \Big\downarrow \lambda_{[G,v]} \\ M\binom{\underline{d}}{\underline{c}} \end{array}$$

for $(\underline{c}; \underline{d}) \in S$ and for strict isomorphism classes of pointed wheeled graphs $[G, v] \in \mathsf{G}_{S*}\binom{\underline{d}}{\underline{c}}$. In particular, for each $G \in \mathsf{G}_S\binom{\underline{d}}{\underline{c}}$ with $\mathrm{Vt}(G) \neq \varnothing$ and each ordering

$$\sigma\colon \mathrm{Vt}(G) \xrightarrow{\cong} \{1, \ldots, m = |\mathrm{Vt}(G)|\},$$

there is a component map

$$\begin{array}{ccc} \left[\bigotimes_{i=1}^{m-1} \mathsf{P}(\widehat{v}_i)\right] \otimes M(\widehat{v}_m) & \xrightarrow{\lambda_{[G,\sigma]}} & M\binom{\underline{d}}{\underline{c}} \\ \text{natural}\Big\downarrow & \nearrow_{\lambda_{[G,v_m]}} & \\ (\mathsf{P}, M)[G, v_m], & & \end{array} \tag{16.23}$$

where the tensor products in the upper left entry are ordered, and $v_i = \sigma^{-1}(i)$ for each i.

For $t = (\underline{c}; \underline{d}) \in S$ and a profile $\underline{s} \in \mathcal{P}(S)$ with $|\underline{s}| \geq 1$, using the notation in (16.13), there is a natural isomorphism

$$\begin{array}{c} \left(\overline{\mathsf{U}}_{\mathcal{G}}, \mathsf{P}, M\right)(\underline{s}; t) \\ \Big\| \\ \overline{\mathsf{U}}_{\mathcal{G}}\binom{t}{\underline{s}} \otimes \mathsf{P}(s_1) \otimes \cdots \otimes \mathsf{P}(s_{m-1}) \otimes M(s_m) \\ \Big\downarrow \cong \\ \coprod_{[G,\sigma]} \left[\bigotimes_{i=1}^{m-1} \mathsf{P}(\widehat{v}_i)\right] \otimes M(\widehat{v}_m), \end{array}$$

where each $v_i = \sigma^{-1}(i)$ has profiles s_i. Therefore, we may assemble the component maps in (16.23) into a structure map

$$\lambda\colon F_{\overline{\mathsf{U}}_{\mathcal{G}}}(\mathsf{P}, M) \longrightarrow M \in \mathcal{E}^{dis(S)}$$

as in Lemma 16.16. With respect to this structure map, the S-colored object M becomes a May P-module. Indeed, the associativity square (16.18) in Lemma 16.16 is commutative because the P-module associativity square (15.4) is also. On the one hand, the map $\mu_{\mathsf{P},M}$ in (15.4) is induced by graph substitution. On the other hand, the $\mu_{\mathsf{P},M}$ in (16.18) is induced by the operad structure map γ of O, which is also induced by graph substitution. One can similarly check the unity diagram (16.19).

To obtain a P-module from a May P-module, simply reverse the above arguments. □

As noted above, consequences of Theorem 16.20 include the existence of (co)limits of modules, as (co)limits of monadic algebras, and an alternative construction of the free module functor over a generalized PROP.

Bibliography

[Bar10] C. Barwick, On left and right model categories and left and right Bousfield localizations, Homology, Homotopy and Applications 12 (2010), 245-320.

[BB] M. Batanin and C. Berger, Homotopy theory for algebras over polynomial monads, arXiv:1305.0086.

[BM14] M. Batanin and M. Markl, Operadic Categories and Duoidal Deligne's Conjecture, arXiv:1404.3886.

[BM03] C. Berger and I. Moerdijk, Axiomatic homotopy theory for operads, Comm. Math. Helv. 78 (2003), 805-831.

[BM06] C. Berger and I. Moerdijk, The Boardman-Vogt resolution of operads in monoidal model categories, Topology 45 (2006), 807-849.

[BM07] C. Berger and I. Moerdijk, Resolution of coloured operads and rectification of homotopy algebras, Contemp. Math. 431 (2007), 31-58.

[BV73] J.M. Boardman and R.M. Vogt, Homotopy invariant algebraic structures on topological spaces, Lecture Notes in Math. 347, Springer-Verlag, Berlin, 1973.

[Bor94] F. Borceux, Handbook of categorical algebra 2: categories and structures, Encyclopedia of Math. Appl. 51, Cambridge Univ. Press, Cambridge, 1994.

[BM08] D.V. Borisov and Y.I. Manin, Generalized operads and their inner cohomomorphisms, Prog. Math. 265 (2008), 247-308.

[CGMV10] C. Casacuberta, J.J. Gutierrez, I. Moerdijk, and R.M. Vogt, Localization of algebras over coloured operads, Proc. London Math. Soc. 101 (2010) 105-136.

[Cha05] D. Chataur, A bordism approach to string topology, Int. Math. Res. Not. (2005), 2829-2875.

[CG04] R.L. Cohen and V. Godin, A polarized view of string topology, London Math. Soc. Lecture Note Ser. 308, p. 127-154, Cambridge Univ. Press, Cambridge, 2004.

[CV06] R.L. Cohen and A.A. Voronov, Notes on string topology, in: String topology and cyclic homology, p. 1-95, Adv. Courses Math. CRM Barcelona, Birkhäuser, Basel, 2006.

[Dwy01] W.G. Dwyer, Classifying spaces and homology decompositions, in: Homotopy theoretical methods in group cohomology, Adv. Courses Math. CRM Barcelona, Birkhäuser, Basel, 2001.

[DS95] W.G. Dwyer and J. Spalinski, Homotopy theories and model categories, in: Handbook of algebraic topology, p. 73-126, North-Holland, Amsterdam, 1995.

[EKMM97] A.D. Elmendorf, I. Kriz, M.A. Mandell, and J.P. May, Rings, modules, and algebras in stable homotopy theory, Math. Surveys and Monographs 47, Amer. Math. Soc., Providence, RI, 1997.

[Far95] E.D. Farjoun, Cellular spaces, null spaces and homotopy localization, Lecture Notes in Math. 1622, Springer, Berlin, 1995.

[FMY09] Y. Frégier, M. Markl, and D. Yau, The L_∞-deformation complex of diagrams of algebras, New York J. Math. 15 (2009), 353-392.

[Fre09] B. Fresse, Modules over operads and functors, Lecture Notes in Math. 1967, Springer-Verlag, Berlin, 2009.

[Fre10] B. Fresse, Props in model categories and homotopy invariance of structures, Georgian Math. J. 17 (2010), 79-160.

[Gan03] W.L. Gan, Koszul duality for dioperads, Math. Res. Lett. 10 (2003), 109-124.

[Ger63] M. Gerstenhaber, The cohomology structure of an associative ring, Ann. Math. 78 (1963), 267-288.

[Get94] E. Getzler, Batalin-Vilkovisky algebras and two-dimensional topological field theories, Comm. Math. Phys. 159 (1994), 265-285.

[Get09] E. Getzler, Operads revisited, in: Algebra, arithmetic, and geometry: in honor of Yu. I. Manin vol. 1, Prog. Math. 269 (2009), 675-698.

[GK98] E. Getzler and M.M. Kapranov, Modular operads, Compositio Math. 110 (1998), 65-126.

[GK94] V. Ginzburg and M. Kapranov, Koszul duality for operads, Duke Math. J. 76 (1994), 203-272.

[GH00] P.G. Goerss and M.J. Hopkins, André-Quillen (co)-homology for simplicial algebras over simplicial operads, in: Contemp. Math. 265, Amer. Math. Soc., Providence, RI, 2000.

[Gut11] J.J. Gutiérrez, Transfer of algebras over operads along derived Quillen adjunctions, arXiv:1104.0584.

[HRY] P. Hackney, M. Robertson, and D. Yau, Infinity Properads and Infinity Wheeled Properads, Lecture Notes in Mathematics, Springer-Verlag, Berlin (to appear).

[Har09] J.E. Harper, Homotopy theory of modules over operads in symmetric spectra, Alg. Geom. Top. 9 (2009), 1637-1680.

[Har10] J.E. Harper, Homotopy theory of modules over operads and non-Σ operads in monoidal model categories, J. Pure Appl. Alg. 214 (2010), 1407-1434.

[Hin97] V. Hinich, Homological algebra of homotopy algebras, Comm. Alg. 25 (1997), 3291-3323.

[Hir03] P. Hirschhorn, Model categories and their localizations, Math. Surveys and Monographs 99, Amer. Math. Soc., Providence, RI, 2003.

[Hov99] M. Hovey, Model categories, Math. Surveys and Monographs 63, Amer. Math. Soc., Providence, RI, 1999.

[HSS00] M. Hovey, B. Shipley, and J. Smith, Symmetric spectra, J. Amer. Math. Soc. 13 (2000), 149-208.

[IJ02] M. Intermont and M.W. Johnson, Model structures on the category of ex-spaces, Top. Appl. 119 (2002), 325-353.

[Ion07] L.M. Ionescu, From operads and props to Feynman processes, JP J. Alg. Number Theory Appl. 7 (2007), 261-283.

[JL01] G. James and M. Liebeck, Representations and characters of groups, 2nd ed., Cambridge Univ. Press, Cambridge, 2001.

[JY09] M.W. Johnson and D. Yau, On homotopy invariance for algebras over colored PROPs, J. Homotopy Related Structures 4 (2009), 275-315.

[Kas95] C. Kassel, Quantum groups, Grad. Texts in Math. 155, Springer-Verlag, New York, 1995.

[Kau07] R.M. Kaufmann, On spineless cacti, Deligne's conjecture and Connes-Kreimer's Hopf algebra, Topology 46 (2007), 39-88.

[KW] R.M. Kaufmann and B.C. Ward, Feynman Categories, arXiv:1312.1269.

[KWZ] R.M. Kaufmann, B.C. Ward, and J.J. Zuniga, The odd origin of Gerstenhaber, BV and the master equation, arXiv:1208.5543.

[Kel82] G.M. Kelly, Basic concepts of enriched category theory, LMS Lecture Notes, Cambridge Univ. Press, Cambridge, 1982.

[KM94] M. Kontsevich and Y. Manin, Gromov-Witten classes, quantum cohomology, and enumerative geometry, Comm. Math. Phys. 164 (1994), 525-562.

[KM95] I. Kriz and J.P. May, Operads, algebras, modules and motives, Astérisque 233, 1995.

[Law63] F.W. Lawvere, Functorial semantics of algebraic theories, Proc. Nat. Acad. Sci. 50 (1963), 869-872.

[Mac63] S. Mac Lane, Natural associativity and commutativity, Rice Univ. Stud. 49(1) (1963), 28-46.

[Mac65] S. Mac Lane, Categorical algebra, Bull. Amer. Math. Soc. 71 (1965), 40-106.

[Mac98] S. Mac Lane, Categories for the working mathematician, Grad. Texts in Math. 5, 2nd ed., Springer, New York, 1998.

[MM92] S. Mac Lane and I. Moerdijk, Sheaves in geometry and logic: A first introduction to topos theory, Universitext, Springer, New York, 1992.

[Man06] M.A. Mandell, Cochains and homotopy type, Publ. Math. IHES. 103 (2006), 213-246.

[Mar96a] M. Markl, Models for operads, Comm. Alg. 24 (1996), 1471-1500.

[Mar96b] M. Markl, Cotangent cohomology of a category and deformations, J. Pure Appl. Alg. 113 (1996), 195-218.

[Mar02] M. Markl, Homotopy diagrams of algebras, Rend. del Circ. Mat. di Palermo, Series II, Suppl. 69 (2002), 161-180.

[Mar04] M. Markl, Homotopy algebras are homotopy algebras, Forum Math. 16 (2004), 129-160.

[Mar06] M. Markl, A resolution (minimal model) of the PROP for bialgebras, J. Pure Appl. Alg. 205 (2006), 341-374.

[Mar08] M. Markl, Operads and PROPs, Handbook of algebra, vol. 5, p.87-140, Elsevier, 2008.

[MMS09] M. Markl, S. Merkulov, and S. Shadrin, Wheeled PROPs, graph complexes and the master equation, J. Pure Appl. Alg. 213 (2009), 496-535.

[MSS02] M. Markl, S. Shnider, and J.D. Stasheff, Operads in Algebra, Topology and Physics, Math. Surveys and Monographs 96, Amer. Math. Soc., Providence, R.I., 2002.

[MV09] M. Markl and A. A. Voronov, PROPped-up graph cohomology, Prog. Math 270 (2009), 249-281.

[May72] J.P. May, The geometry of iterated loop spaces, Lecture Notes in Math. 271, Springer-Verlag, New York, 1972.

[May97] J.P. May, Definitions: operads, algebras and modules, Contemp. Math. 202 (1997), 1-7.

[MS02] J.E. McClure and J.H. Smith, A solution of Deligne's Hochschild cohomology conjecture, Contemp. Math. 293 (2002), 153-193.

[Mer08] S.A. Merkulov, Lectures on PROPs, Poisson geometry and deformation quantization, Contemp. Math. 450 (2008), 223-257.

[Mer09] S.A. Merkulov, Graph complexes with loops and wheels, Prog. Math. 270 (2009), 311-354.

[Mer10a] S.A. Merkulov, Wheeled Pro(p)file of Batalin-Vilkovisky formalism, Comm. Math. Phys. 295 (2010), 585-638.

[Mer10b] S.A. Merkulov, Wheeled props in algebra, geometry and quantization, European Congress of Math., p. 83-114, Eur. Math. Soc., Zürich, 2010.

[MV09] S. Merkulov and B. Vallette, Deformation theory of representations of prop(erad)s. I and II, J. Reine Angew. Math. 634 (2009), 51-106, and 636 (2009), 123-174.

[Qui67] D.G. Quillen, Homotopical algebra, Lecture Notes in Math. 43, Springer-Verlag, Berlin, 1967.

[Rez96] C.W. Rezk, Spaces of algebra structures and cohomology of operads, Ph.D. thesis, MIT, 1996.

[Sch99] S. Schwede, Stable homotopical algebra and Γ-spaces, Math. Proc. Cam. Phil. Soc. 126 (1999), 329-356.

[SS00] S. Schwede and B. E. Shipley, Algebras and modules in monoidal model categories, Proc. London Math. Soc. 80 (2000), 491-511.

[Seg01] G. Segal, Topological structures in string theory, R. Soc. Lond. Philos. Trans. Ser. A Math. Phys. Eng. Sci. 359 (2001), no. 1784, 1389-1398.

[Seg04] G. Segal, The definition of conformal field theory, London Math. Soc. Lecture Note Ser. 308 (2004), 421-577.

[Smi82] V.A. Smirnov, On the cochain complex of topological spaces, Math. USSR Sbornik 43 (1982), 133-144.

[Smi01] V.A. Smirnov, Simplicial and operad methods in algebraic topology, Translations Math. Monographs 198, Amer. Math. Soc., Providence, RI, 2001.

[Sta63] J.D. Stasheff, Homotopy associativity of H-spaces I, II, Trans. Amer. Math. Soc. 108 (1963), 275-312.

[Str10] H. Strohmayer, Prop profile of bi-Hamiltonian structures, J. Noncommut. Geom. 4 (2010), 189-235.

[Val07] B. Vallette, A Koszul duality for props, Trans. Amer. Math. Soc. 359 (2007), 4865-4943.

[Vog03] R.M. Vogt, Cofibrant operads and universal E_∞ operads, Top. Appl. 133 (2003), 69-87.

[Wei94] C.A. Weibel, An introduction to homological algebra, Cambridge Univ. Press, Cambridge, 1994.

Index